Environmental Science and Engineering. Subseries: Environmental Science

Series editors: R. Allan · U. Förstner · W. Salomons

Bernhard Westrich · Ulrich Förstner (Eds.)

Sediment Dynamics and Pollutant Mobility in Rivers

An Interdisciplinary Approach

With 195 Figures and 52 Tables

Springer

Editors

Prof. Dr.-Ing. habil. Bernhard Westrich

University of Stuttgart
Institute of Hydraulic Engineering
Pfaffenwaldring 61
70569 Stuttgart
Germany
Bernhard.Westrich@iws.uni-stuttgart.de

Prof. Dr. Ulrich Förstner

Hamburg University of Technology
Institute of Environmental Technology and Energy Economics
Eissendorfer Straße 40
21071 Hamburg
Germany
u.foerstner@tu-harburg.de

ISSN 1863-5520 **Springer Berlin Heidelberg New York**

ISBN 978-3-642-07110-2 e-ISBN 978-3-540-34785-9

Springer is a part of Springer Science+Business Media
springer.com

Cover design: deblik, Berlin

Printed on acid-free paper 30/2132/AO – 5 4 3 2 1 0

Preface

"Sediment management requires a solid mix of pragmatism and sound science"[1]

The actual discussion on the role of sediments in the European water legislation is a typical example, how the development of environmental policies mixes legal requirements with socio-economic aspects, issues of technical feasibility, and scientific knowledge[2]. In fact, the Sixth Environmental Action programme of 2001 has stipulated, namely, that "sound scientific knowledge and economic assessments, reliable and up-to-date environmental data and information, and the use of indicators will underpin the drawing up, implementation and evaluation of environmental policy"[3].

In the ten chapters of this book, a broad set of practical process knowledge is presented, comprising simulation techniques, laboratory and in-situ studies on the interaction between biological, chemical and hydrodynamic factors as well as models, for solving combined quality and quantity problems of riverine sediments. The underlying research on various types of particulate matter, risk analyses and problem solutions will also contribute to the implementation of the Soil Strategy and – in particular – Marine Strategy Directive in Europe.

Among the 24 original scientific papers of the present book (with 12 review articles and 8 short chapter introductions), 13 stem from the interdisciplinary Joint Program SEDYMO ("Fine *Se*diment *Dy*namics and Pollutant *Mo*bility in Rivers"), funded by the German Federal Ministry of Education and Research (BMBF), May 2002 to July 2006. The respective sub-projects were organized under the three headings "experimental techniques", "processes and properties", and "development and validation of models" (see Sect. 1.2).

The present scientific state-of-the-art overview on the wider area "sediment stability assessment" has been and will further be confronted with the world of policy and management. To characterize these interactions, achievements and deficiencies, the following paragraphs of our preface give a selection of ten sediment-related themes actually under discussion:

---

1 Krantzberg G, Hartig JH, Zarull MA (2000) Sediment management: deciding when to intervene. Environ Sci Technol 34:22A–27A.

2 Quevauviller Ph (2006) Science-policy interfacing in the context of the WFD implementation. J Soils & Sediments 6(4):259–261.

3 Anon (2001) 6th Environmental Action Plan 2001–2010. European Commission, rue de la Loi, B-1040 Brussels.

1. Quantity vs. quality – concept of sustainable sediment management[4]

Sediment management challenges and problems relate to quality and quantity issues. Quality issues relate to contamination, legislation, risk perception and assessment, source control and destination of dredged material. Quantity issues are related to natural processes causing erosion, sedimentation, flooding and the impact of engineering works such as dams, on the river morphology. The concept "sustainable sediment management (SSM)" of the European Sediment Research Network SedNet (www.SedNet.org)[4] is based on the integration of quality and quantity aspects involving typical hydrodynamic issues like erosion stability and transport in a broad spectrum of spatial and temporal scales (Sect. 1.1 *"Quantity and quality issues in river basin" by U. Förstner and P. N. Owens* and Sect. 2.1 *"Hydrodynamics and sustainable sediment management" by B. Westrich*).

2. Hydraulic processes as the driving force for dispersion of contaminated sediments

The quantification of flow rates including the transport of particle aggregates, microorganisms as well as dissolved and adsorbed substances requires an integration of various experimental techniques (flumes, turbulence columns, erosion chambers), to study the combined effects of sediment processes during resuspension, transport and deposition, and to describe these processes by models on different scales for the determination of hydrodynamic, chemical and biological parameters.

Sediment physical parameters are the basis of any risk assessment both on local and river basin scales. In the decision process, sediment stability should be considered a subset of an overall risk-management framework, using a tiered approach[5], which is characterized by a progressive increase in complexity – e.g., definition of key elements at the site, regional geomorphology to understand the sensitivity of the site to flood – associated flows, and definition of needs for sediment sampling, acoustic surveying and current measurements.

Application of the empirical methods used to assess the erosion characteristics of a cohesive deposit would benefit from a demonstration that, despite the relatively small size of most of the available experimental apparatus, the resulting flows represent a reasonable simulation of the flow conditions at the sediment-water interface in the field[5]. The SEDYMO-program included two subprojects using sediment stability tests which could bridge the gap between laboratory and field (no. 1 "Triad System" and no. 3 "Microcosm/Hot film anemometer").

[4] Anon (2002) SedNet Demand driven, European Sediment Research Network, Description of Work, Final Version. EC Proposal No EVK-2001-00058; Salomons W, Brils J (eds) (2004) Contaminated Sediments in European River Basins. SedNet 'Booklet'; SedNet Work Package books (Elsevier Publ 2007): WP2 (PN Owens) Sediment Management at the River Basin Scale; WP3 (D Barcelo) Sediment Quality and Impact Assessment, WP4 (G Bortone) Sediment and Dredged Material Treatment, WP5 (S Heise) Sediment Risk Management and Communication.

[5] Bohlen F, Erickson MJ (2006) Incorporating sediment stability within the management of contaminated sediment sites: A synthesis approach. Integr Environ Assess Manag 2:24–28.

3. Problems with model approaches for fine-grained sediments

Analytical and numerical models are indispensable for both connecting and integrating the interdisciplinary study of individual processes and for transferring the findings from laboratory experiments to a natural aquatic system where processes take place on extremely variable scales both in space and time. Coupling of field data to model formulation requires particular attention when dealing with cohesive sediments; due to the complex nature of the transport processes, designs would benefit from a degree of standardization that does not exist at present[5].

In the SEDYMO-program, special emphasis has been given to fine-grained sediments and to a typical set of factors commonly influencing solution/solid equilibrium conditions. Examples are the delayed release of metals from resuspended anoxic sediments (subproject no. 10) and the relationship between sediment-associated phosphorous entrainment rates and bed shear velocities, which were studied in subproject no. 11 for a series of in-situ experiments for the river Spree, Germany.

4. "Is the site erosive or depositional?" – conceptual site model[6]

The sediment manager's work begins with a conceptual site model (CSM), a description what is known about the contaminant source areas, as well as the physical, chemical, and biological processes that affect contaminant transport from the sources through environmental media to potential environmental receptors[7]. With respect to sediment-associated contaminants, questions that should be asked during selection of management options include: *(i)* Is the site erosive or depositional? *(ii)* Will management options change that, and how will that impact other sites downstream? *(iii)* Can the normal sedimentation process in the area solve the problem through burial and mixing? Or as the principle question[8]: Should we wait for the natural recovery mechanisms to reduce the risk due to sedimentation, degradation or other natural attenuation processes, or does the situation require removal of the contaminated sediments?

5. Sediment management strategies – appropriate selection criteria?

In practice sediment management strategies can be categorized according to typical problem solutions like monitored natural recovery, in-situ containment, in-situ treatment and dredging or excavation[5]. The information required to evaluate or compare each of these options is fundamentally different[9] and the realization of the optimal solution is a difficult, not only technical task. This is because the complexity of sedi-

---

[6] Apitz SE, White S (2003) A conceptual framework for river-basin-scale sediment management. J Soils Sediments 3:132–138; see also Apitz et al. (2005) Integr Environ Assess Manag 1(1):2–8.

[7] Anon (1995) Standard Guide for Developing Conceptual Site Models for Contaminated Sites. In: Annual Book of ASTM Standards, vol. 04.08, E 1689-95 American Society of Testing and Materials. West Conshohocken, Pa.

[8] Wim Salomons, personal communication.

[9] Apitz SE, Power EA (2002) From risk assessment to sediment management. J Soils Sediments 2:61–66.

ment transport processes and associated uncertainties in most cases is fostering application of the precautionary principle, with removal as a presumed conservative but more expensive approach over in-place options.

6. "How best to incorporate sediment stability assessment results when managing risks at contaminated sediment sites?"

In order to give in-situ options such as monitored natural recovery (MNR)[10] a real chance, a shift of emphasis would be needed towards the use and communication of results from the analyses of multiple lines of evidence, e.g., through detailing the potential impacts of large, low-probability events or combinations of probabilities (e.g., the 100-year flood and the probability of erosion to a specific depth) on exposure and risk, and the associated uncertainties.

These lines of evidence should include a range of accepted and independent methods, such as historical review of site characteristics, past usage, and storm events; careful consideration of regional geomorphology and implications regarding the evolution of sediment deposits; detailed geochemical analysis of sediment cores obtained at carefully selected locations; and field measurements to characterize hydrodynamic forces[5]. This approach provides the basis for the development of a conceptual site model – paragraph 4 – that includes a numerical model for prediction potential impacts of rare events on the resuspension of sediments.

7. "Water Framework Directive – the river basin scope also for sediments?"[11]

Since 2000 remediation methodology as well as the preceding risk assessment are part of the holistic river basin approach according to the European Water Framework Directive (WFD), which is aiming to achieve a good ecological potential and a good surface water chemical status in European river basins until the year 2015 by a combined approach using emission and pollutant standards. Although there are immense quantitative and qualitative problems with sediments, the WFD did not take sediments explicitly into its implementation scheme. In particular, lack of information on the role of sediments as a long-term secondary source of contaminants may easily lead to unreliable risk analyses with respect to the – pretended – "good status".

8. Risks due to erosion of contaminated sediments and their potential impacts downstream are not covered by existing regulations …

The Water Framework Directive monitoring objectives require compliance checking with Environmental Quality Standards (EQS) but also the progressive reduction of pollution. The no-deterioration clause implies that trend studies should be foreseen

[10] Magar VS, Wenning RJ (2006) The role of natural recovery in sediment remediation. Integ Environ Assess Manag 2:66–74. Schwartz R, Gerth J, Neumann-Hensel H, Bley S, Förstner U (2006) Assessment of highly polluted fluvisol in the Spittelwasser floodplain based on national guideline values and MNA-criteria. J Soils & Sediments 6(3):145–155.

[11] Förstner U (2002) Sediments and European Water Framework Directive. J Soils & Sediments 2:54.

for sediment and biota; this calls for further guidance under Common Implementation Strategy[12], complementing the existing monitoring guidance. However, *compliance monitoring* for sediment is not yet appropriate because of lack of the definition of valid Environmental Quality Standards ($EQS_{sediment}$) in a European context[13]. One can, however, find certain links of this important issue to present and future activities in the European Water Framework Directive. For example, the strategies against chemical pollution of surface waters (WFD article 16), i.e. establishment of the program of measures until 2009, must consider all potential pollution sources at the catchment scale. Already the first step – screening of all generic sources that can result in releases of priority substances and priority hazardous substances – will include the specific source/pathway "historical pollution from sediment"[14].

9. Pragmatic approach – two case studies on historical contaminated sediments in the Rhine and Elbe River basins

The first study in the Rhine River catchment by Heise et al.[15], on behalf of the Port of Rotterdam, presented a three-step strategy, by the identification of *(i)* substances of concern, *(ii)* areas of concern and *(iii)* areas of risk with regard to the probability of polluting sediments in downstream reaches. The final assessment of the "areas of risk" had to take into account sediment erosion thresholds and the hydrological exceedance probability. The pragmatic approach provided initial evidence, that sediment-associated hexachlorobenzene (HCB) from the Higher and Upper Rhine has a significant effect on the quality of material dredged from Rotterdam Harbor and that this historical HCB source can contribute to a failure of the objectives of the WFD in the Rhine basin and may require additional measures for its control. Similar results were obtained from the study of Heise et al. on the Elbe River basin[16]; here, the transfer of dioxins and furans from the sediments and soils of the former Chemical Triangle at Bitterfeld can be considered as the most critical process involving historical contaminated sediments in this catchment area.

---

[12] Anon (2003) Common Implementation Strategy for the Water Framework Directive (2000/60/EC). Carrying forward the Common Implementation Strategy for the Water Framework Directive. Progress and Work Programme for 2003 and 2004, as agreed by the Water Directors, 17 June 2003, 52 p, Brussels.

[13] Anon (2004a) WFD Expert Group on Analysis and Monitoring of Priority Substances (AMPS), Sediment Guidance Discussion Document, AMPS and SedNet, Draft Version 16 April 2004. See also: Proposal for a Directive of the European Parliament and of the Council on environmental quality standards in the field of water policy and amending Directive 2000/60/EC (WFD-Daughter Directive), Brussels 17.7.2006; e.g., Article 2 (2) "*Member States* shall ensure, that concentrations of substances listed in Parts A and B of Annex I do not increase in *sediment* and biota".

[14] Anon (2004b) Concept Paper on Emission Control from 8 June 2004 of the Expert Advisory Forum (EAF) on Priority Substances and Pollution Control. 7th EAF-Meeting at Brussels, 14–15 June 2004.

[15] Heise S, Förstner U, Westrich B, Jancke T, Karnahl J, Salomons W (2004) Inventory of Historical Contaminated Sediment in Rhine Basin and its Tributaries. On behalf of the Port of Rotterdam. October 2004, Hamburg, 225 p.

[16] Heise S, Claus E, Heininger P, Krämer Th, Krüger F, Schwartz R, Förstner U (2005) Studie zur Schadstoffbelastung der Sedimente im Elbeeinzugsgebiet – Ursachen und Trends. Im Auftrag von Hamburg Port Authority. Abschlussbericht Dezember 2005, 169 S.

10. Emission-immission relationships in the catchment-coast continuum

From their patterns of dioxin/furan congeners characteristic emissions from the Chemistry Triangle at Bitterfeld can be followed 350 km downstream to the sediments of Hamburg Harbour[17]. Such tracers, together with sediment core studies, provide an ideal tool for assessing the spatial and temporal development of pollution at critical sites in the "catchment-coast continuum"[18].

This, in particular, concerns the big river ports such as Rotterdam and Hamburg[19]. On the one hand, as the owners of 'large-sediment traps', they have to pay the expenses of former and actual shortcomings in emission control within their catchment areas. On the other hand, they increasingly tend to get rid of part of their problems by using sea disposal as a relative cheap procedure for less contaminated dredged sediments. In this situation, the question arises on the yardsticks for assessing quality of both types of sediments – ingoing from the characteristic emissions from the catchment area and outgoing into the coastal sea.

A majority of stakeholders at a SedNet round table[20] found environmental quality standards for sediments in the WFD not practicable, due to the highly variable hydrodynamic situations in riverine systems. Different from this, decisions about the disposal of dredged materials could be based on operational target values of key substances or ecotoxicological criteria, as in the case of Rotterdam Harbor, by application of 'Chemical Toxicity Test' (CTT)[21] values. Considering, however, the particular risks from sediment-associated contaminants in the marine environment, sustainable sediment management in coastal zones requires further science-based criteria, e.g., for assessing – in the wider area of this book – *(i)* pollutant loads (quality/quantity relationships) of extremely toxic substances like dioxins/furans, and *(ii)* specific dispersion and sedimentation patterns of "less" contaminated particles following the so-called "relocation" of dredged material.

The planned Marine Strategy Directive in the European Union[22] will follow the Water Framework Directive (WFD) in the stepwise implementation of elements that shall be reviewed every six years after their initial establishment: *(a)* the initial assessment and the determination of good environmental status, *(b)* the environmental targets; *(c)* the monitoring programs; and *(d)* the programs of measures.

---

[17] Götz R, Steiner B, Friesel P, Roch K, Walkow F, Maaß V, Reincke H, Stachel B (1998) Dioxin (PCDD/F) in the river Elbe – investigations of their origin by multivariate statistical methods. Chemosphere 37:1987–2002.

[18] Salomons W (2005) Sediments in the catchment-coast continuum. J Soils & Sediments 5(1):2–8.

[19] Westrich B, Förstner U (2005) Sediment dynamics and pollutant mobility in rivers (SEDYMO). Assessing catchment-wide emission-immission relationships from sediment studies. J Soils & Sediments 5(4):197–200.

[20] "Sediment management – an essential element of river basin management plans", Venice, Nov 22–23, 2006. J Soils & Sediments 7(2):117–132 (2007)

[21] Stronkhorst J (2003) Ecotoxicological effects of Dutch harbour sediments – The development of an effects-based assessment framework to regulate the disposal of dredged material in coastal waters of the Netherlands. Dissertation Vrije Universiteit Amsterdam, 202 p.

[22] Anon (2005a) Proposal for a Directive of the European Parliament and of the Councel establishing a Framework for Community Action in the field of Marine Environmental Policy (Marine Strategy Directive). Brussels, SEC(2005), SEC (2005) 1290. Brussels 24.10.2005.

The preparatory, mostly scientific work in the Marine Strategy[23] focused notably on *(i)* the application of the ecosystem-based approach to management of human activities impacting the marine environment; *(ii)* monitoring and assessment issues; and *(iii)* the particular challenge of hazardous substances. Here, with the objective close to zero for man-made synthetic substances, primarily for dioxins/furans and PCBs, the pressure on emission control can be expected being more stringent than under the WFD.

From the start of the Marine Strategy Directive (MSD), sediments will play a more significant role than in the WFD. Referring to the quality aspects of sediments, the Marine Strategy of 2002 already stated: "as sediments act as sinks for many pollutants, these chemicals continue to be a public health concern and impede the use of marine resources for human use". In Annex II to the MSD the general state of chemical pollution is described, among others, by sediment contamination; the quantitative aspects of sediments refer to their role as bed substrate and potential physical damage due to siltation, abrasion and selective extraction.

In an internet consultation of the Commission[24], the objectives for the Strategy were considered of "high" importance by a large majority of respondents. There was strong support for the dual EU/regional approach and for the proposed methodology on monitoring. Finally, regarding the role of science in this complex approach, while not covered by the statistical evaluation, the statement of an anonymous respondent can be cited (conclusions no. 98):

> "Science-based decision-making must be the leading principle for consistent decisions."

[23] Anon (2002) Towards a strategy to protect and conserve the marine environment (Marine Strategy). Communication from the Commission to the Council and the European Parliament. Brussels 02.10.2002.

[24] Anon (2005b) Thematic strategy on the protection and conservation of the marine environment. Evaluation of the replies to the Internet Consultation, 15 March to 9 May 2005 (133 respondents from 22 Member States).

Acknowledgments

The interdisciplinary research on sediment dynamics and contaminant mobility in the "Sedymo"-program 2002–2006 – nucleus of the present book – was funded by the German Federal Ministry of Education and Research (BMBF) through the program on "Sustainability in Production and Services" formerly supervised by Jürgen Heidborn and it has been successfully managed by Peter Hemberle and Verena Höckele from the Ministry Project Management Agency Forschungszentrum Karlsruhe, Water Technology and Waste Management Division. An interdisciplinary task force committee was constituted under the umbrella of the German Association of Water, Wastewater and Waste (DWA) to which professionals from universities, water authorities, engineering consultancies and other stakeholders were appointed aiming to bridge the gap between research and practice for future sediment management. Similarly, cooperation with the Water Chemical Society, Division of German Chemical Society focused on the quality aspects of river sediments.

For the investigation of contaminated sediment issues sampling campaigns in German rivers were conducted with logistic and technical support provided by various institutions. At the Rhine River it was the State Authority for the Environment Baden-Württemberg Karlsruhe, the German Waterways and Shipping Office Freiburg and the German Federal Waterways Engineering and Research Institute Karlsruhe. In cooperation with the International Commission for the Protection of the Rhine (ICPR), many experiments on sediment erosion were performed to provide basic information on the risk of mobilization of contaminants. Access to chemical and toxicological sediment data was given through the cooperation with the Federal Institute of Hydrology Koblenz thanks to Martin Keller and Peter Heininger. At the river Elbe. support was given by the Hamburg Port Authority, the Free Hanseatic City of Hamburg, the Consulting Centre for Integrated Sediment Management (BIS at the Hamburg University of Technology) and by UFZ Centre for Environmental Research Leipzig-Halle, Department Lake Research, Magedeburg.

First project results have been presented and discussed at the "Sedymo-Workshop" which was held at the University of Stuttgart in October 2004. The final presentation of the joint research project took place at the International Symposium in March 2006 at the Hamburg University of Technology. In addition to the "Sedymo"-Symposium – the proceedings are presented in this book – a special "SedNet"-day was organized in cooperation with the European Sediment Research Network. Two sessions, one focusing on "Sediments in European River Basins" with contributions from Susanne Heise, Axel Netzband, Philip N. Owens, Wim Salomons and Kevin G. Taylor and the other one on "Sediment Risk Assessment" by Marc Babut, Jos Brils and Sue White were devoted to the European perspective of sediment management (titles in the Appendix to this book). We are very grateful to SedNet and to the authors for this multidisciplinary state-of-the-art overview.

Thanks are given to all the reviewers of the submitted contributions, in particular to Johan C. Winterwerp for reviewing the "Transport Modeling" chapter. Special acknowledgment is paid to Dietrich Hammer from the University of Stuttgart for his enduring work on many details of the submitted papers to make them ready for printing. Mrs. Agata Oelschläger and Dr. Christian Witschel from the publisher Springer are gratefully acknowledged for their interest in publishing the present book.

Bernhard Westrich, Ulrich Förstner
Stuttgart, Hamburg, May 2007

Contents

List of Authors

Dr. rer. nat. Gudrun Abbt-Braun
Universität Karlsruhe (TH), Engler-Bunte-Institut, Chair of Water Chemistry,
Engler-Bunte-Ring 1, 76131 Karlsruhe, Germany
Gudrun.Abbt-Braun@ebi-wasser.uni-karlsruhe.de, Sect. 6.2

Dr. Fridbert Ackermann
Bienhornhöhe 27, 56076 Koblenz, Germany
fridbert.ackermann@freenet.de, Sect. 7.4

Dr. habil. Wolfgang Ahlf
Technische Universität Hamburg-Harburg, Institut für Umwelttechnik und
Energiewirtschaft, Eissendorfer Straße 40 (N), 21071 Hamburg, Germany
ahlf@tu-harburg.de, Sect. 10 Intro., 10.1, 10.2

Martina Baborowski
Helmholtz Centre for Environmental Research – UFZ, Department River Ecology,
Brückstraße 3a, 39114 Magdeburg, Germany
martina.baborowski@ufz.de, Sect. 5.3, 8.4

Dr. Jean-Philippe Bedell
LSE, ENTPE, Rue Maurice Audin, 69518 Vaulx-en-Velin, France
bedell@entpe.fr, Sect. 9.3

Dr. Tom Benson
HR Wallingford Ltd., Howberry Park, Wallingford, Oxfordshire, OX10 8BA,
United Kingdom
t.benson@hrwallingford.co.uk, Sect. 8.3

Dr. Reinhard Bierl
Universität Trier – Campus II, Fachbereich VI Geographie/Geowissenschaften,
Abteilung Hydrologie, Behringstraße 21, 54286 Trier, Germany
bierl@uni-trier.de, Sect. 7.1

Dr. Antonio Bispo
Ademe, 2 square Lafayette, 49004 Angers Cedex 01, France
Antonio.Bispo@ademe.fr, Sect. 9.3

Prof. Dr. Ludek Blaha
Masaryk University, RECETOX, Kamenice 3, 62500 Brno, Czech Republic
blaha@recetox.muni.cz, Sect. 10.3

Dr. Björn Bohling
Universität Kiel, Institut für Geowissenschaften, Otto-Hahn-Platz 1, 24118 Kiel, Germany
bohling@gpi.uni-kiel.de, Sect. 8.2

Dr. Hilmar Börnick
Technische Universität Dresden, Institut für Wasserchemie, Helmholtzstraße 10, 01062 Dresden, Germany
Hilmar.Boernick@tu-dresden.de, Sect. 6.3

Prof. Paolo Burlando
ETH Zürich, Institute of Environmental Engineering (IfU), Wolfgang-Pauli-Strasse 15, 8093 Zürich, Swizzerland
paolo.burlando@ifu.baug.ethz.ch, Sect. 5.2

Olaf Büttner
Helmholtz Centre for Environmental Research – UFZ, Department Lake Research, Brückstraße 3a, 39114 Magdeburg, Germany
olaf.buettner@ufz.de, Sect. 5.3

Prof. Dr.-Ing. Wolfgang Calmano
Technische Universität Hamburg-Harburg, Eissendorfer Straße 40, 21071 Hamburg, Germany
calmano@tu-harburg.de, Sect. 6.5

Ekkehard Christoffels
Erftverband, Bereich Gewässer, Paffendorfer Weg 42, 50126 Bergheim, Germany
ekkehard.christoffels@erftverband.de, Sect. 5.1

Guy Collilieux
CNR, 2 rue André Bonin, 69316 Lyon Cedex 04, France
G.Collilieux@cnr.tm.fr, Sect. 9.3

Prof. Dr. William Davison
Lancaster University, Department of Environmental Science, Lancaster, LA1 4YQ, United Kingdom
w.davison@lancaster.ac.uk, Sect. 8.1

Dr. Mike Dearnaley
HR Wallingford Ltd., Howberry Park, Wallingford, Oxfordshire, OX10 8BA, United Kingdom
mpd@hrwallingford.co.uk, Sect. 8.3

Markus Delay
Universität Karlsruhe (TH), Engler-Bunte-Institut, Chair of Water Chemistry, Engler-Bunte-Ring 1, 76131 Karlsruhe, Germany
markus.delay@ebi-wasser.uka.de, Sect. 6.2

Dirk Ditschke
Blue Ridge Numerics GmbH, Espenhausen 10, 35091 Cölbe, Germany
dirk.ditschke@cfdesign.com, Sect. 4.3

Dr. Matthias Dürr
Hygiene Institute Halle, Johann-Andreas-Segner-Straße 12, 06097 Halle (Saale), Germany
matthias.duerr@medizin.uni-halle.de, Sect. 10.3

Dr. Günter Fengler
Ernst-Moritz-Arndt-Universität Greifswald, Institut für Ökologie, Schwedenhagen 6, 18565 Kloster/Hiddensee, Germany, Sect. 3.5

Prof. Dr. Hans-Curt Flemming
University Duisburg-Essen, Aquatic Microbiology, Biofilm Centre, Geibelstraße 41, 47057 Duisburg, Germany
HansCurtFlemming@compuserve.com, Sect. 1.2, 9 Intro., 9.1, 9.2

Prof. Dr. Ulrich Förstner
Hamburg University of Technology, Institute of Environmental Technology and Energy Economics, Eissendorfer Straße 40, 21071 Hamburg, Germany
u.foerstner@tu-harburg.de, Sect. 1.1, 1.2, 2.2

Dr. Kurt Friese
Helmholtz Centre for Environmental Research – UFZ, Dept. Seenforschung, Brückstr 3a, 39114 Magdeburg, Germany
kurt.friese@ufz.de, Sect. 7.3

Prof. Dr. Fritz Hartmann Frimmel
Universität Karlsruhe (TH), Engler-Bunte-Institut, Chair of Water Chemistry, Engler-Bunte-Ring 1, 76131 Karlsruhe, Germany
Fritz.Frimmel@ebi-wasser.uni-karlsruhe.de, Sect. 1.2, 6.2

Annekatrin Fritsche
Sächsisches Serumwerk, Branch of SmithKline Beecham Pharma GmbH & Co. KG, GSK-group company, Zirkusstraße 40, 01069 Dresden, Germany
annekatrin.a.fritsche@gsk.com, Sect. 6.3

Jenna Funnell
University of St. Andrews, Sediment Ecology Research Group, Gatty Marine Laboratory, East Sands, Fife, St. Andrews Scotland, KY16 8LB, United Kingdom, Sect. 3.1

Dr. rer. nat. Sabine Ulrike Gerbersdorf
University of St. Andrews, School of Biology, Gatty Marine Laboratory,
East Sands, Fife, St. Andrews, Scotland, KY16 8LB, United Kingdom
sug@st-andrews.ac.uk, Sect. 3.2, 10.3

Prof. Rémy Gourdon
LGCie, INSA, Bat. S. Carnot INSA LYON, 69621 Villeurbanne, France
Remy.Gourdon@insa-lyon.fr, Sect. 9.3

Dr. Helmut Guhr
Waltherstraße 4, 39116 Magdeburg, Germany
HelmutGMgb@t-online.de, Sect. 5.3

Prof. Dr. rer. nat. Giselher Gust
Hamburg University of Science and Technology, Ocean Engineering 1,
Schwarzenbergstraße 95, 21073 Hamburg, Germany
gust@tu-harburg.de, Sect. 1.2, 3.3, 6.4

Dr.-Ing. Ingo Haag
Beratender Ingenieur Dr.-Ing. Karl Ludwig Wasserwirtschaft – Wasserbau,
Herrenstraße 14, 76133 Karlsruhe, Germany
ingo.haag@ludwig-wawi.de, Sect. 10.3

Dietrich Hammer
Universität Stuttgart, Institut für Wasserbau, Pfaffenwaldring 61, 70569 Stuttgart,
Germany
dietrich.hammer@iws.uni-stuttgart.de, Sect. 2.2

Christian Heise
Universität Karlsruhe (TH), Engler-Bunte-Institut, Chair of Water Chemistry,
Engler-Bunte-Ring 1, 76131 Karlsruhe, Germany, Sect. 6.2

Dr. Susanne Heise
Technische Universität Hamburg-Harburg, Beratungszentrum für
integriertes Sedimentmanagement (BIS), Eissendorfer Straße 40 (N),
21071 Hamburg, Germany
s.heise@tuhh.de, Sect. 10.1

Dr. Klara Hilscherova
Masaryk University, RECETOX, Kamenice 3, 62500 Brno, Czech Republic
hilscherova@recetox.muni.cz, Sect. 10.3

Elke Hinz
ETH Zürich, Institute of Environmental Engineering (IfU),
Wolfgang-Pauli-Strasse 15, 8093 Zürich, Swizzerland, Sect. 5.2

Dr. Henner Hollert
Universität Heidelberg, Department of Zoology, Im Neuenheimer Feld 230,
69120 Heidelberg, Germany
hollert@uni-heidelberg.de, Sect. 10.3

Pei-Chi Hsu
Technische Universität Hamburg-Harburg,
Institut für Umwelttechnik und Energiewirtschaft, Eissendorfer Straße 40 (N),
21071 Hamburg, Germany
pei.hsu@tu-harburg.de, Sect. 10.2

Dr. Michael Hupfer
Leibniz Institute of Freshwater Ecology and Inland Fisheries, Müggelseedamm 301,
12587 Berlin, Germany
hupfer@igb-berlin.de, Sect. 6.4

Nicolas Huybrechts
Universite Libre de Bruxelles, Dept. Water Pollution Control, Campus Plaine
(CP 208), 1050 Brussels, Belgium
nhuybrec@ulb.ac.be, Sect. 4.4

Dr.-Ing. George K. Jacoub
University of Manchester, School of M.A.C.E, Tyndall Centre for Climate
Change Research, Parsier Building, P.O. Box 88, Manchester, M60 1QD,
United Kingdom
George.Jacoub@manchester.ac.uk, Sect. 4.1

Thomas Jancke
Universität Stuttgart, Institut für Wasserbau, Pfaffenwaldring 61, 70550 Stuttgart,
Germany
thomas.jancke@iws.uni-stuttgart.de, Sect. 3.2

Prof. Gerhard H. Jirka, Ph.D.
Universität Karlsruhe, Institute for Hydromechanics, Kaiserstraße 12,
76131 Karlsruhe, Germany
jirka@ifh.uni-karlsruhe.de, Sect. 1.2, 3.4

Joachim Karnahl
Universität Stuttgart, Institut für Wasserbau, Pfaffenwaldring 61, 70550 Stuttgart,
Germany
Joachim@karnahl.com, Sect. 4.2

Dr.-Ing. Ulrich Kern
Erftverband, Bereich Gewässer, Paffendorfer Weg 42, 50126 Bergheim, Germany
ulrich.kern@erftverband.de, Sect. 5 Intro., 5.1

Prof. Dr.-Ing. Michael Kersten
Johannes Gutenberg-Universität, Institut für Geowissenschaften, 55099 Mainz, Germany
michael.kersten@uni-mainz.de, Sect. 8.2

Jörg Kirsch
Björnsen Beratende Ingenieure GmbH, Maria Trost 3, 56070 Koblenz, Germany
j.Kirsch@bjoernsen.de, Sect. 5.2

Dogan Kisacik
Technische Universität Hamburg-Harburg, Institut für Meerestechnik, Schwarzenbergstraße 95, 21073 Hamburg, Germany, Sect. 3.3

PD Dr. Andreas Kleeberg
Leibniz Institute of Freshwater Ecology and Inland Fisheries, Müggelseedamm 301, 12587 Berlin, Germany
kleeberg@igb-berlin.de, Sect. 6.4

Dr. Marion Köster
Ernst-Moritz-Arndt-Universität Greifswald, Institut für Ökologie, Schwedenhagen 6, 18565 Kloster/Hiddensee, Germany
marion.koester@uni-greifswald.de, Sect. 3.5

Claudia Kraft
Technical University Braunschweig, Institute of Environmental Geology, Pockelsstraße 3, 38106 Braunschweig, Germany
c.kraft@tu-bs.de, Sect. 8.4

Jörg Kraft
Friedrich Schiller University of Jena, Institute of Inorganic and Analytical Chemistry, Lessingstraße 8, 07743 Jena, Germany
joerg.kraft@uni-jena.de, Sect. 8.4

Dr. rer. nat. habil. Andreas Krein
Public Research Centre Gabriel Lippmann, Research Unit Geosciences, Department of Environment and Agro-Biotechnologies, Rue du Brill 41, 4422 Belvaux, Luxembourg
krein@lippmann.lu, Sect. 7.1, 7.2

Dr. Frank Krüger
ELANA, Soil Water Monitoring, Dorfstraße 55, 39615 Falkenberg, Germany
frank.krueger@ufz.de, Sect. 5.3, 7.3

Gregor Kühn
Universität Karlsruhe, Institute for Hydromechanics, Kaiserstraße 12, 76131 Karlsruhe, Germany
kuehn@ifh.uni-karlsruhe.de, Sect. 3.4

Andreas Kurtenbach
Universität Trier – Campus II, Fachbereich VI Geographie/Geowissenschaften, Abteilung Hydrologie, Behringstraße 21, 54286 Trier, Germany
kurt6101@uni-trier.de, Sect. 7.1, 7.2

Aurélie Larcy
Universite Libre de Bruxelles, Department of Chemical Engineering, 50 avenue F. D. Roosevelt (CP 165/67), 1050 Brussels, Belgium
Aurelie.Larcy@ulb.ac.be, Sect. 4.4

Dr. Ole Larsen
DHI Wasser und Umwelt GmbH, Wiesenstraße 10 a, 28857 Syke, Germany
ola@dhi-umwelt.de; and
Max Planck Institute for Marine Microbiology, Department of Biogeochemistry, Celsiusstraße 1, 28359 Bremen, Germany
ola@dhi-umwelt.de, Sect. 1.2, 8 Intro., 8.1

Carlos Felipe Leon Morales
University Duisburg-Essen, Aquatic Microbiology, Biofilm Centre, Geibelstraße 41, 47057 Duisburg, Germany
felipe@uni-duisburg.de, Sect. 9.1, 9.2

Dr.-Ing. Chen-Chien Li
National Cheng Kung University, Disaster Prevention Research Center, No. 500, Sec. 3, Anming Rd., 709 Tainan, Taiwan (R.O.C.)
ccli@dprc.ncku.edu.tw, Sect. 2.2

Dr. Ingo Lobe
Helmholtz Centre for Environmental Research – UFZ, Department River Ecology, Brückstraße 3a, 39114 Magdeburg, Germany
ingo.lobe@ufz.de, Sect. 5.3

Margarete Mages
Helmholtz Centre for Environmental Research – UFZ, Department River Ecology, Brückstraße 3a, 39114 Magdeburg, Germany
margarete.mages@ufz.de, Sect. 8.4

Dr. Andy J. Manning
HR Wallingford Ltd., Howberry Park, Wallingford, Oxfordshire, OX10 8BA, United Kingdom
andymanning@yahoo.com, Sect. 8.3

Prof. Mark Markofsky, Ph.D.
Leibniz Universität Hannover, Institute of Fluid Mechanics and Computer Applications in Civil Engineering, Appelstraße 9a, 30167 Hannover, Germany
mark@hydromech.uni-hannover.de, Sect. 1.2, 4 Intro., 4.3

Anett Matthäi
Technische Universität Hamburg-Harburg, Institut für Umwelttechnik und Energiewirtschaft, Eissendorfer Straße 40 (N), 21071 Hamburg, Germany
anett.matthaei@tu-harburg.de, Sect. 10.2

Jennifer L. Mc Connachie
Northgate Minerals Corporation, Kemess Mine, Box 3519, Smithers, BC, V0J 2N0, Canada
jmcconnachie@kemess.com, Sect. 6.1

George Metreveli
Universität Karlsruhe (TH), Engler-Bunte-Institut, Chair of Water Chemistry, Engler-Bunte-Ring 1, 76131 Karlsruhe, Germany
george.metreveli@ebi-wasser.uka.de, Sect. 6.2

Prof. Dr. Lutz-Arend Meyer-Reil
Rothenkirchen 29, 18573 Rambin, Germany, Sect. 1.2, 3.5

Flemming Møhlenberg
DHI Water & Environment, Agern Allé 5, 2970 Hørsholm, Denmark
flm@dhigroup.com, Sect. 8.1

Dr. Peter Molnar
ETH Zürich, Institute of Environmental Engineering (IfU), Wolfgang-Pauli-Strasse 15, 8093 Zürich, Swizzerland
molnar@ifu.baug.ethz.ch, Sect. 5.2

Dr. Peter Morgenstern
Helmholtz Centre for Environmental Research – UFZ, Department Analytics, Permoserstraße 15, 04318 Leipzig, Germany
peter.morgenstern@ufz.de, Sect. 5.3

Dr.-Ing. Volker Müller
Technische Universität Hamburg-Harburg, Institut für Meerestechnik, Schwarzenbergstraße 95, 21073 Hamburg, Germany
v-mueller@tu-harburg.de, Sect. 3.3

Manuelle Neto, Ph.D
LSE, ENTPE, Laboratoire des Sciences de l'Environnement, Rue Maurice Audin, 69518 Vaulx-en-Velin, France
neto.m@hotmail.fr, Sect. 9.3

Prof. Dr. Reinhard Nießner
Technische Universität München, Institute of Hydrochemistry and Chemical Balneology, Marchioninistraße 17, 81377 München, Germany
reinhard.niessner@ch.tum.de, Sect. 1.2

Mihály Óvári
Eötvös University Budapest, Department of Chemical Technology
and Environmental Chemistry, Pazmany Peter stny 1/A, 1117 Budapest,
Hungary
ovari@chem.elte.hu, Sect. 8.4

Dr. Philip N. Owens
University of Northern British Columbia, FRBC Endowed Research, Endowed Chair
of Landscape Ecology, Environmental Sciences Program, 3333 University Way,
Prince George, BC, V2N 4Z9, Canada
owensp@unbc.ca, Sect. 1.1

Prof. David M. Paterson
University of St. Andrews, Sediment Ecology Research Group, Gatty Marine
Laboratory, East Sands, Fife, St. Andrews Scotland, KY16 8LB, United Kingdom
d.paterson@st-and.ac.uk, Sect. 3 Intro., 3.1

Dr. Ellen L. Petticrew
University of Northern British Columbia, FRBC Endowed Research, Endowed Chair
of Landscape Ecology, Geography Program, 3333 University Way, Prince George, BC,
V2N 4Z9, Canada
ellen@unbc.ca, Sect. 6 Intro., 6.1

Dr. Holger Rupp
Helmholtz Centre for Environmental Research – UFZ, Department Soil Physics,
Dorfstraße 55, 39615 Falkenberg, Germany
holger.rupp@ufz.de, Sect. 5.3

James Saunders
University of St. Andrews, Sediment Ecology Research Group, Gatty Marine
Laboratory, East Sands, Fife, St. Andrews Scotland, KY16 8LB, United Kingdom
jes11@st-andrews.ac.uk, Sect. 3.1

Dr. Birgit Schubert
Federal Institute of Hydrology, Qualitative Hydrology, Am Mainzer Tor 1,
56068 Koblenz, Germany
schubert@bafg.de, Sect. 7.4

Dr. René Schwartz
Hamburg University of Technology, Institute for Environmental
Technology and Energy Economics, Eissendorfer Straße 40, 21071 Hamburg,
Germany
schwartz@tu-harburg.de, Sect. 7.3

Andreas Seibel
Germanischer Lloyd AG, Vorsetzen 35, 20459 Hamburg, Germany, Sect. 3.3

Ralf Siepmann
Technische Universität Hamburg-Harburg, Institut für Umwelttechnik und Energiewirtschaft, Eissendorfer Straße 40, 21071 Hamburg, Germany
siepmann@tu-harburg.de, Sect. 6.5

Bryan M. Spears
University of St. Andrews, Sediment Ecology Research Group, Gatty Marine Laboratory, East Sands, Fife, St. Andrews Scotland, KY16 8LB, United Kingdom
bs29@st-andrews.ac.uk, Sect. 3.1

Dr. Kate L. Spencer
Queen Mary University of London, Geography Department,
Centre for Aquatic and Terrestrial Environments, Mile End Road, London, E1 4NS, United Kingdom
k.spencer@qmul.ac.uk, Sect. 8.3

Dr. Martin Strathmann
University Duisburg-Essen, Aquatic Microbiology, Biofilm Centre, Geibelstraße 41, 47057 Duisburg, Germany
strathmann@uni-duisburg.de, Sect. 9.1, 9.2

Dr. Keiko Suzuki
Queen Mary University of London, Geography Department,
Centre for Aquatic and Terrestrial Environments, Mile End Road, London, E1 4NS, United Kingdom
k.suzuki@qmul.ac.uk, Sect. 8.3

Prof. Dr. Wolfhard Symader
Universität Trier – Campus II, Fachbereich VI Geographie/Geowissenschaften, Abteilung Hydrologie, Behringstraße 21, 54286 Trier, Germany
symader@uni-trier.de, Sect. 7 Intro., 7.1

Dr. Jon A. Taylor
Queen Mary University of London, Geography Department, Centre for Aquatic and Terrestrial Environments, Mile End Road, London, E1 4NS, United Kingdom
j.a.taylor@compasshydrographic.co.uk, Sect. 8.3

Dr. Björn Tetzlaff
Research Centre Jülich, Institute of Chemistry and Dynamics of the Geosphere, ICG-IV-Agrosphere, 52425 Jülich, Germany
b.tetzlaff@fz-juelich.de, Sect. 5.4

Kyriakos Vamvakopoulos
Max Planck Institute for Marine Microbiology, Department of Biogeochemistry, Celsiusstraße 1, 28359 Bremen, Germany
kvamvako@mpi-bremen.de, Sect. 8.1

Dr. Andrea van der Veen
ECT Oekotoxikologie GmbH, Böttgerstraße 2–14, 65439 Flörsheim/Main, Germany
a.van-der-veen@web.de, Sect. 8.4

Prof. Jean-Pierre Vanderborght
Universite Libre de Bruxelles, Department of Water Pollution Control, Campus Plaine (CP 208), 1050 Brussels, Belgium, vdborght@ulb.ac.be, Sect. 4.4

Prof. Michel Verbanck
Universite Libre de Bruxelles, Department of Water Pollution Control, Campus Plaine (CP 208), 1050 Brussels, Belgium, mikeverb@ulb.ac.be, Sect. 4.4

Dr. Frank von der Kammer
Universität Wien, Institut für Geologische Wissenschaften, Geologie – Petrologie – Geochemie, Althanstraße 14, 1090 Wien, Österreich
frank.kammer@univie.ac.at, Sect. 1.2, 6.5

Dr. Wolf von Tümpling
Helmholtz Centre for Environmental Research – UFZ, Department of River Ecology, Brückstraße 3a, 39114 Magdeburg, Germany
wolf.vontuempling@ufz.de, Sect. 5.3, 8.4

Dr. Frank Wendland
Research Centre Jülich, Institute of Chemistry and Dynamics of the Geosphere, ICG-IV-Agrosphere, 52425 Jülich, Germany
f.wendland@fz-juelich.de, Sect. 5.1, 5.4

Prof. Dr.-Ing. Bernhard Westrich
University of Stuttgart, Institute of Hydraulic Engineering, Pfaffenwaldring 61, 70550 Stuttgart, Germany
bernhard.westrich@iws.uni-stuttgart.de, Sect. 1.2, 2.1, 2.2, 3.2, 4 Intro., 4.1, 4.2

Jan Wölz
Universität Heidelberg, Department of Zoology, Im Neuenheimer Feld 230, 69120 Heidelberg, Germany
jan.woelz@zoo.uni-heidelberg.de, Sect. 10.3

Prof. Dr. Eckhard Worch
Technische Universität Dresden, Institut für Wasserchemie, Helmholtzstraße 10, 01062 Dresden, Germany
Eckhard.Worch@tu-dresden.de, Sect. 1.2, 6.3

Prof. Werner Zielke
Leibniz Universität Hannover, Fluid Mechanics Institute, Appelstraße 9a, 30167 Hannover, Germany
zielke@hydromech.uni-hannover.de, Sect. 1.2

Authors of Contributions[25] to the Sedymo Symposium 2006 Not Published in This Volume

Dr. Sabine E. Apitz
SEA Environmental Decisions, Ltd., 1 South Cottages, TTie Ford, Little Hadham, SG112AT, United Kingdom
drsea@cvrl.org

Dr. Marc Babut
Cemagref, Freshwater Ecosystems Biology Research Unit, 3 bis Quai Chauveau CP 220, 69336 Lyon Cedex 9, France
marc.babut@cemagref.fr

Dr. Robert Banasiak
Ghent University, Hydraulics Laboratory, Department of Civil Engineering, Belgium
Robert.Banasiak@ugent.be

Jos Brils
TNO, P.O. Box 80015, 3508 TA Utrecht, The Netherlands
jos.brils@tno.nl

Evelyn Claus
German Federal Institute of Hydrology, Am Mainzer Tor 1, 50668 Koblenz, Germany
claus@bafg.de

Pieter J. den Besten
Institute for Inland Water Management and Waste Water Treatment (RIZA), Ministry of Transport, Public Works and Water Management, P.O. Box 17, 8200 AA Lelystad, The Netherlands
p.dbesten@riza.rws.minvenw.nl

Marc Eisma
Port of Rotterdam, World Port Centre, 3002 AP Rotterdam, The Netherlands
m.eisma@portofrotterdam.com

Dr. Ute Feiler
German Federal Institute for Hydrology, Qualitative Hydrology, Am Mainzer Tor 1, 56068 Koblenz, Germany
feiler@bafg.de

[25]See Appendix.

Tom Gallé
Resource Centre for Environmental Technologies (CRTE), CRP Henri Tudor, Box/P.O. Box 144, 4002 Esch-sur-Alzette, Luxembourg
Tom.galle@tudor.lu

Dr. Peter Heininger
German Federal Institute of Hydrology, Qualitative Hydrology, Am Mainzer Tor 1, 50668 Koblenz, Germany
heininger@bafg.de

Prof. Dr. Thilo Hofmann
University of Vienna, Erdwissenschaftliches Zentrum, Umweltgeowissenschaften, Althanstraße 14, 1090 Wien, Austria
Thilo.hofmann@univie.ac.at

Thomas Krämer
German Federal Institute of Technology, Am Mainzer Tor 1, 50668 Koblenz, Germany
thomas.kraemer@bafg.de

Axel Netzband
Hamburg Port Authority, Dalmannstraße 1, 20457 Hamburg, Germany
Axel.netzband@hpa.hamburg.de

Dr. Helga Neumann-Hensel
Laboratory Dr. Fintelmann and Dr. Meyer, Mendelssohnstraße 15D, 22761 Hamburg
hensel@fintelmann-meyer.de

Amy M. P. Oen
Norwegian Geotechnical Institute, Sognsveien 72, 0806 Oslo, Norway
Amy.oen@ngi.no

Prof. Dr. Wim Salomons
Free University Amsterdam, Institute for Environmental Studies, IVM, Kromme Elleboog 21, 9751 RB Haren (GN), The Netherlands
wim.salomons@home.nl

Dr. Marc Stutter
Macaulay Institute, Craigiebuckler, Aberdeen, AB15 8QH, Scotland, United Kingdom
m.stutter@macaulay.ac.uk

Dr. Kevin G. Taylor
Manchester Metropolitan University, Department of Environmental and Geographical Sciences, Manchester, M1, United Kingdom
k.g.taylor@mmu.ac.uk

Tiedo Vellinga
Delft University of Technology, Postbus 5048, 2600 GA Delft, The Netherlands
t.vellinga@portofrotterdam.com

Prof. Sue White
Cranfield University Silsoe, Institute of Water and Environment, MK45 4DT, United Kingdom
sue.white@Cranfield.ac.uk

Prof. Johan Winterwerp
WL I delft hydraulics, PO Box 177, 2600 MH Delft, The Netherlands
han.winterwerp@wldelft.nl

Chapter 1

Introduction

Ulrich Förstner · Philip N. Owens

1.1 Sediment Quantity and Quality Issues in River Basins

1.1.1 Introduction

Sediment is an integral and dynamic part of aquatic systems and it plays a major role in the hydrological, geomorphological and ecological functioning of river basins, defined here to include lakes, reservoirs, estuaries and the coastal zone. In natural and agricultural systems, sediment originates from the weathering of rocks, the mobilization and erosion of soils and river banks, and mass movements such as landslides and debris flows. In most river basins there are also important contributions to the sediment load of organic-rich material from a range of sources such as riparian trees, macrophytes and fish. This inorganic and organic material is susceptible to transportation downstream by flowing water, from headwaters and other source areas towards the outlet of the river basin. Flow rates decline in lowland areas (and areas where flow is reduced) where transported material settles in slack-zones and on the bed of the river, and on river floodplains during overbank events. At the end of the river much of the sediment is deposited in the estuary and on the seabed of the coastal zone.

Sediment provides the substrate for organisms and through interaction with the overlying waters (e.g., nutrient cycling) plays an essential role within aquatic ecosystems. In addition, after flooding, fine-grained sediments are left as a deposit on floodplains, creating fertile soils that are often highly suited for agricultural production. On the other hand, the removal of sediments from harbors, navigation channels, locks, floodplains and river stretches is a high capital cost for authorities and agencies responsible for their maintenance and water quality.

In the 1970s and 1980s, anthropogenic emissions of excessive metals, nutrients, organic pollutants, radionuclides and other substances caused a rapid deterioration of sediment quality. In addition, the hydrodynamic conditions of many rivers were altered: either directly by the construction of hydraulic constructions, such as dykes, dams, seawalls, and drainage; or indirectly by changes in land use, such as deforestation and urbanization. Apart from seasonal flooding of polder areas and floodplains, in recent years there have also been catastrophic events due to extreme rainfall and the failure of dams; examples include the breaching of tailing dams in highly contaminated areas such as the mining districts of Spain (1998) and Romania (2000), which caused considerable and immediate hazards (Hudson-Edwards et al. 2003; Macklin et al. 2003).

Most large river basins throughout the world are highly populated and/or modified by human activities and thus there are many users and uses of sediment within a basin

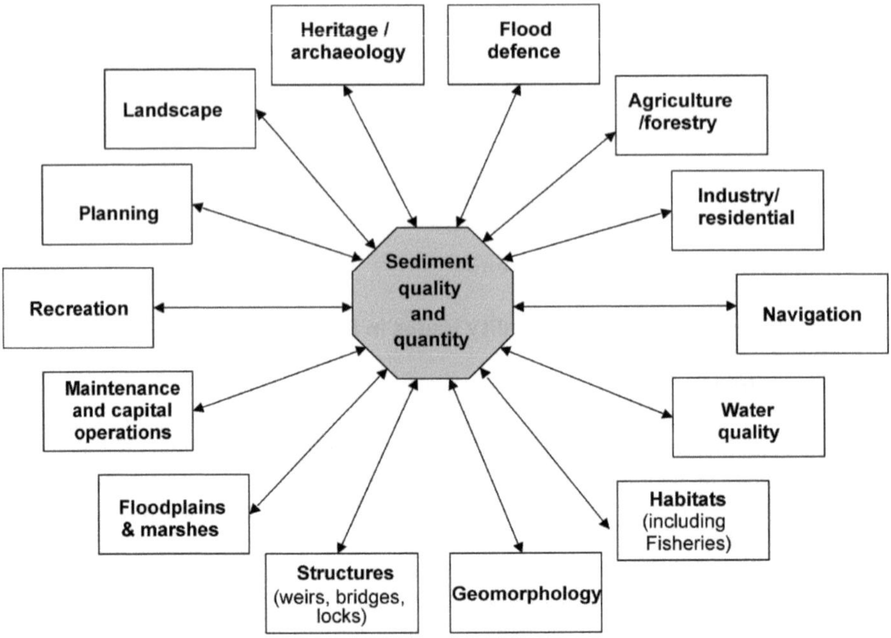

Fig. 1.1. Schematic representation of some of the main influences and impacts on sediment within a river basin (Owens et al. 2004)

(Fig. 1.1). This means that site-specific interventions will have impacts that will likely affect other users and uses of sediment (Owens 2005a), and it is, therefore, necessary to develop ways to consider and evaluate these needs, including involving stakeholders in the decision-making process (Gerrits and Edelenbos 2004).

The river basin represents an appropriate scale for management because in many cases source control will be the optimal long-term solution: environmentally, socially and economically. In particular, the controlling of diffuse sources of sediment and associated contaminants necessitates a river basin approach in order: *(i)* to identify all or most of the sources of the sediment and contaminants; *(ii)* for conducting meaningful risk assessment and evaluation; and *(iii)* to be able to implement remediation and mitigation options that are appropriate for controlling diffuse sources spread over a large area, for example, implementing appropriate land use and land management measures (Owens 2005a). Having extolled the virtues of considering sediment management at the river basin scale, the following sections briefly consider some of the main sediment quantity and quality issues in river basins, and consider some options and future requirements for sediment management at the river basin scale.

1.1.2 Sediment Quantity Issues

Rates and patterns of sediment supply (including soil erosion, bank erosion and mass movements) and sediment transport are a function of three, broad inter-connected groups of factors. First, there are factors that control the inherent susceptibility of the landscape to erode and transport sediment. Factors that control this include the resistance of geology and surficial material, soil type, topography (slope steepness and complexity), and drainage network. Second, there are driving factors that facilitate the physical movement of sediment from source to receptor (such as a river or water body). Factors that control this include the type, amount and intensity of precipitation, and snowmelt. Third, there are anthropogenic factors that modify the inherent susceptibility of the landscape to: *(i)* erode, and this includes land use and management (such as deforestation, cultivation, over-grazing, artificial drainage etc.); and *(ii)* transport sediment, and this includes river use and management (such as channelization, building dams etc.).

Given the large variation in these controlling factors, there is considerable spatial variation in sediment mobilization, sediment supply and sediment transport, both within river basins (especially large basins like the Danube, Yangtze, Mississippi and Amazon) and between river basins. Furthermore, due to temporal variations in the forces that drive sediment production and supply (e.g., rainfall events) there is also considerable temporal variability in sediment movement in river basins. It is these temporal and spatial variations, particularly in terms of the extremes in both maximum and minimum fluxes, that are often of concern for sediment management. At this stage it is important to recognize that sediment quantity issues as part of sediment management tend to apply to specific requirements for society. Large volumes of sediment – both coarse and fine – naturally move through river systems, and it is only when such sediment impinges or impacts on how society uses river basins (see Fig. 1.1) that it becomes an issue and requires some form of management. The following sections consider some of the main issues of sediment quantity that require management. It is convenient to consider the issues of too much and too little sediment, before considering relevant sediment management options and requirements, and also the possible effects of global environmental change.

Too Much Sediment

In most river basins, the main "quantity" issue is too much sediment. Areas within a river basin where there is too much sediment, from a societal perspective, are locations where either the concentration of sediment is too high, such as the intakes of turbines at hydro-electric power plants, and/or where the volume of deposited sediment is too large. In the case of the latter, problem areas include canals, upstream of dams and impoundments, ports and harbors. The specific management options associated with each location vary. In many situations, dredging and subsequent treatment and/or disposal are the preferred management options, especially for canals, ports and harbors (Netzband et al. 2002). For reservoirs, dredging may be an option in some cases, but more often sediment removal by hydraulic flushing, by-passing, or settling ponds are more cost-effective and sustainable (Owens et al. 2005). Too much sediment, especially fine-grained sediment (i.e. clay-, silt- and fine sand-sized particles), can also be a problem for sensitive aquatic habitats such as fish-spawning gravels as the fine sediment blocks the gravels, reducing oxygen supply and increasing fish egg and larval mortality. Unlike canals, ports and harbors, where the amount of sediment requiring dredging can be 1 000s of m^3, the amount of sediment needed to detrimentally influence aquatic habitats can be fairly insignificant, and it is often the type (i.e. fine-grained) and timing (i.e. during important reproductive times) that are important. This underlines that it is not just the amount of sediment that is important but also the timing of sediment transport and storage that are important.

There are numerous examples of the amounts and costs involved with removing sediment. One of the best documented examples is the port of Hamburg, Germany, which spends of the order of 30 million Euros each year to dredge and treat between 2 and 5×10^6 m^3 of sediment, much of which is contaminated thereby increasing the costs (Netzband et al. 2002). Hamer et al. (2005) state that The Netherlands and Germany dredge between 30 and 50×10^6 m^3 of sediment annually. These figures compare to estimates of the annual sediment flux to the estuaries and coastal zones of Europe of $300–700 \times 10^6$ t yr^{-1}, depending of the geographic area of concern (Owens and Batalla, in prep.).

Too Little Sediment

While a significant amount of attention and resources are given to dealing with the removal and treatment/disposal of excessive amounts of sediment, in many situations there are problems associated with too little sediment. Again, it is important to recognize that it is too little from the perspective of society, although in many cases the consequences of too little sediment impact on geomorphological and ecological functioning (Owens et al. 2005). It is also important to note that issues of too much sediment and too little sediment often occur in the same river basin, albeit at different locations. One of the major causes of problems associated with a dramatic reduction in sediment fluxes is the construction of dams and impoundments which reduce the supply of sediment to downstream reaches of a river. The natural supply and flow of sediment along a river system are interrupted and sediment that would normally move through a river network is deposited upstream of the impoundment, often causing problems associated with excessive sedimentation such as the reduction in the lifespan of reservoirs and increased operating costs. Downstream of the impoundment there can be problems due to a dramatic reduction in sediment supply which often leads to "hungry waters" whereby the river compensates for the reduced sediment load by downcutting and lateral erosion (Kondolf 1997). This can result in pronounced changes in the hydromorphology of a river system.

A good example is the Ebro River basin in Spain, where about 190 dams now impound almost 60% of the annual runoff in the basin. Total annual sedimentation in the reservoirs is estimated at 15×10^6 m^3 yr^{-1}. As a result of these impoundments, the total annual sediment load to the coastal zone at present is estimated to be only 3% of the sediment load at the start of the 20th century, and most of this sediment is now derived from channel sources compared to hillslope sources originally (Batalla 2003; Vericat and Batalla 2006). These reductions in sediment supply to downstream river reaches, estuaries and the coastal zone can result in the loss of important habitats such as wetlands, mud flats and salt marshes, which are particularly sensitive to change (Batalla 2003). Downstream of the Hoover Dam on the Colorado River in USA, the riverbed had degraded by 7.5 m within 13 years of dam closure and this erosion had affected 120 km of the river during that period (Williams and Wolman 1984). Downcutting of river channels can undermine structures such as buildings, roads and bridges, and in some cases lead to loss of life, as demonstrated by the collapse of a bridge on the Duero River in Portugal (Batalla 2003). Furthermore, Leopold (1997) describes the situation downstream of the Aswan Dam where 15–19% of the habitable land of the Nile delta could be gone within 60 years due to subsidence resulting from a lack of sediment deposition, which in turn could displace 15% of the population of Egypt.

Towards Solutions for Sediment Quantity Issues

Solutions for the occurrence of excessive amounts of sediment tend to focus on removal/dredging and treatment/disposal/relocation, or on schemes to divert or flush sediments through fluvial systems. The examples described above present information on the amounts of sediment involved and the costs. However, these solutions can be considered as "end-of-pipe" solutions in that they tend to deal with a problem as and when it appears, as opposed to dealing with the root cause. Thus for most situations, controlling the initial mobilization and supply of sediment from various sources (i.e. source control) will often be more sustainable from an ecological, economical and environmental perspective. This requires an assessment of the sources, pathways and fluxes of sediment within a river basin. It also requires an assessment of how much sediment and the type of sediment that are needed in rivers for geomorphological and ecological functioning, and also the timing of sediment fluxes (i.e. not too much at ecologically sensitive times of the year), so that sediment supply can meet geomorphological and ecological requirements. At present we are lacking sufficient information to determine the "requirement" part of the equation.

The sediment budget concept offers considerable potential for sediment management and should form part of the early stages of a river basin plan for sediment management (Owens 2005a). There are a variety of tools that can be used to provide the necessary information needed to establish a sediment budget. Thus, sediment fingerprinting and sediment tracing approaches can be used to identify where sediment originates (Table 1.1) and determine transfer pathways and sediment residence times. Magilligan et al. (2006), for example, used the fallout radionuclide ^{7}Be as a tracer to estimate the sediment transport velocities in regulated and unregulated streams in the USA, so as to assess the impacts of impoundment on sediment transport.

Table 1.1. Load-weighted mean contributions of each source type, derived from agricultural and urban areas of the basin, to the suspended sediment samples collected from two river gauging stations (Beal and Methley) in the downstream reaches of the River Aire basin, UK, during the period November 1997 to January 1999 (from Carter et al. 2003)

Site	Number of samples	Source type contribution (%)					
		Channel bank	Uncultivated topsoil	Cultivated topsoil	Woodland topsoil	Solids from STW[a]	Road dust
Beal	18	33	7	20	0	18	22
Methley	5	18	4	45	0	14	19

[a] *STW:* Sewage treatment works.

Having identified sources, pathways and fluxes (including the identification and estimation of storage elements) (Owens 2005a), there are several options for management, which apply to both issues of too much sediment and too little sediment. At this point, the two-stage risk assessment procedure of Apitz and White (2003) is useful, and this is discussed further in Sect. 1.1.4. On a river basin or regional level, there are several broad-scale options available. One is to implement regional- or basin-scale measures to reduce sediment mobilization and delivery to channels via awareness, education and improved practices. These measures are best achieved voluntarily through realization that benefits are both environmental and economical. A good example is conservation agriculture which has been demonstrated to both reduce adverse environmental impacts, including soil erosion, sediment delivery and diffuse pollution of waters, and also be economically viable (Jones et al. 2006). Another broad-scale approach is through legislation, and national and multi-national policies and agreements such as the Dutch-German Exchange and the International River Protection Commissions. At present, there are fairly limited legislation and policy measures relevant to sediment, particularly from a quantity aspect. A useful review is given in Köthe (2003; see also Salomons and Brils 2004; Owens et al. 2005; Owens and Collins 2006). Two pieces of legislation particularly worthy of mention for sediment quantity for the countries of the EU are the Water Framework Directive (WFD) and the Soil Framework Directive (SFD). The former is discussed further in Sect. 1.1.4, mainly in terms of sediment quality issues, although sediment quantity is also relevant for habitat (i.e. ecology) and hydromorphology goals within the WFD. The SFD is forthcoming legislation that in part will address issues of sediment mobilization and sediment delivery, mainly in the context of reducing soil erosion.

At the basin- or local-scale there are more specific measures that can be used to deal with issues of sediment quantity, many of which are also relevant for sediment quality, that should help to reduce the need for expensive "end-of-pipe" solutions like dredging. Examples include the use of buffering features placed at specific points in the landscape to help reduce sediment delivery to rivers (Owens et al. 2006) and measures to facilitate the more "natural" flow of water and sediments through river channels, which can also have ecological benefits (Batalla et al. 2006). Most of the measures and options identified above agree with the philosophy of the European Sediment Network (SedNet; www.sednet.org).

The Impact of Global Environmental Change

It is important to assess the implications of likely future global environmental change on sediment fluxes in rivers and how this may influence some of the sediment quantity issues described above. Global environmental change mainly refers to biogeochemical processes, including water and sediment fluxes, and how they are modified by human activities, and encompasses global climate change. It is important to stress that at present we are not certain what these changes will be and thus how they will influence sediment fluxes. There is still considerable debate and uncertainty about the direction (i.e. positive or negative, increases or decreases), magnitude and timing of global environmental changes in climate (temperature, precipitation and seasonality) and carbon cycles, for example, and we are not clear how land use and management will change under ever-changing policy and guidance on food production and security. Plus, global environmental change is itself part of a wider concept of global change that includes economic, cultural and geopolitical global issues, among others (Slaymaker and Owens 2004).

Some likely or probable effects of climate change and associated responses in soil erosion and sediment fluxes are described in Owens (2005b). These include a likely increase in sediment supply and fluxes in rivers due to the anticipated increase in wildfires in many parts of the world, especially areas of the Mediterranean and similar bioclimatic regions and the boreal forests of North America and Russia. Although changes in sediment fluxes due to likely changes in climate are no doubt important, some believe that changes in land use and management may be equally or more important (Slaymaker 2001).

In order to determine the likely changes in sediment fluxes in rivers, models can be used to simulate the production and transport of sediment through river basins under certain scenarios of climate and/or land use change. For example, Asselman et al. (2003) developed a suite of GIS-embedded models to simulate the production and transport of fine sediment through the 165 000 km^2 basin of the Rhine River. Table 1.2 presents some of the results of this study based on two different scenarios:

1. CP – present climatic conditions with autonomous changes in land use.
2. CPC – UK Hadley Centre high-resolution atmospheric general circulation model climate conditions (lower, central and upper estimates) with both autonomous and climate-induced changes in land use.

The effects of climate change on the future sediment load of the Rhine River can be estimated from the differences between the sediment load estimated by the combined climate and land use change scenario (CPC scenario) and that produced by the autonomous land use change scenario (CP scenario). The sediment load is estimated to increase by ca. 0.15×10^6 t yr^{-1} near the tidal limit. There are, however, considerable spatial variations in sediment fluxes in this large river basin (Asselman et al. 2003). Other studies have modeled the likely effects of changes in climate and land use on contaminated sediment fluxes and storage in the Rhine River (e.g., Thonon et al. 2006).

Table 1.2.
Expected per cent change in average annual sediment loads under scenario climate and land use conditions in different parts of the River Rhine basin (ranging from Diepoldsau in the headwaters to Rees near the coast) by the year 2100 (from Asselman et al. 2003)

Area	CP	CPC		
		Lower	Central	Upper
Diepoldsau	220	240	257	284
Rheinfelden	112	132	145	167
Maxau	31	31	37	52
Worms	7	5	11	31
Kaub	−16	−17	−12	8
Andernach	−17	−19	−13	8
Rees	−17	−19	−13	9

The effects of global environmental change on sediment fluxes in river basins are at the center of numerous research programs and initiatives, including the International Sediment Initiative (ISI), that has been launched by the United Nations Educational, Scientific, and Cultural Organization (UNESCO), as a major activity of the International Hydrological Programme (IHP; www.unesco.org/water/ihp).

1.1.3 Contaminated Sediments

The perception of aquatic sediments as a very valuable material for humans (e.g., for agriculture) and nature (e.g., as a habitat) rapidly and drastically changed into one of viewing sediments as a legacy of industrialization and related mass consumption, leaving an immense problem for water quality managers and other stakeholders to deal with (Anon. 2002). In fact, the value of today's aquatic sediments is concealed by several negative factors (Förstner 2004a).

Firstly, even without direct and hard dumps, sediments function as a sink for ongoing releases from many sources; these include wet and dry fallout from air emissions, runoff from farms, solid and dissolved inputs from mines, and discharges from landfills, industrial plants, and sewage-treatment plants. Restoration of the quasi-natural state will be a long-lasting process.

Secondly, subsequent to the natural erosive processes described in Sect. 1.1.1, sediment-bound contaminants are dispersed, in an unpredictable manner, on floodplains,

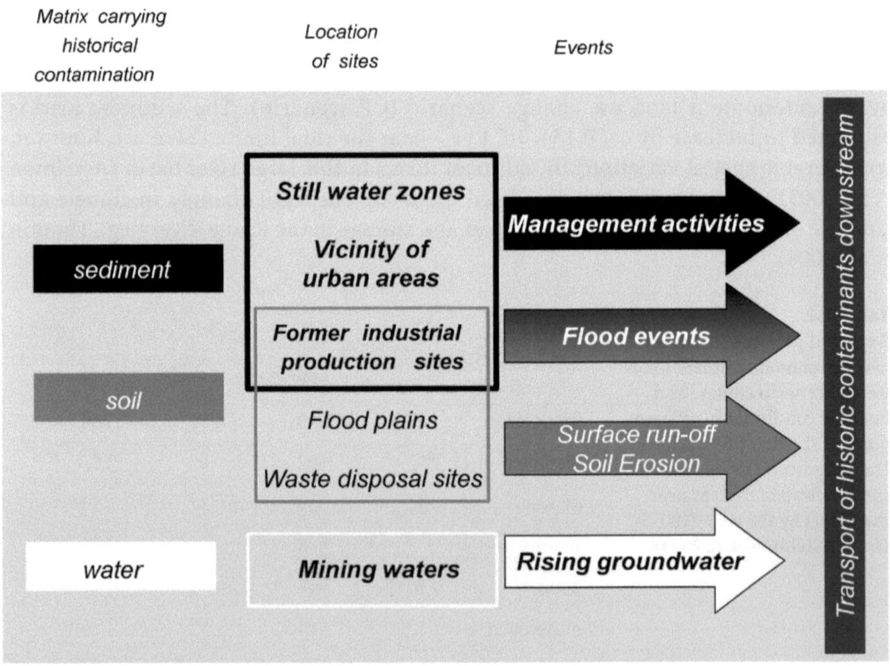

Fig. 1.2. Pathways and processes for the transport of historic contamination downstream (Heise et al. 2004)

dike foreshores, and polder areas. Even moderately polluted solids and pore waters are secondary sources of toxic chemicals, and further accumulation of these substances can then take place in the food chain. Therefore, monitoring and assessment of adverse effects will persist as a priority task in sediment management in the coming decades.

Transport of Historic Contamination Downstream

Today, relatively unpolluted recent sediment surface layers cover older contaminated sediments deposited in areas of low flow in river corridors, such as floodplains, river slack-zones and channel beds, reservoirs, and groyne fields. Nevertheless, there is an increasing risk of the resuspension of old contaminated sediment layers and the transport of the particle-bound pollutants downstream in river systems due to the potential for increasing water discharge associated with both anthropogenic activities (i.e. increased runoff to rivers due to land use changes) and climate change (i.e. increased precipitation). Contaminated material can also be introduced to river systems from contaminated soil and other diffuse sources during surface runoff or erosion events (Fig. 1.2).

The requirements for a river basin-wide sediment management concept will include inventories of interim depots within the catchment area (underground and surficial mining residues, river-dams, lock-reservoirs), integrated studies on hydromechanical, biological and geochemical processes, risk assessments on sedimentary biocoenoses and, last but not least, development of decision tools for sustainable technical measures on a river basin scale including sediment aspects.

Solutions for Contaminated Sediment Problems

Remediation techniques for contaminated sediments are generally much more limited than for most other solid waste materials. The widely diverse sources of contaminants and sediments in large basin areas usually produce a highly complex mixture of pollutants. For most of the sediment derived from maintenance dredging, there are more arguments in support of disposal compared to treatment. Considering the energy input which is needed for the separation of contaminants from their valuable matrices and for the purification of the various gaseous, liquid and solid emissions, then sediment conditioning for reuse – as pushed by waste legislation – will rarely conform to the principles of sustainability. In addition, dubious modes of utilization, such as filling of depressions etc., are often justified by data from test procedures which do not relate to characteristic sediment properties such as the content of redox-sensitive compounds (Förstner 2004a).

At this point, there is an emerging science-based technology for the final storage of sediments, called 'sub-aquatic depot'. The EU Landfill Directive (Anon. 1999) does not refer to waste disposal below the groundwater level, and here the two most promising conditions for a sediment depot can be found: *(i)* a permanent anoxic milieu to guarantee extremely low solubility of metals; and *(ii)* base layers of compacted fine-grained sediments which prevent the advective transport of contaminants to the groundwater (Anon. 2002b). Together with advanced geochemical and transport modeling, such deposits offer the most cost-effective and sustainable problem solutions for dredged sediments.

In the wake of this technology – of which the flagship is the Dutch 'De Slufter' depot – innovative sediment-specific applications can be developed, for example, techniques for active capping to safeguard both depot and in-situ contamination against pollutant release into the overlying surface waters.

Different from the management of dredged sediments, problem solutions for large-scale, complex contamination of floodplains are still in the early stages of development. In the 'intrinsic barrier' concept – presumably one of the very few realistic approaches to deal with contaminated soils and sediments on floodplains – soil and sediment components not only act as substrates ('habitats') for organisms to biodegrade substances, but also as media for supporting chemical and mechanical stability ('geochemical engineering approaches'; Förstner 2003).

1.1.4 Risk Assessment at the River Basin Scale

Many of the sediment issues and problems described above are mostly located at the mouth of the large rivers, such as deposition of sediment in harbors and ports, and the problem owners, such as the port authorities, are in a rather uncomfortable situation as they have to pay the costs for all former, actual and future shortcomings in the emission control within their catchment area. According to available information, there should be many more 'interim owners' of sediment problems in the upstream river basin. Many of them, however, ignore their problems or claim to follow a procedure called 'sediment relocation'. In the case of the latter, management problems and costs for the 'end owners' can be exacerbated further (Förstner 2002). In fact, it is not really a relocation or transfer of contaminated sediment to its original site, but rather a down-locating (as re-cycling of waste materials mostly is a down-cycling), and with the dispersion of pollutants there is an unecological increase of entropy. Although the large-scale effect of natural and technical resuspension processes is well-known – for example, in the Elbe basin typical patterns of dioxin congeners from the Bitterfeld area can be detected in the sediments of the Port of Hamburg more than 300 km downstream (Götz et al. 1998) – sediment problems in river basins are still regulated locally, sometimes by means of dubious threshold values.

Sediments and the EU Water Framework Directive

Here, a clear deficiency of the European Water Framework Directive (WFD) becomes evident. The WFD aims at achieving a good ecological potential and good surface water chemical status in European river basins by the year 2015, using a combined approach of emission and pollutant standards. These consider priority pollutants from diffuse and point sources, but neglect the role of sediments as a long-term secondary source of contaminants. Such a lack of information may easily lead to unreliable risk analyses with respect to the, apparent, 'good status' of waters.

The WFD monitoring objectives require compliance checking with Environmental Quality Standards (EQS) but also require the progressive reduction of pollution. However, *compliance monitoring for sediment* is not yet appropriate because of: a lack of definition of valid Environmental Quality Standards for sediment ($EQS_{Sediment}$) in a European context; analytical limitations; and the anticipated high costs required to obtain full spatial coverage. Sediment *trend monitoring* may be both spatial and temporal, and may be related to the chemical and ecological status of a water body. Sediment monitoring may also play a part in *risk-assessment*, for example in cases where the good-ecological-status/potential is not met or where water quality is adversely affected by the channel bed and/or resuspended sediment, and also in order to prioritize sites where actions can take place, and/or where monitoring should be intensified with respect to its effects along the river basin (Anon. 2004).

In principle, it has been recognized that harmonization of sediment monitoring is particularly relevant at a river basin level. Different objectives (trend monitoring, compliance monitoring, risk assessment and source control) will be involved and subsequently also different sampling strategies. However, technical issues such as sediment collection, sample treatment, sediment analysis and reporting results will have to follow a common level of quality requirements. An example is the application of the traceability concept in chemical sediment analysis (see Sect. 10.1, Förstner 2004b).

Catchment-Wide Assessment of Hazards and Risks

A basin-scale framework for sediment management should be comprised of two main levels of decision-making: the first for basin-scale evaluation (site prioritization) and the second for site-specific assessment (risk ranking) (Apitz and White 2003). Prioritization, among other things, needs the development of appropriate indicators for sediment mobility at a catchment scale and determination of the sediment dynamics and budget in a river basin (Heise 2003; Owens 2005a; Babut et al. 2007). In practice, a catchment-wide assessment of historical contaminated soil and sediment should apply a three-step approach (Heise et al. 2004): *(i)* identification of substances of concern (s.o.c.) and their classification into 'hazard classes of compounds'; *(ii)* identification of areas of concern (a.o.c.) and their classification into 'hazard classes of sites'; *(iii)* identification of areas of risk (a.o.r.) through consideration of erosion and transport processes and their assessment relative to each other with regard to the probability of polluting the sediments in the downstream reaches.

In a study of the historical contaminated sediments of the Rhine River (Heise et al. 2004), the target area was the Port of Rotterdam, with respect to a possible exceedance of those sediment quality criteria (based on the "Chemical Toxicity Test" (CTT) values of key substances) that decide the fate of dredged material: open water or the more expensive upland disposal. The probability of exposure was determined by calculations of erosion thresholds and indications that resuspension occurred. Probability of exposure was the most difficult parameter to quantify, as very little information existed about critical erosion thresholds and shear stresses for different flood situations.

In a catchment scale view, i.e., assessing the risk for downstream areas such as harbors or the coastal zone, inclusion of mechanical effects (e.g., resuspension of contaminated sediments) will significantly increase uncertainties with respect to the interpretation of combined erosion risk and chemical mobilization data, due to the large variability of granulometric and compositional parameters in the hydraulic term (Westrich, in Heise et al. 2004). Modeling pollutant transport on a river-basin scale requires broad information on water volumes, sediment dynamics and processes at the sediment-water interface. Apart from the quantification of anthropogenic activities (e.g., dredging, reservoir flushing) which should be dealt with when addressing advanced watershed management, prediction of the effects of large storm events on flow and the accompanying sediment load – even more pronounced because of its exponential increase – are among the most challenging tasks. In the three-step-approach of the Rhine River study, the hazards of "substances of concern" and of "areas of concern" could be determined with higher certainty than the risks of polluting sediments within the Port of Rotterdam. However, the combined information from critical erosion thresholds and indication that resuspension took place, as well as the differentiation of four risk classes with regard to the exceedance of well-defined target values, provided "evidence for high risk" for the Port of Rotterdam from historical contamination of sediments contained in the barrages of the Higher and Upper Rhine, even at annual flood situations (Heise et al. 2004).

In the future, an increase in the precision and accuracy of the term "indication that resuspension occurred" should significantly increase the weight of evidence for risks on downstream target areas. Here, sediment core profiles merit special attention (Westrich and Förstner 2005). The best locations for such historical records are within or close to the critical target areas (harbor basins, lakes, depressions, lowlands, floodplain soils and sediments, etc.). Additional information on the source areas of specific pollutants that are analyzed in the target sediment cores can be gained from indicator substances or from typical isotopes (e.g., lead isotopes) and patterns of congeners (e.g., for dioxins/furans; Götz et al. 1998).

1.1.5 Integrated River Basin Strategies

Both for establishing sediment-related quality objectives and for developing and implementing technical problem solutions, practical process-based knowledge is needed that uses a wide range of simulation techniques and models in different spatial and temporal scales (for example see Table 1.2). In practice, specific information on the interactions of hydromechanical, biological and geochemical processes is required for two reasons:

1. *Sediment quality assessment* is considerably more complex than water quality assessment due to the many site-specific parameters that need to be considered, which is not necessarily the case for water. Bioavailability of a contaminant is not only species specific, but also depends on sediment characteristics and sediment deposition and erosion. The methodologies developed to date do not adequately deal with the complex nature of sediments.
2. *Remediation techniques* for contaminated sediments are generally more limited than for other solid waste materials. Considering world-wide dredging activities, only a very small percentage of these materials can undergo "treatment" in the true sense – solvent extraction, bioremediation, thermal desorption, etc. Here, geochemical mechanisms such as stabilization and other forms of long-term, self-containing barriers could reduce the mobility and biological availability of critical pollutants.

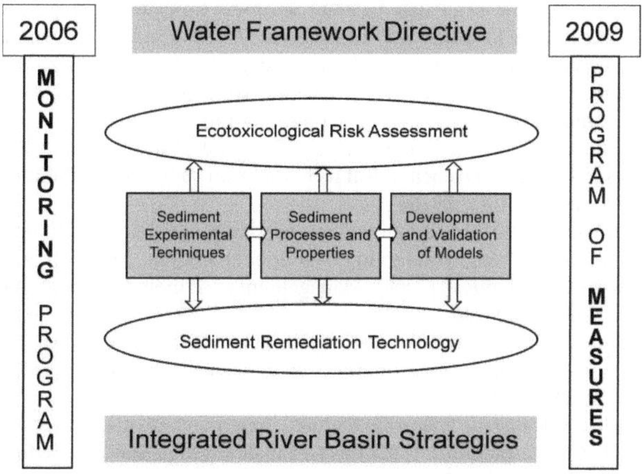

Fig. 1.3. Embedding of the themes and aims of the Priority Research Project SEDYMO (Sediment dynamics and pollutant mobility in rivers', Sect. 1.2) in the EU Framework Directive and other integrated river basin strategies

The position of integrated process studies, for example in the SEDYMO Program (Sect. 1.2), between ecotoxicological risk assessment and remediation technologies in the management of aquatic sediments and dredged materials is presented in Fig. 1.3, which also explains the position of this multidisciplinary research program in the context of the WFD and other integrated river basin strategies. As mentioned above, the WFD, which focuses on the catchment scale, does not consider sediment quality and quantity as a major issue. However, the strategies against chemical pollution of surface waters (WFD article 16) – i.e. implementation of monitoring programs until 2006 and establishment of the program of measures until 2009 – have to consider sediment quality (and quantity) at the catchment scale. With respect to the latter date, the first step – screening of all generic sources that can result in releases of priority substances and priority hazardous substances – will already include an assessment of the specific source/pathway of 'historical pollution from sediment'.

References

Anon (1999) Council Directive 1999/31/EC of 26 April 1999 on the landfill of waste

Anon (2002a) SedNet Demand driven, European Sediment Research Network, Description of Work (DoW), Final Version. EC Proposal no. EVK-2001-00058. 31 p

Anon (2002b) DEPOTEC, Amersfoort, and Strom- & Hafenbau, Hamburg, Sub-aquatic Depots of Dredged Material in the Netherlands (in German), Report, January 2002. Hamburg

Anon (2004) Expert Group of Analysis and Monitoring of Priority Substances (AMPS). Discussion Document, 13 January 2004, 8 p, Ispra

Apitz S, White S (2003) A conceptual framework for river-basin-scale sediment management. J Soils Sediments 3:132–138

Assleman NEM, Middelkoop H, van Dijk PM (2003) The impact of changes in climate and land use on soil erosion, transport and deposition of suspended sediment in the river Rhine. Hydrol Processes 17:3225–3244

Babut M, Oen A, Hollert H, Apitz SE, Heise S, White S (2007) Prioritization at river basin scale, risk assessment at local scale: suggested approaches. In: Heise S (ed) Sustainable Management of Sediment Resources: Sediment Risk Management and Communication. Elsevier Amsterdampp, 107–151

Batalla RJ (2003) Sediment deficit in rivers caused by dams and instream gravel mining. A review with examples from NE Spain. Cuaternario y Geomorfología 17:79–91

Batalla RJ, Vericat D, Palau A (2006) Sediment transport during a flushing flow in the lower Ebro River. In: Rowan JS, Duck RW and Werrity A (eds) Sediment Dynamics and the Hydromorphology of Fluvial Systems, IAHS Publication 306, IAHS Press, Wallingford, UK, pp 37–44

Carter J, Owens PN, Walling DE, Leeks GJL (2003) Fingerprinting suspended sediment sources in a large urban river system. Sci Total Environ 314–316:513–534

Förstner U (2002) Sediments and the European Water Framework Directive. J Soils Sediment 2:54

Förstner U (2003) Geochemical techniques on contaminated sediments – river basin view. Environ Sci Pollut Res 10(1):58–68

Förstner U (2004a) Sediments – resource or waste? J Soils Sediments 4(1):3

Förstner U (2004b) Traceability of sediment analysis. Trends Anal Chem 23(3):217–236

Gerrits L, Edelenbos J (2004). Management of sediments through stakeholder involvement. J Soils Sediments 4:239–246

Götz R, Steiner B, Friesel P, Roch K, Walkow F, Maaß V, Reincke H, Stachel B (1998) Dioxin (PCDD/F) in the river Elbe – investigations of their origin by multivariate statistical methods. Chemosphere 37:1987–2002

Hamer K, Hakstege P, Arevalo E (2005) Treatment and disposal of contaminated dredged sediments. In: Lens P, Grotenhuis T, Malina G, Tabak H (eds), Soil and Sediment Remediation, IWA Publishing, London, UK, pp 345–369

Heise S (2003) Sediment working group on risk management: The current discussion status. J Soils Sediments 3(3):129–131

Heise S, Förstner U, Westrich B, Jancke T, Karnahl J, Salomons W (2004) Inventory of Historical Contaminated Sediment in Rhine Basin and its Tributaries. On behalf of the Port of Rotterdam. October 2004, Hamburg, 225 p

Hudson-Edwards KA, Macklin MG, Jamieson HE, Brewer PA, Coulthard TJ, Howard AJ, Turner JN (2003) The impact of tailings dam spills and clean-up operations on sediment and water quality in river systems: the Ríos Agrio-Guadiamar, Aznacóllar, Spain. Appl Geochem 18:221–239

Jones CA, Basch G, Bayliss AD, Bazzoni D, Biggs J, Bradbury RB, Chaney K, Deeks LK, Field R, Gomez JA, Jones RJA, Jordan VWL, Lane MCG, Leake A, Livermore M, Owens PN, Ritz K, Stury WG, Thomas F (2006) Conservation Agriculture in Europe: An Approach to Sustainable Crop Production by Protecting Soil and Water? SOWAP, Jealott's Hill, Bracknell, UK

Kondolf GM (1997) Hungry water: effects of dams and river mining on river channels. Environ Manage 21:533–351

Köthe H (2003) Existing sediment management guidelines: an overview. J Soils Sediments 3:139–143

Leopold LB (1997) Waters, Rivers and Creeks. University Science Books, California, USA

Macklin MG, Brewer PA, Balteanu D, Coulthard T, Driga B, Howard AJ, Zahari S (2003) The long term fate and environmental significance of contaminant metal released by the January and March 2000 mining tailings dam failures in Maramure County, upper Tia Basin, Romania. Appl Geochem 18:241–257

Magilligan FJ, Salant NL, Renshaw CE, Nislow KE, Heimsath R, Kaste JM (2006) Evaluating the impacts of impoundments on sediment transport using short-lived fallout radionuclides. In: Rowan JS, Duck RW, Werrity, A (eds) Sediment Dynamics and the Hydromorphology of Fluvial Systems, IAHS Publication 306, IAHS Press, Wallingford, UK, pp 159–165

Netzband A, Reincke H, Bergemann M (2002) The river Elbe A case study for the ecological and economical chain of sediments. J Soils Sediments 2:112–116

Owens PN (2005a) Conceptual models and budgets for sediment management at the river basin scale. J Soils Sediments 5:201–212

Owens PN (2005b) Soil erosion and sediment fluxes in river basins: the influence of anthropogenic activities and climate change. In: Lens P, Grotenhuis T, Malina G, Tabak H (eds) Soil and Sediment Remediation, IWA Publishing, London, UK, pp 418–433

Owens PN, Collins AJ (2006) Soil erosion and sediment redistribution in river catchments: summary, outlook and future requirements. In: Owens PN, Collins AJ (eds) Soil Erosion and Sediment Redistribution in River Catchments: Measurement, Modelling and Management, CABI Publishing, Wallingford, UK, pp 297–317

Owens PN, Apitz S, Batalla R, Collins A, Eisma M, Glindemann H, Hoonstra S, Kothe H, Quinton J, Taylor K, Westrich B, White S, Wilkinson H (2004) Sediment management at the river basin scale: synthesis of SedNet Working Group 2 outcomes. J Soils Sediments 4:219–222

Owens PN, Batalla RJ, Collins AJ, Gomez B, Hicks DM, Horowitz AJ, Kondolf GM, Marden M, Page MJ, Peacock DH, Petticrew EL, Salomons W, Trustrum NA (2005) Fine-grained sediment in river systems: environmental significance and management issues. River Res Appl 21:693–717
Owens PN, Duzant JH, Deeks LK, Wood GA, Morgan RPC, Collins AJ (2006) The use of buffer features for sediment and phosphorus retention in the landscape: implications for sediment delivery and water quality. In: Rowan JS, Duck RW, Werrity A (eds) Sediment Dynamics and the Hydromorphology of Fluvial Systems, IAHS Publication 306, IAHS Press, Wallingford, UK, pp 223–230
Salomons W, Brils J (eds) (2004) Contaminated Sediments in European River Basins. SedNet Document
Slaymaker O (2001) Why so much concern about climate change and so little attention to land use change? The Canadian Geogr 45:71–78
Slaymaker O, Owens PN (2004) Mountain geomorphology and global environmental change. In: Owens PN, Slaymaker O (eds) Mountain Geomorphology. Arnold, London, pp 277–300
Thonon I, Middlekoop H, Van der Perk M (2006) The impact of changes in climate, upstream land use and flood plan topography on overbank deposition. In: Rowan JS, Duck RW, Werrity A (eds) Sediment Dynamics and the Hydromorphology of Fluvial Systems, IAHS Publication 306, IAHS Press, Wallingford, UK, pp 480–486
Vericat D, Batalla RJ (2006) Sediment transport in a large impounded river: the lower Ebro, NE Iberian Peninsula. Geomorphology 79:72–92
Westrich B, Förstner U (2005) Assessing catchment-wide emission-immission relationships from sediment studies. J Soils Sediments 5(4):197–200
Williams GP, Wolman MG (1984) Downstream Effects of Dams on Alluvial Rivers. USGS Report, US Government Printing Office, Washington DC, USA

Ulrich Förstner · Hans-Curt Flemming · Fritz Hartmann Frimmel · Giselher Gust Gerhard H. Jirka · Frank von der Kammer · Ole Larsen · Mark Markofsky Lutz-Arend Meyer-Reil · Reinhard Nießner · Bernhard Westrich · Eckhard Worch Werner Zielke

1.2 Sediment- and Pollutant-Related Processes – Interdisciplinary Approach

1.2.1 Introduction

Particulate pollutants represent a multi-disciplinary challenge for prediction and strategies to prevent their distribution in the environment. Four aspects, in an overlapping succession, reflect the development of knowledge in particle-associated pollutants during the past thirty years:

- the identification, surveillance, monitoring and control of sources and the resulting pollutant distribution;
- the evaluation of solid/solution relationships for contaminants in surface waters;
- the assessment of the environmental impact of particle-bound contaminants, i.e. the development of sediment quality criteria;
- the study of processes and mechanisms of pollutant transfer in various compartments of aquatic ecosystems.

For simulating real world effects on various temporal and spatial scales, an integrated approach has been developed combining hydrodynamic and chemical/biological factors which influence pollutant mobilization and transfer – the interdisciplinary SEDYMO approach ('*SE*diment *DY*namics and pollutant *MO*bility in rivers').

The following sections give an overview of process interactions which influence contaminant mobilization in rivers (Sect. 1.2.2); summarizes the state of knowledge in the three major SEDYMO themes: 'experimental techniques', 'processes and properties' and 'development and validation of models' (Sect. 1.2.3), describes three pre-SEDYMO examples for combined hydrodynamic and chemical/biological process studies (Sect. 1.2.4) and, finally, introduces the structure of the SEDYMO program (2002–2006) and its sub-projects (Sect. 1.2.5).

1.2.2 Sediment- and Pollutant-Related Processes

In Table 1.3 the major processes influencing the cycling of contaminants in aquatic systems are arranged according to the primary research disciplines involved, and pollutant phase (dissolved or particulate). Sediment bioturbation is a process in which benthic fauna contributes to the mixing and resuspension of different sediment layers. Bioturbation stimulates many of the characteristic interactions between chemistry and biology and between chemistry and photo-degradation. Biological activity is involved in physical cycling of particulate matter both in the water column and at the sediment/water interface. Organic excretions may produce fecal pellets and may enhance aggregation and thus accelerate the settling of particles. There are well-documented effects of reworking and resuspension of sediments by benthic organisms such as tubificid worms, but also by amphipods, shrimps, and clams. Bioturbation is a major post-sedimentation process, affecting the fate of particle-associated toxic metals and persistent organic chemicals, which are not primarily affected by volatilization, photolysis or bio- and photo-degradation (Allan 1986). In fluvial systems, the cycling of pollutants is dominated by the processes of resuspension, settling and the burial of particulate matter.

System of Interacting Natural Processes in Rivers

Due to their particular dynamics, three characteristic features of sedimentary processes in rivers should receive special attention:

- The dramatic effects of flood events on particle transport,
- the rapid and far-reaching effects of sulfide oxidation, and
- the biological accumulation and potential release of toxic compounds.

In practice, emphasis has to be put on the role of fine-grained sediments and suspended matter, since these materials exhibit large surface areas and high sorption capacities. Organic materials are highly reactive. Degradation of organic matter will cause depletion of oxygen and may enhance formation of flocs and biofilms.

Within the system of substrates and processes, three scientific disciplines and three study objects can be discerned from Table 1.3: suspended matter, sediment and porewater/bulk water; the formation of aggregates in turbulent water, flocs and biofilms from biological transformations; and the formation of new surfaces for re-adsorption of dissolved pollutants. The main focus is on the degradation of organic matter, which affects both hydrodynamic processes – here erosion vs. sedimentation – and (bio)geochemical redox

Table 1.3.
Processes affecting the cycling of pollutants (organics, metals) in aquatic systems

	Aqueous species	Particulate species
"Chemical"	Dissolution Desorption Complexation	Precipitation Adsorption Aggregation
	Species transformation	
"Biological"	Decomposition Sorption, release Cell wall exchange	Food web transfer Filtering, digestion Pellet generation
	Bioturbation	
"Physical"	Advection Diffusion Colloidal transfer	Resuspension Settling Burial

cycles. The crucial question, after all possible interactions between both existing and newly formed solid and dissolved phases, leads to the net release of dissolved organic carbon (DOC), nutrients and pollutants, including metals, into the open water.

1.2.3 Interdisciplinary Approach

Figure 1.4 reflects the basic concept of the coordinated research program "Fine Sediment Dynamics and Pollutant Mobility in Rivers". An interdisciplinary approach has been derived from an evaluation of the state-of-the-art technology with three major themes: "experimental techniques", "processes and properties", and "development and validation of models".

Experimental Techniques

The quantification of flow rates, including the transport of particle aggregates, microorganisms as well as dissolved and adsorbed substances, needs an integration of various experimental and modeling techniques for the determination of hydrodynamical, chemical and (micro)biological parameters. In these fields, research at various institutions has led to new perceptions and a working hypothesis, which allow quantification of the above-mentioned relocation and flocculation/aggregation processes. The development of new systems can be based on the following research work:

- Clarification of the correlation between mineralogical, microbiological and pore water parameters in erosion devices with the precise control of bottom hydrodynamics (Amos et al. 1992; Booij et al. 1994; Wiltshire et al. 1998).
- Determination of the dependence of erosion stability on consolidation and mechanical properties of bottom sediments, including biogenic stabilization by microorganisms (Spork 1997; Tolhurst et al. 2002).
- Studies on the types of aggregates in the water column under *(a)* turbulent, *(b)* oscillating and *(c)* slowly cyclic variable (tidal) flow (Perkins et al. 2004).

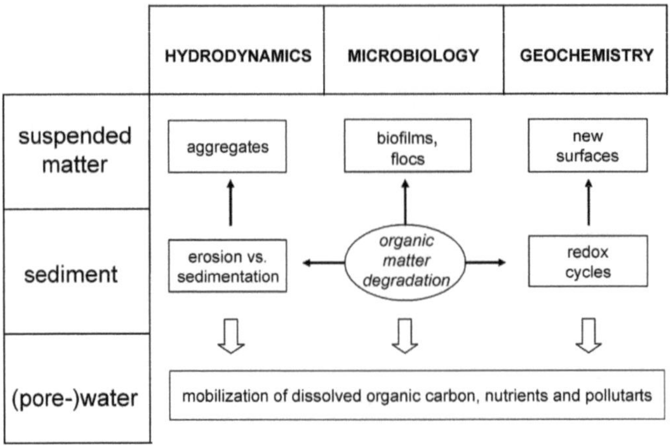

Fig. 1.4.
Scientific disciplines and study areas in a priority research program of the German Federal Ministry of Education and Research, 2002–2006, on sediment dynamics and pollutant mobility in rivers (SEDYMO)

- Investigations of the erosion and deposition behavior of particles and the respective vertical mass fluxes in different hydrodynamic flow field simulators, such as erosion chambers, erosion flumes, differential turbulence columns, etc.(Gust and Müller 1997; Porter et al. 2004)

Processes and Properties

The second main topic comprises the micro-biological and geochemical studies, which are needed as a basis for the evaluation of priority pollutants relative to quality standards for sediments and suspended matter. The *biological sub-projects* are mainly aimed at investigating the quantity and quality of particulate and dissolved organic substances, their microbial metabolism using various electron acceptors, and the role of biofilms and colloids on the behavior of pollutants in sedimentary systems. The major objectives of the *geochemical sub-projects* are realistic descriptions of nutrient and pollutant transfer from the particulate into dissolved phases (pore water and open water body); the dynamic and subsequent micro-scale heterogeneity of material dispersion within the sediment; and the diffusive transfer of pollutants across the sediment/water interface. Most of the micro- and mesoscale geochemical information will be used in the development of models.

Riverine flocs have a complex composition and may be dynamic in both structure and function due to manifold interactive processes which operate between various physical, chemical and biological factors. Most flocculated natural aquatic sediments commonly have a living and active biological component in conjunction with inorganic and non-living biological particles. Flocculation alters the hydrodynamics of particles and therefore influences the fate and effect of sediment-associated contaminants. In a review of these processes, Droppo et al. (1997) suggest that fibrils consisting of extracellular polymers are the dominant agent for both the development and stabilization of flocculated materials. This does not exclude electrochemical flocculation completely, but rather it appears to be less significant than the biological flocculation in natural systems.

Biofilms

Biofilms are very heterogenous. They consist of various microorganisms, which develop on various surfaces under multiple conditions (Characklis 1990). In many natural cases, mineral surfaces are at least partially covered by a biofilm. Dissolved substances will thus cross this biofilm first and be possibly sorbed there before they reach the original mineral surface (Chap. 9).

The sorption properties of the insulating layer are, therefore, of importance for the dissolved and the sorbed state of pollutants (Flemming et al. 1995; Flemming and Leis 2002). Active binding occurs through the excretion of binding-, chelating- or precipitation-cell products in response to the presence of the dissolved substance. In addition, active transport systems may allow the uptake of (e.g.,) metal ions into the cytoplasm. Metal binding by bacterial surfaces is considered largely as a passive phenomenon within the process of electrostatic interaction between cationic metals and anionic cell surface groups (Flemming and Leis 2002). In the presence of metabolic activity, however, microorganisms can bring about metal precipitation indirectly through the production of inorganic ligands, such as, sulfide and phosphate among others and directly by changing metal redox state, e.g., oxidation of reduced iron.

Colloids

In the transport of pollutants in surface, subsurface and porewater of soils and sediments, the colloidal phase, often defined as particles between 0.001 μm and 1 μm, can play a major role due to their high specific surface area and high mobility (Buffle and Leppard 1995). From the available data, it appears that sub-micron particles in oxygenated river water mainly consist of organic matter (fulvics, humics, polysaccharides, proteins), silica, iron oxyhydroxides and possibly small clay particles. While representing only a small fraction (<10%) of the total particle mass, their number increases with decreasing particle size (Buffle and Van Leeuwen 1992), and thus represent a considerable portion of the overall surface area available for interaction with pollutants. There are indications that pollutants bound to colloids can increase toxicity in aquatic systems (Vignati et al. 2005).

Key questions are the mobility of colloidal particles and hence their stability against aggregation and capture by larger flocs and the colloid-pollutant interaction, which can significantly differ between different colloidal phases (e.g., iron-oxide colloids and natural organic matter, NOM). Other points of interest are colloidal particle release from sediments and pore water, and generation by aquatic biota. The colloidal part of the NOM seems to play a mayor role in fresh water due to its high mobility. One key observation to understand the behavior of inorganic colloids was reported by Wilkinson et al. (1999) where the aquatic organic matter (mainly polysaccharides) promoted aggregation of inorganic colloids, while soil-derived organic matter stabilized the inorganic colloids against aggregation. Hence not only the general water chemistry as a whole plays a mayor role but also the origin of NOM can change colloidal behavior. Today a thorough quantitative investigation of the role of different colloidal carriers for pollutant transport on the river scale is still rare (Vignati et al. 2005). This is mainly due to the lack of suitable methods to quantify the colloidal particles and the pollutants associated with them. On a routine basis, sequential filtration is often applied to quantify colloidal components and although it is comparably easy to perform it is prone to artifacts (Morrison and Benoit 2004). Also the choice of cut-off diameters and molecular weights seems to be arbitrary and complicates comparisons (Babiarz et al. 2001; Vignati et al. 2005; Rostad and Leenheer 1997; Wen et al. 1999). Studies applying advanced methods for colloid analysis are tedious (Lyven et al. 2003; Lead et al. 2005; Lead and Wilkinson 2006) and therefore limited in their application in terms of temporal and spatial extension. Hence the available data is often only a snapshot, from which it is difficult to derive information about processes. Recently Lyven et al. (2003) identified colloids in the size range between 1 and 10 nm in diameter in a small Swedish creek. While iron (and many other main and trace metals) were found to be associated with carbon-rich colloids (NOM), it was also present with larger colloids (5 nm) of potential inorganic origin. Lead was mainly bound to this second fraction. Investigations of Baalousha et al. (2006) on the river Loire confirmed the absence of lead in the carbon-dominated fraction. Stolpe et al. (2004) sampled the same Swedish creek under different seasonal conditions and the qualitative partitioning of lead seems to be unaffected. This points out the need for investigations which account for the continuum-like size distribution of aquatic colloids, as well as the further development, harmonization and validation of methodologies.

Sensor Techniques

Micro-electrodes are useful tools for high spatial resolution assessment of relevant parameter distributions in the immediate vicinity of micro-organisms. For example, the development of a microbial biosensor now allows micro-scale determination of the bio-available fraction of organic carbon in sediments and bio-films (Neudörfer and Meyer-Reil 1997). Considerable progress in the simulation of metabolic processes in sediments has been made possible through the development of particle-oriented sensor techniques, e.g., by the Max Planck Institute for Marine Biology in Bremen (Jørgensen 1994). Apart from glass-based micro-electrodes for measuring oxygen (Glud et al. 1998), hydrogen sulfide (Kühl et al. 1999) and carbon dioxide (De Beer et al. 1997), there are fiber-optical sensors (optodes) for measuring the time-decay of analyte-specific fluorescence receptors of dissolved substances such as oxygen, nitrate, nitrite and ammonia. This allows an analysis on a micro-scale, e.g., at the sediment-water interface (Kühl et al. 1997).

A new photo-acoustic sensor was developed for in-situ and online concentration monitoring of aromatic substances in water (Mohacsi et al. 2001). Selective detection is accomplished by transferring the analyte of interest into the gas phase through a permeable membrane. Unlike the conventional approach, where sampling and detection units are separated, here the membrane is inserted directly into the central part of the photo-acoustic resonator, thereby eliminating the need for various gas-handling components and purging gases. In this way, the system becomes simpler, more compact, has a response time of 40 min and a potential for fully automated operation. The use of a near-IR (1.67 μm) diode laser coupled to the PA cell by an optical fiber yields detection limits of 350 μg (for benzene) and 1.1 mg (for toluene) per liter of water.

When placed directly into water wells or sediments, the proposed sensor can serve as a warning system for long-term automatic observations.

A fiber-optic sensor system for the online detection of heavy metal ions in water was presented by Prestel et al. 2000 and Zhang et al. 2000. This is based on the laser-induced fluorescence spectroscopy of suitable metal-ligand complexes. The sensor system is designed to measure heavy metal ions in the field. Flow injection analysis (FIA) is coupled with the sensor system to overcome problems associated with the slow diffusion rates of heavy metals through the membrane of an in-situ sensor head. First experiments have shown that the new FIA system has good reproducibility, a high sample analysis rate and that one can measure heavy metal ions (Cu(II), Ni(II), Cd(II) and Zn(II)) at the ppb level, if the appropriate ligands are used.

Pore Water Studies

The composition of pore water is a highly sensitive indicator for reactions between chemicals on solid substrates and the aqueous phase which contacts them. It must be remembered that the chemical composition of the pore water is controlled primarily by microbial processes working at significantly higher spatial resolution. High resolution techniques applied to pore water studies contain among other microsensor studies (see above). Experiments have clearly demonstrated that the surface structure of the sediment plays a key role for the advective oxygen transport (e.g., Huettel et al. 1998).

The turn-over of other redox sensitive elements has turned out to be more difficult to analyze and the results less clear in their interpretation. (Hydr)oxides of iron(III) and manganese(III,IV) are recognized to be very important sinks of pollutants as well as quantitative important electron acceptors in anoxic sediments. The pore water concentration of Fe^{2+} and Mn^{2+} correspond to a very small fraction of the total iron and manganese in sediments and extreme analytical care must be taken using traditional analytical approaches. Direct measurements of the transition metals include voltammetric assessment (Nuester and Larsen) and sampling the metals in gels after diffusive equilibrium has been obtained (Davison, see Sect. 8.1).

Given the fact that most processes involving sediment particles and pore water are fast and that the concentrations merely reflects partial equilibria concentrations the concentrations alone tell very little about the actual rates of turn-over in sediments and transport of over the sediment-water interface. Various methods have been developed for the quantification of rates based on radio tracers and stable isotopes (Elsgaard and Jørgensen 1992).

For instance, the rates of sulfate reduction and the turn-over of electron donors like volatile fatty acids can be directly measured using radiotracer techniques. Incubating sediments (e.g., cores or slurries) added radio labeled reactants allow after measuring the partitioning of the radio tracer between product and reactant as well as the reactant concentration a direct quantification of the turn-over rates (Cranfield et al. 1993). Using landers incubation of deep-sea sediments under in-situ conditions is even possible (Greef et al. 1998).

In riverine sediments advective transport plays a large role and the transport regime may in the estuaries change direction and force within few hours. Hence, technology can be transferred from the extensive marine experimental basis but most techniques must be adopted in order to function satisfactorily in rivers (Larsen, Sect. 8.1 in this book).

Development and Validation of Models

Analytical and numerical models are indispensable for both connecting and integrating the interdisciplinary study of individual processes and for transferring the results of laboratory experiments to natural aquatic systems, where processes take place on extremely variable spatial and temporal scales.

Numerical models can be applied to schematize and simulate physical, chemical and biological processes:

- Transport and reaction modeling, considering advective, dispersive and diffusive transport mechanisms as well as ad- and desorption processes (e.g., CoTReM; Landenberger 1998).
- Hydrodynamic (Johnson and Tezduyar 1997; Boivin et al. 1998; Ling et al. 1998), statistical (Lick et al. 1992) and/or stochastic (Hesse and Tory 1996) models operating on the particle level are best suited for the study of fine-scale aggregation/segregation processes and, in addition, may include biological and chemical processes.
- Macroscale long-term simulation can only be performed using 1- or 2-D model approaches (including particle-tracking models (Wollschläger 1996)) due to limited computing capabilities.
- Three dimensional continuum-models (Le Normant et al. 1998; Malcherek 1995 and 2001) are particularly efficient for locally concentrated emissions and for conducting short to medium-term simulations.

Available material transport models are still restricted mainly to the description of transport and dispersion processes of suspended sediments as well as dissolved and particulate substances. The hydrodynamic interactions between turbulent flow with suspended and bottom sediment are still not totally resolved and process descriptions and numerical simulations of biological and chemical influences on particles and pollutants, especially in the near bottom-zone – are still in an early stage of development.

Diversity and uncertainty complicate the determination and parameterization of bio-chemical parameters affecting binding in heterogeneous systems. Sorption sites have variable affinity for the adsorbents and there is highly variable solubility of the amorphous solid phase. In particular, for describing the transport of inorganic and organic substances, high priority has to be given to the effects of competing adsorption and replacement desorption; here, the influence of dissolved organic matter (DOM) on sorption processes has not adequately been considered as yet. It was shown that DOM can reduce the sorption of hydrophobic organic pollutants due to complex formation (Rebhun et al. 1996), but it seems that the strength of complex binding is substance-specific (Amiri et al. 2005) and therefore of different significance for sorption modeling.

Modeling Hydrodynamic and Biogeochemical Data

The calculation of equilibrium speciation in aqueous systems, using computer programs like MINEQL (Westall et al. 1976) or PHREEQC (Parkhurst 1995), requires the exact knowledge of the formation constants of all species under consideration as well as the total mass of some selected components which are derived from chemical analy-

Table 1.4. Development of models coupling hydrodynamic and biogeochemical data for the prediction of pollutant transport in rivers (after Kern 1997)

Numerical description	Components in water body	Solute/solid interaction	Formulation of transport equation
1. Distribution coefficent	Dissolved + particulate	K_d = constant	2 coupled linear differential equations
2. Extended K_d-concept	Dissolved + particulate + milieu factors	$K_d = f$(pH, pe, complexing agents, competing ions)	Additional n linear differential equations for milieu factors
3. Chemical multi-component model	Chemical individual species	Dissociation and binding constants, solubility products	Coupled differential-, algebraic equation system
4. Biochemical multi-component model	Chemical species + biota	Additional growth and decay rates	Differential algebraic equation system

ses. Imprecision may arise from uncertainties in experimental parameter determination as well as from inconsistencies in published data, sometimes differing by orders of magnitude. These effects are particularly apparent in heterogeneous systems such as sediment and suspended matter.

Models for predicting pollutant transport in rivers are dominated by hydromechanical parameters, including bio-chemical terms using constant distribution coefficients (Table 1.4, Kern 1997). A first step for extending these models involves consideration of typical ecosystem factors such as competing ions, complexing agents, redox conditions and pH values when metals are considered. The next level of sophistication would be the inclusion of binding constants, solubility products and other factors, which can describe solid/solution interactions of critical chemicals in a multi-component system. An quantum step would extend the mechanical-chemical model into biology. Such biochemical multi-component models should at least consider rates of growth and decay of organisms and organic matter.

1.2.4 Pre-SEDYMO Integrated Process Studies

First integrated process studies on sediment dynamics and pollutant mobility in river sediments date back to a coordinated research project of the German Research Council "Interactions between abiotic and biotic processes in the tidal Elbe River" (1986–1996; Kausch and Michaelis 1996). This included:

- studies on the temporal and spatial fluctuations of suspended matter discharges via sound ranging, which provide data on the fractioning and remixing of solids in various stages of tidal river flow (Seibt-Winckler and Schirmer 1996);
- model development on the hydrodynamics of suspended matter, allowing simulation of three-dimensional transport (including interstitial scenarios) and variable bottom topographies in the tidal Elbe River (Rolinski 1997);

- studies of the increased enzymatic activities on suspended matter, and the dependence of microbial inhibition on the sediment load (Neumann-Hensel and Ahlf 1995);
- investigations on exopolymeric substances, which both provide highly sorptive matrices for particulate/dissolved substances and may induce stabilization of aggregates (Kies 1995; Humann 1995);
- the mutually influencing groups of variables – such as "driving forces" (C-, N-, S-, Fe/Mn-cycles) and capacity controlling properties of the solid matrix – for the redistribution of trace elements (Hong 1995; Förstner 1996; Petersen et al. 1996).

Two experimental devices, the suspension cell, consisting of a thermostatic bioreactor with a 3 liter water volume in a gas cycle, and the LABOratory SImulation Apparatus (LABOSI), where six sediment cores can be inserted and processed at the sediment-water interface, were developed by the GKSS Research Centre in Geesthacht (Petersen et al. 1995). In his dissertation, Hennies (1997) provided detailed insights on the exchange processes at the sediment/water boundary layer and on the release of previously accumulated individual substances during biological degradation of phytoplankton in the water column. Suspension cell experiments demonstrated the effect of typical estuarine conditions, i.e. light deficiency, increasing salinity, on the decay of limnic algae during a vegetation period. About 30–50% of the particle-bound copper, cadmium, zinc and lead was remobilized, presumably due to interactions with organic matter. Using the LABOSI apparatus in which up to six sediment cores are in contact with laminar water flow in a close water cycle (Schroeder et al. 1992); Hennies (1997) showed that pore water transfer of chemicals in the upper 1 cm sediment layer is controlled by molecular diffusion and benthic activity. In the deeper sediment advection along voids caused by methane gas dominates.

SETEG Flume Experiments – Hydrodynamic vs. Chemical/Biological Stability in Sediment Core Profiles

Since 1994, a sequence of individual projects was carried out at the Institute of Hydraulic Engineering at the University of Stuttgart, combining channel experiments and field studies on the lock-regulated Neckar River in Southern Germany (Westrich and Kern 1996). At a later stage, the experiments were performed on layered sediments at different flow velocities (Westrich and Kern 1996). Transport equations involving terms for convection and longitudinal dispersion, and information on either sedimentation or erosion were combined with a set of equations including data on adsorption and desorption, pollutant degradation and evaporation. With regard to pollutant transport, the actual model is more or less based on a simple K_d-approach. This involves the assumption of a uniform distribution coefficient for each pollutant, disregarding the specific conditions of the liquid and solid phases. Further developments are mainly aimed at overcoming the limitations which result from this gross simplification.

Using this model of Westrich and Kern (1996), the hazard potential arising from resuspension of contaminated sediments can be estimated from the product of hydraulic mobilization and chemical mobilization (Fig. 1.5). Mobility is the net result of stabilizing and mobilizing effects in both sectors. For example, hydraulic mobility can be assessed by measuring sediment coverage, critical shear stress and the bottom shear

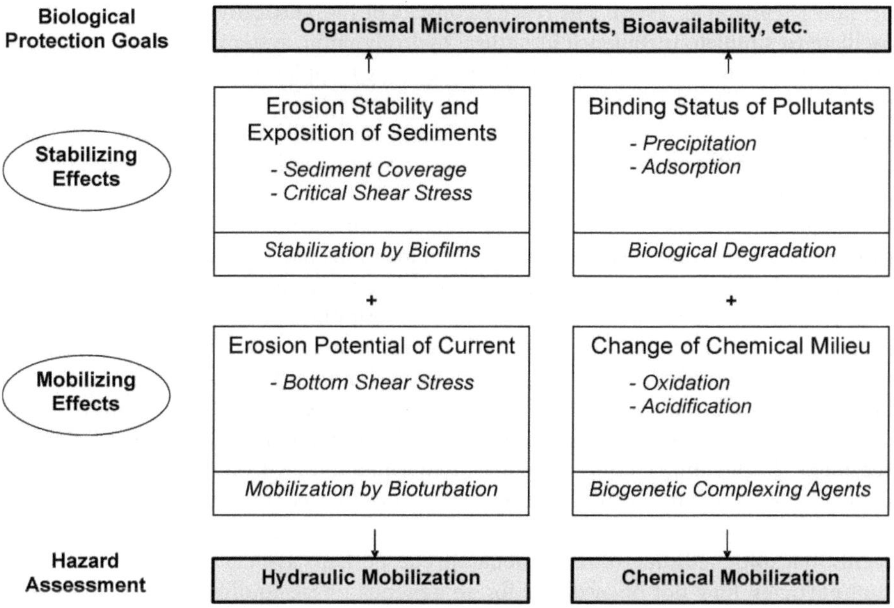

Fig. 1.5. Physical and chemical factors for contaminated sediment mobility assessment (after Westrich and Kern 1996)

stress. Geochemical mobility, e.g., for heavy metals, can be estimated from solubility and desorption data, which may, however, be strongly influenced by changes in the chemical milieu. For heavy metals, acidity plays a dominant role, and protons may be produced from the oxidation of metal sulfides.

At this time, major deficiencies were identified in biological information. Boxes in Fig. 1.5 indicate typical examples for biological implications both for stabilizing and mobilizing sediments and their associated pollutants. Biofilms, for example, will stabilize sediment deposits mechanically. However, some of their exopolymers may become effective as complexing agents for the mobilization of heavy metals. With regard to organic pollutants, biological degradation can be considered as a stabilizing effect, when degradation reaches its completion. However, it is possible that metabolic reactions will produce more mobile and toxic intermediates than the original pollutant. In many cases, biofilms will provide a temporary sink for sorbed pollutants which will turn into a source when either the sediment which supports the biofilms is disrupted by high shear forces or if nutrient depletion leads to biofilm decomposition. As a consequence, sorbed substances which are not fully degraded or non-degradable (e.g., metals) will be remobilized.

Simulating Natural Hydrodynamics and Behavior of Organic Contaminants with a Differential-Turbulence Column

The Differential-Turbulence Column was developed at Cornell University in the framework of a Ph.D. thesis by Brett Brunk (1995) and was initially used to simulate the homogeneous turbulent kinetic energy and sediment-loading profiles for open channel flow (Brunk et al. 1996). The reactor consists of five vertically spaced grids, which oscillate to simulate turbulence in natural hydrodynamic systems. The spatial distribution of turbulence is measured using an acoustic Doppler velocimeter. In-situ sampling can be made by introducing chemicals and monitoring chemical dynamics. Under homogeneous and open channel flow turbulent conditions, sediment profiles obtained in the differential-turbulence column accurately followed conventional theory (Brunk et al. 1996).

Using this device, Brunk et al. (1997) studied the enhancement of sorption of phenanthrene to particles in an estuary, since certain locations had been reported to be sinks for hydrophobic pollutants, and sorption has been commonly considered to be an important mechanism for the observed pollutant trapping. The sorption enhancement caused by 'salinity effects' and dissolved organic matter (DOM) coatings were both measured and modeled. The polycyclic aromatic compound phenathrene, an extracellular polymer from a soil bacterial isolate, and a low organic carbon kaolinite were used respectively as models for the hydrophobic pollutant, DOM, and for suspended sediment. Both salinity effects and DOM coatings induced increased sorption, the former ~50%, the latter ~10%. These experiments showed, that equilibrium sorption of phenanthrene cannot explain the full extent of pollutant trapping in estuaries. It seems that some sediment-bound phenanthrene, perhaps associated with atmospheric soot particles, may not be available for an aqueous phase equilibrium distribution.

'Microcosm' Experiments – Metal Transfer during Sediment Resuspension in Rivers and Estuaries

Due to the capacity of sediments to store and immobilize toxic chemicals in so-called 'chemical sinks', direct effects of pollution may not be directly manifested. This positive function of sediments does not guarantee, however, that the chemicals are safely stored for ever. Factors influencing the storage capacity of sediments or the bio-availability of the stored chemical can change and indirectly cause sudden and often unexpected mobilization of chemicals in the environment (Stigliani 1988). From the discussions on the 'Chemical Time Bomb' (CTB) concept during the early 1990s, it became apparent, that it is imperative to know what sediment properties will control the toxicity levels of a chemical and how sensitive the chemical toxicity is to changes of these properties. *Acidity*, as suggested by Stigliani (1991), is the most important driving force in chemical time bomb effects. In river sediments, acidity can be produced from the process 'split of sulfate' (Van Breemen 1987): During organic degradation, iron sulfide and calcium bicarbonate are formed and the latter is removed with running water. The (solid) acid producing potential can come into action during resuspension and oxidation. With each cycle of deposition and erosion, a certain proportion of buffer capacity in the sediment is consumed. In cases when buffer substance – mainly calcium carbonate – is no longer available, a breakthrough of acidity and heavy metals can be expected (Förstner 1995).

In a project funded by the German Research Foundation (Förstner and Gust 1996), resuspension experiments on cohesive harbor sediments at defined shear stress and related parameters was undertaken using an erosion chamber device ('microcosm'). The principle operational settings of the erosion chamber designed by Gust (1991) are chosen to generate a spatially homogenous skin friction at the sediment surface. This is achieved *(i)* by the variable rotational speed of a stirring disk placed 8 cm over the sediment-water interface, and *(ii)* by the volume of recirculated suspension sucked concurrently through the rotating axis. The resulting suspension is monitored on-line in the concentration range 0–20 g l^{-1} by using an infrared two-channel turbidity meter and a mass flow meter. Concentration is indicative of the phase with erosion/deposition cycles. Investigations on the release of trace metals from model sediments by Akkiparambath (1999) revealed four typical effects:

1. during the simulation of skin friction velocity over tidal cycles, the initially resuspended particles of organic bottom sediment typically exhibit higher concentrations of copper and cadmium compared to the subsequent resuspension phases.
2. an increase in erosion stability was realized in a sequence of resuspension events. As a result, a decrease of metal mobilization by resuspension took place, particularly in organic-rich model sediments.
3. during the resuspension experiments with model sediments with close-to-real sediment composition, an increase in inorganic carbon was observed. This was accompanied by a decrease of dissolved calcium. This result suggested that the precipitation of calcium carbonate occurred.
4. During oxidation of sulfide components, i.e. during resuspension of anoxic sediments, the carbonate content steadily decreased. Oxidation of ammonia to nitrate during a batch experiment using artificial sediments, induced a drastic pH decrease and resulted in a continuous increase of dissolved cadmium concentrations during three redox cycles.

In Fig. 1.6, results are presented from a microcosm experiment by Fengler et al. (1999) on the release of calcium and cadmium from a poorly buffered harbor sediment from the Elbe River at Hamburg during an erosion period of more than 1 000 hours. There is a significant difference in the metal release at later stages – after approx. 500 hours depending on whether there was bottom sediment or not. In the former situation, there is practically no pH change and no metal release from the resuspended sediment. This can be explained by an ongoing supply of buffer capacity.

The sequence of factors and processes controlling cadmium mobility has been clarified by Peiffer (1997). Cadmium is relatively mobile and is affected by exchange processes with calcium. In the case of a well-buffered neutral sediment, the addition of

Fig. 1.6.
Mobilization of calcium and cadmium in resuspension experiments with poorly buffered harbor sediment from the Elbe River at Hamburg. Open signature: no bottom sediment; full signature: with bottom sediment (Fengler et al. 1999)

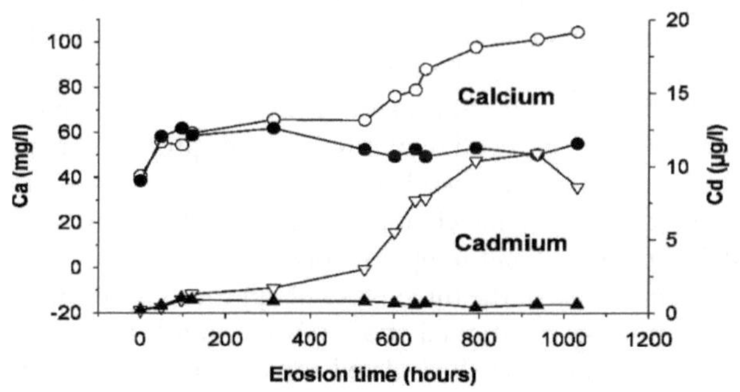

oxygen leads to the oxidation of sulfides, ammonia and organic matter. Acidity, in the form of carbonic acid and protons, is then consumed within the system with matrix-bound Cd^{2+} by the dissolution of calcium carbonate and the exchange of released Ca^{2+} and protons. Further input of protons is provided from the oxidation of Fe(II) to Fe(III), which then mediates further oxidation of iron sulfides. It is important to note that the exchange of calcium (or magnesium) is the major mechanism for the release of cadmium into the water phase in such buffered systems.

At this stage, the cause of observed *delayed* changes in pH and metal release was not fully resolved. Reports of such effects date back as far as the mid-eighties. Interference with microbial activity (Prause et al. 1985) seems to provide more probable explanations than inorganic complexation following slow oxidation (Salomons et al. 1987).

1.2.5 Sedymo Priority Program 2002–2006

The approaches used to understand sediment-contaminant mobility and resuspension-deposition tend to be either *(i)* detailed laboratory simulation and/or analytical experiments, *(ii)* field-based measurement programs, or *(iii)* numerical and/or GIS-based modeling approaches. The three types of approach offer different but complementary information (Owens and Petticrew 2006).

On the basis of the concept outlined in Sect. 1.2.2, a proposal for an interdisciplinary research project on sediment processes was initiated in 1999. During discussions concerned with the new EU water directive and with sediment removal by hydraulic dredging – a highly controversial issue on a national level – the theme of interacting sediment processes included also practical aspects. The resulting coordinated research program comprising 13 sub-projects is funded by the Federal Ministry of Education and Research (BMBF); the first seven sub-projects (nos. 1/13, 2, 3, 5, 14, 18b in Fig. 1.7; see also Table 1.5) started in May 2002; Phase 2, comprising six projects (7, 8, 10, 11, 15, and 19 in Fig. 1.7) started in May 2004. A short description of the 13 sub-projects was presented in a pre-conference overview of the International Sedymo Symposium, held at Hamburg University of Technology, March 26–29, 2006 (Förstner and Westrich 2005). Sections representing contributions of SEDYMO sub-projects in the present book are listed in Table 1.5.

In the center of the first phase was the optimization and application of devices for the study of erosion behavior, i.e., sub-project nos. 1, 2 and 3 (Fig. 1.7), and the development and validation of sediment transport models (sub-projects nos. 13 and 14 in Fig. 1.7). The second phase added the sub-projects focusing on chemical and biological parameters that control the mobility and transport of river sediments and their pollutants. Of particular interest were several subjects studying the potential function and influence of biofilms, such as extracellular polymeric substances (EPS), on sediment stability and pollutant mobility. Sub-project 19 aims at a better understanding of the relationship between key bio-chemical processes and bio-availability of contaminants; the toxicity and chemical data of sediment and water samples were studied under different spatial and temporal conditions along the Elbe River (Sect. 10.2).

In the sense of the three types of approaches for understanding sediment-contaminant mobility mentioned above (Owens and Petticrew 2006), the first enables specific controlling factors to be isolated and examined. Typical Sedymo examples are the sub-projects no. 1 (SETEG flume) and no. 2 (turbulence column), the latter closely cooperating with sub-project 8, where the main physico-chemical factors determining the resuspension and deposition behavior of fine sediments was studied. Similar cooperation between the sub-projects no. 10 (metal transfer processes) and no. 15 (sorption of hydrophobic organic pollutants) was based on Gust's 'erosion chamber' (sub-project 3).

The second ('ii'), field-based approach provides a more realistic assessment of conditions within a river, often at larger spatial scales; as an example, the combined work of sub-projects no. 3 (erosion chamber plus field devices) and no. 11 (phosphorus re- and immobilization) shows the difference in sediment-associated phosphorus entrainment rates between laboratory experiments and in-situ studies at the Spree River (Sect. 6.4).

With regard to the third ('iii') approach, modeling often uses the information and understanding obtained from the previous two approaches to develop numerical models and to extrapolate over time and space. This is demonstrated through the validation of the numerical model in sub-project 14 with the laboratory data from sub-project 2 and the two sub-projects no. 1 and no. 13 of Westrich's research group. The relative

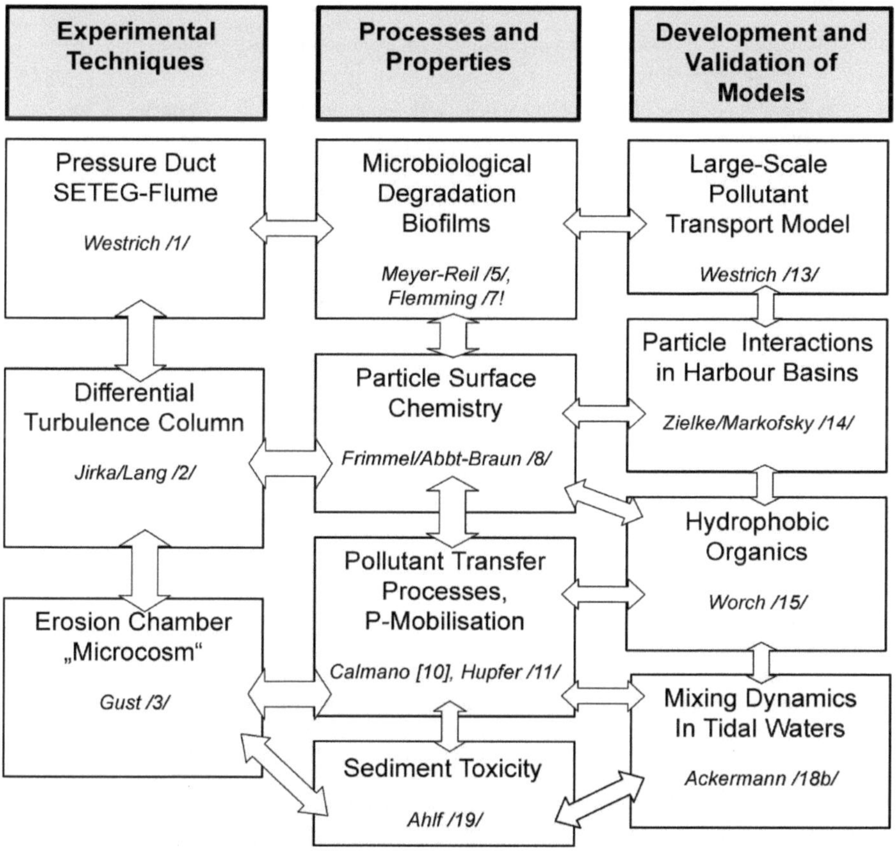

Fig. 1.7. Structure of the SEDYMO program. 13 sub-projects are involved in the coordinated research project 2002–2006

Table 1.5. Titles of the SEDYMO sub-projects and location of SEDYMO contributions in the present book

No	SEDYMO project title	Leader	Section
1	Parameters determining the sediment stability/erosion of cohesive riverine sediments	Westrich	3.2
2	Fine sediment dynamics and flocculation unter the influence of turbulence/salt stratification	Jirka	3.4
3	Erosion and deposition cycles of cohesive sediments	Gust	3.3
5	Microbial mediated processes on resuspended sediments	Meyer-Reil	3.5
7	Role of biofilms on the mobility of pollutants in rivers	Flemming	9.2
8	Influence of the particle charge on their erosion and deposition behavior	Frimmel	6.2
10	Forecast of contaminant transfer from sediment and suspended matter	Calmano	6.5
11	Microbial re- and immobilisation of phosphorus at alternating environmental conditions	Hupfer	6.4
13	Numerical modeling of the mobility of pollutants originating from riverine sediments	Westrich	4.2
14	Modeling fine sediment dynamics in tidal harbor basins and harbor entrances with consideration of the fine structure of both the flow and transport process	Zielke, Markofsky	4.3
15	Sorption equilibrium and desorption kinetics of hydrophobic organics	Worch	6.3
18b	Trace metals as indicators for the dynamics of suspended particulate matter in the tidal reach of the River Elbe: Model calculations and measurements	Ackermann, Keller	7.4
19	Evaluation of the toxicity of sediments from different river basins	Ahlf	10.2

close contact to practice of this approach is shown from the involvement of sub-project 13 in the Iffezheim barrage case (Sect. 4.3; case study 'Upper Rhine' in Sect. 2.3 and from the cooperation of sub-project 14 with the Hamburg Port Authority (Sect. 4.4).

References

Akkiparambath A (1999) Investigations on Metal Mobilization from Sediments under Conditions Close to Nature (in German). Diploma Thesis Technical School Hamburg. 194 p

Allan RJ (1986) The Role of Particulate Matter in the Fate of Contaminants in Aquatic Ecosystems. National Water Research Institute, Scientific Series no. 142, 128 p. Burlington: Canada Centre for Inland Waters

Amiri F, Börnick H, Worch E (2005) Sorption of phenols onto sandy aquifer material: the effect of dissolved organic matter (DOM). Water Res 39:933–941

Amos CC, Daborn GR, Christian HA, Atkinson A, Robertson A (1992) In-situ erosion measurements of fine-grained sediments from the Bay of Fundy. Mar Geol 108:175–196

Baalousha M, v d Kammer F, Motelica-Heino M, Baborowski M, Hofmeister C, Le Coustumer P (2006) Size-based speciation of natural colloidal particles by flow field flow fractionation, inductively coupled plasma-mass spectroscopy, and transmission electron microscopy/X-ray energy dispersive spectroscopy: Colloids-trace element interaction. Environ Sci Technol 40:2156–2162
Babiarz CL, Hurley JP, Hoffman SR, Andren AW, Shafer MM, Armstrong DE (2001) Partitioning of total mercury and methylmercury to the colloidal phase in freshwaters. Environ Sci Technol 35:4773–4782
Boivin M, Simonin O, Squires KD (1998) Direct numerical simulation of turbulence modulation by particles in isotropic turbulence. J Fluid Mech 375:235–263
Booij K, Sundby B, Helder W (1994) Measuring flux of oxygen to a muddy sediment with a cylindrical microcosm. Neth J Sea Res 32:1–11
Brunk BK, Jirka GH, Lion LW (1997) Effects of salinity changes and the formation of dissolved organic matter coatings on the sorption of phenanthrene: Implications for pollutant trapping in estuaries. Environ Sci Technol 31:119–125
Brunk BK, Weber-Shirk M, Jensen-Lavan A, Jirka GH, Lion LW (1996) Modeling natural hydrodynamic systems with a differential-turbulence column. J Hydraulic Engin July 1996:373–380
Buffle J, Leppard GG (1995) Characterization of aquatic colloids and macromolecules. 1. Structure and behaviour of colloidal material. Environ Sci Technol 29:2169–2175
Buffle J, Van Leeuwen HP (1992) Foreword. In: Buffle J, Van Leeuwen HP (eds) Environmental Particles. Lewis Publ. Chelsea Mich
Characklis WG (1990) Biofilm processes. In: Characklis WG, Marshall KC (eds) Biofilms. John Wiley, New York, pp 195–232
Canfield DE, Jørgensen BB, Fossing H, Glud R, Gundersen J, Ramsing NB, Thamdrup B, Hansen JW, Nielsen LP, Hall POJ (1993) Pathways of organic carbon oxidation in three continental margin sediments. Mar Geol 113:27–40
De Beer D, Glud A, Epping E, Kühl M (1997) A fast responding CO_2 microelectrode for profiling sediments, microbial mats and biofilms. Limnol Oceanog 43:1590–1600
Droppo IG, Leppard GG, Flannigan DT, Liss SN (1997) The freshwater floc: A functional relationship of water and organic and inorganic floc constituents affecting suspended sediment properties. Water Air Soil Pollut 99:43–54
Elsgaard L, Jørgensen BB (1992) Anoxic transformations of radiolabeled hydrogen sulfide in marine and freshwater sediments. Geochim Cosmochim Acta 56:2425–2435
Fengler G, Förstner U, Gust G (1999) Verification experiments on delayed metal release from sediments using a hydrodynamically controlled erosion apparatus (in German). Abstract Annual Meeting German Society of Water Chemistry, Regensburg, pp 240–243
Flemming H-C, Leis A (2002) Sorption properties of biofilms. In: Flemming H-C (ed) Biofilms. In: Bitton G (ed) Encyclopedia of Environmental Microbiology vol. 5:2958–2967
Flemming H-C, Schmitt J, Marshall KC (1995) Sorption properties of biofilms. In: Calmano W, Förstner U (eds) Sediments and Toxic Substances. Springer-Verlag, Berlin, pp 115–157
Förstner U (1984) Effects of salinity on the metal sorption onto organic particulate matter. In: Laane DN, Wolff WJ (eds) The Role of Organic Matter in the Wadden Sea. Neth J Sea Res 10/84:195–209
Förstner U (1995) Non-linear release of metals from aquatic sediments. In: Salomons W, Stigliani WM (eds) Biogeodynamics of pollutants in soils and sediments – risk assessment of delayed and non-linear responses. Springer-Verlag, Berlin, pp 247–307
Förstner U (1996) Solutes/solids interaction of metals in estuaries (keynote lecture). In: Kausch H, Michaelis W (eds) Particulate Matter in River and Estuaries. Arch Hydrobiol Spec Issues Advanc Limnol 47:271–287
Förstner U, Gust G (1996) Development and Verification of a Model for Heavy Metal Transfer from Close-to-Nature Mobilized Aquatic Sediments (in German). Proposal to the German Research Foundation (DFG), Project FO 95/26-1 (Feb 1, 1997 until Jan 31, 1999)
Glud RN, Santegoeds CM, Beer DD, Kohls O, Ramsing NB (1998) Oxygen dynamics at the base of a biofilm studied with planar optodes. Aquat Microbiol Ecol 114:223–233
Greef O, Glud RN, Gundersen JK, Holby O, Jørgensen, BB (1998) A benthic lander for tracer studies in the sea bed: in situ measurements of sulfate reduction. Contin Shelf Res 18:1581–1594
Gust G (1976) Observations on turbulent drag reduction in a dilute suspension of clay in sea water. J Fluid Mech 75:29–47

Gust G (1991) Fluid velocity measurement instrument. U.S. Patent no. 4,986,122
Gust G, Müller V (1997) Interfacial hydrodynamics and entrainment functions of currently used erosion devices. In: Burt N, Parker R, Watts J (eds) Cohesive Sediments. John Wiley & Sons, Chichester, pp 149–174
Hennies K (1997) Biogeochemical Process Studies on the Transport Behavior of Trace Elements in the Tidal Elbe River (in German) Dissertation Hamburg University of Technology, Hamburg-Harburg, 288 p
Hesse CH, Tory EM (1996) The stochastics of sedimentation. Adv Fluid Mech 7:199–240
Hong J (1995) Characteristics and Mobilization of Heavy Metals in Anoxic Sediments of the Elbe River during Resuspension/Oxidation. Doctoral Thesis at Hamburg University of Technology, Hamburg-Harburg, 157 p
Humann K (1995) The Influence of Microphytobenthos on Sediment Stability and Formation of Suspended Particulate Matter from Sediments of the Elbe River Estuary (in German). Doctoral Thesis, University of Hamburg
Huettel M, Ziebis W, Forster S, Luther GW (1998) Advective transport affecting metal and nutrient distribution and interfacial fluxes in permeable sediments. Geochim Cosmochim Acta 62:613–631
Johnson AA, Tezduyar TE (1997) 3D simulation of fluid-particle interactions with the number of particles reaching 100. Comput. Meth Appl Mech Eng 145:301–321
Jørgensen BB (1994) Diffusion processes and boundary layer processes in microbial mats. In: Stal LJ, Caumette P (eds) Microbial Mats – Structure, Development and Environmental Significance. Springer Verlag, Berlin, pp 243–253
Kausch H, Michaelis W (eds, 1996) Suspended Particulate Matter in Rivers and Estuaries. Advances in Limnology 47. Schweizerbart'sche Verlagsbuchhandlung (Nägele u. Obermiller), Stuttgart, 573 p
Kern U (1997) Transport of Suspended Matter and Pollutants in Lock-regulated River – Example of Neckar River (in German). Dissertation University of Stuttgart
Kies L (1995) Algal snow and the contribution of algae to suspended particulate matter in the Elbe Estuary. In: Wiessner W, Schnepf E, Starr R (eds) Algae, Environment and Human Affairs. Biopress, Bristol, pp 93–121
Kühl M, Lassen C, Revsbech NP (1997) A simple light meter for measurements of PAR (400 to 700 nm) with fiberoptic microprobes: application for P vs E_0 (PAR) measurements in a microbial mat. Aquat Microbial Ecol 13:197–207
Kühl M, Steuckart C, Eickert G, Jeroschewski P (1999) A H_2S microsensor for profiling biofilms and sediments: Application in an acidic lake sediment. Aquatic Microbial Ecol 15:201–209
Landenberger H (1998) CoTReM – A Multicomponent Transport- and Reaction Model (in German). Department of Geosciences, Section of Geochemistry and Hydrogeology, University of Bremen, 142 p
Lead JR, Wilkinson KJ (2006) Aquatic colloids and nanoparticles: Current knowledge and future trends. Environ Chem 2006(3):159–171
Lead JR, Muirhead D, Gibson CT (2005) Characterization of freshwater natural aquatic colloids by atomic force microscopy (AFM). Environ Sci Technol 39:6930–6936
Le Normant C, Peltier E, Teisson C (1998) Three dimensional modelling of cohesive sediment transport in estuaries. In: Dronkers J, Scheffers MBAM (eds) Physics of Estuaries and Coastal Seas. AA Baklema Publ, Rotterdam, pp 65–72
Lick W, Lick J, Ziegler CK (1992) Flocculation and its effect on the vertical transport of fine-grained sediments. In: Hart BT, Sly PG (eds) Sediment/Water Interactions V. Kluwer Academic Publ, Dordrecht, pp 1–16
Ling W, Chung JN, Troutt TR, Crowe CT (1998) Direct numerical simulation of a three-dimensional temporal mixing layer with particle dispersion. J Fluid Mech 358:61–85
Lyven B, Hassellov M, Turner DR, Haraldsson C, Andersson K (2003) Competition between iron- and carbon-based colloidal carriers for trace metals in a freshwater assessed using flow field-flow fractionation coupled to ICPMS. Geochim Cosmochim Acta 67:3791–3802
Malcherek A (1995) Mathematical Modeling of Hydraulic Flow and Transport Processes in Estuaries (in German). Doctoral Thesis at the Institute of Fluid Mechanics and Electronic Computation in Civil Engineering. University of Hannover, Report no. 44
Malcherek A (2001) Hydromechanik der Fließgewässer (in German). Habilitation at the Institute of Fluid Mechanics and Electronic Computation in Civil Engineering. University of Hannover, Report no. 61

Mohacsi A, Bozoki Z, Niessner R (2001) Direct diffusion sampling-based photoacoustic cell for in situ and on-line monitoring of benzene and toluene concentrations in water. Sensors and Actuators B79(2–3):127–131

Morrison MA, Benoit G (2004) Investigation of conventional membrane and tangential flow ultrafiltration artifacts and their application to the characterization of freshwater colloids. Environ Sci Technol 38:6817–6823

Neudörfer F, Meyer-Reil L-A (1997) A microbial biosensor for the microscale measurement of bioavailable organic carbon in oxic sediments. Mar Ecol Prog Ser 147:295–300

Neumann-Hensel H, Ahlf W (1995) Fate and effect of copper and cadmium in a sediment-water system and effect on chitin degrading bacteria. Acta Hydrochim Hydrobiol 23:72–75

Owens PhN, Petticrew EL (2006) Sediment dynamics and pollutant mobility in river basins – Sedymo 2006 Symposium, Hamburg University of Technology, Germany, 26–29 March 2006. J Soils and Sediments 6(2):122–124

Parkhurst DL, Appelo CAJ (1999) User's guide to PHREEQC (Version 2) – a computer program for speciation, batch-reaction, one-dimensional transport and inverse geochemical calculations. Wat Resour Invest US Geol Surv Report 99–4259

Peiffer S (1997) Environmental Geochemical Significance of Formation and Oxidation of Pyrite in Aquatic Sediments (in German). Bayreuth Forum Ecology vol. 47, University of Bayreuth, 105 p

Perkins RG, Sun H, Watson J, Player MA, Gust G, Paterson DM (2004) In-line laser holography and video analysis of eroded floc from engineered and estuarine sediments. Environ Sci Technol 38:4640–4648

Petersen W, Hong J, Willamowski C, Wallmann K (1996): Release of trace contaminants during reoxidation of anoxic sediment slurries in oxic water. In: Kausch H, Michaelis W (eds) Particulate Matter in River and Estuaries. Arch Hydrobiol Spec Issues Advanc Limnol 47:295–305

Petersen W, Wallmann K, Li P, Schroeder F, Knauth H-D (1995) Exchange of trace elements at the sediment-water interface during early diagenesis processes. Mar Freshwater Res 46:19–26

Porter ET, Sanford LP, Gust G, Porter FS (2004) Combined water-column mixing and benthic boundary-layer flow in mesocosms: key for realistic benthic-pelagic coupling studies. Mar Ecol Prog Ser 217:43–60

Prause B, Rehm E, Schulz-Baldes M (1985) The remobilisation of Pb and Cd from contaminated dredged soil after dumping in the marine environment. Environ Technol Lett 6:261–266

Prestel H, Gahr A, Niessner R (2000) Detection of heavy metals in water by fluorescence spectroscopy: On the way to a suitable sensor system. Fresenius J Anal Chem 368 (2–3):182–191

Rebhun M, De Smedt F, Rwetabula J (1996) Dissolved humic substances for remediation of sites contaminated by organic pollutants. Binding-desorption model predictions. Water Res 30:2027–2038

Rolinski S (1997) On the Dynamics of Suspended Particulate Matter in the Tidal Elbe River – Numerical Simulation Using a Lagrangean Procedure (in German). Doctoral Thesis, University of Hamburg, Reports of the Center of Marine and Climate Research no. 25, 117 p

Rostad CE, Leenheer JA (1997) Organic carbon and nitrogen content associated with colloids and suspended particulates from the Mississippi River and some of its tributaries. Environ Sci Technol 31:3218–3225

Salomons W, de Rooij NM, Kerdijk H, Bril J (1987) Sediments as a source for contaminants. In: Thomas RL, Evans R, Hamilton A, Munawar M, Reynoldson T, Sadar H (eds) Ecological Effects of In Situ Sediment Contaminant. Hydrobiologia 149:13–30

Schroeder F, Klages D, Blöcker G, Vajen-Finnern H, Knauth H-D (1992) The application of a laboratory apparatus for the study of nutrient fluxes between sediment and water. Hydrobiol 235/236:545–552

Seibt-Winckler A, Schirmer F (1996) Measurements with a three frequency echo sounder for the detection of suspended matter in a river estuary. In: Kausch H, Michaelis W (eds) Particulate Matter in River and Estuaries. Arch Hydrobiol Spec Issues Advanc Limnol 47:497–506

Spork V (1997) Erosion Behavior of Fine-Grained Sediments and Their Biogenic Stabilisation (in German). Communications of the Institute for Hydraulics and Water Management, RWTH Aachen, vol. 114

Stigliani WM (1988) Changes in the values 'capacities' of soils and sediments as indicators of nonlinear and time-delayed environmental effects. Environ Monit Assessm 10:245–307

Stigliani WM (1991) Chemical Time Bombs: Definition, Concepts, and Examples. Executive Report 16 (CTB Basic Document). IIASA Laxenburg

Stolpe B, Hassellöv M, Andersson K, Turner DR (2005) High resolution ICPMS as an on-line detector for Flow Field-Flow Fractionation; multi-element determination of colloidal size distributions in a natural water sample. Analytica Chimica Acta 535:109–121

Tolhurst TJ, Gust G, Paterson DM (2002) The influence of an extracellular polymeric substance (EPS) on cohesive sediment stability. In: Winterwerp JC, Kranenburg C (eds) Fine Sediment Dynamics in the Marine Environment. Proceedings in Marine Science 5:409–425

Van Breemen N (1983) Acidification and alkalinization of soils. Plant and Soil 75:283–308

Vignati DAL, Dworak T, Ferrari B, Koukal B, Loizeau J-L, Minouflet M, Camusso MI, Polesello S, Dominik J (2005) Assessment of the gochemical role of colloids and their impact on contaminant toxicity in freshwaters: An example from the Lambro-Po system (Italy). Environ Sci Technol 39:489–497

Wen LS, Santschi PH, Paternostro C, Gill G (1999) Estuarine trace metal distributions in Galveston Bay I: Importance of colloidal forms in the speciation of the dissolved phase. Mar Chem 63:185–212

Westall JC, Zachary JL, Morel FMM (1976) MINEQL: A compact program for the calculation of chemical equilibrium composition of aqueous systems. R.M. Parsons Lab Techn Note no. 18. MIT, Cambridge, Mass

Westrich B, Kern U (1996) Mobility of Contaminants in the Sediments of Lock-Regulated Rivers – Field Experiments in the Lock Reservoir Lauffen, Modeling and Estimation of the Remobilisation Risk of Older Sediment Deposits (in German). Final Report no. 96/23, Institute for Hydraulics, University of Stuttgart, 186 p

Wiltshire K, Tolhurst T, Paterson DM, Davidson I, Gust G (1998) Pigment fingerprints as markers of erosion and changes in cohesive sediment surface properties in simulated and natural erosion events. In: Black KS, Paterson DM, Cramp A (eds) Sedimentary Processes in the Intertidal Zone. Geological Society London, Spec Publ 139:99–114

Wilkinson KJ, Balnois E, Leppard GG, Buffle J (1999) Characteristic features of major components of freshwater organic matter revealed by transmission electron and atomic force microscopy. Colloids Surf A 1999, 155:287–310

Wollschläger A (1996) A Random-Walk-Model for Heavy Metal Particles in Natural Waters (in German). Doctoral Thesis at the Institute for Flow Mechanics and Electronic Computation in Civil Engineering, University of Hannover, Report no. 49

Zhang J, Prestel H, Gahr A, Niessner R (2000) Development of a flow injection analysis (FIA) system for the measurement of heavy metals using a fiber optic chemical sensor based on laser-induced fluorescence. Proc Inter Soc Optical Engineering, 4077 Sensors and Control Techniques, pp 32–39

Chapter 2

Managing River Sediments

Bernhard Westrich

2.1 Hydrodynamics and Sustainable Sediment Management

2.1.1 Introduction

Sediments play an important role in river engineering and water resources management. In the past, many rivers in developed countries have been engineered by training and regulation works for navigation, hydropower generation and flood protection. In the past decades, municipal and industrial wastewater discharge and various diffusive sources from agriculture have caused a widespread contamination of river sediments by heavy metals, organic toxicants and agrochemicals. Meanwhile, many historically contaminated sites in rivers are localized and identified as a severe latent hazard for the river ecosystem (see Sect. 1.1.3). Most of the contaminated sites have been detected in low flowing water bodies which are either permanently or temporarily connected to the main river channel such as near bank groyne fields in waterways or harbors, river dead arms, flood plains and last not least flood retention reservoirs (Fig. 2.1). Many deposits are most likely to be resuspended and transported over a long distance by extreme discharges causing contamination of not yet polluted surface water bodies and unpolluted soils subject to flooding.

High discharges in rivers may cause the mobilization of contaminants deposited in such low flowing zones of river channels. The recent flood events in the river Odra in 1997, river Rhine in 1999 and river Elbe in 2002 have illustrated not only the devastating power of floods by damaging hydraulic structures and breaching dams but also the enormous erosion capacity of flowing water associated with the mobilization, transport and partial deposition of contaminated sediments in tidal harbors, estuaries and coastal areas. The precautionary as well as the nondeterioration principle calls for the development and implementation of an integrated sediment management aiming to reduce the risk of contaminated sediment mobilization and their impact on the environment according to the EU water framework directive.

Integrated Risk Assessment

Sustainable sediment management aims to reduce the risk of adverse impact and ecological damage by sediment associated toxicants and to improve the ecological status of surface water bodies. A comprehensive risk assessment, which is an essential contribution to the challenging task, requires an interdisciplinary approach to cope with the interacting physical, chemical and biological processes occurring on extremely differ-

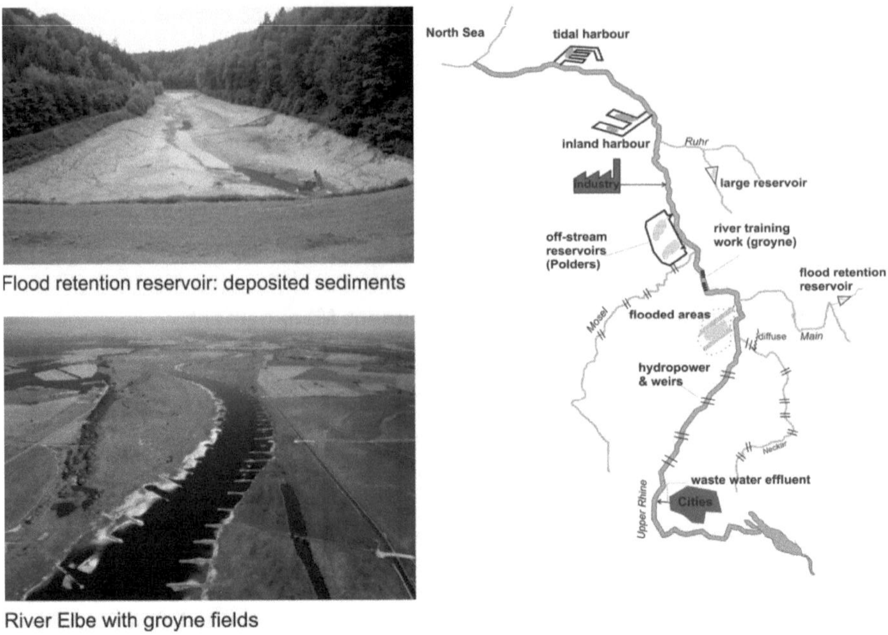

Fig. 2.1. Sources, sinks and pathways of contaminants in a large river basin

ent space and time scale (Kern 1997). Management strategies must include river engineering issues, and environmental problems and economic aspects on the local, regional and river basin scale (Carlon et al. 2000). Contaminated sediments deposited in groyne fields, harbors and water reservoirs can be mobilized in many ways, for instance by floods, maintenance dredging, partial or total emptying of reservoirs, revision or technical inspection of structures. After resuspension, fine sediments are mostly transported over long distances through the whole river system while simultaneously various other processes occur, such as mixing, dilution and loading by tributaries, fractional sedimentation, pollutants repartitioning as well as chemical and biological transformation and degradation.

Beside the quantitative aspects of sediment transport, such as river bed stabilization, habitat improvement, flood protection and navigation, the mobility and transport behavior of sediment bound contaminants and nutrients are emerging key issues of vital importance to future sediment management and surface water quality improvement.

Contaminant immission at a downstream site in terms of concentration and load depends on the catchment characteristics and the hydrological situation such as

- location and connectivity of contaminated sites in the catchment
- actual hydraulic conditions in the river channel network
- in-situ toxicity and total amount of contaminants mobilized upstream.

Floods play a dominant role in sediment erosion risk assessment because of their extreme erosion and transport capacity. Hence, there is a high probability that histori-

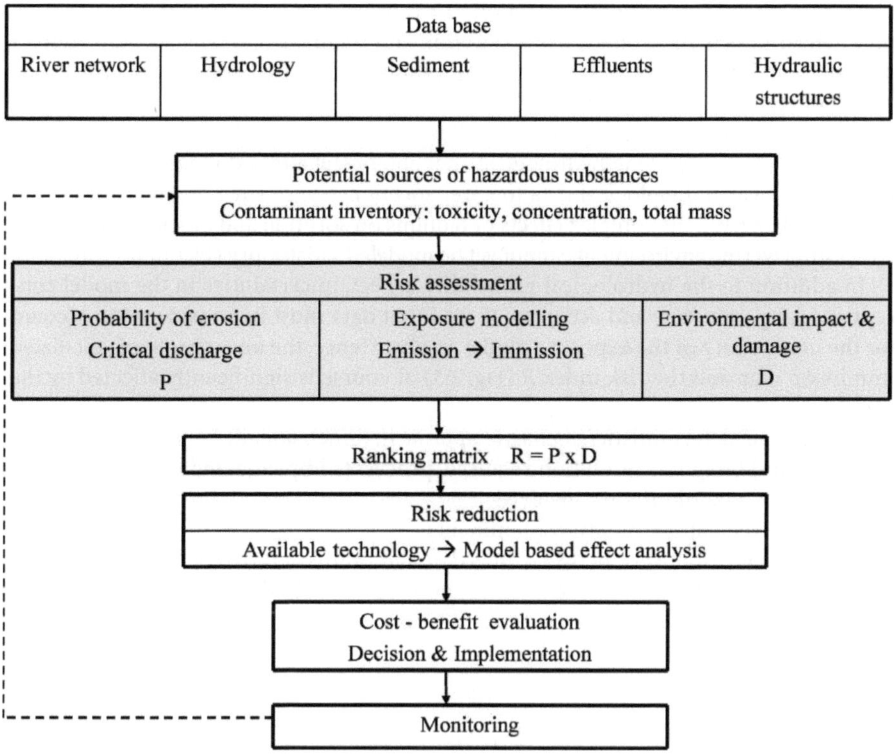

Fig. 2.2. Catchment related integrated sediment management concept

cally contaminated sediments in deeper layers can be resuspended and transported through the river basin to the estuarine and coastal waters. Another key factor is the sediment erosion stability because it controls the mobility and contaminant mass flux and hence the initial conditions for the subsequent transboundary transport process (Fig. 2.2). The site specific relationship between discharge, bed shear stress and sediment mass flux can be completely described by hydraulic modeling. Discharge statistics are directly transformed by hydrodynamics into bed shear stress statistics. Finally, erosion probability results from the convolution of the probability density function of the hydrodynamic bed shear stress and the sediment specific erosion resistance.

After the exploration of polluted sites and their contaminant inventory, a risk analysis must be performed to quantify the risk index R as the product of erosion probability and environmental damage or impact (Carlon et al. 2000).

Numerical exposure models are useful tools to describe the pathway and fate of mobilized contaminants and aim to quantify the spatial and temporal distribution of dissolved and particulate substances in the water column and the river bed as well, and to identify sedimentation zones in the river system (Baart et al. 2001).

Numerical modeling allows us to integrate different results and experience from engineering and natural science, and to simulate processes differing by some orders of magnitude both in space and time (Kern 1997). Individual scenario modeling provides data on the intensity and duration of exposure and, with statistical input data, they deliver information on exposure duration and frequency causing accumulation of deposited pollutants which can be used for a statistically based effect model (de Zwart 2005; Öberg and Bergbäck 2005).

The application of a contaminated sediment transport model requires a comprehensive data base including hydrological, morphological and sedimentological data as well as chemical and biological data to cover sorption, transformation and degradation processes. However, the complex effects of biofilms on sorption and biodegradation processes in a riverine environment cannot yet be modeled satisfactory (Flemming, this vol.).

In addition to the hydrological probability aspect, uncertainties in the model concept, model parameters and accuracy of the input data must be considered to account for the uncertainty of the exposure model results. Hence, the uncertainty of calculated immission data and the risk index *R* (Fig. 2.3) of course is significantly affected by the

- model type, deterministic/stochastic approach, dimensionality
- spatial and temporal resolution required, processes implemented
- number of hazardous sites involved
- quality of data about in-situ contaminants
- pathway and tributaries between emission and immission site
- physical and in particular, biochemical processes involved
- data base for model calibration and validation.

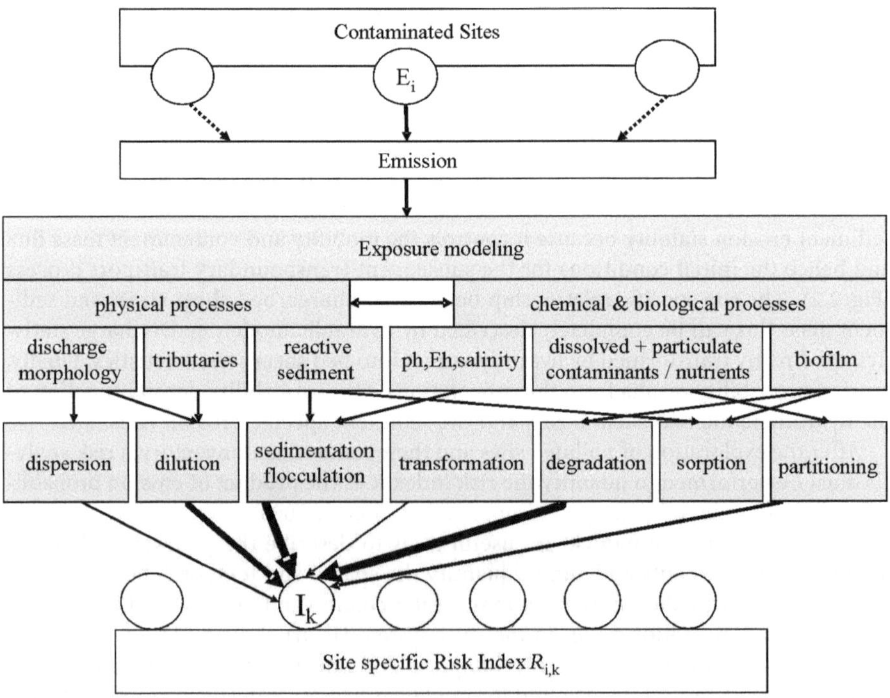

Fig. 2.3. Exposure modeling for site specific risk index evaluation

Based on the numerical results, an emission or immission related site specific risk index $R_{i,k}$ can be obtained by evaluating the impact of each emission site, marked by the index i, to a defined immission site or vise versa, by ranking the immission load of an individually considered site at risk, expressed by the index k. The matrix enables an optional ranking of the damage potential of contaminated sites as well as for receiving water bodies or flooding areas at risk. Both ranking figures provide useful information for remediation planning.

The ecotoxicological aspects of a comprehensive risk analysis can be supported and substantially improved by using effect models which allow us to describe the dose/effect relationship, e.g., by application of Artificial Neural Network (Lek and Guegan 2000) or Fuzzy Logic approach (Ahlf, this vol.). The exposure model can, of course, be used for a risk reduction analysis investigating the effect of alternative remediation measures. A risk based sustainable sediment management strategy must, of course, try to find a source oriented solution instead of an end of pipe solution. Hence, the source related risk index is of first priority. After a cost-benefit evaluation a prioritization of remediation action can be performed as a rational basis for the decision on a cost effective solution for sediment improvement.

Experimental Methods

Because of the great variety of river characteristics, water chemistry, quality and biology it is evident that the sediment stability is very site specific and subject to seasonal variation (Paterson 1997). Therefore, experimental results cannot be transferred directly from one site to another. Because of a lack of a conclusive generic description of cohesive sediment erosion processes, experimental investigations on undisturbed sediments either in the laboratory or the field seem to be indispensable to gain site and river specific data (Lick et al. 1994; Zreik et al. 1998).

Sediment erosion stability controls not only the onset and source strength of particle mass flux but also the flux of the dissolved and colloidal components associated with the pore water. Immediately after erosion larger aggregates are exposed to strong turbulent shear forces disrupting the eroded lumps and generating new reactive surface for the exchange and transfer of adsorbed pollutants (Worch, this vol.). During the hydraulic transport, concentrations, grain size spectrum and most probably the chemical and biological milieu condition will also change and hence reactions and interactions between the particulate and dissolved phase accordingly.

To investigate some key processes with regard to fine sediment mobility and particulate contaminant behavior specific experimental equipment has been developed and applied as follows (Fig. 2.4):

- *SETEG system:* Depth profile of erosion threshold and erosion rate in a pressurized channel, sediment testing area 150 cm^2, bed shear stress up to 25 N m^{-2}, sediment core length up to 150 cm (Witt and Westrich 2003)
- *Differential Turbulence Column:* Concentration profile of different particle fractions, flocculation, desorption and remobilization of sediment bound contaminants, pollutants partitioning (Kühn, this vol.)
- *Mesocosm:* Erosion, sedimentation cycles under tidal like conditions, sorption process under controlled chemical conditions in the water column (Gust, this vol.)

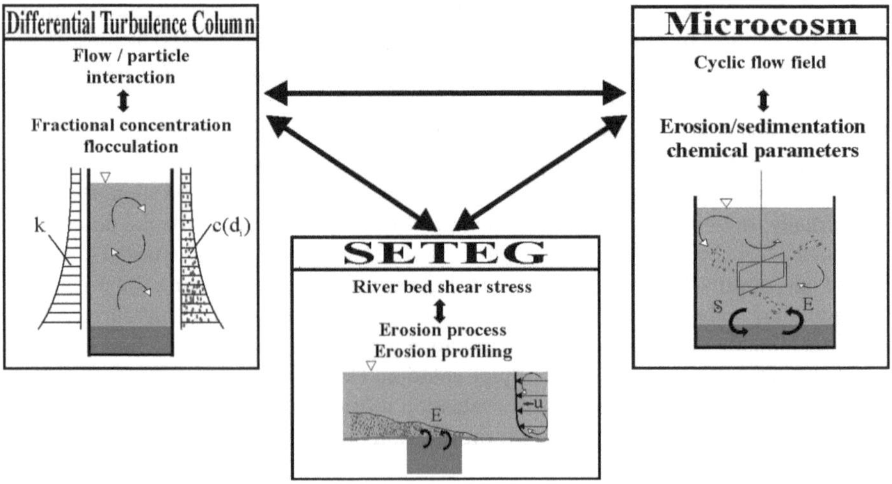

Fig. 2.4. Experimental techniques for investigating contaminated sediment stability and suspended particle/turbulent flow interaction

Complementary sediment stability tests can be performed to quantify scale effects of erosion and hence to facilitate the up-scaling of experimental laboratory data to the field and to compare laboratory data with in-situ measurements (Westrich and Förstner 2005).

- *Box Sampler:* Erosion tests in a flume, sediment testing area 30×70 cm^2, sediment depth 28 cm, maximum bed shear stress 20 N m^{-2}.
- *EROMOB:* Mobile in situ erosion testing equipment; sediment testing area 30×70 cm^2, maximum bed shear stress 10 N m^{-2} (Westrich and Schmid 2004)

Both instruments have a testing area ten times larger than the abovementioned SETEG equipment. The intercomparison of the abovementioned erosion testing methods with the inclusion of other methods like the CSM method (de Deckere et al. 2002) or the in-situ flume (Debnath et al. 2006) has not yet been concluded.

Experimental investigations have been performed using two parallel undisturbed sediment samples from the same spot with the aim of providing sediment depth profiles with a resolution of about 2 cm as follows:

- one sediment sample is used for erosion profiling after physical properties profiling such as grain size, bulk density, water and gas content,
- the other sample is used for chemical and biological parameter profiling.

For practical application it is advisable to restrict the analytical effort on sediment exploration and to model the erosion process by using a limited number of variables representing a significant percentage of the total variance of the sediment parameters.

A comprehensive indepth investigation was conducted to identify and quantify the relevant parameters by multivariate statistical analysis in order to find a relationship between sediment erosion behavior and measurable sediment properties (Gerbersdorf et al. 2005).

Spatial Variability of Sediments

Beside the physical properties, contaminated sediments exhibit a great spatial variability not only in the horizontal but also in the vertical direction which mainly indicates the history of the pollution. The deposition pattern reflects the spatial and temporal variation of the flow field. Fine reactive sediments with a fall velocity of the order of magnitude of some 10^{-4} m s^{-1} can only be found in zones where, for some time, the local bed shear stress was below the critical value of sedimentation to build up a certain sediment layer thickness which was protected by overlaying sediments or could withstand due to its erosion resistance. Because of the variation in discharge, suspended sediment inflow and river water pollution, large gradients of sediment properties can be detected only by a respectively high vertical resolution of a few centimeters. The critical erosion shear stress may change by a factor of 3 within sediment layers of 5 to 10 cm as shown in Fig. 2.5a, 5c and 5d for different sites.

Particulate contaminant profiles are not simply correlated to a single sediment parameter profile. Therefore, mobilization modeling must refer to depth profile data of both the sediment erodibility parameter and the contamination to ensure that the depth dependent contaminant source strength is captured by the model.

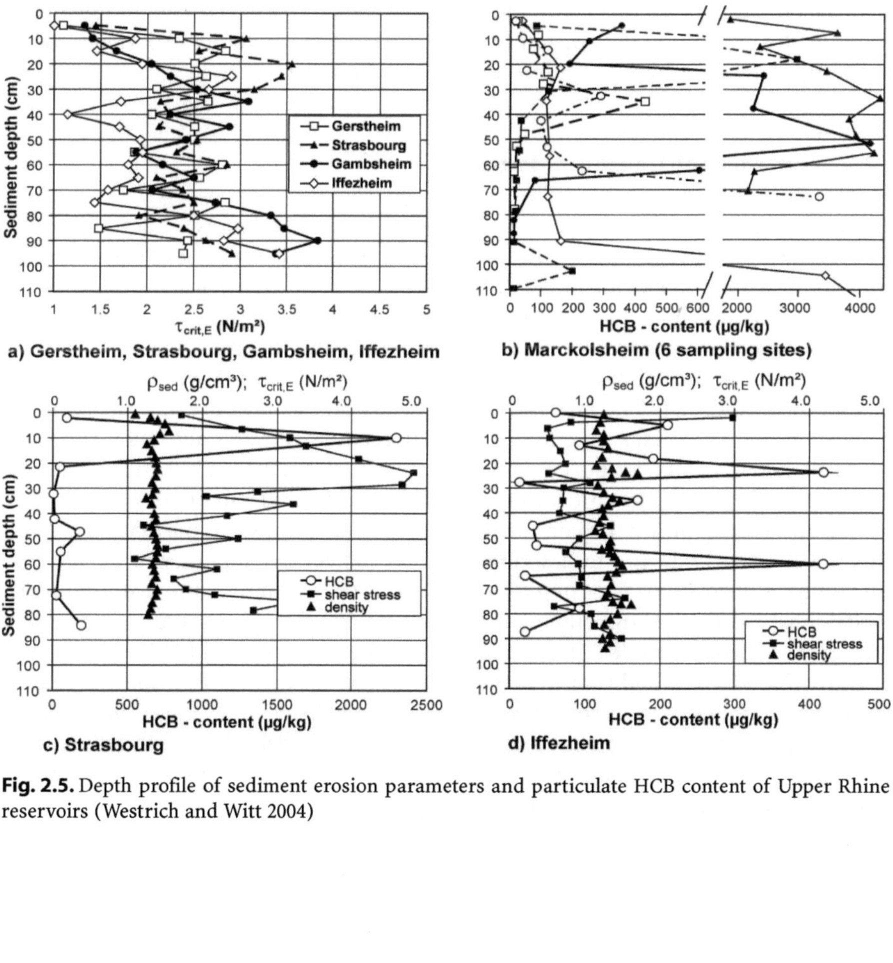

Fig. 2.5. Depth profile of sediment erosion parameters and particulate HCB content of Upper Rhine reservoirs (Westrich and Witt 2004)

With a sufficiently high number and spatial density of sediment samples a geostatistical analysis is advisable in order to improve the reliability of the model input data, to enhance the efficiency of sediment monitoring and to reduce the costs of maintenance dredging (Winkler and Stein 1997). In the case of a poor data base of sediment properties and contaminant inventory, simple interpolation and extrapolation can be applied which, of course, increases the uncertainty of the model output.

Apart from the variability of the critical shear stress the sediment samples from Marckolsheim, Strasbourg an Iffezheim (Fig. 2.5b, 5c and 5d) are typical for contaminated sites in river reservoirs as they illustrate the high spatial variability of the contamination of organic toxicants, e.g., *Hexachlorobenzene* (HCB). Similar large gradients of grain size, bulk density and gas content were also detected. Depth profiles of neighboring samples always exhibit some small difference, known as the nugget effect, in the semi-variogram (Asselmann 1997), which must be considered when combining erosion and biochemical measurements of two sediment cores. The large micro- and mesoscale heterogeneity of sediment parameters underlines the necessity of a high spatial resolution of samples for reducing the uncertainty of model input data.

Freshly deposited fluffy sediments show a small timeindependent erosion rate of some 10^{-6} kg m^{-2} s^{-1}, whereas consolidated sediments exhibit a linear progression of initial erosion rate caused by enhancing the erosive potential of the local disturbance of turbulent boundary layer (Witt and Westrich 2004). This phenomenon illustrates the difficulty of defining the erosion rate from smallscale laboratory experiments and for transferring the data to nature or to numerical models.

2.1.2 Contaminant Transport Modeling

Physically based numerical models have proven to be powerful tools for describing the pathway and fate of contaminants in surface water and hence the relationship between emission and immission (Onishi 1981). Moreover, predictive numerical transport models are used to anticipate the environmental impact of hydrological mobilization scenarios and to analyze the effect of optional remediation measures as well. They provide necessary information for assessing alternative riskreducing measures and estimating their efficiency. The model choice depends on the objective and the requirements in terms of spatial and temporal resolution and accuracy. Calibration and validation of contaminant transport models is a crucial task because of lack of appropriate data, especially on sorption processes and biochemical transformation.

The aim is to quantify the concentration field of dissolved and particulate pollutants in the water body and to describe the areas subject to temporary or permanent contaminant deposition. Focusing on transport in river channels, the hydrodynamic model must be supplied directly by discharge data from gauging stations or by a hydrological catchment model. Deep water bodies like estuaries and tidal harbors very often require 3-D flow and transport modeling (Ditschke, this vol.). In contrast, many transport processes in lowland rivers can be described by depth averaged flow velocities and suspended sediment concentrations with 2-D models. Assuming fully mixed conditions, 1-dimensional advection/dispersion models can be applied to investigate large scale far field transport and long term processes.

Input Data Base

Beside the hydrological, hydraulic and sediment data for the hydrodynamic model part, the sorption parameters of the reactive particle fraction must be available in a data base for the description of the interaction between the aqueous and solid phase. A comprehensive data base must include information on different subjects, such as the following:

- discharge statistics at gauging stations
- river network, channel bathymetry
- digital terrain model of the flood plains
- hydraulic structures, location and operation
- suspended sediment concentration
- grain size/fall velocity spectrum of suspended matter
- fall velocity and sorption parameter of contaminated fraction
- site specific erosion threshold and erosion rate parameters
- sediment and pollutant specific sorption parameter
- biochemical degradation parameter

Model input data are of varying quality with respect to accuracy and density in space and time. Statistical information should be available, such as expected value and statistical variance, especially for chemical and biological parameters, to perform a sensitivity analysis and to facilitate an uncertainty assessment. Consistent field data of extreme events are very poor and, as a result, hard data for contaminant transport model calibration and validation and in particular model prediction are uncertain (Karnahl, this vol.).

Coping with Uncertainties

There are various sources of uncertainties: uncertainty in the model concept, the model parameters and the data itself. The impact of model parameter uncertainties on the prediction of reservoir sediment erosion by floods has been investigated by Li (2004). The boot strap method was applied to gain the mean value and statistical variance of the critical shear stress and the erosion rate from the experimental data. The numerical analysis was performed with a 1-D model (Kern 1997) using the Monte Carlo method. It reveals the predominant influence of the peak discharge and the flood duration. Referring to historical floods it has been shown that with a critical bed shear stress ranging from 3.5 N m^{-2} ±0.5 N m^{-2} with a variance of 0.12 the eroded sediment volume was higher when assuming spatially uncorrelated erosion data.

Sedimentation of fine suspended material is primarily dominated by the mean bed shear stress, fall velocity and concentration of the contaminated fraction. Modeling fractional sedimentation under natural conditions in flood plains is difficult and uncertain because of the influence of roughness elements, like vegetation, on the near bottom turbulence. However, in most cases, large flood plains show low flow velocities and hence are a significant sink of contaminants as experienced by the Elbe flood in August 2002.

2.1.3 Case Study: Upper Rhine

Site Description

The lower six hydropower stations on the Upper Rhine, built in the years 1961 to 1977, show a characteristic sedimentation pattern related to the individual layout of the hydropower channel, the weir channel and the ship lock (Fig. 2.6). A sustainable sediment management must be established to keep the required freeboard of the embankment for safety reasons and, in particular, to reduce the risk of erosion of highly contaminated sediments in deeper layers. The dynamic behavior of fine suspended sediments is very much controlled by the fact that the hydropower capacity is limited and, in the case of a flood, the river discharge is split and the surplus is thereby directed to the weirs which serve as a spillway. The operation rules are as follows:

- The discharge capacity of the hydropower station is $Q_{Turbine} = 1\,400\ m^3\,s^{-1}$ and $1\,100\ m^3\,s^{-1}$. Most of time there is no discharge through the weir section except about $15\ m^3\,s^{-1}$ for ecological purpose.
- If the river discharge exceeds $1\,400\ m^3\,s^{-1}$ and $1\,100\ m^3\,s^{-1}$ the surplus discharge, i.e. the difference between Q_{River} and $Q_{Turbine}$, is directed to the weir channel.
- The headwater at the dam is kept at a constant level.

Study Objective

On the one hand, the main objective was to estimate the future risk of resuspension of historically deposited contaminants, mainly *Hexachlorobenzene* (HCB) and, on the other hand, to perform a retrospective analysis of the erosion and sedimentation of HCB during the last flood in May 1999 during which a substantial amount of particulate HCB was mobilized. The study was conducted using a 2-D numerical flow and transport model (Jacoub, this vol.). Each of the six reservoirs was investigated to estimate the HCB mass eroded and to quantify the cumulative contribution of the respective reservoirs to the total particulate HCB load released to the Lower Rhine and monitored at the German/Dutch border. Unfortunately, the pre-flood data of contaminated sediment zones were scarce. Water samples were taken only in front of the turbine section at the lowest hydropower station Iffezheim (Fig. 2.6) during the flood event. The inflow discharge hydrograph, the suspended sediment outflow concentration and the associated daily HCB load are given and used as boundary conditions (Fig. 2.7).

Since the flow velocities approaching the hydropower section are most of the time high enough, no deposition of fine contaminated sediment fraction takes place whereas, in the headwater of the weir section, the flow velocities are small to allow fractional sedimentation except the short period of erosive flood event. At the beginning of the rising hydrograph, fresh sediments can be deposited in the weir branch but shortly afterwards they are resuspended together with older sediments and flushed through the weirs. After the peak flow, when the flow velocities in the weir branch again become small enough, the inflowing suspended sediments and particulate contaminants start settling and remain deposited until the next flood event. Hence, the flood removes the previously deposited sediment top layer of some 10 cm and causes an in-

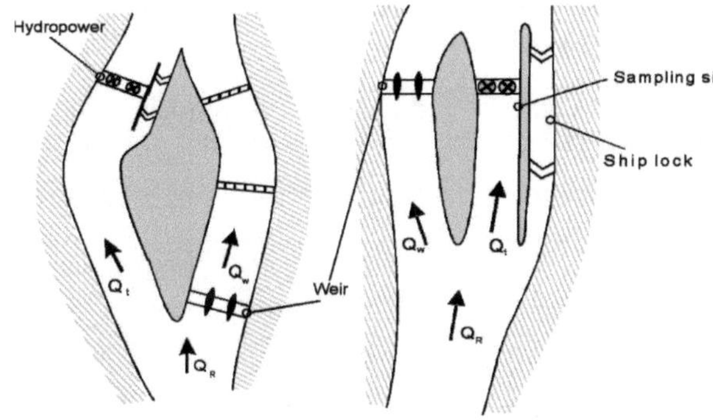

Fig. 2.6. Layout scheme of two representative Upper Rhine reservoirs: Marckolsheim (*left*) and Iffezheim (*right*)

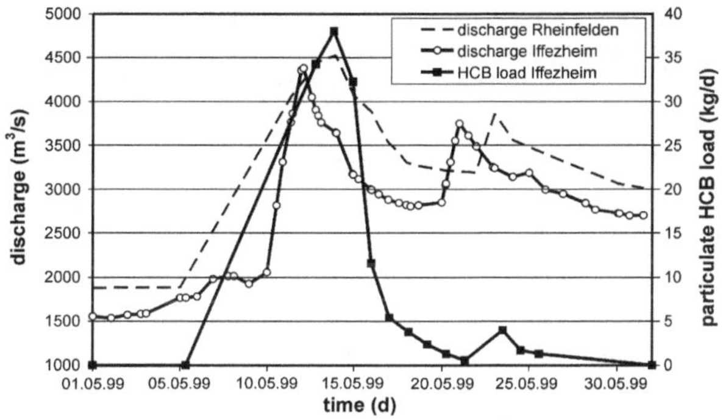

Fig. 2.7. Discharge hydrograph and daily particulate HCB load during the flood in May 1999 (BfG in Heise et al. 2004)

put of currently mobilized contaminants which are almost completely deposited in the weir branch.

The reservoir Marckolsheim completed in 1961 was selected for an intensive field investigation based on 25 sediment sampling spots to cover an estimated total contaminated area of about 54×10^3 m^2. The spatial sampling density at Iffezheim was 12 samples representing a total area of 35×10^3 m^2 whereas in the Straßburg reservoir only 7 samples were taken of a total contaminated area of an estimated size of 23×10^3 m^2. Simple techniques were applied such as interpolation, total averaging under exclusion of extreme values etc. to assign to each node of the computational mesh the required sediment parameters for critical erosion shear, erosion rate and particulate contaminant concentration. The latter value was averaged over the erosion depth of some 12 cm resulting from the calibrated 2-D flow model. In the Upper Rhine reservoirs, the sediment bulk density varied from 1.1 to 1.7 g cm^{-3}, the critical erosion shear stress between 0.5 to 10 N m^{-2} and erosion rates between 10^{-3} and 10^{-5} kg m^{-2} s^{-1} were measured and used for numerical modeling accordingly (Witt 2004; Jacoub and Westrich 2006).

Despite the extensive field investigation at Marckolsheim, the uncertainty of the calculated eroded particulate HCB was unsurprisingly high. The main reason is the lack of pre-flood data on sediments and the high spatial variability of sediments and contaminants of the post-flood samples. The estimated HCB mass eroded during the flood ranges from 2.4 to 17 kg (Table 2.1). The latter value is far beyond the ICPR (International Commission for the Protection of the river Rhine) target value of 1.3 kg referring to a maximum permissible sediment contamination of 40 μg kg^{-1} for HCB.

Table 2.1. One-year sediment and HCB mass balance of the reservoir Iffezheim from 1 May 1999 to 30 April 2000 (Jacoub 2004)

Particle size (µm)	Input (10^3 t)	Output (10^3 t)	Suspension (10^3 t)	Deposited (10^3 t)	Eroded (10^3 t)	Volume (10^3 m^3)
20	3674	3670	1.9	29.5	27.2	1.95
60	1775	1689	2.5	103	18.9	69.78
100	593	508.7	2.0	101	18.5	68.83
Total deposited sediment volume:				140 560 m^3		
Averaged annual deposited sediment to be dredged:				120 000–140 000 m^3		
Particulate HCB concentration of 60 µm fraction in the river bed: Total averaged value:				18 µg kg^{-1}		
Averaged value of the weir section only:				290 µm kg^{-1} (220 µg kg^{-1} measured)		

Additional effort was spent on Iffezheim to present a detailed diagnosis of the transport dynamics of different suspended sediment fraction during low flow periods and in particular, during the flood from 5 to 30 May 1999 (Fig. 2.7). During low flow period ($Q_{Rhine} < 1\,500$ m^3 s^{-1}), no inflow of particulate HCB was measured; however, fine suspended particles can be transported by lateral dispersion into the left weir branch and deposited at different rates according to the fractional fall velocity as shown in Fig. 2.8 for the grain size: 20 µm, 60 µm and 100 µm (SS20, SS60 and SS100). If the partial discharge through the left branch exceeds about 1 500 m^3 s^{-1}, which corresponds to a total river discharge of about 2 900 m^3 s^{-1}, erosion starts reaching a maximum at the flood peak as depicted in Fig. 2.8 (lower right) and decreasing with falling hydrograph. The numerical results reveal that the inflowing particulate HCB which was assumed to be associated with the 60 µm sediment fraction starts settling after the beginning of the closure of the weir.

Even though the HCB-concentration of the riverbed before the flood was unknown and assumed to be zero, the numerical model results show good agreement when comparing the calculated value of 220 µg kg^{-1} with the HCB contamination of 290 µg kg^{-1} measured in the years 2001 to 2003 and averaged over the erodible top layer of about 10 cm.

According to the numerical results of the individual reservoirs investigated the total mass of HCB mobilized during the flood in all six reservoirs amounts to some 61 kg, which must be considered an underestimation because the computation was performed with the erosion depth averaged contamination measured after the flood as mentioned above (Table 2.2). It is also evident that the measured value of 145 kg HCB must be considered too low because the sampling site in front of the turbine section on the right hand side (Fig. 2.6) could not capture the sediments eroded in the weir channel. It rather represents the fractional load of HCB through the hydropower branch at Iffezheim.

The results of the post-flood retrospective study on the transport dynamics of particulate HCB, in conjunction with the volume of sediments deposited in the subsequent 12 months, provide useful information for future sediment management and contaminant mobilization risk assessment.

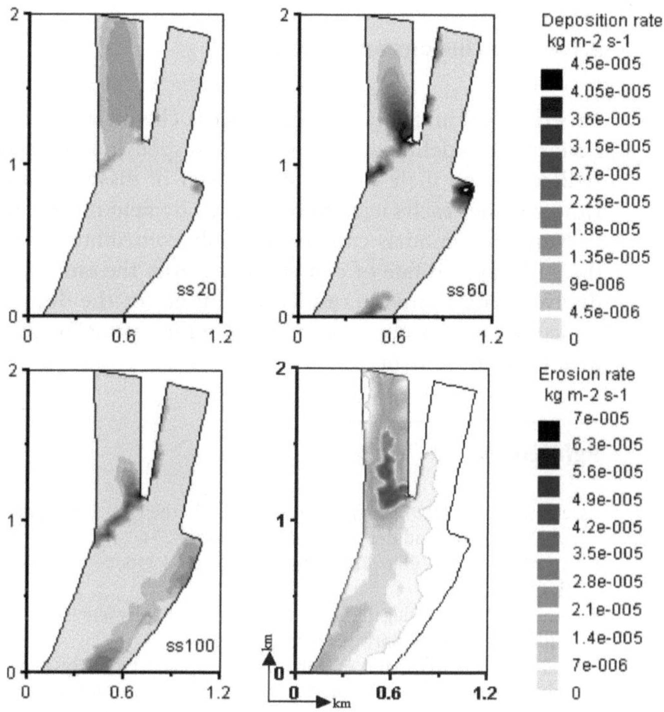

Fig. 2.8. Numerical results of the spatial distribution of the deposition rates of three grain size fractions at low river discharge ($Q = 1\,500\ m^3\,s^{-1}$) and erosion rates (*lower right*) at flood peak ($Q = 4\,250\ m^3\,s^{-1}$) for the Iffezheim reservoir (Jacoub 2004)

Table 2.2. Calculated total HCB mass eroded in the Upper Rhine reservoirs during the flood in May 1999 (Westrich and Witt 2004)

Hydropower station/reservoir	Contaminated area ($10^3\ m^2$)	Particulate HCB (kg)	Specific HCB mass content ($g\ m^{-2}$)	Uncertainty assessment
Marckolsheim	54	5	0.11	Underestimated
Rheinau	13	10	0.08	Estimated*
Gerstheim	178	14	0.08	Plausible
Strasbourg	230	23	0.10	Plausible
Gambsheim	68	6	0.09	Highly underestimated
Iffezheim	36	3	0.08	Highly underestimated
Total:		61 (145 measured)		Highly underestimated

The total amount of deposited sediments from 5 to 30 May 1999 was calculated to be 140 000 m^3 which is close to the averaged annual volume necessary for re-establishing the original channel geometry by maintenance dredging. The good agreement of the deposited sediment volume also confirms the computational results of the contaminated mass budget and hence the applicability of the contaminant transport model.

2.1.4 Conclusions and Outlook

Experimental investigations of physical, chemical and biological parameters must be performed to define erosion process, suspension and sedimentation of fine cohesive sediments and their role in the transfer of dissolved and particulate contaminants. However, the results need to be verified by field measurements.

Numerical models can considerably contribute to risk assessment by describing the pathway and fate of contaminants from the emission to the immission site. Predictive models provide results to be used for the design and analysis of alternative remediation measures and as basic information for further cost effective sediment quality improvement.

References

Asselmann NEM (1997) Suspended sediment in the river Rhine, the impact of climate change on erosion, transport and deposition ISBN 90-6809-254-5 (NGS)

Baart AC, Boon JG, Villars MT (2001) Generic model for contaminants (GEMCO), Z2725, Delft Hydraulics, Delft

Boudreau BP (1997) Diagenetic models and their implementation. Modeling transport and reactions in aquatic sediments, Springer-Verlag, Berlin

Carlon C, Norbiato C, Critto A, Marcomini A, (2000) Risk analysis applied to a contaminated site. Determination of risk based remedial targets. Ann Chim 90(5-6):349–358

Debnath K, Nokora V, Aberle J, Westrich B, Muste M (2006) Erosion of cohesive sediments: resuspension, bed load and erosion patterns from field experiments, to be published by Journal Hydraulic Research, ASCE

De Deckere, EM, van der Waal P, Andree S (2002) A flume study on the effect of Corophium volutator on the sediment stability. J Experimental Marine Biology and Ecology

De Zwart D (2005) Impact of toxicity on species, composition of aquatic communities: Concordance of predictions and field observations, Netherlands National Institue for public helth and the Environment (RIVM)

Formica SJ, Baron JA, Thibodeaux LJ, Valsaraj KT (1988) PCB transport into lake sediments: conceptual model and laboratory simulation. Environmental Sc. Technology, 22:1435–1440

Gerbersdorf S, Jancke T, Westrich B (2004) Physico-chemical and biological sediment properties determining erosion resistance of contaminated riverine sediments – Temporal and vertical pattern at the Lauffen reservoir river Neckar, Germany. Limnologia 5:132–144

Haag I, Westrich B (2001) Erosion investigation and sediment quality measurements for a comprehensive risk assessment of contaminated aquatic sediments. The Science of The Total Environment 266:249–257

Jacoub G (2004) Development of a 2-d numerical code of particulate contaminant transport in impounded rivers and flood retention reservoirs. Institut für Wasserbau, Universität Stuttgart, Mitteilungen Heft 133

Jacoub G, Westrich B (2006) Modelling transport dynamics of contaminated sediments in the headwater of a hydropower plant at the Upper River Rhine. Acta Hydrochimica and Hydrobiologica 33(3)

Kern U (1997) Transport von Schweb – und Schadstoffen in staugeregelten Fließgewässern am Beispiel des Neckars. Institut für Wasserbau, Universität Stuttgart, Mitteilungen, Heft 93, ISBN -3-921694-94-9 (in German)

Lek S, Guegan JF (2000) Artifical Neuronal Networks: application to ecology and evolution. Springer-Verlag

Li, Chien Chen (2004) Deterministisch-stochastisches Berechnungskonzept zur Beurteilung der Auswirkungen erosiver Hochwasserereignisse in Flussstauhaltungen. Universität Stuttgart, Institut für Wasserbau, Mitteilungen Heft 129

Lick W, Lick J, Ziegler CK (1994) The resuspension and transport of fine-grained sediments in Lake Erie. J Great Lakes, vol. 20, pp 599–612
Öberg T, Bergbäck B (2005) A review of probabilistic risk assessment of contaminated land. J Soils and Sediments 5(4):213–224
Onishi Y (1981) Sediment contaminant transport model. American Society of Civil Engineering. Journal of Hydraulic Division 107:1089–1107
Paterson DM (1997) Biological mediation of sediment erodibility: ecology and physical dynamics. In: Burt N, Parker R, Watts J (eds) Cohesive Sediments. Wiley and Sons, pp 215–229
Westrich B, Förstner U (2005) Sediment dynamics and pollutant mobility in rivers (SEDYMO), assessing catchment-wide emission–immission relationship from sediment studies. J Soil and Sediments 5(4):194–200
Westrich B, Schmid G, Entwicklung und Einsatz eines mobilen Gerätes zur in-situ Bestimmung der Erosionsstabilität kontaminierter Feinsedimente. Institut für Wasserbau, Universität Stuttgart, Report TB 2004/05-VA 49
Westrich B, Witt O (2004) Untersuchungen zum Resuspensionsrisiko von Sedimentablagerungen in ausgewählten Staustufen des Rheingebiets. IKSR Report, Institut für Wasserbau, Universität Stuttgart (in German, not published)
Winkler H, Stein A (1997) Optimal sampling for monitoring and dredging contaminated sediments. International Conference on Contaminated Sediments, vol. II, pp 1019–1028
Witt O (2004) Erosionstabilität von Gewässersedimenten mit Auswirkungen auf den Stofftransport bei Hochwasser am Beispiel ausgewählter Staustufen am Oberrhein. Institut für Wasserbau, Universität Stuttgart, Mitteilungen Heft 127, ISBN 3-933761-30-1
Witt O, Westrich B (2003) Quantification of erosion rates for undisturbed cohesive sediment cores by image analysis. Hydrobiologia 494 (1–3):271–276
Zreik DA, Krishnappan G, Germaine JT, Madson OS, Ladd C (1998) Erosional and mechanical Strength of deposited cohesive sediments. Hydraulic Engineering 124:1076–1085

Bernhard Westrich · Chen-Chien Li · Dietrich Hammer · Ulrich Förstner

2.2 Requirement on Sediment Data Quality – Hydrodynamics and Pollutant Mobility in Rivers

2.2.1 Introduction

Three principal media can be used for aquatic monitoring: water, particulate matter and living organisms. With respect to particulate matter, characteristics have been noted such as: *(1)* good specificity to a given pollutant, *(2)* high sensitivity to low levels of pollution, *(3)* medium to low sample contamination risk, *(4)* short (suspended matter) or long to very long (deposited sediment) time span, respectively of information obtained.

The objectives of an assessment program for particulate matter quality can be (Thomas and Meybeck 1992):

- to assess the *present concentrations* of substances including pollutants found in the particulate matter and their variations in time and in space (*basic surveys*), particularly when pollution cannot be accurately and definitely shown from water analysis;
- to estimate *past pollution levels* and events (e.g., for the last 100 years) from the analysis of deposited sediments (*environmental archive*);
- to determine the direct or potential *bioavailability* of substances or pollutants during the transport of particulate matter through rivers and reservoirs (*bioavailability assessment*);

- to determine the *fluxes* of substances and pollutants to major water bodies (i.e., regional seas, oceans) (*flux monitoring*); and
- to establish the *trends* in concentrations and fluxes of substances and pollutants (*trend monitoring*).

The use of particulate matter as an assessment medium has several advantages, at least compared to the water phase, mainly due to the high sensitivity to low levels of pollution and the medium to low sample contamination risk. However, considering the complex system of a large river basin, a closer look is necessary both with respect to state-of-the-art of quality control and quality assurance in these water quality assessment procedures (Sect. 2.2.3) and specifically to quality requirements in relation to hydraulic sediment data (Sect. 2.2.4).

The European Water Framework Directive (WFD) monitoring objectives require compliance checking with Environmental Quality Standards (EQS) but also the progressive reduction of pollution (Sect. 1.1.3). The no-deterioration clause implies that trend studies should be foreseen for sediment and biota. However, *compliance monitoring for sediment* is not yet appropriate because of lack of the definition of valid Environmental Quality Standards (EQSediment) in a European context, analytical limitations and anticipated costs involved to obtain full spatial coverage (Anon 2004a, 2006).

Sediment *trend monitoring* may be both spatial and temporal, and may be related to the chemical and ecological status of a water body. Sediment monitoring may also play a part in *risk-assessment* (see Sect. 10.1).

In principle, it has been recognized that harmonization of sediment monitoring is particularly relevant at a river basin level. In particular, technical issues such as sediment collection, sample treatment, sediment analysis and reporting results will have to follow a common level of quality requirements. Major problem areas have been identified and discussed by the European thematic framework "Metropolis" (Metrology in Support of Precautionary Sciences and Sustainable Development Policies; Anon. 2004b), and comprise lack of representativeness, a high level of uncertainty, lack of metadata, and lack of traceability.

The concept of traceability (Quevauviller 2002) implies that measurement data are *(1)* linked to stated references *(2)* through an unbroken chain of comparison, *(3)* all with stated uncertainties. In the following the implications of the traceability concept for the quality control of sediment analysis will be demonstrated with special reference to the sampling of sediments (Sect. 2.2.2) and the combination of pollutant mobility data (Sect. 2.2.3) and hydrodynamic information (Sect. 2.2.4) at the river basin scale. In Sect. 10.1 "Quality assurance of ecotoxicological sediment analysis" these findings were extended on the influences arising from biological factors.

2.2.2 Sediment Sampling

Natural sediment formed during weathering processes may be modified quite markedly during transportation and deposition by chemicals of anthropogenic origin. Firstly, it must be noted that anthropogenic chemicals may be scavenged by fine sediment particles at any point from their origin to the final sink or their deposition. Secondly, to

Table 2.3. Development of particulate matter quality assessment in rivers in relation to increasing levels of monitoring sophistication (after Thomas and Meybeck 1992)

	Monitoring level		
	A	B	C
Suspended matter (SPM)	Survey of SPM quantity throughout flood stage (mostly when rising)	Survey of SPM quality at high flow (filtration or concentration)	Full cover of SPM quality throughout flood stage
Deposited sediment	Grab sample at station (end of low flow period)	Longitudinal profiles of grab samples (end of low flow period)	Cores at selected sites where continuous sedimentation may have occurred

Level A: Simple monitoring, no requirement for special field and laboratory equipment.
Level B: More advanced monitoring requiring special equipment and more manpower.
Level C: Specialised monitoring which can only be undertaken by fully trained and equipped teams of personal.

compute a geochemical mass balance for sediment-associated elements, it is imperative to derive, by measurement, a mass balance for the sediment in the system under evaluation.

To establish background levels of particulate matter composition, samples of bottom sediment should be taken in the upper reaches of the river basin. The effects of tributaries on the main river should be covered by sampling tributaries close to their junction with the main river. In practice different levels of monitoring sophistication can be distinguished (Table 2.3).

Study of Dated Sediment Cores

The study of dated sediment cores has proven particularly useful as it provides a historical record of the various influences on the aquatic system by indicating both the natural background levels and the man-induced accumulation of elements over an extended period of time. Marine and, in particular, lacustrine environments have the ideal conditions necessary for the incorporation and permanent fixing of metals and organic pollutants in sediments: reducing (anoxic) and non-turbulent environments, steady deposition, and the presence of suitable, fine-grained mineral particles for pollutant fixation. Various approaches to the dating of sedimentary profiles have been used but the isotopic techniques, using ^{210}Pb, ^{137}Cs and $^{239+240}Pu$, have produced the more unambiguous results and therefore have been the most successful (see review on "Historical Monitoring" by Alderton 1985).

Sampling and Filtration of Suspended Matter

Suspended-sediment sampler fall into three general categories (Anonymous 1982; Ongley and Blachford 1982; Horowitz 1991): *(i)* integrating samplers that accumulate a water-sediment mixture over time, *(ii)* instantaneous samplers that trap a volume of whole water by sealing the ends of a flow-through chamber, and *(iii)* pumping samplers that collect a whole-water sample by pump action. Integrating samplers usually are preferred because they appear to obtain the most representative fluvial cross-sectional samples. Cross-sectional spatial and temporal variations in suspended sediment and associated trace elements and their causes are discussed by Horowitz (1991).

Filtration may be carried out under positive pressure or vacuum; excessive pressure or vacuum should be avoided because this may cause rupture of algal cells and release of their intracellular contents into the filtered sample (Hunt and Wilson 1986). Filters having different structures, pore sizes, and composition are available (Brock 1983); the effective pore size of depth filters – having a complex system of channels – changes as the filter becomes more loaded with particles, whereas the effective pore size of screen filters is not affected by filter loading (Apte et al. 2002). Filtration and ultrafiltration can be used for size fractionation of aquatic particles, colloids, and macromolecules (Buffle et al. 1992).

Uncertainties. Handling of suspended sediments includes medium to high contamination risk, similar to the sampling and processing of water samples. Beside problems with filtration techniques (see above), it is important to minimize the time between sample collection and filtration because adsorption/desorption reactions involving particulates and bacterial activity can lead to changes in sample composition.

Handling, Preparation and Storage of Sediment Samples

A review of Mudroch and Azcue (1995) covers the major operations such as *(i)* measurement of pH and Eh (including a detailed description of equipment and solutions used in the measurements), *(ii)* subsampling for determination of cation exchange capacity, *(iii)* subsampling under oxygen-free atmosphere, *(iv)* sample mixing and subsampling into prepared containers and *(v)* sampling hazardous sediments and safety requirements.

Wet Samples

A general scheme for handling samples for tests and analyses on wet sediments is presented in Fig. 2.9 (Mudroch and Bourbonniere 1994). Samples for determination of *particle size-distribution* should not be frozen but stored at 4 °C. Tightly sealed plastic bags, glass jars, or other containers can be used to store samples prior to particle size analyses. Sediments with a high iron content should be stored in air-tight containers to avoid precipitation of iron oxides on particle surfaces and should be analyzed as soon as possible after collection.

Sediment samples for *geotechnical studies* can be stored at 4 °C in a humidity-controlled room, without any large changes in sediment properties for several months. Long cores, such as those collected by piston coring, can be cut into lengths suitable for storage, wrapped to preserve their original consistency, and stored in a refrigerated room.

Freezing has long been an acceptable preservation method for sediments collected for the determination of organic and inorganic constituents. It has been shown that rapid and deep-freezing can best maintain sample integrity and thus enable investigation for concentrations of contaminants. The lower the temperature of deep-freezing the better: a temperature of –80 °C is the suggested maximum.

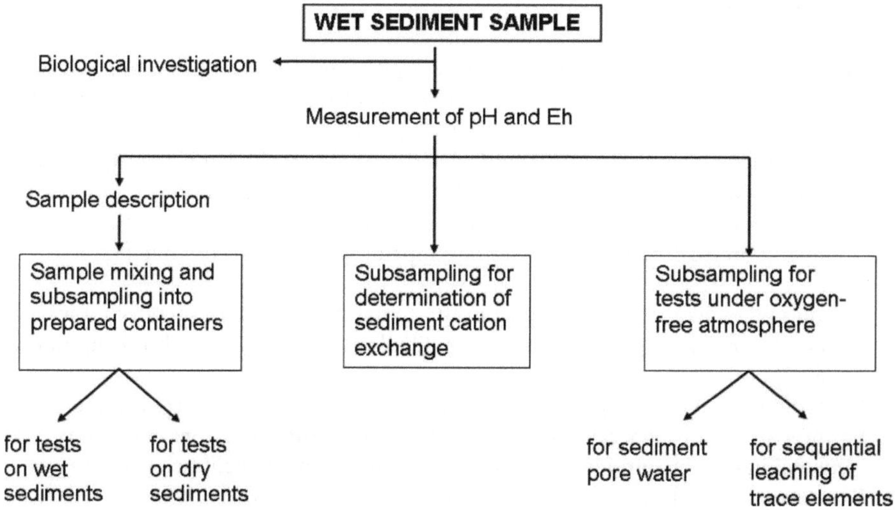

Fig. 2.9. Handling samples for tests and analyses on wet sediments (after Mudroch and Bourbonniere 1994)

Samples collected for investigations of *benthic organisms* are usually processed in the field by wet sieving through different size sieves. If, for any reason, the samples cannot be processed in the field, they should be stored at 4 °C in the dark and processed in the laboratory as soon as possible.

Dry Sediment Sample Preparation

Handling operations of dry sediments include drying, sieving, grinding, mixing, and homogenization. Three types of drying are commonly used to prepare solid samples prior to analysis (Mudroch and Bourbonniere 1994):

- *Air-drying* is rarely used for the preparation of sediments for pollution studies, since it may generate undesirable changes in sediment properties. For example, changes in metal availability and complexation were shown for samples that were air-dried.
- *Oven-drying* of sediments is usually carried out on samples collected for the determination of inorganic components, such as major and trace elements. Oven-drying is not acceptable for sediments which contain any volatile or oxidizable components, whether they be organic or inorganic, and may contribute to the alteration of even non-volatile organics.
- *Freeze-drying* can be used for drying sediments collected for the determination of most organic pollutants as well as for analyses of inorganic components, such as the major and trace elements. The principal advantages of freeze-drying for sediments are *(i)* low temperatures avoid chemical changes in labile components, *(ii)* loss of volatile constituents, including certain organic compounds, is minimized, *(iii)* most particles of dried sediments remain dispersed, *(iv)* aggregation of the particles is minimized, *(v)* sterility is maintained, and *(vi)* oxidation of various minerals or organic compounds is minimized or eliminated.

Anoxic Sediment Treatment

Anoxic sediment samples require different sampling preservation techniques such as oxygen exclusion. Drying and freezing (also freeze-drying) of the samples should be avoided for material designated for extraction procedures. If total analyses or strong acid digestion is planned, the sediment is dried at 60 °C, crushed and stored; for mass calculations, reweighing after drying at 105 °C may become necessary. For a more differentiated approach, in particular for solid speciation studies on anaerobic samples, the following pretreatment scheme was developed (Kersten and Förstner 1987):

- Samples were taken immediately from the center of the material (collected with a grab or corer) with a polyethylene spoon, filled into a polyethylene bottle up to the surface.
- Immediately after arriving at the laboratory, sediments were inserted into a glove box prepared with an inert argon atmosphere. Oxygen-free conditions in the glove box were maintained by purging continuously with argon under slight positive pressure.
- Extractants were deaerated prior to the treatment procedure.

Quality Control

Containers and other equipment used in handling sediment samples after retrieval can be a significant source of contamination (Mudroch and Azcue 1995). For example, plastics contain plasticizers that can be potential contaminants in the determination of organic compounds. Glass, porcelain, stainless steel, Teflon, or Teflon-coated instruments should be used in handling sediment samples to be analyzed for organic components. Wide-mouth amber or clear glass jars and bottles with aluminum foil or Teflon-lined caps are the best containers, but certain compounds (e.g., phenols) can adsorb to these surfaces. Metal containers, spoons, or other equipment may contaminate samples that will be analyzed for metals and trace elements. If both organic and metal analysis are required for a given sediment sample, a Teflon container is recommended.

Since standard sampling and preparation techniques are not available for sediments, results from sediment analyses and in particular their application for sediment quality criteria (SQC), depend in a special way from a high level of quality control (QC) and quality assurance (QA) both in field and laboratory (Keith 1991). QC in *planning* includes choice of *(i)* sampling locations, *(ii)* sampling procedures, and *(iii)* material; quality control in *field sampling* covers *(i)* sample collection, *(ii)* sample handling, *(iii)* cleaning procedures, *(iv)* transport, *(v)* preservation, and *(vi)* storage.

Two techniques can be used for QC in sediment sampling (Mudroch and Azcue 1995):

1. Collection of more than one sediment sample at selected sampling sites using identical sampling equipment, such as multicorers, as well as using identical field subsampling procedures, handling and storage of the samples, and methods for sediment analyses.
2. Subdivision of the collected sample into a few subsamples and treatment of each subsample as an individual sample. The results of chemical analyses of all subsamples indicate the variability due to the sampling and analytical techniques and sediment heterogeneity within a single collected sample.

The control samples used in sediment studies include sampling, transport, sampling equipment, etc., and control samples for laboratory procedures. Contrary to water sampling, sediment sampling generally does not require the use of blanks.

2.2.3 Traceability in Chemical Sediment Analysis[1]

Chemical analyses on sediments, including suspended particulate matter and porewater, are efficient tools in water-quality management (surveillance, survey, monitoring); in this context they refer – in order of increasing complexity – to different objectives, such as preliminary site characterization, identification of chemical anomalies, establishment of references and identification of time changes (chemical, biological), calculation of mass balances, and process studies (Golterman et al. 1983). Chemical analyses are also used to directly characterize contaminated in situ sediments and dredged materials in relation to various treatment techniques.

In the view of the traceability concept, a 'basic sequence' of measurements consists of three steps, which can be considered as an unbroken chain of comparisons (Sect. 2.2.1):

1. *Sampling and sample preparation.* Project planning, sampling stations, sampling devices, handling and storage, and quality control are not standardized, but well-documented in all aspects (Mudroch and Azcue 1995).
2. *Grain size as a characteristic sediment feature.* Sampling on fine-grained sediment (Horowitz 1991) and grain size normalization with 'conservative elements' such as Cs, Sc, Li and Al (reflecting clayey material content) is recommended as standard approach (Förstner 1989).
3. *Analytical procedures.* Reference sediment materials are commercially available. While direct species analysis is still limited, standardized extraction schemes for metals and phosphorus in sediments as well as certified reference materials for comparisons were developed under the auspices of BCR/IRMM.

Further steps in chemical sediment analysis are split up with regard to specific purposes – sediment quality assessment including biological effects (see Sect. 10.1), coupling of sediment quality data with erosion risk evaluation (*(4)* and *(5)* below); chemical changes following resuspension of anoxic sediments (*(6)* and *(7)* below); and modeling of chemical sediment data (*(8)* below).

Due to the particular dynamics of fluvial processes, hydraulic parameters such as the critical shear stress of erosion processes form the primary input factors for investigating and predicting large-scale dispersion of contaminants in flood-plains, dike foreshores and polder areas. Unlike problems related to conventional polluted sites, the risks here are primarily connected with the depositing of contaminated solids on soils in downstream regions. Short-term issues include the fate of sediment associated contaminants when sediment is deposited upland and a better understand-

[1] Based on Förstner U (2004) Trends in Analytical Chemistry 23:217–236.

ing of the impact on ground water, water and soil ecosystems. Medium/long-term issues will focus on integration of quality and quantity aspects, and to determine the sediment transport processes at the river basin scale as a function of land and water use and hydrological (climate) change (see Sect. 1.1). A schematic view of the combined assessment of chemical and hydrodynamic effects in river sediments is presented in Fig. 2.10.

4. *Erosion effects.* Sediment physical parameters and techniques form the basis of any risk assessment in this field. Sampling of flood-plain soils and sediments is affected by strong granulometric and compositional heterogeneities arising from the wide spectrum of flow velocities at which the sediments were eroded, transported and deposited. Standardized fractionation schemes and respective reference materials can be useful for studying ecotoxicological aspects of resuspended sediments. Sediment quality issues should include experimental designs for the study of chemical and biological effects during erosion and deposition (Kosian et al. 1999). Coupling of erosion experiments with investigations on aging effects as well as on the mobilization of chemicals from porewater and labile sediment phases (below) could provide a valuable tool in the decision-making process for remediation techniques.
5. *Aging effects.* 'Diagenetic' effects, which apart from chemical processes (sorption, precipitation, occlusion, incorporation in reservoir minerals and other geosorbents such as char, soot and ashes) involve an enhanced mechanical consolidation of soil and sediment components by compaction, loss of water and mineral precipitations in the pore space, may induce a quite essential reduction of the reactivity of solid matrices. The methodologies developed so far (Corbisier et al. 1999; Thompson et al. 1999; Verbruggen et al. 2000; Zhang et al. 2001) could influence the traceability aspects in the field of ecological/chemical risk assessment (Sect. 10.1) and in relation to erosion stability/pollutant mobility both in situ and river basin wide, and these informations will also affect the decision-making process for remediation techniques (Sect. 1.1.3).

Fig. 2.10.
Assessment of combined chemical and hydrodynamic effects

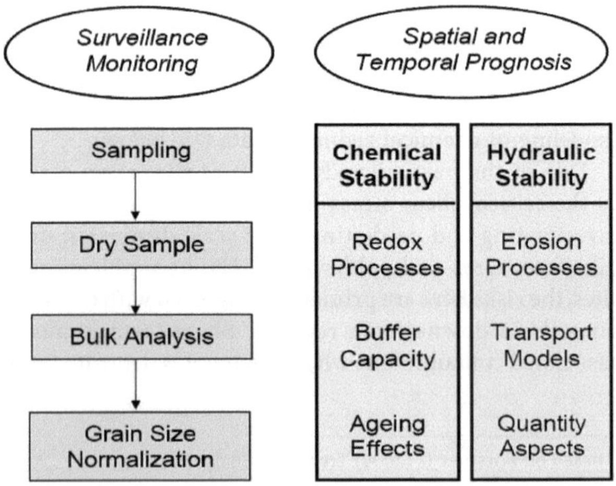

The dramatic effects of stormwater events on particle transport can coincide with rapid and far-reaching chemical changes, in particular, by the effects of sulfide oxidation on the mobilization of toxic metals. The objectives of this research fall under the category 'process studies', usually involving a relative high degree of complexity (Sect. 1.2). Such field and laboratory studies as well as the models using these data are indispensable for long-term prognosis of erosion and chemical mobilization risks arising from subaqueous deposition and capping, both favorable technologies for dredged material and in situ sediments (Förstner 2003).

6. *Anoxic/oxidized samples.* Changes of the forms of major, minor and trace constituents cannot be excluded, when the sediment is transferred from its typical anoxic environment to chemical analysis via normal sample preparation. On the other hand a comparison of extraction data from the original and oxidized samples could be used for worst-case considerations in respect to potential metal release during sediment resuspension or subsequent to up-land deposition of dredged material.
7. *Capacity controlling properties.* Both pH and redox potential in sediment/water systems are significant parameters for mobilization and transformation of metals or phosphorus. Criteria for prognosis of the middle- and long-term behavior of these and other substances should, therefore, include the abilities of sediment matrices for producing acidity and for neutralizing such acid constituents (Sect. 6.5).
8. *Modeling.* The data of critical trace metals and matrix components, as determined from original samples, can also be used in models and in this way, sequential extractions can serve as effective conformational tools to reduce the complexity of the natural system (Wallmann et al. 1993). Pore water analytical data can be applied in geochemical models for short-, medium- and long-term predictions (Parkhurst and Postma 1999). Transport and reaction models consider advective, dispersive and diffusive transport mechanisms as well as ad- and desorption processes (Landenberger 1998).

In total, the traceability of 'further steps' (4–8 above) is less pronounced than that of the three steps (1–3 above) of the 'basic sequence'. However, in the light of the economic value of these analyses for developing and executing far-reaching management plan, coordinated efforts should be undertaken to improve this situation. Short-term measures should range from organized propagation of results from on-going research ('aging effects'), official documentation of techniques and instruments in a relative new field ('erosion effects') and state-of-the-art procedures ('modeling', e.g., analytical data from pore water), via extension of standardized extraction schemes and reference materials (prescription for handling 'anoxic sediments' for fractionation, certification of specific constituents like Ca, S and Fe(II) for the study of 'capacity controlling properties'), up to the development of new reference materials ('pore water'). With regard to the quantity aspect of contaminated sediments in river basin scale, chemical inventories of interim deposits like mining residues, river bank, polder and flood plain deposits, fillings of river-dams and lock-reservoirs, should be given high priority.

2.2.4 Hydraulic Data Quality

Hydrological, hydraulic and sediment data are usually collected from different sources, such as water authorities, institutes and agencies. In many cases, the original measuring data are not accessible; they have already been processed and aggregated so to present them as daily, monthly or annually averaged values in terms of discharge, concentration, load etc. Very often, they are communicated without any technical specification of the sampling site, sampling technique, data-processing method and uncertainty assessment. Little is known about the representativity of the measuring points in flow cross-sections and only sparse information is given on the sampling frequency which is especially necessary for flood events. Gauging stations must provide calibrated discharge rating curves, which normally do not cover the upper range of extreme discharges with overbank flow.

With the aim of quantifying transport rates, flow velocities in the cross-section and the respective fractional concentrations of contaminated sediment must be measured along vertical transects to capture the entire water depth and to allow the calculation of the total transport rate in a given cross-section as shown by the following relationship

$$Q_s = \int_A \int u \cdot c \cdot \mathrm{d}A \qquad (2.1)$$

$$Q_s = Q \cdot C^A + \int_A \int u'' \cdot c'' \cdot \mathrm{d}A \qquad (2.2)$$

$$u'' = U^A - u \qquad (2.2i)$$

$$c'' = C^A - c \qquad (2.2ii)$$

$$Q_s = \alpha \cdot Q \cdot C_1 \qquad (2.3)$$

with the discharge Q in $m^3\,s^{-1}$, the total suspended sediment transport rate Q_s in $kg\,s^{-1}$, the total flow cross-section A, the cross-section averaged suspended sediment concentration C^A, the cross-section averaged flow velocity U^A which equals Q/A, the local flow velocity u in $m\,s^{-1}$ and the respective suspended sediment concentration c in $kg\,m^{-3}$. When only the cross-section averaged data Q and C^A is referred to, a systematic error is obtained where the second term in Eq. 2.2 which represents the differential advection is neglected. This term can be determined only by extensive measurements over the whole cross-section of the flow. The differential advection term vanishes only if there is a uniform flow velocity or a homogeneous concentration of suspended matter, i.e. if u'' or c'' equals zero. In the case of a fixed monitoring station the measured concentration C_1 must be weighted by a site-specific and time-dependent factor α to account for the non-homogeneous distribution of the suspended sediments.

The sampling frequency must be adjusted to the discharge in order to provide an appropriate temporal resolution of the process and to account properly for the discharge as a key parameter for the evaluation of the transport rate and mass balance. The evaluation of single flood events often requires a sampling frequency of the order of magnitude of hours. The spatial density of sampling points depends on the local gradients of both the flow field and the contaminated suspended sediment concentration.

The weakest link in the information chain seems to be the sediment. There are only a few data available on the grain size of suspended sediment, on fall-velocity distribution and sediment erosion stability. Various instruments have been developed and applied for in-situ or on-site particle fall-velocity measurements. Advanced experimental techniques including optical methods with digital image processing are presented and discussed by Eisma et al. (1997). A variety of experimental methods has been developed for cohesive sediment erosion tests. A comprehensive overview of in-situ erosion measurement devices is reported by Cornelisse et al. (1997), the performance of selected erosion devices is described by Gust and Müller (1997). However, a detailed comparison of the various methods is not yet available (see Sect. 2.1).

Vertical profiles of sediment properties and contaminant concentration, including dissolved and colloidal substances in the pore water, must be available to quantify the total mass flux of deep erosion processes. Therefore, the variability in the vertical sediment parameters has to be taken into account, which increases the variance of the computational output. The more processes involved along the river pathway, the higher the variance of the transport quantities involved and the larger the gap between the best- and worst-case assumption for sediment management.

Model Parameter Uncertainty Assessment

Numerous physically based models have been developed to describe the effect of flood events on river morphology and sediment transport, but most of them are deterministic and do not account for the uncertainties involved in the input variables and model parameters. Therefore, it is necessary to apply statistical methods to assess and improve the reliability of model results. In most stochastic approaches, probabilistic distributions of the input variables and model parameters are used for an uncertainty assessment. However, in most cases, the data set is not sufficient to determine the statistics in a conventional way. The integration of the stochastic concept into a deterministic model provides a useful alternative to cope with such uncertainties.

For the risk assessment of contaminated sediment resuspension, various sources of uncertainties must be considered. The most significant contribution to the uncertainty is due to discharge hydrology, which is known as the hydrological risk. Additional uncertainties originate from the imperfection of the model concept and, in particular, from the erosion-related sediment properties including the erosion threshold and erosion rate. Each of the quantities exhibits a specific measuring inaccuracy and shows a high spatial variability in nature. For an environmental impact assessment, the in-situ sediment contamination, sorption, transformation and degradation processes must be described based on chemical and biological parameters, which are subject to significantly higher uncertainties compared to physical parameter uncertainties. Therefore, any quantity calculated at the far downstream end of the contaminant pathway is subject to all the uncertainties included and hence exhibits the cumulative effect of the uncertainties involved (see Fig. 2.11). The impact of the uncertainties of physical, geochemical and biological parameters on the results is linked to and superimposed on the hydrological occurrence probability. Assuming statistical independence of the

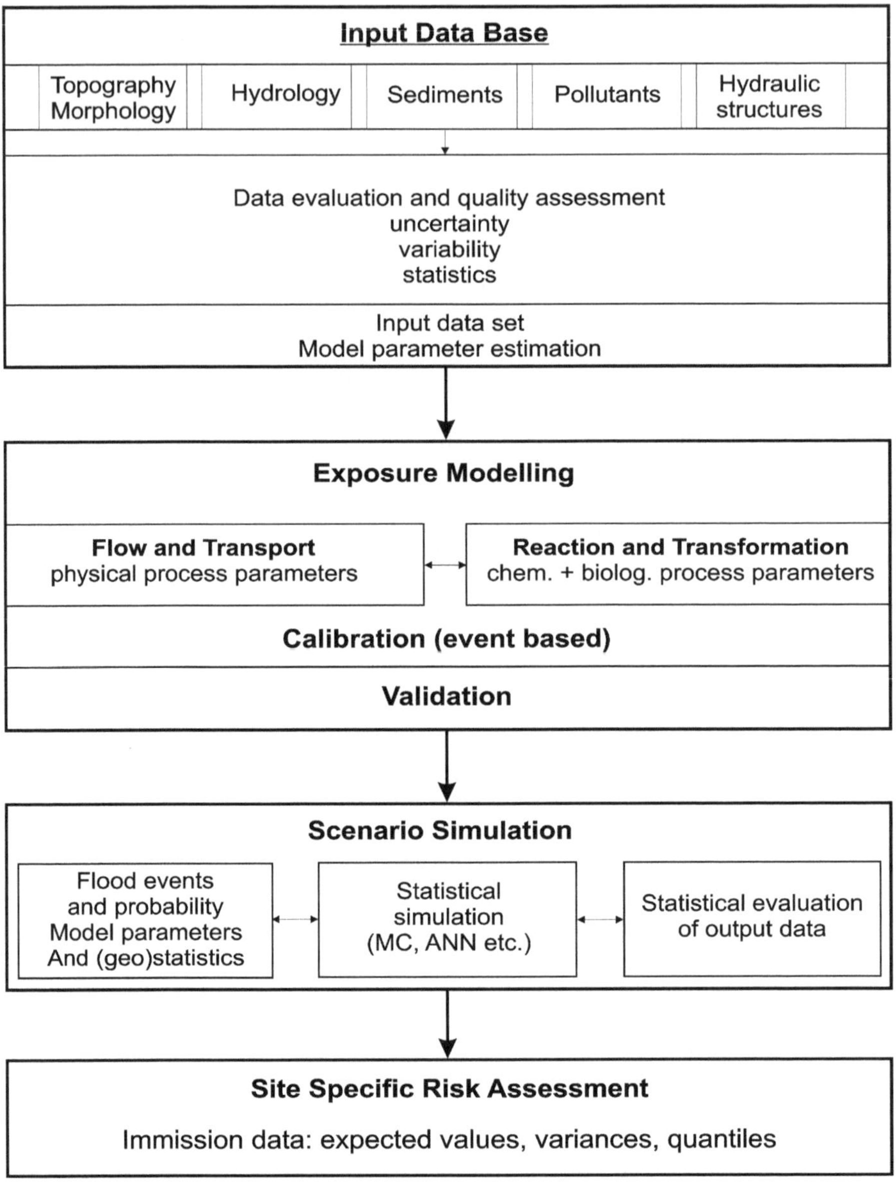

Fig. 2.11. Origin and transmission of uncertainties

parameters, the Gaussian law of error transmission can be applied for estimating the final uncertainty of the quantities. It clearly shows that the uncertainty of the target quantity can be much larger than any of the input parameters.

To cope with the uncertainty of measurement based model parameters and the effect of the variability of hydrological input variables, the use of a stochastic concept for assessing the uncertainty and improving the reliability of the model results is advisable. Stochastic concepts can be applied and integrated into a deterministic hydrodynamic transport model (Fig. 2.11).

The statistical components of the problem can be treated, for instance, by the Monte Carlo method which allows a statistical evaluation of the output. The bootstrap method (Efron and Tibshirani 1993) is an effective method for evaluating field data, especially when the amount of available data does not allow a conventional statistical analysis.

Since the hydrological component makes the dominant contribution to the sediment resuspension risk assessment, the probability, i.e. the uncertainty with respect to time, of any model based quantity increases substantially with the complexity and size of the catchment area. Large catchments consisting of different individual subcatchments with contaminated sites subject to potential resuspension are difficult to capture because of their weakly correlated regional or local flood events. Therefore, one can hardly predict the probability of an immission in terms of concentration of polluted sediments deposited at a site which is far down-stream from the emission site.

Case Study: Erosion Capacity of Floods

To demonstrate the influence of the sediment erosion parameters numerical computations with a 1-dimensional transport model based on the Monte Carlo simulation method were carried out for the 11-km stretch of the river Neckar barrage Lauffen, focusing on floods with various discharge hydrographs and a spatial variability of the sediment erosion properties (Li 2004). Historical flood events were selected from the data series covering the last 50 years to show the effect of the peak flood hydrograph characterized by the discharge and the flood volume on the erosion capacity of the flood. In addition, the uncertainty of the eroded sediment mass caused by the statistical variation of the sediment erosion parameters is quantified.

Two field studies on sediment erosion were made in 1997 and 1998. Altogether, 29 sediment cores were taken for experimental sediment erosion tests. In total, 460 data on the critical shear stress of sediment erosion are then available.

Each value from the collected data is regarded statistically independently. Using the non-parametric bootstrapping method (Efron and Tibshirani 1993) a certain number of bootstraps of the sample mean of the collected field data was produced. Each bootstrapped sample mean is assumed to be the mean critical erosion shear stress of the whole river reach, and was put into the function (Eq. 2.4) suggested by Kuijper et al. (1989) for calculating the erosion rate (E)

$$E = M\left(1 - \tau_0 / \tau_{\text{crit. E}}\right)^n \tag{2.4}$$

where M is the erosion coefficient, n the erosion exponent, τ_0 the actual bed shear stress, and $\tau_{\text{crit. E}}$ the critical erosion shear stress.

A data set of discharges covering almost 50 years from 1950 onwards was available for numerical simulations. The concentrations of inflowing suspended sediment as a function of the discharge were calculated using an experimentally determined power law function (Kern 1997). A field study of the flood event from 28 October 1998 to

4 November 1998 provided another data set of discharges and corresponding suspended sediment concentration (Haag et al. 2002). In the study of Li and Westrich (2004), the influence of all three sediment parameters in Eq. 2.4 was systematically investigated by a local sensitivity analysis. In the following, the discharge hydrograph and the critical erosion shear stress are considered stochastic. The influence of the two variables on the sediment erosion capacity of the flood is estimated by applying a 1-dimensional flow and sediment transport model (Kern and Westrich 1996) to the 11-km lock-regulated stretch of the river Neckar. The erosion parameters M and n were regarded as constant and set at 7.5×10^{-4} kg m^{-2} s^{-1} and 3.2 respectively.

Four historical flood events between 1950 and 1994 with different peak discharge and duration are chosen to demonstrate the effect of the shape of the hydrograph and the variability of the critical erosion shear stress on the sediment erosion potential of floods. The duration of each flood event is 10 days. 50 long term simulations were carried out for a period of 45 years using the measured discharges from 1950 to 1994. At the end of the 45-year simulation period, the impact of the respective flood event was investigated. For each simulation following the Monte Carlo method, the critical erosion shear stress was statistically determined and assumed to be constant in the entire river reach.

Figure 2.12 shows the calculated quantiles of eroded sediment for four historical flood events. Q_{max} stands for the flood peak discharge. Flood (a) differs from flood (b) in the larger flood volume and longer erosion duration. Both floods (a) and (b) show higher peak flow rate than floods (c) and (d) and hence exhibit much higher eroded sediment mass. The slope of the line indicates the effect of the variability of critical erosion shear stress. Among the four flood hydrographs in Fig. 2.12, flood (a) has the largest spreading of eroded sediment mass. A comparison between floods (a) and (b) shows that the larger flood volume of the hydrograph (a) seen in Fig. 2.13 results in a significant larger erosion capacity than flood hydrograph (b). The sediment mass eroded by flood (a) is about 53 000 tons larger than that of flood (b). The calculated results in case (a) spread over 2 000 tons, i.e. ±1 000 tons, whereas the spreading of

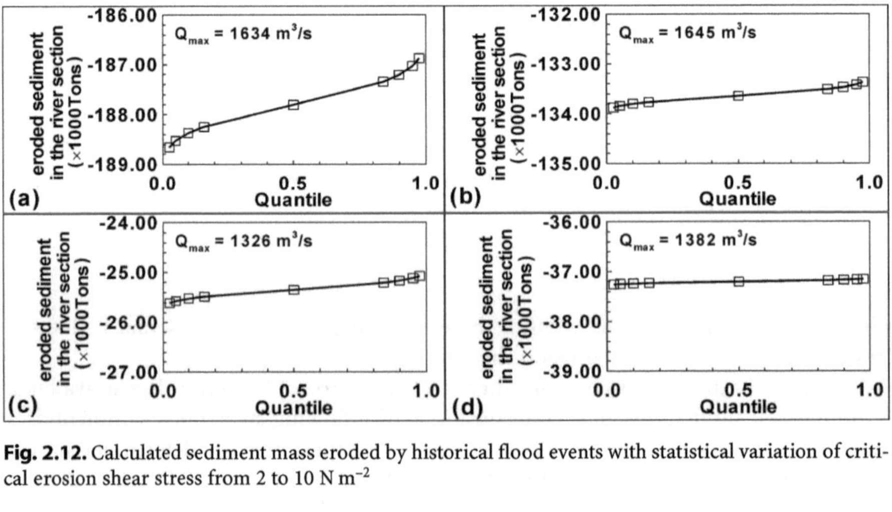

Fig. 2.12. Calculated sediment mass eroded by historical flood events with statistical variation of critical erosion shear stress from 2 to 10 N m^{-2}

flood (c) is, as expected, only about 1 000 tons, i.e. ±500 tons maximum. In comparison to both flood events (c) and (d) in Fig. 2.12 which have about the same peak discharge of 1 634 and 1 645 $m^3 s^{-1}$ respectively, more sediment is eroded by the flood in 1998 even though it had a lower peak discharge of 1 055 $m^3 s^{-1}$. The effect of the variability of the hydrograph, that is the peak discharge and duration of erosive discharge is evident.

One can conclude from the numerical investigation that each flood event has its own individual erosion capacity. Therefore, in order to assess the impact of flood events with a given return period, the whole statistical ensemble must be simulated to provide the required statistical answer. For instance, when assessing the impact of a 100-year flood event on the resuspension of contaminated sediments in a confined river stretch, a series of synthetic hydrographs with varying peak discharge and duration must be investigated and analyzed to provide a probabilistic answer to how much sediment can be eroded by such an event. The joint effect of peak flow rate and the duration of the erosive flood discharge determines the concentration and load of resuspended sediments and, in conjunction with the contamination of the sediments, it also controls the event-related total contaminant load.

Fifty simulations were carried out for the flood event from 28 October to 4 November 1998 using the measured hydrograph with a peak discharge of 1 055 $m^3 s^{-1}$ and 50 samples of critical erosion shear generated using bootstrap sampling. The value of the critical erosion shear stress was given by a random process for each simulation. The critical erosion shear stress was assumed to be constant both in space and time.

The impact of the spatial variability of critical shear stress on the erosion process can be demonstrated by the flood in 1998 (Fig. 2.14). Assuming the critical erosion shear stress varies in space, the usual case in the field, the range of the statistical results is significantly increased and amounts to 8 000 tons of eroded sediments. The maxi-

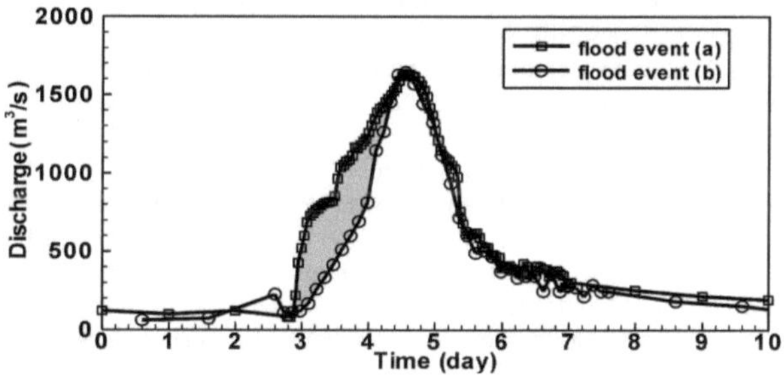

Fig. 2.13.
Hydrograph of two selected historical flood events with equal peak flow (Li 2004)

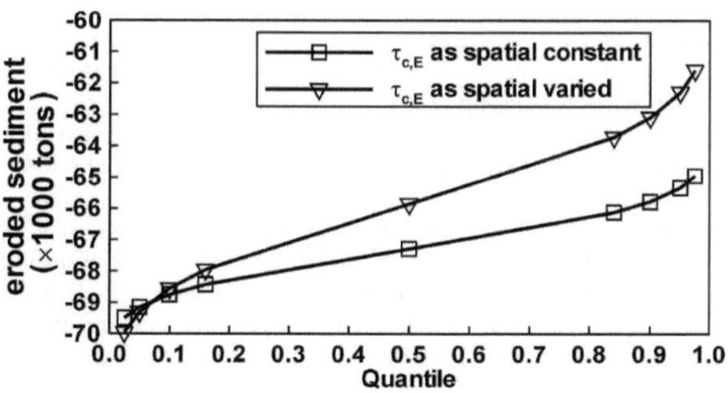

Fig. 2.14.
Quantile of calculated sediment mass eroded during flood event 1998 with a peak flow rate of 1 055 $m^3 s^{-1}$:
a spatially constant and
b spatially varied (Li 2004)

mum deviation from the mean value is about ±4 000 tons, which means a maximum uncertainty of about ±6% related to the expected quantity of 65 000 tons. The graph indicates that the statistical expected value of the eroded sediment mass is increased by about 1 500 tons due to only the spatial variability of the critical erosion shear stress. Moreover, the spatial variability of the sediment parameters does not only increase the variance but also shifts the expected mean value of the eroded sediment mass to a higher level. Only if a constant concentration of contaminants in $\mu g\,kg^{-1}$ across the sediment layers is assumed is the variance of the mass of resuspended contaminants the same as it is for the total eroded sediment mass.

References

Alderton DHM (1985) Sediments. In: Historical Monitoring, pp 1-95. MARC Technical Report 31. Monitoring and Assessment Research Centrer, University of London, UK

Anonymous (1982) National Handbook of Recommended Methods for Water-Data Acquisition. Chapter 3 – Sediment pp 3-1 – 3-100. U.S. Geological Survey

Anonymous (2004a) Expert Group on Analysis and Monitoring of Priority Substances (AMPS). WFD AMPS Sediment Monitoring Guidance Discussion Document, AMPS and SedNet, Draft Version 1 from 16 April 2004

Anonymous (2004b) Evaluation of Current Gaps and Recommendations for further Actions in the Field of Environmental Analysis and Monitoring. METROPOLIS (Metrology in Support of EU Policies) Position Paper, March 2004, 8 p. Verneuil-en-Halatte/France

Anonymous (2006) Proposal for a Directive of the European Parliament and of the Council on Environmental Quality Standards in the Field of Water Policy and Amending Directive 2000/60/EC. Commission of the European Communities, Brussels, 17.7.2006. http://ec.europa.eu/environment/water/water-dangersub/pdf/com_2006_397_en.pdf

Apte SC, Batley G, Maher WA (2002) Monitoring of trace metals and metalloids in natural waters. In: Burden FR, McKelvie I, Förstner U, Guenther A (eds) Environmental Monitoring Handbook, Chapter 6. McGraw-Hill, New York

Brock TD (1983) Membrane Filtration: A User's Guide and Reference Manual. Springer-Verlag, Berlin

Buffle J, Perret J, Newman J (1992) The use of filtration and ultrafiltration for size fractionation of aquatic particles, colloids, and macromolecules. In: Buffle J, Van Leeuwen HP (eds) Environmental Particles, vol. I, Chap 5. Lewis Publ Chelsea MI

Corbisier P, van der Lelie D, Borremans B, Provoost A, de Lorenzo V, Brown N, Lloyd J, Hobman J, Csöregi E, Johannsson G, Mattiasson B (1999) Whole cell- and protein-based biosensors for the detection of bioavailable heavy metals in environmental samples. Anal Chim Acta 387:235–244

Cornelisse JM, Mulder HPJ, Houwing EJ, Williamson HJ, Witte G (1997) On the development of instruments for in-situ erosion measurements. In: Burt N, Parker R, Watts J (eds) Cohesive Sediments. John Wiley & Sons, pp 175–186

Efron B, Tibshirani R (1993) An Introduction to the Bootstrap. Chapman and Hall, New York

Eisma D, Dyer KR, van Leussen W (1997) The in-situ determination of the settling velocities of suspended fine-grined sediment – a review. In: Burt N, Parker R, Watts J (eds) Cohesive Sediments. John Wiley & Sons, pp 17–44

Förstner U (1989) Contaminated Sediments. Lecture Notes in Earth Sciences 21, Springer-Verlag, Berlin, 157 p

Förstner U (2003) Geochemical techniques on contaminated sediments – river basin view. Environ Sci Pollut Res 10:58–68

Förstner U (2004) Traceability of sediment analysis. Trends Anal Chem 23(3):217–236

Golterman HL, Sly PG, Thomas RL (1983) Study of the Relationship between Water Quality and Sediment Transport. Technical Papers in Hydrology no. 26, 231 p. UNESCO, Paris

Gust G, Müller V (1997) Interfacial hydrodynamics and entrainment functions of currently used erosion devices. In: Burt N, Parker R, Watts J (eds) Cohesive Sediments. John Wiley & Sons, pp 149–174

Haag I, Hollert I, Kern U, Braunbeck T, Westrich B (2002) Flood event sediment budget for a lock-regulated river reach and toxicity of suspended particulate matter. Proc 3rd Intern Conf on Water Resources and Environment Research (ICWRER), Dresden, Germany, July 2002

Horowitz A (1991) A Primer on Sediment-Trace Element Chemistry. 2nd ed. 136 p. Lewis Publ. Chelsea/Mich

Hunt DTE, Wilson AL (1986) The Chemical Analysis of Water: General Principles and Techniques. 2nd ed. Royal Society of Chemistry London

Keith L (1991) Environmental Samples and Anayses: A Practical Guide. 143 p. Lewis Publ. Chelsea/Mich

Kern U (1997) Transport von Schweb- und Schadstoffen in staugeregelten Fließgewässern am Beispiel des Neckars. Mitteilung des Intituts für Wasserbau, Heft 93, Universität Stuttgart

Kern U, Westrich B (1996) Mobilität von Schadstoffen in den Sediment staugeregelter Flüsse – Natur versuche in der Stauhaltung Lauffen, Modellierung und Abschätzung des Remobilisierungsrisikos kontaminierter Altsedimente. Wissenschaftlicher Bericht Nr. 96/23 (HG 237), Institut für Wasserbau, Universität Stuttgart

Kersten M, Förstner U (1987) Effect of sample pretreatment on the reliability of solid speciation data of heavy metals – implication for the study of diagenetic processes. Mar Chem 22:299–312

Kosian PA, West MS, Pasha CW, Cox JS, Mount DR, Huggett RJ, Ankley GT (1999) Use of nonpolar resin for reduction of fluoranthene bioavailability in sediment. Environ Toxicol Chem 18:201–206

Kuijper C, Cornelisse JM, Winterwerp JC(1989): Research on erosive properties of cohesive sediments. J Geophys Res 94(C10):341–350

Landenberger H (1998) CoTReM – A Multi-component Transport and Reaction Model (in German). Section of Geochemistry and Hydrogeology, University of Bremen, Germany

Li CC (2004) Deterministisch-stochastisches Bemessungskonzept zur Beurteilung der Auswirkungen erosiver Hochwasserereignisse in Flußstauhaltungen. Mitteilungen des Instituts für Wasserbau, Heft 129, Universität Stuttgart, ISBN 3-933761-32-8

Li CC, Westrich B (2004) Modeling reservoir sediment erosion by floods for assessing the effect of uncertainties and variability of input parameters. Presentation held at 6th Inter Conf on Hydroscience and Engineering (ICHE-2004), May 30–June 3, Brisbane, Australia

Mudroch A, Azcue JM (1995) Manual of Aquatic Sediment Sampling. 219 p. Lewis Publ, Boca Raton

Mudroch A, Bourbonniere RA (1994) Sediment preservation, processing, and storage. In: Mudroch A, MacKnight SD (eds) Techniques for Aquatic Sediments Sampling. 2nd ed., pp 131–169, Lewis Publ. Boca Raton

Ongley E, Blachford N (1982) Application of continuous-flow centrifugation to contaminant analysis of suspended sediment in fluvial systems. Environ Technol Letts 3:219–228

Parkhurst DL, Postma CAJ (1999) User's Guide to PHREEQC (Version 2) – A Computer Program for Speciation, Batch-Reaction, One-Dimensional Transport, and Inverse Geochemical Calculations. Water-Resources Investigations Report 99-4259. U.S. Geol. Survey, Denver, Colorado, USA

Quevauviller Ph (ed) (2002) Methodologies for Soil and Sediment Fractionation Studies. 180 p. The Royal Society of Chemistry Cambridge UK

Thomas R, Meybeck M (1992) The use of particulate material. In: Chapman D (ed) Water Quality Assessments. A Guide to the Use of Biota, Sediments and Water in Environmental Monitoring. Chapter 4, pp 121–170, Chapman & Hall, London

Thompson HA, Parks GA, Brown GE jr (1999) Dynamic interaction of dissolution, surface adsorption and precipitation in an aging cobalt(II)-clay-water-system. Geochim Cosmochim Acta 63: 1767–1779

Verbruggen EMJ, Vaes WHJ, Parkerton TF, Hermens JLM (2000) Polyacrylate-coated SPME fibers as a tool to simulate body residues and target concentrations of complex organic mixtures for estimation of baseline toxicity. Environ Sci Technol 34:324–331

Wallmann K Kersten M, Gruber J, U. Förstner U (1993) Artifacts in the determination of trace metal binding forms in anoxic sediments by sequential extraction. In: Rauret G, Quevauviller P (eds) Sequential Extraction of Trace Metals in Soils and Sediment. Int J Environ Anal Chem 51:187–200

Zhang H, Zhao F-J, Sun B, Davison W, McGrath SP (2001) A new method to measure effective soil solution concentration predicts copper availability to plants. Environ Sci Technol 35:2602–2607

Chapter 3

Hydrodynamics

David M. Paterson

Over the last decades, fluid dynamics has undergone a revolution in the understanding of the importance of flow to the ecology of aquatic ecosystems. This advance has been driven in two ways: firstly as a by-product of improved system analysis that allows the determination of flow patterns at ever decreasing scales of space and time; Secondly, many new methods have been specifically developed to examine questions in ecology that are related to flow dynamics and the behavior of organism in response to flow. Early work was fairly predictable, employing standard fluid dynamics approaches to test biological theory under laboratory conditions (e.g., "flume and flora" experiments), then a period began in which many laboratories produced devices to measure the stability of sediment under field conditions, including fully submerged marine systems. Examples range from linear flumes, carousel or circular chamber systems, to suction or jet devices. These systems all produce useful data but rarely on a comparative scale across ecosystems. The size of the imprint, the nature of the flow created and the method of determining the point of incipient erosion all vary. These are ongoing problems that cannot easily be addressed but of further interest is the imbalance in the applications of these techniques across the fields of research that have developed and applied these technologies. Far more information in terms of relative sediment stability and biogenic interactions is available in the literature concerning brackish (estuarine) and marine environments than fresh water systems. This is perhaps surprising since the overarching characteristic of fluvial systems is the continual longitudinal displacement of the medium ... the river flow. The sediment of river system is continually in motion and undergo catastrophic redistribution during extreme episodic events. These systems are therefore vulnerable to hydrodynamic forcing and the complex control of sediment dynamics is worthy of further consideration. The understanding of the stability of natural fluvial sediments and their behavior on erosion is a fundamental part of the suite of research. The sections in the following chapter help to redress this balance. In the first paper, Spears et al. make an explicit the comparison of sediment stability across different systems, providing an unusual direct comparison between fresh water and brackish habitats. For this work, a single stability device was applied across the varied system which allows for confidence in their comparative approach. In contrast, Gerbersdorf et al. consider the "strength with depth" of natural sediment systems. The novelty of their approach is in determining depth profiles of variables, such as bed shear stress, but linked to the biological and physicochemical properties of the sediment. They argue that this is important in predicting the nature of the bed failure and consequent transport of material, which may also include contaminants from the river system. The behavior of material after erosion is also of critical importance and Mueller et al. continue their well-known ability to produce novel engineering methods and approaches by introducing the *Benthic Water Column Simulator* (BWCS). This engineering device allows experimental investigation of particles and floc throughout the simulated cycle of erosion, transportation, deposition and consolidation of fine sediments. A further engineering approach is presented by Kuehn and Jirka with the objective of creating a system in which turbulence can be controlled and varied over the depth of a water column, essentially separating advection from turbulent intensity regime. This is an innovative approach that will allow the introduction of biological variables into the design system of great engineering precision. Each of these contributions is innovative in its own way and outlines the great potential in combining engineering approaches to answer current questions in the field of fluvial sediment transport.

Bryan M. Spears · Jenna Funnell · James Saunders · David M. Paterson

3.1 On the Boundaries: Sediment Stability Measurements across Aquatic Ecosystems

3.1.1 Introduction

Few studies in the literature compare the sediment stability of depositional habits across marine, freshwater and brackish ecosystems. This is partly because there is conceptual difficulty in comparing different erosional devices but also because scientist often focus on specific habitats. In addition, many field devices generate shear stresses over the 0–1 N m^{-2} range, with few capable of generating erosive forces beyond this level (Tolhurst et al. 2000). However, habitats such as intertidal deposits and salt marshes are often quite resistant to hydrodynamic forcing and are considered to provide an "ecosystem service" of coastal protection. Most existing measurements have been made within a "measurement comfort zone" (Fig. 3.1), usually where a bed shear stress between approximately 0.1 and 1 N m^{-2} surpasses the critical threshold. However, the study of a wider range of habitats is fundamental to the understanding of ecosystem dynamics in aquatic environments.

Resistance to sediment disturbance can be enhanced via physicochemical (particle-particle attraction/resistance; Lerman 1979; Lick and Huang 1993), bio-physical (e.g., increased structural integrity within root beds; Kenworthy et al. 1982; Benoy and Kalff 1999) and bio-chemical processes (the production of extracellular polymeric substances: EPS) (Paterson 1994; Yallop et al. 2000; Deco 2000). While the majority of field based sediment stability studies have been conducted within estuarine mudflat ecosystems, the relevance of these findings across different sediment ecosystems remains largely unknown. Variation should be expected since significant variations in ecosystem processes, known to affect sediment stability, occur within, and between, different ecosystems (e.g., lakes, rivers, and estuaries). These include exposure/submersion cycles, the deposition/removal of fine particulate matter (FPM), the community structure of benthic fauna and flora, and the shear stress exerted across the sediment surface (O'Sullivan and Reynolds 2004; Nedwell and Raffaelli 1999).

In estuarine systems, the presence of EPS, secreted mainly by benthic diatoms (Goto et al. 1999; Yallop et al. 2000), has been shown to enhance sediment stability (see Paterson

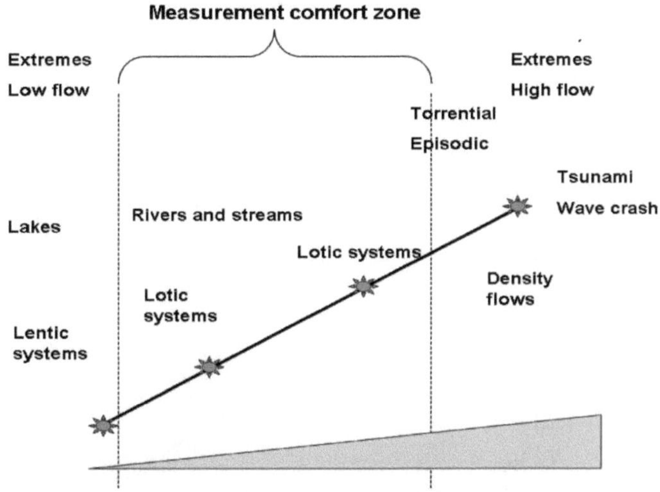

Fig. 3.1. Schematic diagram of the measurement comfort zone for many erosion devices

1997; Tolhurst 1999; Decho 2000). Recent studies have also highlighted the role of vascular plant root-systems in enhancing sediment stability and accumulation by increasing the structural integrity and, therefore, the bed shear strength of colonized riverine and estuarine sediments (Grady 1981). Similar ecosystem engineering may occur in shallow freshwater lakes in which light regimes favor the colonization of macrophytes and microphytobenthos. However, a number of differences occur between ecosystems. For example, tidal cycles drive diurnal fluctuations in estuarine sediment exposure/submersion and compaction. In contrast, lacustrine sediments are continually submerged, with fewer disturbances and a higher depositional flux of fine particulate matter. Another key difference between freshwater and estuarine systems is the strength of the electrostatic attractions (cohesion) between sediment particles as influenced by salinity (Packman and Jerolmack 2004).

A relatively simple comparative assessment of sediment stability across estuarine and freshwater habitats, using the same measurement system, was undertaken. Additionally, spatial variation within each ecosystem was assessed (lake depth, intertidal position). The specific objectives of the study were to *(1)* assess the spatial variation (i.e. inter- and intra-site) of sediment stability and related sediment characteristics; *(2)* identify the key variables regulating stability in each site; and *(3)* identify trends in stability regulation across ecosystem types.

3.1.2 Methods

Study Sites

Loch Leven is a shallow eutrophic loch situated in the southeast of Scotland (56°10' N, 3°30' W), with a mean depth of 3.9 m, covers an area of 13.7 km^2 and is generally dominated by *Potamogeton* sp. Two sample sites were chosen at 2.1 m (shallow Loch Leven site) and 4.3 m (deep Loch Leven site) depth (Fig. 3.2) to represent macrophyte and microphytobenthos dominated sediment, respectively.

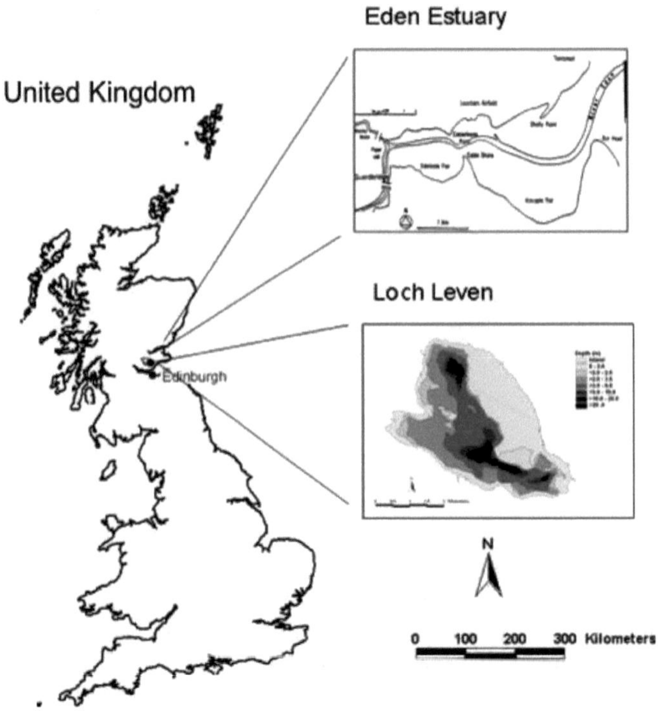

Fig. 3.2. Map showing sample sites within Loch Leven and the Eden Estuary

The Eden Estuary is situated in southeast Scotland (56°22' N, 2°50' W) and is composed of both mudflat (8 km^2 surface area) and saltmarsh (0.11 km^2). The mudflat is generally dominated by microphytobenthos (mainly diatoms) and the saltmarsh mainly by *Puccinellia maritima*. Three sample sites were chosen on the southern shore near the mouth of the estuary to represent low shore and mid shore mudflat, and saltmarsh sediment areas (Fig. 3.2). Sediment is predominantly sandy mud.

Sample Collection

20 cores were collected from each site (i.e. 100 cores in total). At each site, 10 cores were used for stability analysis and 10 cores for the quantification of other sediment characteristics. In the latter, the upper 2 mm of sediment was cryogenically preserved using the contact coring technique (HIMOM 2005). Cores (8 cm internal diameter and 5 cm deep) were collected manually (7–15 February, 2006) from within a 2 m^2 area of exposed sediment at both mudflat sites and the saltmarsh site. To maintain the natural structure of the sediment, macrophytes present in the saltmarsh cores were not removed. Freshwater sediment cores (6.7 cm internal diameter) were collected (21 March 2006 by boat) at each site, within a 10 m^2 area of submerged sediment using a Jenkin surface-sediment sampler. The cores consisted of 20 cm sediment overlain by 30 cm water. The overlying water was either left (for the analysis of sediment stability) or carefully siphoned off to leave an exposed sediment surface as required for contact coring.

Sediment Stability

Sediment stability was measured using a Mark IV Cohesive Strength Meter (CSM) (Paterson et al. 1989). The CSM operates by firing a sequence of water jets with incremental increases in pressure onto the sediment surface inside a small (2 cm diameter) erosion chamber filled with filtered seawater. Eroded sediment is detected by an infrared transmission across the inside of the erosion chamber. The critical erosion threshold is deemed to have passed when suspended sediment levels result in a 10% drop in the original transmission value encountered at the beginning of the test (Tolhurst et al. 2000). The test program "Fine 1" was selected due to its high resolution at low critical erosion thresholds which were expected after initial trials upon the sediments. The CSM was calibrated and results expressed as the stagnation pressure upon the sediment surface expressed as $N\,m^{-2}$.

Cores ($n = 10$) from the saltmarsh and mudflat sites were analyzed in the laboratory. The erosion chamber was set flush with the exposed sediment surface and manually filled with filtered saline water by syringe. Stability in freshwater cores was analyzed immediately after collection and with the overlying water column intact. Filtered saline water was used as the erosive medium in saltmarsh and mudflat cores.

Other Sediment Characteristics

At each site, 10 contact cores were collected for the analysis of wet bulk density, water content, and organic content (HIMOM protocols 2005). Bound and colloidal carbohydrates were separated following the addition of 5 ml distilled water to 50 mg freeze dried sediment and subsequent centrifugation (1 500 rpm for 15 minutes). The resultant pellet was analyzed for bound (attached), and the supernatant colloidal, carbohydrates (Underwood et al. 1995). Both colloidal and bound carbohydrate concentrations were determined using Dubois assays as described within HIMOM protocols (2005) and are expressed against a glucose standard curve as glucose equivalents ($\mu g\,g^{-1}$ dry weight glucose equivalents). Chlorophyll *a* was extracted from freeze dried sediment in 90% acetone. The extraction was conducted at –4 °C in an ultrasound bath for 90 min. Following this, samples were stored in the dark at –80 °C for 24 hours before being vortexed and stored for a further 24 h in the same conditions. Samples were centrifuged (2000 rpm) for 3 min prior to spectrophotometric pigment analysis according to HIMOM (2005).

Statistical Analysis

Statistical variation was assessed using one-way analysis of variance (ANOVA) followed by Fisher's least significant difference post-hoc analysis (critical value = 2.014, $\alpha = 0.05$), on observation of a significant ANOVA result. An analysis of the relationships between variables within each site (intra-site relationships) was conducted using correlation analysis ($n = 10$). An analysis of the relationships between variables across the five sites (inter-site relationships) was also conducted using correlation analysis ($n = 5$). All error bars shown represent the standard error of the mean ($n = 10$) taken from each variable.

3.1.3 Results and Discussion

Variation in Sediment Stability Properties between Ecosystem Types

Significant intra-site variation (Table 3.1 and Fig. 3.3; ANOVA analyses; $p < 0.05$) was observed in sediment stability, colloidal carbohydrate concentration, bound carbohydrate concentration, chlorophyll *a* concentration, organic matter content, water content, and bulk density. Sediment stability was lowest in the freshwater sites, intermediate in the mudflat sites and highest in the saltmarsh site. Stability was observed to increase with depth in the freshwater sites and with increasing distance up the shore in the estuary. The saltmarsh site was more stable than the intertidal low shore site, the shallow and deep freshwater sites. Sediment stability was significantly higher for the mid shore intertidal site than it was in the shallow freshwater site (Fig. 3.3a). Colloidal carbohydrate and bound carbohydrate concentrations were significantly higher in the deep freshwater site than at all other sites (Fig. 3.3b). The bound carbohydrate concentration in the shallow freshwater site was higher than for mid and low shore mudflat sites but similar to the values observed in the saltmarsh site (Fig. 3.3c). Chlorophyll *a* concentration was significantly higher in the deep freshwater site than any other site (Fig. 3.3d). The salt marsh and the shallow freshwater site had similar chlorophyll *a* concentrations, both being higher than the mudflat sites. Sediment organic content was similar for salt marsh and the deep freshwater sites which were both significantly higher than all other sites (Fig. 3.3e). The water content of the mudflat sites were both low and did not differ from each other. Intermediate water content was observed in the saltmarsh site followed by the shallow freshwater site with the highest water content observed in the deep freshwater site (Fig. 3.3f). The bulk density of the intertidal sites were greatest, followed by the salt marsh site and then the freshwater sites. Bulk density in the mid shore site was significantly greater than all other sites with the exception of the low shore tidal site (Fig. 3.3g).

Intra-Site Relationships between Sediment Stability Properties

A positive correlation between sediment stability and colloidal carbohydrate concentration was found for the deep freshwater site (Table 3.2). Negative correlations were observed between sediment stability and chlorophyll *a*, and sediment stability and organic content in the mid shore tidal site and between sediment stability and bulk density in the salt marsh site. Sediment stability was not significantly correlated with any of the measured variables in the shallow freshwater site or the low shore site.

Inter-Site Relationships between Sediment Stability Properties

No significant relationships were observed between sediment stability and any of the other measured variables across the sites. Colloidal carbohydrate, bound carbohydrate and chlorophyll *a* concentrations were significantly positively inter-correlated. The measured sediment characteristics varied significantly both across and within the five sites. The majority of these differences can be accounted for by considering the key ecosystem processes acting within each. Within the estuarine environment,

Table 3.1. Summary of results taken from Fishers post hoc analysis. Direct comparisons of primary variables versus all other variables included where '+' represents instances where primary variables are higher than other variables and '–' represents instances where the primary variable is less than other variables

Primary variable	2.1 m	4.3 m	Low	Mid
Sediment stability				
2.1 m				–
4.3 m				–
Low				
Mid			+	
Salt	+	+	+	+
Colloidal carbohydrates				
2.1 m		+		
4.3 m				
Low		+		
Mid		+		
Salt		+		
Bound carbohydrates				
2.1 m			+	+
4.3 m	+		+	+
Low				
Mid				
Salt	–	–	+	+
Chlorophyll *a*				
2.1 m			+	+
4.3 m	+		+	+
Low				
Mid				
Salt		–	+	+
Organic matter content				
2.1 m				
4.3 m	+		+	+
Low				
Mid				
Salt	+		+	+
Water content				
2.1 m			+	+
4.3 m	+		+	+
Low				
Mid			+	
Salt	–	–	+	+
Bulk density				
2.1 m			–	–
4.3 m			–	–
Low				
Mid				
Salt	+	+	–	–

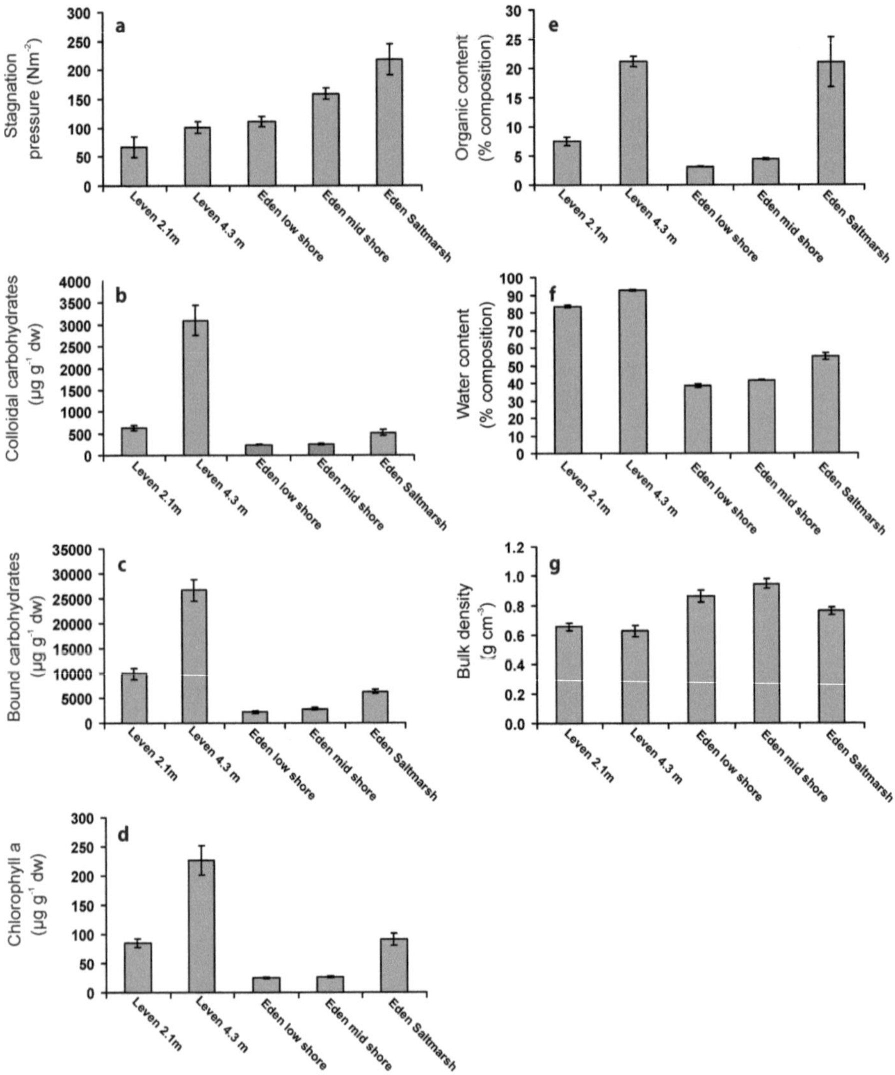

Fig. 3.3. Average cross site trends in **a** stagnation pressure that corresponds to the critical erosion threshold, **b** colloidal carbohydrate concentration, **c** bound carbohydrate concentration, **d** chlorophyll *a* concentration, **e** organic content, **f** water content, and **g** wet bulk density. Error bars represent the standard error of the mean ($n = 10$)

bulk density was lowest in the saltmarsh site and similar in both low and mid shore sediment. This is probably the result of a number of factors including bioturbation, the effect of vegetative cover on the salt marsh and the consequent retention of fine sediments. In comparison, the bulk density values of both freshwater sites were significantly lower than all the estuarine sites. This is due to the depositional environment in productive freshwater lakes where high fluxes of FPM and flocs are combined

Table 3.2. Summary of correlation analysis ($n = 10$; $p < 0.05$) for each site. Only significant correlations reported. *Coll. carb:* Colloidal carbohydrates; *Bound carb.:* bound carbohydrates; *Chl. a:* chlorophyll a; *Org.:* Organic content; *Bulk den.:* bulk density.

Relations	Variable *X*	Variable *Y*	*P*-value	R^2
Loch Leven 2.1 m	Coll. carb.	Chl. *a*	0.043	0.65
	Coll. carb.	Bulk den.	0.021	−0.71
	Bulk den.	Water	0.025	−0.70
	Chl. *a*	Org.	0.043	0.64
Loch Leven 4.3 m	Stability	Coll. carb.	0.032	0.675
	Coll. carb.	Water	0.002	0.85
	Bound carb.	Chl. *a*	0.009	−0.77
	Chl. *a*	Water	0.015	0.74
Eden low shore	Chl. *a*	Org.	0.007	0.79
	Water	Chl. *a*	0.014	0.74
	Water	Org.	0.004	0.81
Eden mid shore	Stability	Chl. *a*	0.044	−0.64
	Stability	Org.	0.016	−0.73
	Coll. carb.	Bound carb.	0.016	0.73
	Chl. *a*	Org.	0.011	0.76
	Chl. *a*	Water	0.034	0.67
	Org.	Water	0.008	0.78
	Bulk den.	Water	0.013	−0.75
Eden saltmarsh	Stability	Bulk den.	0.033	−0.67
	Coll. carb.	Chl. *a*	0.028	0.69
	Coll. carb.	Org.	0.013	0.75
	Coll. carb.	Water	0.025	0.70
	Chl. *a*	Org.	0.000	0.90

with low sediment flushing rates (Hilton et al. 1986; Weyenmeyer at al. 1995). These processes result in the accumulation of fine, unconsolidated organic matter (Nedwell et al. 1999).

Water content was expected to decrease up the shore line as a result of increased exposure and evaporation. However, the opposite trend was observed, most likely resulting from biologically-mediated ecosystem engineering where biofilms and rooted plant beds can trap moisture and prevent desiccation (Paterson and Black 1999; Yallop et al. 2000).

Chlorophyll a, Organic Matter, and Carbohydrate Concentrations

Similar trends in chlorophyll *a* concentration, organic matter content, and bound and colloidal carbohydrate concentrations were observed across the sites. The correlations between chlorophyll *a* concentration and bound and colloidal carbohydrate concentrations within freshwater and saltmarsh sediments suggest that the main source of carbohydrates in these systems are derived from autotrophic processes (Table 3.2). The absence of a significant correlation between chlorophyll *a* and carbohydrate concentrations in the mudflat sites agrees with the observations of Perkins et al. (2003) in which carbohydrate concentration was shown to be more sensitive to a simulated tidal cycle than chlorophyll *a* concentration as a result of the greater solubility of carbohydrates in water. Thus, an uncoupling of the two variables can be expected in sediment ecosystems exposed to severe tidal processes. Both bound and colloidal carbohydrate concentrations were higher in freshwater than they were in estuarine sediments. Coupled with the chlorophyll *a* concentrations, this suggests that the highest autotrophic production of carbohydrates occurred in freshwater sediment which may be due to the relatively high autotrophic (i.e. combination of phytoplankton, microphytobenthos, epiphytes etc.) production in these high nutrient systems.

The higher chlorophyll *a* values in the freshwater sites compared to the mudflat sites are likely due to a higher accumulation of phytoplanktonic detritus. This process is especially important in eutrophic lakes where movement of phytoplankton between the sediment and the water column is regulated by environmental variables and can account for major portions of total phosphorus and nitrogen partitioning (Head et al. 1999). The high organic content and chlorophyll *a* concentrations observed in the saltmarsh, in comparison to the mudflat sites, is likely due to an increase

Table 3.3. Correlation analysis of sediment characteristics across five sites (Loch Leven 2.2 m, Loch Leven 4.3 m, Eden low shore, Eden mid shore, and Eden saltmarsh). *Coll. carb:* Colloidal carbohydrates; *Bound carb.:* bound carbohydrates; *Chl. a:* chlorophyll *a*; *Org.:* Organic content; *Bulk den.:* bulk density. Significant correlations in bold with *p* values in parenthesis

	Stability	Coll. carb.	Bound carb.	Chl. *a*	Org.	Water
Coll. carb.	−0.309 (0.613)					
Bound. carb.	−0.371 (0.538)	**0.984** **(0.002)**				
Chl. *a*	−0.223 (0.718)	**0.966** **(0.008)**	**0.983** **(0.003)**			
Org.	0.360 (0.552)	0.662 (0.223)	0.673 (0.213)	0.797 (0.106)		
Water	−0.525 (0.363)	0.779 (0.120)	0.878 (0.050)	0.856 (0.064)	0.551 (0.336)	
Bulk den.	0.436 (0.462)	−0.690 (0.197)	−0.815 (0.093)	−0.815 (0.093)	−0.631 (0.254)	**−0.936** **(0.019)**

in *P. maritime* in combination with microphytobenthos biomass as a result of increased exposure time (Austen et al. 1999).

Stability Regulation: an Ecosystem Comparison

Significant spatial variation was observed in sediment stability between freshwater, estuarine mudflat, and estuarine saltmarsh sites. Stability was highest in the saltmarsh site and lowest in the shallow freshwater site. Sediment stability was observed to increase with depth in the freshwater sites and with distance up the shore in the estuarine sites. In general, sediment stability was higher in the estuary than it was in the lake. No significant correlations were observed between sediment stability and any of the measured variables across the five sites indicating stability regulation varies between sites or is controlled by variables not considered in this study (Table 3.3). One example of the former is the influence of electrostatic and physico-chemical particle-particle attractions that increase with sodium chloride concentration (i.e. salinity, Lerman 1979; Packman and Jerolmack 2004). Thus, cohesive aggregation increases across the salinity gradient such that estuarine sediments are more 'cohesive' than freshwater sediments. However, this would represent a sedimentological view and it is almost certain that extracellular polymers are also influenced by the ionic nature of their surrounding and contribute more to sediment cohesion under saline conditions. This hypothesis suggests that physico-chemical processes are the main drivers of sediment stability across ecosystem types with bio-physical and bio-chemical processes also being affected by the physicochemical changes. This also explains why, although levels of extracellular polymers are high in the deep freshwater sites, the enhanced stability is less than for the intertidal sites. This possibility requires further research.

Stability was significantly correlated with a number of variables within sites. The positive correlation between sediment stability regulation and colloidal carbohydrate concentration in the deep freshwater site was in agreement with conventional diatom biostabilization theory (e.g., Madsen et al. 1993; Paterson 1994). The absence of this relation in the shallow freshwater site may be the result of excessive wind mixing reducing microphytobenthos biomass, and hence the production of carbohydrates.

3.1.4 Conclusions

The variation in sediment stability was not explained by a single variable across ecosystems. Instead, stability was found to be system specific and therefore care must be taken when defining general rules of stability across ecosystem types. The stability of the bed was higher in the estuarine sites than it was in the freshwater system, probably as a result of increasing electrostatic effects on particle-particle attraction and polymer related adhesion. Stability was higher at 4.3 m overlying water depth compared to 2.1 m overlying water depth in the freshwater site most likely as a result of higher autotrophic production of carbohydrate concentration. Colloidal carbohydrate concentrations were higher in the freshwater ecosystem than in the estuarine ecosystem but less effective at stabilization.

Acknowledgments

We thank the organizers of the SEDYMO conference and acknowledge NERC Doctoral funding to Bryan Spears (NER/S/A/2003/11324) and James Saunders (NER/S/A/2003/11890). The authors acknowledge the support by the MarBEF Network of Excellence 'Marine Biodiversity and Ecosystem Functioning' which is funded by the Sustainable Development, Global Change and Ecosystems Programme of the European Community's Sixth Framework Programme (contract no. GOCE-CT-2003-505446). This publication is contribution number MPS-06056 for MarBEF.

References

Austin I, Andersen TJ, Edelvang K (1999) The influence of benthic diatoms and invertebrates on the erodibility of an intertidal mudflat, the Danish Wadden Sea. Estuarine and Coastal and Shelf Science 49:99–111

Benoy GA, Kalff J (1999) Sediment acuumulation and Pb burdens in submerged macrophyte beds. Limnology and Oceanography 44(4):1081–1090

Decho W (2000) Microbial biofilms in intertidal systems: an overview. Continental Shelf Research 20:1257–1273

Grady JR (1981) Properties of seagrass and sand flat sediments from the intertidal zone of St Andrews Bay, Florida. Estuaries 4(4):335–344

Goto N, Kawamura T, Mitamura O, Terai H (1999) Importance of extracellular organic carbon production in the total primary production by tidal-flat diatoms in comparison to phytoplankton. Marine Ecology Progress Series 190:289–295

Head RM, Jones RI, Bailey-Watts AE (1999) Vertical movements by planktonic cyanobacterial and the translocation of phosphorus: implications for lake restoration. Aquatic Conservation: Marine and Freshwater Ecosystems 9:111–120

Hilton J, Lishman P, Allen V (1986) The dominant processes of sediment distribution and focussing in a small, eutrophic, monomictic lake. Limnology and Oceanography 31:125–133

HIMOM (2005) Heirarchical Monitoring Methods. European commission fifth framework programme. Contract: EVK3-CT-2001-00052

Kenworthy WJ, Ziemen JC, Thayer GW (1982) Evidence for the influence of seagrasses on the benthic nitrogen cycle in a coastal plain estuary near Beaufort, North Carolina (USA). Oecologia 54(2):152–158

Lick W, Huang H (1993) Flocculation and the physical properties of flocs. In: Mehta AJ (ed) Nearshore and estuarine cohesive sediment transport. AGU, Washington, DC, pp 21–39

Lerman A (1979) Geochemical processes: water and sediment environments. John Wiley and Sons Publishers, New York

Madsen KN, Nilsson P, Sundbäck K (1993) The influence of benthic micro-algae on the stability of a subtidal sediment. Journal of Experimental Marine Biology and Ecology 170:159–177

Nedwell DB, Raffaelli DG (eds) (1999) Advances in Ecological Research Estuaries 29. Academic Press, San Diego, CA

Nedwell DB, Jickells TD, Trimmer M, Sanders R (1999) Nutrients in estuaries. In: Nedwell DB, Raffaelli DG (eds) Advances in Ecological Research: Estuaries 29. Academic Press, San Diego, CA

Packman AI, Jerolmack D (2004) The role of physicochemical processes in controlling sediment transport and deposition in turbidity currents. Marine Geology 204:1–9

Paterson DM (1989) Short-term changes in the erodibility of intertidal cohesive sediments related to the migratory behaviour of epipelic diatoms. Limnology and Oceanography 24:223–234

Paterson DM (1994) Microbiological mediation of sediment structure and behaviour. In: Stal LJ, Caumette P (eds) Microbial Mats. NATO ASI Series vol. G35, Springer-Verlag, Berlin Heidelberg

Paterson DM (1997) Biological mediation of sediment erodibility, ecology and physical dynamics. In: Burt N, Parker R, Watts I (eds) Cohesive Sediments. pp 215–229, Wiley Interscience, Chichester

Paterson DM, Black KS (1999) Water flow, sediment dynamics, and benthic biology. In: Raffaelii D, Nedwell D (eds) Advances in Ecological Research. pp 155–193, Oxford University Press, Oxford
Perkins RG, Honeywill C, Consalvey M, Austin HA, Tolhurst TJ, Paterson DM (2003) Changes in microphytobenthic chlorophyll a and EPS resulting from sediment compaction due to de-watering: opposing patterns in concentration and content. Continental Shelf Research 23:575–586
Tolhurst TJ (1999) Microbial mediation of intertidal sediment erosion. Ph.D. thesis. University of St Andrews
Tolhurst TJ, Black KS, Paterson DM, Mitchener H, Termaat R, Shayler SA (2000) A comparison and measurement standardisation of four in situ devices for determining the erosion sheer stress of intertidal sediments. Continental Shelf Research 20:1397–1418
Underwood GJC, Paterson DM, Parkes RJ (1995) The measurement of microbial carbohydrate exopolymers from intertidal sediments. Limnology and Oceanography 40:1243–1253
Weyenmeyer GA, Meili M, Pierson DC (1995) A simple method to quantify sources of settling particles in lakes: resuspension versus new sedimentation of material from planktonic production. Marine and Freshwater Research 46:223–231
Yallop ML, Paterson DM, Wellsbury P (2000) Interrelationships between rates of microbial production, exopolymer production, microbial biomass and sediment stability in biofilms of intertidal sediments. Microbial Ecology 39:116–127

Sabine Ulrike Gerbersdorf · Thomas Jancke · Bernhard Westrich

3.2 Determination of Sediment Stability by Its Physico-Chemical and Biological Properties: Considering Temporal and Vertical Gradients at Different Contaminated Riverine Sites

3.2.1 Introduction

In the past, hazardous contaminants were discharged in large quantities into water bodies where they subsequently accumulated within the bottom sediments. This legacy of the past is found world-wide in rivers, lakes, harbors, estuaries and near-shore areas which are often buried at depths of up to several meters (summarized in Foerstner et al. 2004; Ziegler 2002). However, under certain hydraulic conditions, these former immobilized contaminants might be eroded and become bioavailable and toxic. To prevent ecological disasters and to achieve a good ecological status of the aquatic habitats, as required in the Water Framework (2000/60/EC) Directive, the potential erosion risk of areas of concern has to be evaluated to recommend further treatments such as remediation or sub-aqueous capping. Although several contaminated riverine sites and lakes have been investigated for their sediment erosion behavior world-wide (Ziegler 2002), only a few studies are known from German rivers with emphasis on sediment stability (Haag et al. 2001). In this context, most studies concentrated on the physico-chemical sediment properties such as bulk density, particle size classes and mineralogy that are characterizing erosion resistance (Jepsen et al. 1997; McNeil and Lick 2004). Over the last years, the biogenic mediation of sediment erosion has received increasing attention, namely through stabilizing effects by the mucilaginous matrix of extracellular polymeric substances (EPS) produced by macrofauna, microalgae and bacteria (de Brouwer et al. 2000; de Deckere et al. 2001; Flemming and Wingender 2001). However, it is still rare that physico-chemical *and* biological sediment properties are investigated simultaneously with regard to sediment stability, and only a few properties are generally considered in total (de Brouwer et al. 2000). Considering biostabilization, most investigations originate from tidal flats and thus concentrate on biofilms at the sediment surface with a focus on microalgae as the main producers of EPS. However, bacteria are also known to excrete significant amounts of polymeric substances such as carbohydrates as well as proteins (Flemming and Wingender 2001), and the bacterial production is likely to dominate in deeper sediment layers below the biofilm/photic zone. So far, studies on bacterial biomass and bacterial EPS production have not been related to sediment stability. Hence, the present study aimed to determine a wide range not only of physico-chemical (water content, organic content, grain size, cation exchange capacity, liquid and plastic limits, bulk density) but also biological parameters including macrofauna abundance, microalgal biomass, bacterial cell numbers and EPS fractions (water- and resin-soluble carbohydrates and proteins). Covariance pattern of the different sediment properties were considered by simultaneously sampling and statistical evaluation. In parallel, the critical erosion shear stress for mass erosion was determined in a rectangular pressure duct called the SETEG (*S*trömungskanal zur *E*rmittlung der *t*iefenabhängigen *E*rosionsstabilität von *G*ewässersedimenten) system (Kern et al. 1999). Hence, master-variables for a reliable, efficient and economically viable erosion risk assessment of contaminated riverine sites can be derived. The transferability of the data obtained was addressed by including spatial (different sampling sites) and temporal (different seasons) gradients. Since the erosion properties of sediments may vary in a non-uniform and non-predictable manner as a function of depth, the present study investigated sediment stability and its sediment properties within natural sediments spanning the zone between surface and 50 cm depth.

3.2.2 Material and Methods

Site description. The lock-regulated river Neckar is a major tributary to the river Rhine, draining a highly industrialized and agrarian affected catchment area of about 14 000 km^2 in the Southwest of Germany. Sediment samples were taken in the reservoirs of Deiziau (km 200), Hofen (km 176), Poppenweiler (km 165) and Lauffen (km 137), near-bank in a water depth of less than 2 m. These study sites are known for their high sediment contamination with heavy metals and polychlorinated biphenyls as well as organic stannous compounds (Haag et al. 2001). The Iffezheim barrage is located in the river Rhine (catchment area of 165 000 km^2) at km 334. Sediments were withdrawn in front of the weir section in a water depth of around 2 m. The sediments of this reservoir are heavily contaminated with hexachlorobenzene (HCB, Zipperle and Deventer 2003). At the river Elbe (catchment area of 148 268 km^2), sampling took place at water depths of 1–2.5 m in front of the weir section (Pœlouc, Czech site, km 223) and in the center of groyne fields: Coswig (km 235), Stecky (km 280), Magdeburg (km 318), Fahlberg-List (km 319), Havelberg (km 419) and Hamburg (km 607). The groyne fields are polluted with heavy metals and arsenic originating from mines by the discharge of the tributary rivers Mulde and Saale (Foerstner et al. 2004). Sampling campaigns took place in different seasons: Deizisau, Hofen, Poppenweiler: April, June, September 2004; Lauffen: January–December 2003 (river Neckar); Iffezheim: March 2004 (river Rhine); Prelouc: April 2005; Coswig, Steckby: June 2005; Fahlberg-List: June, August, November 2005; Magdeburg, Havelberg, Hamburg: August 2003 (river Elbe).

Sampling and proceeding. At each study site, sediment samples were withdrawn using two sizes of cylindrical coring tubes of 150/100 cm in length and 13.5/11 cm inner diameter, respectively. The first two sediment cores were used to determine bulk density non-intrusively by a gamma-ray densitometer (Dreher 1997) as well as the critical shear stress for mass erosion in the rectangular pressurized SETEG-flume (Kern et al. 1999; Haag et al. 2001) in cm steps to depths of usually 50 cm. Two more sediment cores with similar bulk density profiles were sectioned into different layers at intervals of 0.5–2 cm and the appropriate layers from both cores were pooled and mixed thoroughly to overcome intrinsic patchiness. The sediment properties were determined within each of these layers. The fifth sediment core was used for sieving the top 10 cm sediment by a 500 µm mesh sieve, retaining all organisms which are referred to as macrofauna.

Sediment properties. The following physico-chemical sediment properties were determined: *water content, mineral composition, soil particle size classes* (sand 2–0.063 mm, silt 0.002–0.063 mm, clay <0.002 mm), *TOC* (*t*otal *o*rganic *c*arbon), *CEC* (*c*ation *e*xchange *c*apacity), *liquid* and *plastic limits*. The biological sediment properties included: abundances of *macroinvertebrates*, concentrations and content of *chlorophyll a* (proxy for algal biomass) and *pheopigments* (indicator for degraded microalgal pigments), *bacterial cell numbers* (DAPI staining, Epifluorescence Microscope) and *EPS* (*e*xtracellular *p*olymeric *s*ubstances) fractions. Briefly, EPS were extracted from dried and homogenized sediment samples over a fixed period with distilled water as well as with CER (*c*ation *e*xchange *r*esin), modified after de Brouwer and Stal 2001; Frølund et al. 1996. Within the supernatants, the water-extractable (colloidal) carbohydrates, the CER-extractable carbohydrates as well as the CER-extractable proteins – corrected for humic acids – were determined spectrophotometrically (protocols in Frølund et al. 1996). In pre-experiments, the extraction procedure was optimized regarding the yield and ratio of the different EPS fractions by varying extractants, extracting volume and sediment dry weights. For more details on the determination of sediment properties, especially the EPS fractions see also Gerbersdorf et al. (2005).

Statistics. Multivariate Statistics, PCA (*P*rincipal *C*omponent *A*nalysis), was conducted by SPSS 12.0 for Windows to address the covariance pattern of the physico-chemical and biological sediment properties and their impact on sediment stability (Haag and Westrich 2002). For more details see also Gerbersdorf et al. (2005).

3.2.3 Results and Discussion

Sediment stability – vertical profiles of $\tau_{crit.\,E}$ and ρ. On the basis of bulk density profiles, horizons of differing consolidation and particles sizes could be selected. Using this data it was ensured, that the appropriate horizons of replicate sediment cores were pooled after sectioning. At the sediment surface of all stations, the bulk density was relatively low (1.1–1.3 g cm^{-3}) along with high water content (up to 71%) (Fig. 3.4). Although the bulk density tended to increase towards deeper sediment layers, in most cases, clear consolidation effects could not be visualized in the top 25 cm. Pronounced

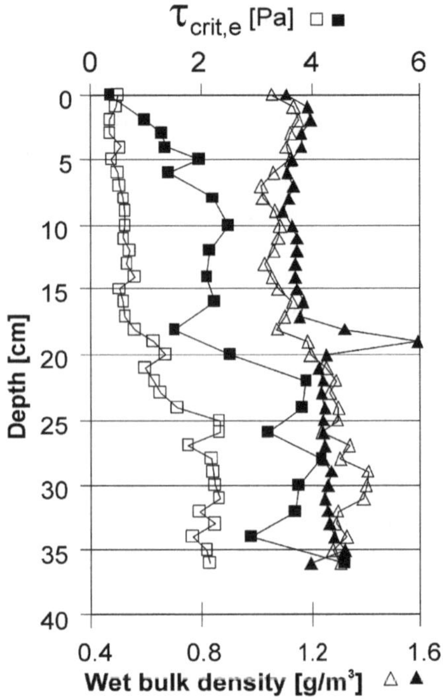

Fig. 3.4. Vertical depths profiles of critical erosion shear stress ($\tau_{crit.\ E}$, squares) and wet bulk density (ρ, *triangles*), examples shown for Prelouc reservoir (*open symbols*) and Fahlberg groyne field (*black symbols*), April and June 2005, respectively

peaks of bulk density values usually corresponded to sandy horizons along with low organic content and decreasing sediment stability (e.g., 18 cm depth, Fahlberg, Fig. 3.4). The critical shear stress determined within the top 25 cm sediment by the SETEG flume, were also relatively low at all reservoirs (mostly below 1 Pascal (Pa), range of 0.3–2.9 Pa, Table 3.4, Fig. 3.4) in accordance with data from active sedimentation/resuspension areas such as the intertidal flats (0.13–1.0 Pa, Austen and Witte 2000). Although older, more consolidated horizons showed higher critical shear stress values compared to younger, freshly deposited sediment layers on top, the stability of all horizons was surprisingly low. The measured critical shear stress of the sediments was opposed to different scenarios of naturally occurring bottom shear stress, for instance calculated by the 1-D flow and transport model COSMOS for the Lauffen Reservoir/river Neckar (Kern 1997). Thus, it was indicated that a three to five fold mean discharge (around 500 $m^3 s^{-1}$ instead of 80 $m^3 s^{-1}$) would erode the top 25 cm sediment layers with shear stress values of 1–2.5 Pa and a five years flood event (corresponds to 9 Pa) could remove all sediments at depths below 50 cm (Haag et al. 2001). Haag et al. (2001) detected polluted sediment layers in the Lauffen Reservoir/river Neckar at only 20–70 cm depth but some contaminations were even directly located at the sediment surface due to local heterogeneity in erosion/deposition pattern. Chemical and ecotoxicological studies revealed the occurrence and bioavailability of contaminations within the top 15 cm sediment at different study sites in river Neckar and Rhine (Zipperle and Deventer 2003). Thus, the risk of erosion was severe at the contaminated reservoirs of river Neckar, even under moderate bottom shear stresses. The same applies for the study sites of river Rhine and Elbe (data not discussed).

Table 3.4. Mean values of critical shear stress (Pa) given for different study sites and at different depths

	$\tau_{crit.E}$ (Pa)		
	Surface	1–5 cm	16–20 cm
Deizisau	0.52	0.61	0.70
Hofen	0.45	0.59	0.86
Poppenweiler	0.64	0.64	1.04
Lauffen	0.69	1.07	1.67
Iffezheim	0.38	0.44	0.78
Prelouc	0.47	0.41	0.90
Coswig	0.28	0.40	8.40
Steckby	0.33	8.90	8.90
Fahlberg	0.44	1.11	4.85
Magdeburg	0.80	0.66	0.43
Havelberg	0.52	0.72	2.50

Biogenic Mediation of Sediment Stability

Macrofauna. The impact of macroorganisms on sediment stability can be negative due to bioturbation and grazing activities but we could not prove such an effect in the present study. On the other hand, certain worms and midge larvae such as *Tubificidae* and *Chironomidae* are known to live in tubes which act as sediment traps and which are coated with polymeric substances (Datry et al. 2003; Olafsson and Paterson 2004). For all river Neckar data ($n = 21$), a positive relation between the critical shear stress values and the abundances of *Tubificidae* (Pearson's $r = 0.63$) as well as *Chironomidae* (Pearson's $r = 0.61$) could be shown for the top 10 cm sediment. Since there were no correlation between polymeric substances and macrozoobenthos abundance, the tubes itself must have contributed to sediment stabilization. It remains to be investigated, which proportion of biostabilization might be due to macrofauna tubes formation, when compared to other biological impact factors like EPS produced by microalgae and bacteria.

Microalgae. At all riverine sites, especially within the groyne fields, the turbidity and light attenuation ($k = 2–3\ m^{-1}$) within the water column was high, calculated by vertical underwater light profiles. However, the resulting light regime at the sediment surface (minimum range 8–30 $\mu E\ m^{-2}\ s^{-1}$) still allowed photosynthetic activity by benthic microalgae and thus, possible EPS excretion (Smith and Underwood 2000). Microscopy revealed the occurrence and dominance of benthic diatoms (Bacillariophyceae, Pennales), and around 70 species were determined in total at all sites (Gerbersdorf et al. unpubl.). In the sediments of the reservoirs, only a few deposited phytoplankton species were found. In the groyne fields sediments, pelagic diatoms (Bacillariophyceae, Centrales) and pelagic green algae (Chlorophyta, Chlorococcales/Chlorellales) were more dominant, thereby reflecting the two main components of the phytoplankton in the river Elbe (Gerbersdorf et al. unpubl.). Presumably the deposited pelagic species contrib-

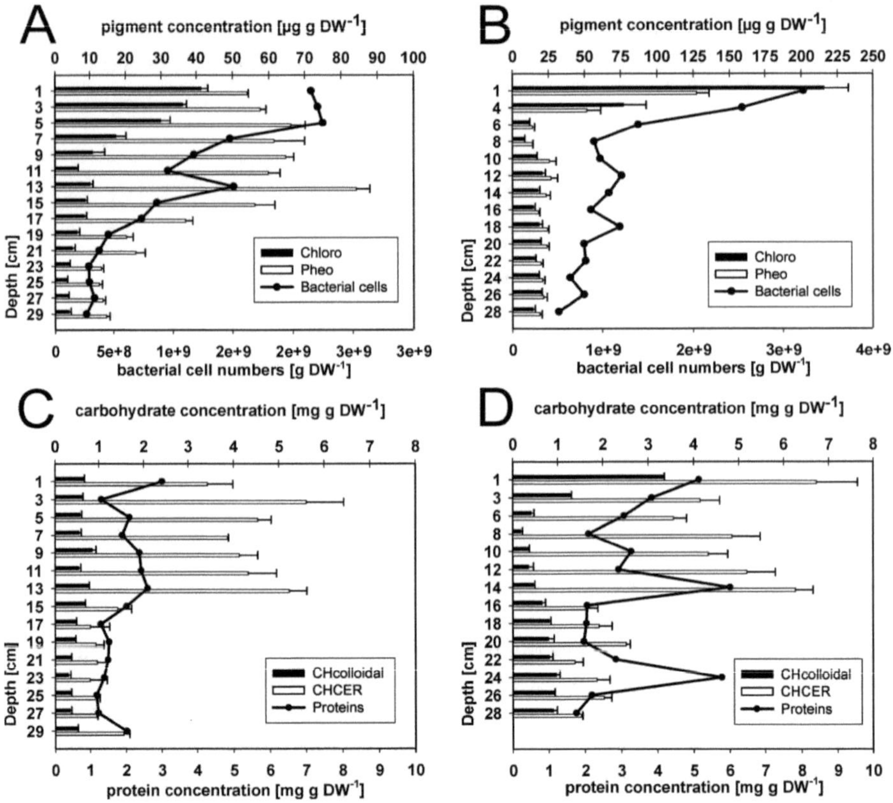

Fig. 3.5. Vertical depth profiles of chlorophyll *a* (*black bars*), pheopigment (*white bars*), bacterial cell numbers (*black line with circles*) (*A* and *B*) and polymeric substances such as colloidal carbohydrates (*black bars*), resin-extractable carbohydrates (*white bars*) and proteins (*black line with circles*) (*C* and *D*). Examples shown for Prelouc reservoir (*A* and *C*) as well as for the Fahlberg groyne field (*B* and *D*), April and June 2005, respectively

uted considerably to the chlorophyll *a* concentration in the sediments indicated by higher microalgal biomass in the groyne fields (river Elbe: Fahlberg 24, UFZ Magdeburg 36, Havelberg 75, Hamburg 60), compared to the reservoirs (river Neckar: Deizisau 27, Hofen 9, Poppenweiler 12, Lauffen 12, river Elbe: Prelouc 11) (µg cm^{-3}), mean value of surface, fresh sediment, Fig. 3.5a and b). Still, the microalgal biomass in all riverine sediments investigated was surprisingly high and even in the range of highly photosynthetic active habitats such as intertidal flats or tidal sandy beaches (de Brouwer and Stal 2001).

Bacteria. Similar to the vertical distribution of the microalgal biomass, the bacterial cell numbers were highest in the top sediment layers and decreased over depth (Fig. 3.5a and b). This indicates the well-known mutual profit of bacteria and microalgae by their excudates (Yallop et al. 2000). The bacterial cell numbers within the sediment surface ranged from $4–9 \times 10^8$ bacterial cells cm^{-3} fresh sediment at all study sites.

EPS fractions. The concentrations of water-extractable (colloidal) carbohydrates in the sediment surface layer were in the range of 0.11–0.49 mg g DW^{-1} with maximum values of 1.52 and 1.82 mg g DW^{-1} (groyne field Hamburg and Havelberg/river Elbe, respectively) and 2.37 mg g DW^{-1} (Deizisau Reservoir/river Neckar). While the colloidal carbohydrates represented only about 10% of the total carbohydrate pool in most reservoirs (exception Deizisau: 20–40%), their contribution was much higher in the groyne fields (9–59%, mostly around 30%, Fig. 3.5c and d). Although the concentrations of colloidal carbohydrates seemed low at first, they were well in the range of data known from pronounced biofilms with high photosynthetic activity and EPS excretion (de Brouwer et al. 2000). The concentrations of proteins in the sediment surface layer were similar to the CER-extractable carbohydrates and mostly in the range of 1.80–5.15 mg g DW^{-1} with some higher values in the reservoirs of the river Neckar (Deizisau 12.90, Hofen 9.60 and Poppenweiler 10.20 (mg g DW^{-1})). Thus, the quantity of polymeric substances could be sufficient to have a significant impact on sediment stability.

Which organisms are excreting polymeric substances? Microalgae are known to produce significant amounts of carbohydrates (Underwood et al. 2004), while bacteria could excrete additionally proteins (Flemming and Wingender 2001). Plotting the data from all study sites at all seasons together, excretion of colloidal and CER-extractable carbohydrates as well as proteins by microalgae and by bacteria, could be shown at all depths (except for the sediment surface, no relation of bacteria and protein, Table 3.5). The positive relation of pheopigments and polymeric substances hints towards the release of EPS from senescent algal material (Table 3.5).

Table 3.5. Potential EPS producers (*Chlorophyll a* as proxy for microalgal biomass, *Pheophytin* as proxy for microalgal degradation products, *Bacteria* = bacterial cell numbers) versus EPS components (*CH colloidal* = water-extractable Carbohydrates, *CH CER* = resin-extractable carbohydrates, proteins) and their relation to water content. The correlation coefficients are given for the top layers (*Surface*, 0.5 cm) and deeper layers (*Deep*, to depth of 10 cm). The different significance levels are indicated by $p < 0.05$ (a) and $p < 0.001$ (b) for Pearson's r

	Water Content		Chlorophyll *a*		Pheophytin		Bacteria	
	Surface	Deep	Surface	Deep	Surface	Deep	Surface	Deep
Chlorophyll *a*	0.81[b]	0.76[b]						
Pheophytin	0.96[b]	0.88[b]	0.77[b]	0.79[b]				
Bacteria	0.92[b]	0.57[b]	0.53[a]	0.58[b]	0.55[a]	0.34[b]		
CH colloidal	0.68[b]	0.72[b]	0.62[b]	0.57[b]	0.69[b]	0.73[b]	0.66[b]	0.25[a]
CH CER	0.41	0.45[b]	0.45[a]	0.46[b]	0.17	0.39[b]	0.48	0.37[b]
Proteins	0.17	–0.11	0.53[b]	0.13	0.41[a]	–0.06	0.01	0.11

[a] $p < 0.05$.
[b] $p < 0.001$.

Correlation of sediment stability and sediment properties

Physico-chemical properties. The critical shear stress showed positive relations with the clay and silt fraction (<63 µm), as well as with TOC and CEC (Table 3.6, Gerbersdorf et al. 2005, 2007). Along with decreasing grain size, the cation exchange capacity (CEC) is increased due to the enhanced number of active adsorption sites in fine-grained sediments. In muddy sediments, the organic content is much higher compared to quartz sand, and these polymeric compounds with active binding sites will add to sediment stability by adsorption to the small particles (de Brouwer et al. 2000). Moreover, the sediment will consolidate over time and depth (Table 3.4), mostly a dewatering effect (Perkins et al. 2003), which is often reflected in increasing bulk densities. In the present data set, sandy layers with high bulk density and low erosion resistance were not excluded from the data set. Accordingly, the relation between critical shear stress and bulk density as well as water content showed some inherent pattern (Table 3.6). Usually, the shear stress was positively correlated with bulk density (e.g., 0.65, $p < 0.001$ for top layers of Fahlberg), while the relation with water content was inverse as could be shown for the top 10 cm sediment from the groyne field Fahlberg (Table 3.6).

Table 3.6. Critical shear stress versus sediment properties (*CH colloidal* = water-extractable Carbohydrates, *CH CER* = resin-extractable Carbohydrates, $d < 63$ µm = grain sizes, *TOC* = Total Organic Carbon, *CEC* = Cation Exchange Capacity). The correlation coefficients are given for the upper layers (*Top*, from surface to 10 cm depth) as well as for deeper layers (*Deep*, below depths of 10 cm), shown exemplary for one reservoir (Lauffen, River Neckar) and one groyne field (Fahlberg, River Elbe). The significance levels are indicated by $p < 0.05$ (a) and $p < 0.001$ (b) for Pearson's r

	Lauffen reservoir		Fahlberg groyne field	
	Top	Deep	Top	Deep
Chlorophyll *a*	–0.45	0.20	–0.45	0.09
Pheophytin	–0.08	0.11	–0.57[a]	0.24
Bacteria	–0.95[b]	–0.19	–0.44	0.10
CH colloidal	0.30	0.54[b]	–0.63[a]	0.54[a]
CH CER	0.02	0.44[a]	0.73[b]	0.02
Proteins	0.33	0.43[a]	–0.21	0.43[a]
Bulk density	0.38	–0.49[b]	0.65[b]	–0.08
Water content	–0.57	0.29	–0.65[b]	0.29
$d < 63$ µm	0.49	0.68[b]	0.41	0.30
TOC	–0.34	0.23	–0.10	0.53[b]
CEC	0.91[a]	0.35	0.30	0.49[a]
Depth	0.53[a]	0.56[b]	0.85[b]	0.48[a]

[a] $p < 0.05$.
[b] $p < 0.001$.

Biological properties. The interrelations between the critical shear stress of erosion and the microorganisms as well as their excreted EPS components have been highly variable in spatial, temporal and vertical terms (see also Gerbersdorf et al. 2005, 2006a,b). At times of high photosynthetic/metabolic activity, huge concentrations of polymeric substances were produced in the top sediment layers, closely related to the biomass of their producers: microalgae and bacteria (Table 3.5). Plotting sediment surface (0–0.5 cm) data from different study sites and seasons together, positive relations between the sediment stability and microalgal/bacterial biomass (Pearson's $r = 0.28 / 0.42$) as well as the resin-extractable carbohydrates (Pearson's $r = 0.44$) could be ruled out (Gerbersdorf et al. 2006b). With depth, the biomass of the microorganisms as well as their excretion products were decreasing (Fig. 3.5a and b) while sediment stability was increasing, resulting in some negative interrelations within the top sediment layers (Table 3.6). In deeper layers, a positive relation between all biologically produced EPS components and sediment stability could be ruled out, independent of the biomass of the EPS producers (Table 3.6, Yallop et al. 2000). Migratory activity of buried microalgae might have maintained the pool of polymeric substances by permanent excretion of colloidal carbohydrates during movements (de Brouwer and Stal 2002; Underwood et al. 2004). Secondly, the bacterial EPS production might have gained more and more importance in layers below the photic zone. However, increasing ratios of carbohydrates and proteins over microorganisms biomass indicated rather irreversible binding of the biologically produced polymeric material to sediment particles and thus, accumulation over depth. Although it is known that colloidal carbohydrates are easily dissolved in water, the large adsorptive surface area in fine-grained sediments might retain even loosely bound carbohydrates, followed by a possible conversion into stronger attached fractions (de Brouwer et al. 2000). Consequently, not only the more recalcitrant CER carbohydrates with higher partition coefficient (indicating higher adsorption to sediment particles, de Brouwer and Stal 2001), but also the colloidal carbohydrates as well as proteins (presumably exoenzymes, Flemming and Wingender 2001) accumulate over depth or in distinctive sediment horizons. Presumably, the variations in the relation between biological properties and sediment stability were not only due to different quantities at the different study sites, seasons or depth (Table 3.6). The quality of the polymeric substances might also be crucial for binding sediment particles (Gerbersdorf et al. 2005). The quantity as well as the stabilizing potential of EPS might vary due to the physiological status and the taxonomic composition of their producers, influenced by abiotic parameters such as nutrients and light (de Brouwer and Stal 2001; Smith and Underwood 2000).

Combined influence of sediment properties on sediment stability. Although the correlations between the critical shear stress and single sediment properties have been significant, the correlation coefficients were low (up to 0.54 only, Table 3.6). Hence, PCA (Principal Component Analysis) was conducted in order to assess the covariance pattern of the sediment properties and their impact on sediment stability. Due to the high variability of the biological data, this multivariate statistic approach was performed for limited data sets of similar origin (e.g., study site, season): Lauffen January–December 2003, Deizisau/Hofen/Poppenweiler April 2004 (river Neckar), Fahlberg June–November 2005 (river Elbe). All other study sites were sampled only once, thus their data set was too small to conduct PCA. Congruently, the combined influence of grain size, TOC, CEC and polymeric substances, constituting the interparticle forces, was most important for sediment stability. In all cases, the critical shear stress correlated best to the main components in which these appropriate sediment properties were combined. Thereby, the magnitude of the correlation coefficient was significantly enhanced compared to the correlations between shear stress and single sediment properties only. The correlation coefficients applying the PCA approach were $R = 0.88$ for Lauffen (Gerbersdorf et al. 2005), $R = 0.70$ and 0.91 for Deizisau, Hofen, Poppenweiler (Gerbersdorf et al. 2007) and $R = 0.81$ for Fahlberg (biplot of critical shear stress to main component I in Table 3.7, biplot not shown).

3.2.4 Conclusions

The present study evaluated the covariance pattern of physico-chemical and biological sediment properties and their impact on sediment stability, investigating natural sediments from contaminated riverine sites over depth. Biogenic mediation was mainly due to the mucilaginous matrix of extracellular polymeric substances (EPS), excreted by microalgae and bacteria. These polymeric substances such as proteins and carbohydrates contributed significantly to the inter-particle forces, along with grain size, TOC (total organic carbon) and CEC (cation exchange capacity). This interplay between fine-grained sediment, offering high binding capacities and charge densities, as

Table 3.7.
Loading matrix showing three extracted main components (= vectors, vertical) explaining 39%, 21% and 17% of the variance in the data set, respectively. The loadings corresponds to Pearson's *r* and loadings > 0.35 ($n < 200$) are considered to be significant (*Chlorophyll a* as proxy for microalgal biomass, *Pheophytin* as proxy for microalgal degradation product, *Bacteria* = bacterial cell numbers, *CH colloidal* = water-extractable carbohydrates, *CH CER* = resin-extractable carbohydrates, *TOC* = Total Organic Carbon, *CEC* = Cation Exchange Capacity)

	Component		
	1	2	3
Chlorophyll *a*	0.917	0.235	0.038
Pheophytin	0.890	0.016	0.113
Bacteria	0.173	0.233	0.722
Protein	0.486	0.068	0.344
CH colloidal	0.317	0.524	−0.494
CH CER	0.232	0.075	0.872
Bulk density	−0.561	−0.241	0.540
Water content	0.908	0.136	0.064
Clay	0.477	0.706	−0.154
Silt	−0.113	0.897	0.135
Sand	−0.189	−0.961	−0.007
TOC	0.832	0.074	−0.271
CEC	0.181	0.181	−0.651
Depth	0.559	0.307	−0.485

well as polymeric substances, permeating the void space and coating particles with their active binding sites, is most crucial for erosion resistance. Hence, the influence of both, physico-chemical *and* biological properties could be shown, even over depth, where mostly sedimentological impact was considered before. This overall sediment protection by particle-organisms interaction may be more effective in resisting erosion forces than pronounced surface biofilms, which have been the focus of investigations so far. In assessing the collective influence of sedimentological and biological properties, a better correlation between sediment stability and the master-variables could be achieved in comparison to single correlations. The need for such a comprehensive risk assessment of contaminated riverine sites became again evident in the present study, because at all study sites, low erosion resistance was determined within the sediments to depths of 50 cm. Compared to naturally occurring bottom shear stress, a severe erosion risk of these polluted sediment horizons could be established, even for medium-energy events.

Acknowledgments

The authors would like to thank for their enthusiastic and excellent support: J. Stork and the crew of 'Max Honsell' (sampling facilities); T. Eder, T. Fimpel, R. Ninov (EPS determination); T. Basta, M. Eder, A. Kuhm, W. Wen (bacterial cell numbers); L. Taeuscher, U. Mueller (microalgae species composition) and I. Haag (statistics). The investigations presented here were part of the project SEDYMO, (*Se*diment *Dy*namic and *Mo*bility) funded by the German Federal Ministry of Education and Research (BMBF).

References

Austen I, Witte G (2000) Comparison of the erosion shear stress of oxic and anoxic sediments in the East Frisian Wadden Sea. In: Flemming BW, Delafontaine MT, Liebezeit G (eds) Muddy coast dynamics and resource management, Proc Mar Sci, vol. 2. Elsevier, Amsterdam, pp 75–84

Datry T, Malard F, Niederreiter R, Gibert J (2003) Video-logging for examining biogenic structures in deep heterogeneous subsurface sediments. C. R. Biologies 326:589–597

De Brouwer JFC, Stal LJ (2001) Short-term dynamics in microphytobenthos distribution and associated extracellular carbohydrates in surface sediments of and intertidal mudflat. Mar Ecol Prog Ser 218:33–44

De Brouwer JFC, Stal LJ (2002) Daily fluctuations of exopolymers in cultures of benthic diatoms *Cylindrotheca closterium* and *Nitzschia sp.* (Bacillariophyceae). J Phycol 38:464–472

De Brouwer JFC, Bjelic S, de Deckere, EMGT, Stal, LJ (2000) Interplay between biology and sedimentology in a mudflat (Biezelingse Ham, Westerschelde, The Netherlands). Cont Shelf Res 20:1159–1177

De Deckere EMGT, Tolhurst TJ, de Brouwer JFC (2001) Destabilisation of cohesive intertidal sediments by infauna. Estuar Coast Shelf Sci 56:665–669

Dreher T (1997) Non intrusive measurement of particle concentration and experimental characterization of sedimentation. Sonderforschungsbericht 404, Universitaet Stuttgart

Flemming HC, Wingender J (2001) Relevance of microbial extracellular polymeric substances (EPS) – Part I: Structural and ecological aspects. Water Sci Technol 43(6):1–8

Foerstner U, Heise S, Schwartz R, Westrich B, Ahlf W (2004) Historical Contaminated Sediments and Soils at the River Basin Scale. J Soil and Sediments 4:247–260

Frølund B, Palmgren R, Keiding K, Nielsen PH (1996) Extraction of extracellular polymers from activated sludge using a cation exchange resin. Wat Res 30:1749–1758

Gerbersdorf SU, Jancke T, Westrich B (2005) Physico-chemical and biological sediment properties determining erosion resistance of contaminated riverine sediments – temporal and vertical pattern at the Lauffen reservoir / river Neckar, Germany. Limnologica 35:132–144
Gerbersdorf SU, Jancke T, Westrich B (2007) Sediment properties for assessing the erosion risk of contaminated riverine sites. Journal of Soils and Sediments: 7(1):25–35
Haag I, Westrich B (2002) Process governing river water quality identified by principal component analysis. Hydrol Process 16:3113–3130
Haag I, Kern U, Westrich B (2001) Erosion investigation and sediment quality measurements for a comprehensive risk assessment of contaminated sediments. Sci Total Environ 266:249–257
Jepsen R, Roberts J, Lick W (1997) Effects of bulk density on sediment erosion rates. Water Air Soil Poll 99:21–31
Kern U (1997) Transport von Schweb- und Schadstoffen in staugeregelten Fließgewaessern am Beispiel des Neckars. Mitteilungen des Instituts fuer Wasserbau 93. Universitaet Stuttgart
Kern U, Schuerlein V, Holzwarth M, Haag I, Westrich B (1999) Ein Strömungskanal zur Ermittlung der tiefenabhängigen Erosionsstabilität von Gewässersedimenten: das SETEG-System. Wasserwirtschaft 89:72–77
McNeil J, Lick W (2004) Erosion rates and bulk properties of sediments from the Kalamazoo River. J Great Lakes Res 30:407–418
Olafsson JS, Paterson DM (2004) Alteration of biogenic structure and physical properties by tube-building chironomid larvae in cohesive sediments. Aquatic Ecology 38:219–229
Perkins RG, Honeywill C, Consalvey M, Austin HA, Tolhurst TJ, Paterson DM (2003) Changes in microphytobenthic chlorophyll *a* and EPS resulting from sediment compaction due to de-watering: opposing patterns in concentration and content. Cont Shelf Re 23:575–586
Smith DJ, Underwood GJC (2000) The production of extracellular carbohydrates by estuarine benthic diatoms: the effects of growth phase and light and dark treatment. J Phycol 36:321–333
Underwood GJC, Boulcott M, Raines CA, Waldron K (2004) Environmental effects on exopolymer production by marine benthic diatoms: dynamics, changes in composition, and pathways of produc tion. J Phycol 40:293–304
Yallop ML, Paterson DM, Wellsbury P (2000) Interrelationships between Rates of Microbial Production, Exopolymer Production, Microbial Biomass, and Sediment Stability in Biofilms of Intertidal Sediments. Microb Ecol 39:116–127
Ziegler CK (2002) Evaluating sediment stability at sites of historic contamination. Environ Manage 29:409–427
Zipperle J, Deventer K (2003) Wirkungsbezogene Sedimentuntersuchungen zur Ableitung von Qualitätsmerkmalen und Handlungsempfehlungen, Teilprojekt 1: Entwicklung und Erprobung einer Strategie zur Beurteilung der Sedimentbeschaffenheit auf der Basis von Wirktests. LFU Karlsruhe

Volker Müller · Andreas Seibel · Dogan Kisacik · Giselher Gust

3.3 Simulation of Water Column Hydrodynamics by Benthic Chambers

3.3.1 Introduction

Water quality in coastal and inland water depends mainly on pollutant mobilization and dispersal. The processes of changing water quality are closely connected to the dynamics of fine sediments. Therefore the understanding of water quality requires a deeper insight in the processes of erosion, transport, deposition and consolidation of particulate matter under various physical, chemically and biologically conditions. Not ignoring the importance of chemical and biological controlled dynamics of particulate matter, it is obvious that all processes determining the water quality basically are affected by hydrodynamics.

Laboratory tests of the various interlinked processes of water quality dynamics are better reproduced under hydrodynamic conditions equivalent to the natural environment. These tasks call for a suite of different experimental devices to reproduce the specific hydrodynamic conditions *(a)* in the benthic zone as well as *(b)* in the water body. In principle, a set of different experimental devices exist to fulfill the specific hydrodynamic conditions in these regions. With most of these only a single hydrodynamic parameter, considered as the most important one, for a selected sedimentological, chemical or biological process can be well controlled.

Within the SEDYMO project we developed an experimental apparatus to control more than one hydrodynamic parameter which affect processes on the benthic zone and in the water body occurring synchronously. The main task for the *Benthic-Water-Column-Simulator* (BWCS) was the ability to perform investigations on the full cycle of erosion, transportation, deposition and consolidation of fine sediments at different physical, chemical and biological loading. Here the design and main characteristic of the BWCS are presented. BWCSs or the parent system – the Gust microcosm (Gust and Müller 1997; Tengberg et al. 2004) – were used for various projects within SEDYMO and the reader is referred to the corresponding contributions (Kleeberg 2006; Paterson 2006; Siepmann 2006).

3.3.2 Design of the Benthic Water-Column-Simulator

The design of the BWCS is determined by the requirement to control the most important hydrodynamic parameter for small scale processes occurring in the benthic zone as well as the most important hydrodynamic parameter controlling small scale processes in the pelagic region. In the benthic zone where exchange processes of particles and solutes at the interface between the fluid and the bed are dominant, the dynamic flow characteristics of the boundary layer are most relevant although the hydrostatic pressure is also important. For an equilibrium boundary layer flow, the most important hydrodynamic parameter is the bottom shear stress τ_b. For chemical and/or particulate reactions in the pelagic flow field then turbulence, characterized by u' (= RMS velocity) or, assuming isotropy, the turbulent kinetic energy (TKE)

$$\text{TKE} = \frac{3}{2}u'^2 \tag{3.1}$$

is the most important hydrodynamic parameter.

A review of the existing erosion devices, channels or flow chambers (Brunk et al. 1996; Gust and Müller 1997; Tengberg et al. 2004) shows that in most cases either the bottom shear stress or the flow turbulence is well controlled. The principles of flow generation used in these different devices to adjust the hydrodynamic parameters are of various kind. Rotating disks, cylinders or special shaped bodies are mostly used for generating the desired values of bottom stress. For generating turbulence in the pelagic zone, moving grids or rotating paddle systems are used. But sometimes these systems are also used for generating bottom stress and vise versa. The Gust microcosm (rotating disk with central suction) and the EROMES system (propeller system) are widely used devices with a long history of investigation (Gust 1990; Müller et al. 1995; Gust and Müller 1997). A combination of these two systems could establish an experimental apparatus which allows the control of mean flow, bottom stress and flow turbulence within the same experimental system.

Fig. 3.6.
Benthic Water-Column-Simulator (BWCS) equipped with an optical window for LDA measurements (*left side*)

The newly developed Benthic-Water-Column-Simulator is a dual-propeller system with adjustable pitch and shaft length (Fig. 3.6). The propeller system itself is housed inside a cylinder (diameter $D = 290$ mm, overall height = 420 mm), with the first propeller plane at a height of 180 mm above the bed. The pitch angle α_P of this four-bladed propeller (diameter $D_{p1} = 200$ mm) is adjustable between +30 deg and –30 deg. The propeller blades terminate in a skirt of 70 mm height. The second propeller (diameter $D_{p2} = 50$ mm) is a fixed pitch ($\alpha_P = 40$ deg) three-bladed propeller. The propeller blades also terminate in a skirt (30 mm height). Both propellers are fixed on a single shaft and rotate at the same frequency. The distance of both propellers on the shaft relative to each other and the bottom of the BWCS is adjustable. Consequently, by adjusting the pitch of the uppermost propeller and changing the direction of rotation, four different operational modes can be realized – two propellers with opposite or same pitch and clockwise or counter clockwise rotation.

Due to the mainly rotational character of the BWCS flow there is a difference in comparison to the open channel flow of a river flow. In the BWCS the main flow component is the peripheral flow component u. In addition the mean radial flow component and the mean vertical flow component which are negligible in-situ are present in the BWCS. However, these components should be negligible compared to the peripheral flow component in general. The radial and vertical dependence of all velocity

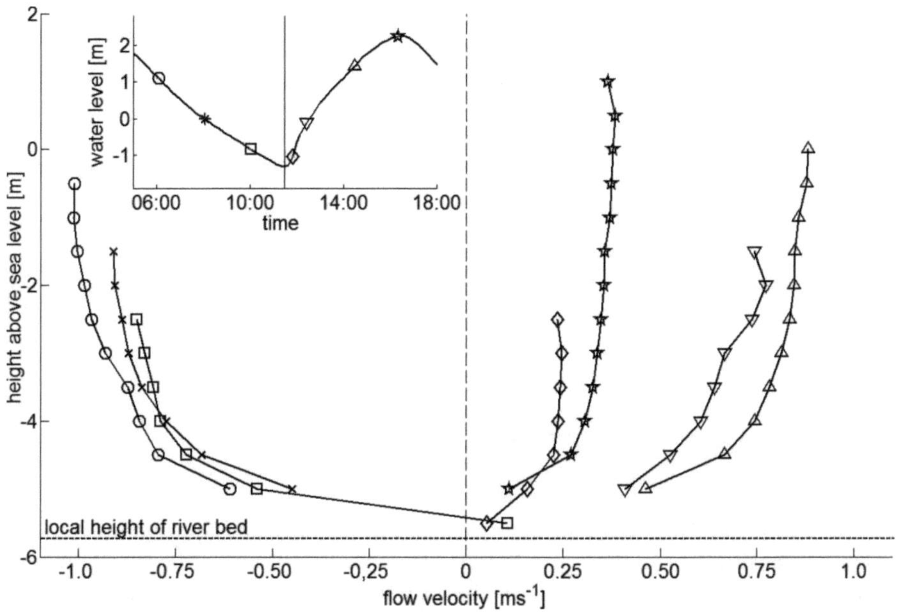

Fig. 3.7. Representative results showing flow velocity (m s^{-1}) against height (m) of the normal flow component through a cross-section at the river Elbe (Süderelbe, near Oortkaten, at September 25, 2003) obtained with a Broadband-ADCP during one tidal cycle

components also leads to anisotropy of turbulence and a radial distribution of bottom stress. Consequently, the design and especially the operational modes of the BWCS have to ensure, that these device-dependent differences will not influence the small scale processes to be investigated.

The design of the BWCS has to meet requirements arising from the variability of bottom stress and turbulence in natural flow. One regional focus of the SEDYMO project was the river Elbe, thus the BWCS need to fulfill the in-situ conditions of the tidal dominated estuary of this river at least. Synchronous measurements with a Broadband-ADCP system and a lowered Single-Point-ADV system were carried out during one tidal cycle to obtain characteristic in-situ data on flow turbulence and bottom shear stress. The mean flow characteristic (Fig. 3.7) was continuously monitored by the ADCP-system and the ADV-system recorded the depth-dependent 3-D turbulence characteristic (Fig. 3.8). Data were sampled at a frequency of 25 Hz. Using the ADV capability to measure the near-bottom distance with high precision, the bottom shear stress was determined by

$$\tau_\mathrm{b} = \mu \lim_{z \to -h} \left(\frac{\partial u}{\partial z} \right) , \quad u_* = \sqrt{\frac{\tau_\mathrm{b}}{\rho}} \tag{3.2}$$

where μ = dynamic viscosity, u = depth-dependent mean flow velocity, h = depth, ρ = density and u_* = friction velocity. The origin of the coordinate system used is located at the (time-dependent) water surface with the vertical axis directed upward.

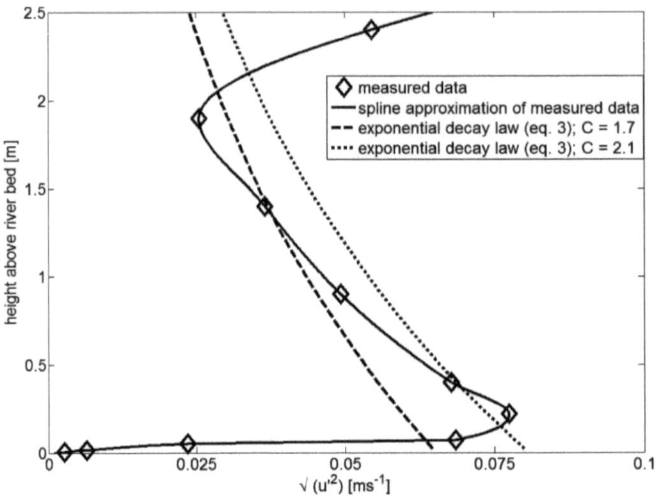

Fig. 3.8. Representative results for depth-dependent flow turbulence obtained with a lowered Single-Point-ADV system (same location and time as data from Fig. 3.7)

The acquired field data from river Elbe established the range and upper values of turbulence intensities, mean flows and bottom stress for the design of the BWCS. In absence of the water–air boundary layer the field data on turbulence intensity fits the exponential decay law

$$\frac{u'}{u_*} = C\exp\left(-\frac{z+h}{h}\right), \quad C \cong 2.3 \tag{3.3}$$

as proposed by Nezu and Nakagawa (1993). We found that over a tidal cycle the value of the constant C varies in range between 1.7 and 2.1. For the design of the BWCS the value of $C = 2.3$ was taken as the upper limit for realistic turbulence intensities coupled to the friction velocity u_* as the main design parameter.

3.3.3 Characteristics of the Benthic Water-Column-Simulator

Compared with existing but more simple flow chambers and erosion devices, where only one parameter (e.g., frequency of the rotating disk) can be adjusted, the variety of adjustable parameters (frequency, pitch, distance of the both propeller and operational modus) leads to a large amount of calibration details. For calibration of the BWCS temperature-compensated hot film technique (Hensse et al. 1997) and a DANTEC 2-D LDA system were applied. The hot film technique provides the friction velocity (mean and fluctuation) and the LDA system was used to record velocity and turbulence of the flow at selected locations in the water column of the BWCS.

Results from first measurements confirmed the expectation that two optimal operational modes exist, one a so-called 'propeller mode' and the other a so-called 'microcosm mode'. The propeller mode is characterized by an orientation of pitch and direction of rotation such that both propeller generate a rotational flow which is directed towards the bottom of the BWCS. The principle of continuity leads to an upward directed rotating flow within the gap between housing and skirt of the four-bladed propeller. The microcosm mode is characterized by an orientation of pitch and direction of rotation such that the four-bladed propeller (the upper one) operates in the same mode as at the propeller mode, but the three-bladed propeller (lower one) operates in the opposite rotation. A flow inside the BWCS similar to that inside the Gust microcosm (rotational flow with a central suction part and an outer region where the secondary flow is directed towards the bottom) is thus generated. A BWCS with a four-bladed propeller with a developed area ratio DAR = 1.0 and pitch $\alpha_P = 0$ deg produces a similar flow field to the Gust microcosm.

The experimental work to calibrate the flow inside the BWCS was supported by a flow simulation model using the Navier-Stokes-Equation-Solver COMET. Due to the sensitivity of the flow to small geometrical changes a validation of the numerical simulation method was executed by comparison of results from simulating the Gust microcosm with extensive LDA measurements of the overall flow. The results shows a good agreement except for the highly turbulent central region where either simulation or experimental data are discordant. The next level of flow simulation simulated a propeller inside a cylinder and we again compared the results with measured data. The established numerical flow model is suitable for design of single-propeller BWCS and similar flow chambers or erosion devices such as the EROMES system.

The simulation results are used on one hand to establish an effective grid of measuring points which allows a timesaving calibration procedure but on the other hand to guarantee that any overlooking of a flow sensitive region cannot occur. Flow sensitive regions are characterized by unstable flows which could appear at the central axis, the regions of strong flow interaction between the both propellers (esp. at the microcosm mode) and in the region with the radius $D_{p1}/2$ of the uppermost propeller.

Control Parameter: Bottom Shear Stress

To calibrate the BWCS for the control parameter bottom shear stress or friction velocity u_* this was measured by the hot film technique at radius $r = 0, 54, 92, 110$ and 130 mm. Data were sampled at a frequency of 25 Hz over 6 minutes to obtain a sufficient sample for statistical evaluation. For comparison of the results at different operational modes and/or changed geometric configuration an area-weighted value of friction velocity

$$u_* = \frac{1}{\left(\frac{D}{2}\right)^2} \sum_{i=1}^{N} \left[\left(\left(\frac{r_{i+1} + r_i}{2} \right)^2 - \left(\frac{r_i + r_{i-1}}{2} \right)^2 \right) u_{*_i} \right] \tag{3.4}$$

was calculated from the time-averaged values u_{*i} at the radii r_i. The pitch α_P was set to $\alpha_P = 10$, 20 and 30 deg and the distance d between both propellers was set to $d = 155$, 130, 100 and 65 mm. The distance H of the uppermost propeller to the bottom always remained at $H = 180$ mm. The rpm-values ranged from 20 to 180 rev. min^{-1}. Measurements were performed under all four different operational modes but only results for the most different modes are presented here.

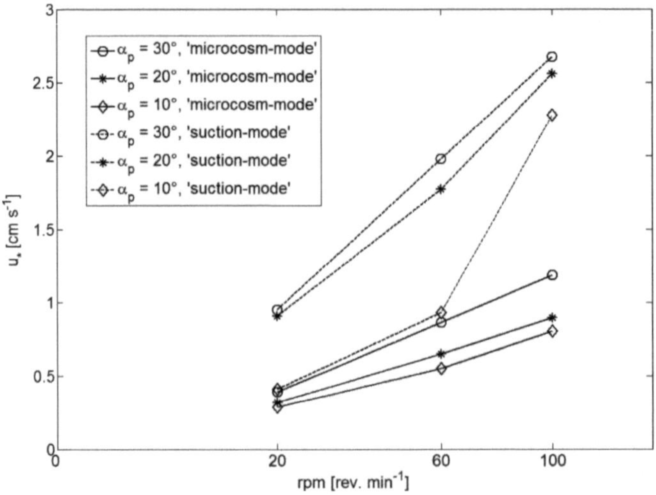

Fig. 3.9. Characteristics of the BWCS for area-weighted friction velocity (cm s^{-1}) depending on operational mode, pitch angle of the uppermost propeller and revolutions per minute (rev. min^{-1})

The selection for the two operational modes (suction mode and microcosm mode) is on one hand due to the different flow stability these modes provide and on the other hand due to the very different hydrodynamic characteristics for each (see Fig. 3.9). In the microcosm mode and the propeller mode a very sensitive investigation of fine sediment dynamics at low friction velocities is possible. At the same rpm the area-weighted values of friction velocity generated in the BWCS are lower than at a comparable setting in the Gust-microcosm.

In the suction mode (area-weighted) bottom shear stress up to $\tau_b = 1$ N m^{-2} ($u_* \cong 3.2$ cm s^{-1}) were generated. Here an extension of the operational range of the Gust-microcosm is achieved and investigations on the erosion-deposition cycle of more erosion-resistant sediments can be performed. The influence of pitch angle changes can clearly be seen in Fig. 3.9. At the pitch angle $\alpha_P = 10$ deg the flow characteristics is more similar to the original Gust-microcosm, especially at low rpm. To generate stable flow conditions, especially at the suction mode, we recommend pitch angles $\alpha_P \geq 20$ deg.

Changing the distance between the propellers had only a weak influence on the area-weighted bottom shear stress. In the microcosm mode, which has the greatest sensitivity to this distance, the values of the area-averaged friction velocity at rpm = 180 rev. min^{-1} differ by 15% between the minimal and maximal distance. In agreement with the numerical model, the greatest friction velocities were measured for maximum distance. At rpm = 20 rev. min^{-1} no significant difference due to changed propeller distance d was measured.

Control Parameter: Flow Turbulence

The synonym 'flow turbulence' is here used for turbulent phenomena as caused by turbulent fluctuations of the bottom shear stress τ_b' (resp. u_*') and by turbulent fluctuations u' in the water column of the BWCS. This differentiation is required because of the two measuring principles used for calibration of the BWCS. With the used hot film skin friction probes, the measurement of bottom shear stress and fluctuations

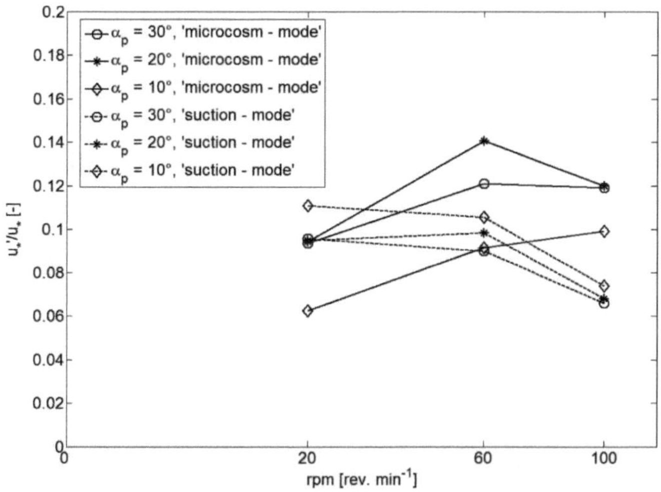

Fig. 3.10.
Characteristics of the BWCS for turbulence of friction velocity (ratio of fluctuations and mean; area-weighted values) depending on operational mode, pitch angle of the uppermost propeller and revolutions per minute (rev. min^{-1})

at the head of the BWCS are feasible, but not the measurement of the internal flow. With the LDA technique, the internal flow velocity and its fluctuations are measured. There is a connection between these measurements where a relationship similar to Eq. 3.3 exists for the very complex flow inside the BWCS.

From the hot film measurements the turbulent fluctuations of the bottom shear stress were obtained (Fig. 3.10). Both of the presented operational modes shows a slightly different behavior, yet, the values remain around $u_*'/u_* \sim 0.1$ over the full range of rpm. This parameter shows clearly an advantage of the BWCS design. This value lies at the same range achieved with the Gust microcosm and is similar to that of open-channel flow (Gust and Müller 1997). The high bottom shear stress fluctuations which were observed for the EROMES system (simple stirrer principle) (Müller et al. 1995) were effectively reduced but the advantage of a propeller system was maintained – reaching high values of friction velocity at moderate rpm. The characteristic feature of a Gust microcosm – rising friction velocity turbulence with growing rpm – as well as the constant or slightly decreasing turbulence of a propeller system with growing rpm were found in a less pronounced manner for the BWCS.

Execution of the LDA measurements was not simple. By using a 2-D LDA system a coincident measurement of the 3-D flow velocity vector at the desired measuring points inside the BWCS was not possible. The measurement of the 3-D velocity was executed by a twofold 2-D measurement under the fulfillment of matching conditions for the peripheral velocity. The measurements covered the water column from near bed up to 60 mm above the height of the uppermost propeller plane. The first result is that the mean radial and vertical velocities are mostly 10 percent lower than the local peripheral velocity, except for the gap between the skirt of the uppermost propeller and the housing of the BWCS. Here a significant rising vertical velocity was measured (up to 50 percent of the peripheral velocity), leading to a rapid transport of particles and solutes to the top of the BWCS. Downward transport will occur through the propeller plane.

The characteristic of flow turbulence in the water column up to 250 mm height above bottom of the BWCS is shown (Fig. 3.11). This depth-dependent distribution of

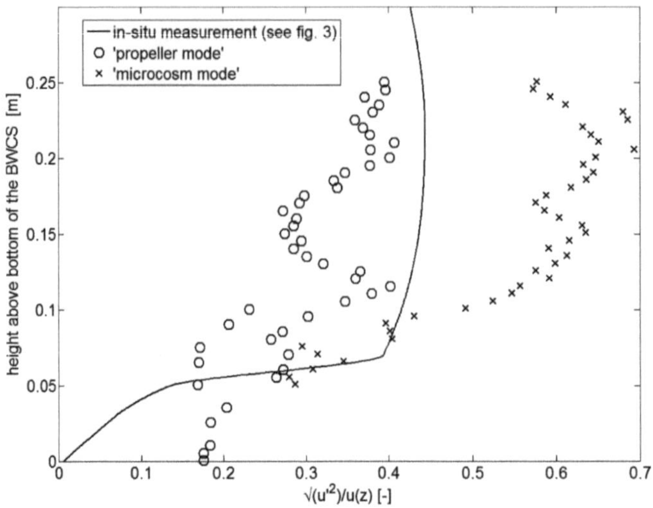

Fig. 3.11. Representative results for flow turbulence against height (m) inside the BWCS depending on the operational mode compared to the in-situ measurement (see spline approximation of measured data at Fig. 3.8); the absolute value of the pitch of the upper propeller at both modes is $|\alpha_P| = 30$ deg

turbulence was measured at a radius of $r = 119$ mm. Vertical profiles at other radii look similar. The measured differences between microcosm mode and propeller mode as documented (Fig. 3.11) are caused by flow interaction of both propellers. Due to the opposite action of the two propellers in the microcosm mode the turbulence intensity is larger in this mode. More detailed tuning of the propeller design should allow an exact matching of an in-situ measured near-bed depth-dependence.

3.3.4 Conclusions

The newly developed *B*enthic-*W*ater-*C*olumn-*S*imulator (BWCS) is an effective tool for laboratory investigations of fine sediment dynamics. It merges the advantages both of the Gust-microcosm and the EROMES system. Synchronous reproduction of the small scale flow and bottom shear stress characteristics of natural flows similar to an open channel flow can be achieved segment-wise for the whole water column with the BWCS filled with artificial or natural sediments. At the present development stage, the system is calibrated up to bottom shear stress $\tau_b = 1$ N m^{-2}. The construction of the system is simple and robust. Within the SEDYMO project this device was applied for erosion and entrainment measurements. The reader is referred to the corresponding contributions (Kleeberg 2006; Paterson 2006; Siepmann 2006).

Acknowledgments

We would like to thank the Hamburg Port Authority and the Project Management Group at the Federal Ministry of Education and Research for a very successful cooperation. This work was funded by the Federal Ministry of Education and Research under contract no. 02WF0318.

References

Brunk B, Weber-Shirk M, Jensen A, Jirka G, Lion LW (1996) Modeling natural hydrodynamic systems with a differential-turbulence column. J Hydraulic Engineering 122(7):373–380

Gust G (1990) Method of generating precisely-defined wall shearing stresses. US Patent Number: 4,973,165,1990

Gust G, Müller V (1997) Interfacial hydrodynamics and entrainment functions of currently used erosion devices. In: Burt N, Parker R, Watts J (eds) Cohesive sediments – Proc. 4th nearshore and estuarine cohesive sediment transport conference INTERCOH '94, Wallingford 1994. Wiley, Chichester, UK:149–174

Hensse J, Müller V, Gust G (1997) Dynamic temperature compensation for hot film anemometry in turbulent flows – necessity and realisation. In: Shen X, Sun X (eds) Modern techniques and measurements in fluid flows – Proceedings of the 3rd conference on fluid dynamic measurement and its applications, Beijing 1997, Int. Academic Publishers, Beijing, P.R. of China, ISBN 7-80003-407-0/TB

Kleeberg A, Hupfer M, Gust G (2007) Phosphorus Entrainment due to Resuspension, river Spree, NE Germany. This volume

Müller V, Vorrath D, Werner A, Witte G (1995) Schubspannungscharakteristik des EROMES-Systems – Messungen zur Hydrodynamik und Erosionsversuche mit Kaolinit. report GKSS 95/E/43, Geesthacht, Germany, ISSN 0344–9629

Nezu I, Nakagawa H (1993) Turbulence in open-channel flow. IAHR Monograph Series, A. A. Balkema Publishers, Rotterdam, Netherlands

Paterson D (2007) On the Boundaries: Measurements of Extreme Systems. This volume

Siepmann R, von der Kammer F, Calmano W (2007) Mobility of Heavy Metals from Resuspended Anoxic Sediments – Close to Nature Approach in Benthic Chambers. This volume

Tengberg A, Stahl H, Gust G, Müller V, Arning U, Andersson H, Hall POJ (2004) Intercalibration of benthic flux chambers I. Accuracy of flux measurements and influence of chamber hydrodynamics. Progress in Oceanography 60:1–28

Gregor Kühn · Gerhard H. Jirka

3.4 Fine Sediment Behavior in Open Channel Turbulence: an Experimental Study

3.4.1 Introduction

Water quality in natural rivers is strongly related to the occurrence of fine sediment particles in the water body. Fine sediment particles offer large surface areas relative to their volume and a high adsorption potential leading to electrostatic agglomeration of contaminants at the particle surface. Furthermore, chemical reactions that occur at the particle surface modify the contaminants. To study contaminant transport in rivers the behavior of fine particles under natural conditions has to be investigated. The important processes are aggregation and disaggregation, leading to floc sizes of different orders of magnitude.

The driving mechanical force in the system is induced by the flow. The flow velocity is composed of a mean flow and the turbulent velocity fluctuations around the mean. The velocity fluctuations are responsible for shear forces and collision of single particles (van Leussen 1994). Therefore, the transformation of the floc size is strongly related to the turbulence conditions in the open channel water column as described by Nezu and Nakagawa (1993).

In an improvement of the differential turbulence column (Brunk et al. 1996) the turbulence is well-controlled and the particle size and the concentration profile can be observed in order to obtain a relationship between mass concentration and shear forces affecting aggregation and disaggregation. This system was used to simulate these processes for different conditions like in nature.

3.4.2 Experimental System

The system was designed to simulate the turbulence profile of open-channel flow using grids oscillating at different frequencies distributed over the vertical column. Thus, the particle behavior over time can be studied locally under-well-controlled turbulence without advection taking place. In open-channel flow particles are advected with the mean flow and therefore, not traceable with stationary measurement equipment on their path in the river though changing environmental conditions. Based on the studies of Brunk et al. (1996) a differential turbulence column was constructed to reproduce a turbulence profile typical for open channel flows, such as rivers (Nezu and Nakagawa 1993).

Sridic et al. (1996) showed the advantage of using oscillating grid pairs instead of single grids to produce a larger area of homogeneous turbulence and amplifying the turbulent fluctuations additionally. In the present study, five pairs of oscillating grids

Fig. 3.12.
Sketch of the differential turbulence column

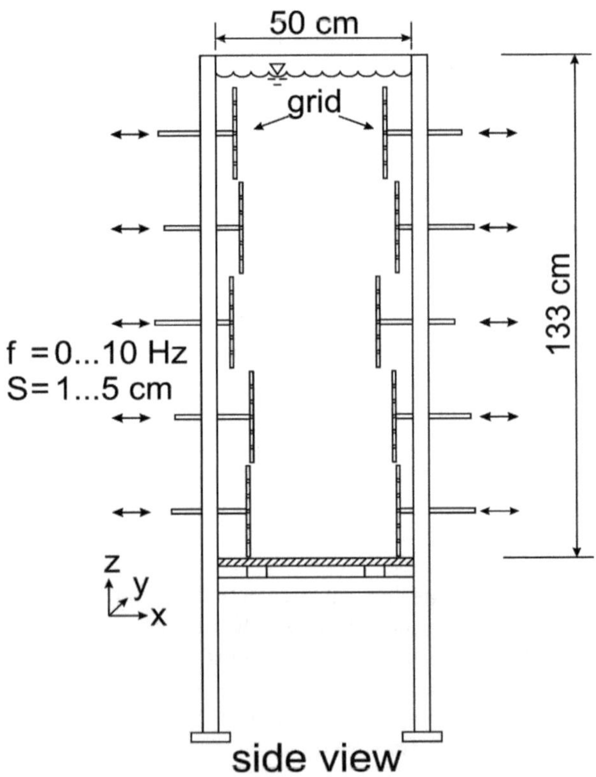

were arranged vertically in a perspex tank (Fig. 3.12). The plan dimensions were 50 × 35 cm, while the height was 133 cm. Each grid was controlled separately via PC and could be adjusted to different frequency and strokes, leading to turbulence characteristics as given by Hopfinger and Toly (1976).

To determine the flow characteristics in the tank different methods for velocity measurement were used. A Laser Doppler Velocimeter (TSI Corp.) was adapted to an automatic traversing system to measure the mean flow and the turbulence intensities in vertical profiles. To get an impression of the complete flow field in the middle region of the tank Particle Image Velocimetry (PIV) was used. Therefore an PIV-System (LaVision) with an 20 mJ Nd:YAG Laser and a CCD camera was installed.

To investigate changes in the particle size distribution of the suspended sediment particles under given turbulence conditions, an in-line microscope (Aello 7000) was used. The microscope consists of an stainless steel tube with a diameter of 38 mm which has a 8 mm wide gap in the mid-section, where the CCD sensor of a fire wire camera and illumination unit are located. The camera can resolve particles between 4 and 500 μm which are located in the focal plane behind the sapphire observation window. The in-line microscope could be inserted into the differential turbulence column at different positions through process inlets, while the system was running. Seven of these inlets were located on the back of the tank at different levels. The images were analyzed with image recognition software to detect particles and to determine their size in the focal plane. For that purpose a two-step analysis was used. In the first step, sharp edges were detected. In the second step contiguous areas were analyzed. Regions with contiguous area and sharp edges are counted as particles.

The measurements of the particle size were completed by measuring the water column turbidity continuously to record the mass distribution of sediment. A single turbidity sensor (Honeywell APMS-10GRCF-KIT) with a automatic sampling unit was positioned at seven sample positions with height over the chamber. The sample positions were measured concurrently. At the end of each cycle a reference fluid was measured and between each change of position the sensor was flushed. This cycle was continued for the complete time of an experiment. The sensor was also individually calibrated to each type of sediment.

3.4.3 Results

Turbulent Flow Field

A key precondition for conducting experiments on floc size distribution is a well controlled turbulence distribution in the tank. For the first phase it was necessary to determine the turbulent conditions in the differential turbulence column and optimize it to represent the profile of the turbulent intensities of an open channel flow. For these measurements the LDV was used, with a measurement location in the center of the column. The agitation conditions of the oscillating grids were based on the result of Brunk et al. (1996). The frequencies of the grids were varied between 1 Hz to 6 Hz. The stroke was varied between 2 cm and 5 cm. Finally the stroke was fixed at 3 cm to span a wide range of useful turbulent fluctuations.

PIV measurements at 14 different positions were taken (Fig. 3.13). The bottom grid frequency was set to 4 Hz. The mean flow shows clear secondary flow cells with flow velocities less than 5 cm s^{-1}. This flow is result of the geometry and production of turbulence by the oscillating grids. The size of these cells correspond with the size of the grids.

The turbulence intensity can be expressed as turbulent kinetic energy $k^{1/2}$, defined as

$$k = \text{TKE} = \frac{1}{2}\left(u'^2 + v'^2 + w'^2\right) \tag{3.5}$$

in which u', v' and w' are the velocity fluctuations in x-, y- and z-direction (Fig. 3.13b).

With the assumption of isotropy, the turbulent kinetic energy can be written as

$$\sqrt{k} = \sqrt{\frac{3}{4}\left(u'^2 + w'^2\right)} \tag{3.6}$$

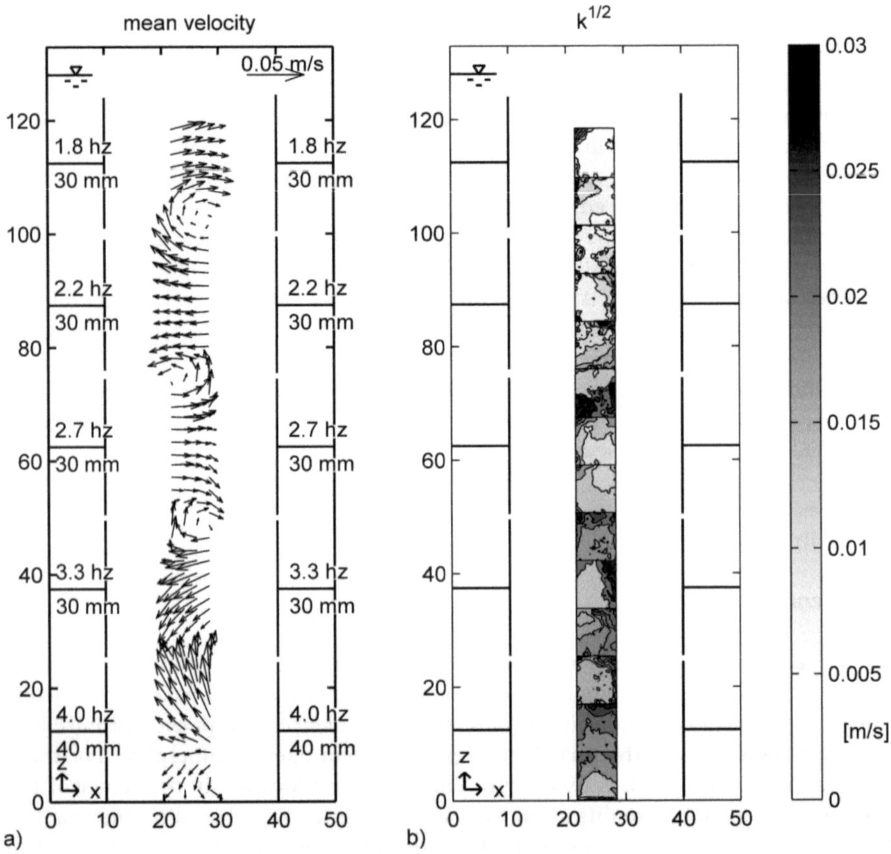

Fig. 3.13. a Mean velocity and **b** turbulent kinetic energy in the center region of the differential turbulence column, measured by PIV

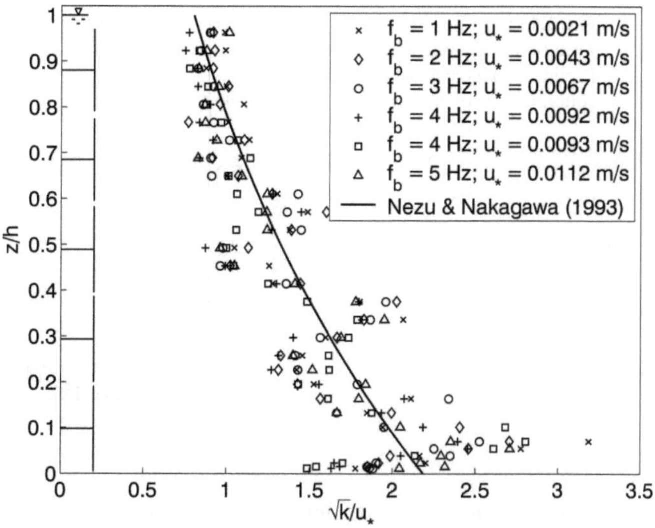

Fig. 3.14. Distribution of the turbulent kinetic energy for different turbulent conditions

The distribution of k in the center region is given (Fig. 3.13). The turbulent kinetic energy decreased towards the surface, while in the center region it stays laterally constant. Close to the bottom the fluctuations were damped to zero. The turbulent fluctuations represented by the RMS-velocities were slightly stronger than the secondary flow.

The LDV measurements were obtained to visualize the distribution of the turbulent fluctuations (Fig. 3.14). The result of different agitation conditions from 1 Hz to 5 Hz of the bottom grid frequencies, scaled by the water depth and a corresponding shear velocity u_* were recorded (Fig. 3.14). In comparison the vertical profile of the turbulence intensities in a natural river as determined by Nezu and Nakagawa (1993) was also plotted. In detail, the effect of the oscillating grids was seen in the distribution of the turbulence intensities. Due to the different frequencies of the grids and their resultant movement in opposite directions, an increase in turbulent fluctuations at the gap between two pairs of grids was observed. This produces strong peaks of turbulent intensities between pairs of grids. In the center of a pair of grids the production of turbulence is decreased because of the lesser fringe effects. However, the average of turbulent fluctuations in the differential turbulence column fell within the natural range of conditions of an natural river.

Particle Size Distribution and Development

The second phase of work was to determine the behavior of fine sediment under turbulent conditions. In the present case kaoline (Dorfner H III GF) was used to represent the sediment and to remove the influence of the complex mixture of different ingredients in natural sediment. Kaoline is one of the largest fractions in natural fine sediments. The sediment concentration was in a range from 100 to 1 000 mg l^{-1} under different turbulent conditions. At the beginning of the experiment, the sediment was completely mixed into the water column. During the experiment, the particle size distribution was measured at a constant height with the Aello microscope. After a steady state

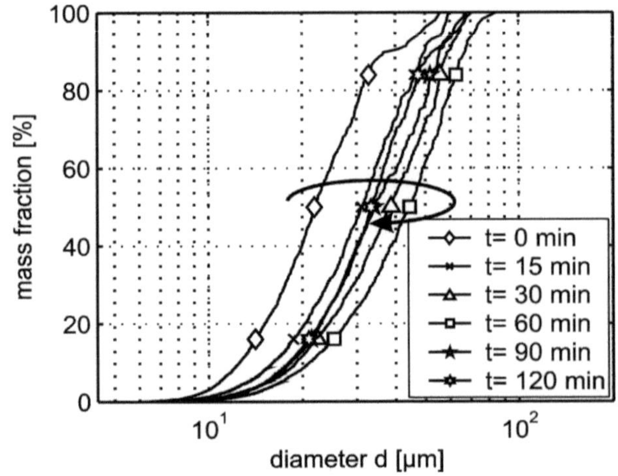

Fig. 3.15.
Evolution of the particle size of kaoline at $z/h = 0.68$ with a mean concentration of 500 mg l^{-1}

of the particle size distribution was reached, the microscope was used to measure the sediment size distribution at seven different heights. Simultaneously, the turbidity was measured at the same seven positions during the complete experiment.

The evolution of the volumetric particle size distribution over time was recorded (Fig. 3.15). The initial concentration of kaoline was 500 mg l^{-1}. The frequency of the bottom grid was 1 Hz. The measuring plane of the microscope was positioned in the center of the column at $z/h = 0.68$.

While the experiment was running an increase of the particle size could be observed from $t = 0$ min to $t = 60$ min. After reaching a maximum particle size at $t = 60$ min, the particle size decreased reaching a steady state after approximately 90 min. This phenomenon can be explained by the development of a mass concentration profile, starting from fully mixed conditions. This will be shown later using the turbidity measurements. The particle size in the steady state is larger than the initial particle size, which means that under the given turbulence conditions particle aggregation is taking place.

After achieving the steady state particle size distribution, the position of the microscope was changed over height, so that a volumetric size distribution profile could be measured (Fig. 3.16).

The particle size increased towards the surface in accordance with decreasing turbulence. In the upper region, the turbulence and therefore, the shear forces are small enough to allow aggregation. The largest particle size distribution was found at the second level at $z/h = 0.68$. This can be explained using the mass concentration profile (Fig. 3.18). In the top level the amount of sediment material is so small, that the random conjunction of two particles is relatively rare and the fabric of the aggregated particles is not very strong. The effect of doubling the turbulence intensities on particle size distribution was investigated (Fig. 3.17). The initial particle size distribution was the steady state of the described experiment. The particle size distribution was measured as before at $z/h = 0.68$ in the center of the column and the volumetric particle size distribution at different time steps was recorded (Fig. 3.17). At time $t = 0$ min the particle distribution shows larger particles due to aggregation that occurred under

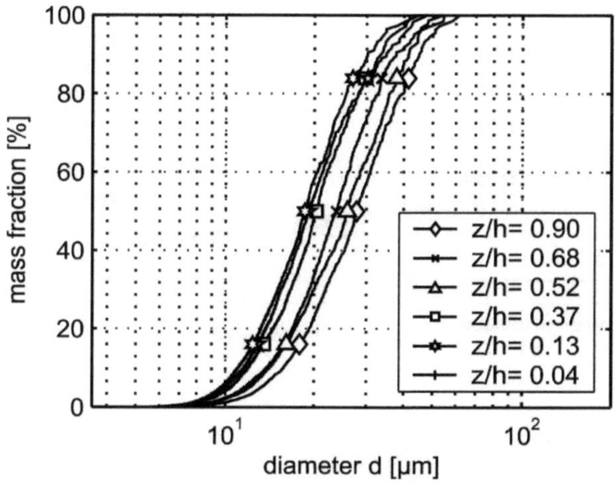

Fig. 3.16.
Distribution of floc size of kaoline over depth at $t = 240$ min with a mean concentration of 500 mg l^{-1}

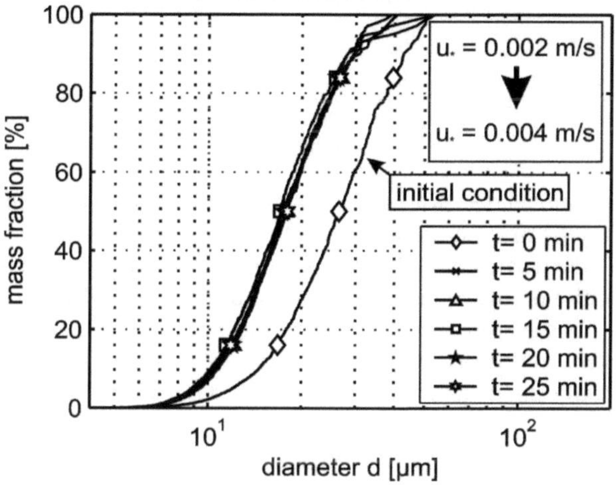

Fig. 3.17.
Segregation of kaoline flocs with a mean concentration of 500 mg l^{-1} by doubling the turbulent intensity at $z/h = 0.68$

the previous low turbulence regime. After $t = 5$ min with the turbulent intensity doubled, most of the aggregates were destroyed and the distribution again reached a steady state, with smaller particle sizes similar to the initial conditions at the beginning.

Additional to the investigation of particle size, the mass concentration in the differential turbulence column was also measured with a turbidity sensor at the same heights. Scaled by a calibration, the data can be transformed into mass concentration profiles. The artificial sediment Kaolin was used with an initial concentration of $c_0 = 500$ mg l^{-1}, fully mixed into the differential turbulence column. The calibrated mass concentration at different levels for different turbulent conditions was recorded (Fig. 3.18). Initially, the turbidity in the top region started to decrease after 20 min followed by a decrease in turbidity at lower levels. This diagram represents pure particle settling (Fig. 3.18a). In the next case (Fig. 3.18b) the bottom grid frequency was 1 Hz, with lower frequencies in the upper layers to simulate an open channel turbulence profile. A slight settling tendency in the complete water body was noted. However, in the two upper regions an

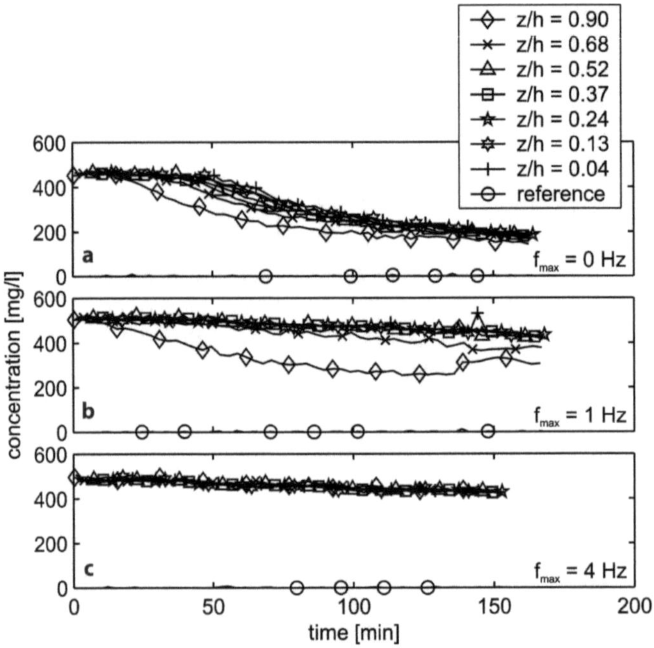

Fig. 3.18.
Mass distribution evolution of kaoline at various turbulent conditions, **a** no agitation and pure settling, **b** low turbulence with 1 Hz bottom grid frequency, **c** stronger turbulence with 4 Hz bottom grid frequency

additional decrease of mass was observed at the beginning. The timescale for developing the mass stratification is about $t = 90$ min, the same as for reaching a steady state of the particle size distribution. In the lower region the suspension remains fully mixed with homogeneous concentration due to the higher turbulence level.

In the last case shown, the turbulence of the bottom grid frequency was increased to 4 Hz. In this case only a slight sedimentation was observed. The suspension in the water body remained fully mixed and the mass distribution over depth is constant (Fig. 3.18c).

3.4.4 Conclusion and Outlook

For investigating the behavior of fine sediment particles under turbulence conditions similar to open channel flow a differential turbulence column was designed. Oscillating pairs of grids were used to produce a turbulence profile. Velocity measurements using LDA and PIV demonstrate that the turbulence conditions in the column can be well controlled. The turbulence profile can be adjusted to represents average conditions of open channel river flow. Secondary flow in the column due to the geometry and the agitation has been suppressed as much as possible so that its influence on the particles can be neglected.

The evolution of the particle size of the particle ensemble and of the concentration profile was observed by an in-situ microscope and by continuously sampling the turbidity at 7 different levels. The investigation showed a tendency for single particles to build aggregates in regions of low turbulence. The size of the aggregates was limited by the shear forces due to the turbulent conditions and the amount of mass.

The investigations on the particle size distribution and the concentration profile will be used for calibrating a numerical model (Ditschke and Markofsky 2006). For that purpose additional experiments are planned with other artificial and natural sediments, including those containing biological activity.

Acknowledgments

This research is supported by the German Federal Ministry of Education and Research in the framework of the joint project Sedymo (Fö-Nr: 02WF0317).

References

Brunk B, Weber-Shirk M, Jensen A, Jirka G, Lion L W (1996) Modeling natural hydrodynamic systems with a differential-turbulence column. J Hydr Eng 122:373–380
Brunk B (1997) Turbulent Coagulation of Particles Smaller Than the length Scales of Turbulence and equilibrium Sorption of Phenantrene to Clay. Ph.D. Thesis, Cornell University, New York
Ditschke D, Markofsky M (2006) A non-equilibrium, multi-class flocculation model. Proc. SEDYMO International Symposium 2006, Hamburg
Fengler G, Köster M, Meyer-Reil L-A (2006) Sediment erodibility in an intertidal groyne field of the Elbe River: Impact on microbial mediated processes. Proc. SEDYMO International Symposium, Hamburg
Hopfinger EJ, Toly JA (1976) Spatially decaying turbulence and its relation to mixing across density interfaces. J Fluid Mech 78:155–175
Nezu I, Nakagawa H (1993) Turbulence in Open-Channel Flow. Rotterdam, Brookfield: A. A. Balkema
Srdic A, Fernando HJS, Montenegro L (1996) Generation of nearly isotropic turbulence using two oscillating grids. Exp Fluids 20:395–397
Van Leussen W (1994) Esturaine Macreoflocs and their Role in Fine-Grained Sediment Transport. Ph.D. Thesis, University of Utrecht, Utrecht

Marion Köster · Lutz-Arend Meyer-Reil · Günter Fengler

3.5 Influence of Microbial Colonization on the Sediment Erodibility in an Intertidal Groyne Field of the River Elbe

3.5.1 Ecological Relevance of Erosion and Resuspension

The accumulation of fine grained cohesive sediments and their enhanced concentrations of inorganic and organic nutrients as well as pollutants is a wide spread problem in aquatic environments. For the estimation of the eutrophication risk of an aquatic ecosystem, detailed knowledge of the mobilization of polluted sediments by erosion is required for water resource management (Westrich and Förstner 2005). Erosion may be caused by natural events, such as water movement and biological activities, as well as anthropogenic impacts, such as dredging. During the last decades, the erosion of cohesive sediments and the resulting resuspension of particles in the water column have been identified as processes of high ecological relevance. This applies to the remobilization of nutrients required for pelagic primary production causing potential eutrophication as well as to the liberation of pollutants.

The erosion and subsequent resuspension of sediment particles is controlled by the current dynamics and the sediment stability. Benthic organisms which colonize benthic habitats have a great impact on sediment stability (Graf and Rosenberg 1997; Widdows et al. 2000, 2006). Among the biological effects, the secretion of extracellular polymeric substances (EPS) by (micro)organisms (e.g., de Brouwer et al. 2000, 2005) and bioturbation by meio- and macrofauna (e.g., Widdows et al. 2000; Roast et al. 2004) seem to be most important mechanisms of sediment stabilization and destabilization, respectively. Resuspension of sediments leads to the liberation of organisms (Shimeta et al. 2004) and particles as well as inorganic and organic matter into the water column. The suspended particles are modified and degraded through microbial activities thus causing the liberation of nutrients and pollutants. Since suspended matter can be transported over long distances, erosion and resuspension events may influence regions remote from the origin of the events.

The goal of this study was to investigate the influence of microbial communities on the sediment stability in an intertidal groyne field of the river Elbe (stream km 607.5). It was of special interest to investigate the sediment stability in relation to the composition and microbial colonization of resuspended sediments, and the prevailing hydrodynamic conditions.

Experimental Approach

Sediments were collected with a manually operated corer from the groyne field of the tidal Elbe during the sampling period from June 2003 to September 2005. Undisturbed sediment columns (inner diameter of 10 cm, length of 40 cm) were exposed to increasing shear stress in a particulate entrainment simulator (PES, Tsai and Lick 1986) comprising the range of shear stress expected in the field (>0 to 0.23 N m^{-2}). After 2 min exposure intervals, eroded sediment particles were taken with a 50 ml syringe from the overlying water and filtered on Whatman GF/C filters (for details see Fengler et al. 2006).

Sediment-specific erosion curves were obtained by plotting resuspended sediments (total particulate matter, TPM) versus shear stress. The critical erosion shear stress τ_{crit} was derived from erosion curves based on log transformed TPM data. The value τ_{crit} was defined as the threshold value above which the initial motion of sediment particles occurred (for discussions of the different approaches for the determination of the critical shear stress compare for example Grant and Gust 1987; Paterson et al. 1999).

The structure and chemical composition of eroded particles was described by their size spectra, contents of organic carbon and nitrogen (Köster et al. 1997), and trace metal contents (Petersen et al. 1996). For the (micro)biological characterization of the eroded particles, defined subsamples of suspended sediments were analyzed for total bacteria (Meyer-Reil 1983), photoautotrophic cells, total and photoautotrophic microbial biomass (Findlay et al. 1989; HELCOM 1988), and EPS (de Brower and Stal 2001). Enzymatic microbial degradation activity and fluxes of oxygen, ammonia, and phosphate were analyzed in time-course experiments according to Köster et al. (1997) and Meyercordt and Meyer-Reil (1999). Additionally, roller tank experiments (Shanks and Edmondson 1989; Ziervogel 2003) were performed to study properties of eroded sediment particles during transport processes (for details see Fengler et al. 2006).

3.5.2 Sediment Specific Erosion Curves and Critical Erosion Shear Stress

With increasing stabilization of the sediments, the slope of the sediment-specific erosion curves decreased (Fig. 3.19). The stability of sediments from the intertidal groyne field of the river Elbe varied spatially and seasonally with values for the critical shear stress between >0 and 0.07 N m^{-2}. Exposed sandy mud sediments at the edges of the groyne field showed higher stability as compared to the more muddy sediments in the center of the groyne field, which are less influenced by the currents. The stability of the sediments seemed to increase in summer and to decrease in autumn and winter during the investigation period from 2003 to 2005.

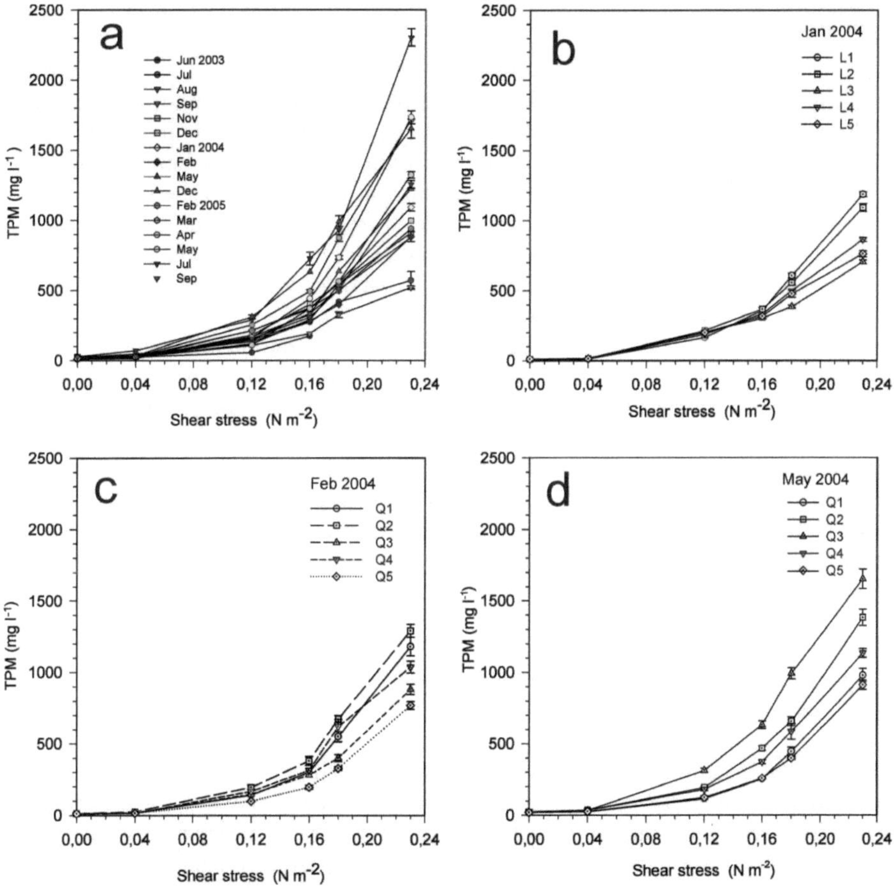

Fig. 3.19. Sediment-specific erosion curves obtained from sediments sampled at different sites in the groyne field of the river Elbe (**a** central station Z during the investigation period 2003–2005, **b–d** stations of the transects L and Q in January, February, and May 2004, respectively; *TPM:* total particulate matter). Error bars represent standard deviation ($n = 3$)

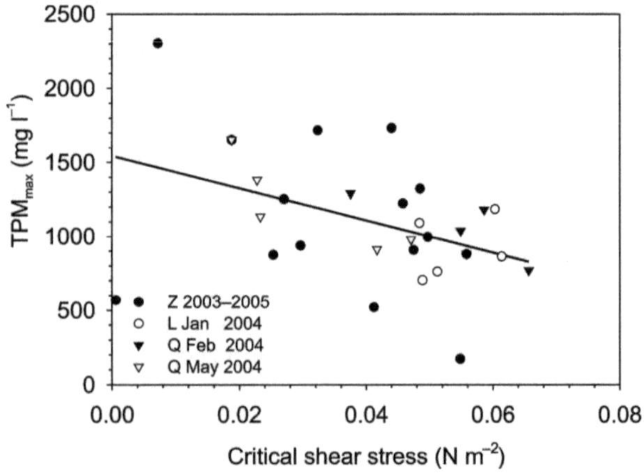

Fig. 3.20. Relationship between maximum concentration of eroded particles (*TPM:* total particulate matter) and critical shear stress in sediments sampled at different sites in the groyne field of the river Elbe ($f(x) = -10\,865x + 1\,543$, $r = 0.427$; *Z:* central station during the investigation period 2003–2005; *L*, *Q:* stations of the transects *L* and *Q* sampled in January, February, and May 2004, respectively). The straight line indicates the linear regression curve

Critical shear stress of the sediments that was limited to the maximum applied stress in the system correlated significantly and inversely with the maximum concentration of eroded sediment particles. The higher the critical shear stress, the lower the number of eroded particles (Fig. 3.20). This significant relation underlines the validity of the determination of the critical shear stress derived from the sediment-specific erosion curves by extrapolation to the threshold value of particle liberation. A comparable relation between shear stress and liberation of particles was described by Widdows et al. (1998) for intertidal sediments.

Observations of the erosion process in the particle entrainment simulator showed that above the critical shear stress, small particles were randomly liberated from the sediment. The particles moved randomly on the sediment surface or were suspended in the overlying water. With increasing shear stress, larger particles and aggregates were released from the sediment (Fig. 3.21). The erosion was a highly dynamic process: smaller particles stuck together, and larger aggregates broke into smaller units. Beside erosion, deposition of aggregates on the sediment surface occurred. A layer of fluffy material on the sediment surface ("Sedimentauflage", Meyer-Reil 2006) was easily resuspended even at low shear stress.

3.5.3 Sediment Stability, Composition and Microbial Colonization of Resuspended Particles

The stability of sediments of the river Elbe was reflected in the number and size composition of resuspended particles. A high number of small particles (<30 µm) was released from sediments that were easily eroded. A comparable dominance of small particles was observed by Ploug et al. (2002) in the highly dynamic upper Elbe Estuary. Investigations with resuspended coastal sediments of the southern Baltic Sea showed that fine particles are main contributors to nutrient load (carbon, nitrogen) and sites of enhanced microbial colonization and activity (Rieling 2000).

The erosion experiments revealed that with increasing sediment stability, the number of particles released decreased. Their size spectrum was shifted to larger

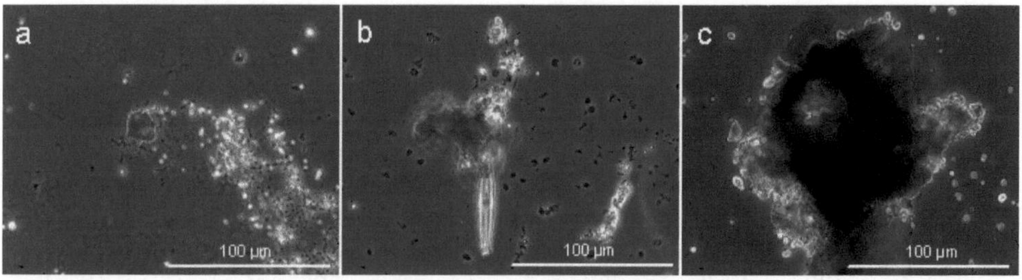

Fig. 3.21. Phase contrast micrographs of sediment particles eroded at different shear stress (**a** 0.50 N m^{-2}; **b** 1.00 N m^{-2}; **c** 1.25 N m^{-2}) from sediments of the river Neckar

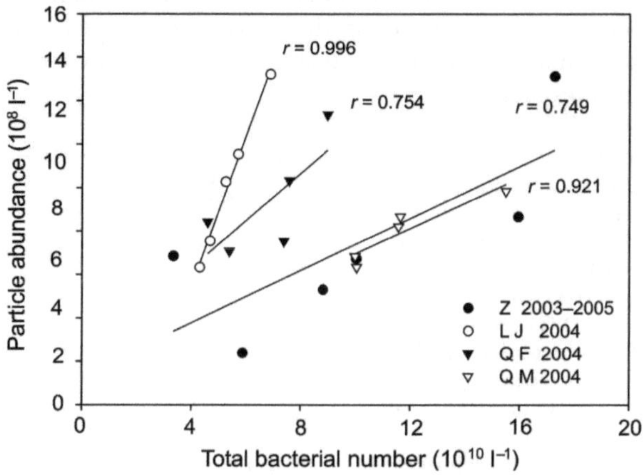

Fig. 3.22. Relationships between particle abundance and total bacterial number of eroded particles in sediments investigated at different sites in the groyne field of the river Elbe (*Z:* central station during the investigation period 2003–2005; *L, Q:* stations of the transects *L* and *Q* sampled in January, February, and May 2004, respectively). The *straight lines* indicate the linear regression curves

particles and aggregates. This applied especially to the exposed sandy mud sediments at the edges of the groyne field of the river Elbe. Sutherland et al. (1998) discussed that with increasing stability (time of non-disturbance) of the sediments, the size of the eroded particles increased. Resuspended particles from the groyne field of the river Elbe were intensively colonized by bacteria. The number of particles released was significantly correlated with the number of bacteria, although the relationship patterns differed depending upon season and location in the groyne field (Fig. 3.22). This may be explained by differences in the chemical composition of the particles and their colonization by different physiological and taxonomic groups of microorganisms.

Roller table experiments showed that the composition, microbial colonization and activity of resuspended particles altered during transport processes (Fengler et al. 2006). The particle size spectrum shifted from smaller (0–30 µm) to larger particles (31–150 µm) during a 7-d period. The ratio of organic carbon to nitrogen increased due to intense microbial metabolism in the dark. In the light the ratio decreased due to the formation of fresh organic matter by primary production as indicated by the increase of chlorophyll *a*. Pronounced increases in bacterial numbers, enzymatic microbial activity, and EPS were found within the first 2 d during light incubation.

3.5.4 Sediment Stabilization and Extracellular Polymeric Substances

Microorganisms (microalgae, bacteria, cyanobacteria) secrete extracellular polymeric substances (EPS), which consist of carbohydrates, proteins and a variety of other organic compounds (e.g., Decho 1990; Nichols et al. 2005). Because of their gel structure (Decho et al. 2003) and their high adsorption capacity, the EPS fulfill key functions for the organisms and in their environment (Köster and Meyer-Reil 2002). Beside their role in motility and adhesion of organisms to surfaces, EPS are important for binding organic and inorganic matter and pollutants, concentrating extracellular enzymes and supporting genetic transfer throughout the EPS matrix among the organisms (Decho 1990; Flemming and Wingender 2001). By the mediation of the properties of EPS, sediment particles are glued together. Because of their variable and complex chemical composition and function, the analysis of EPS and their relation to different functions is still a problem.

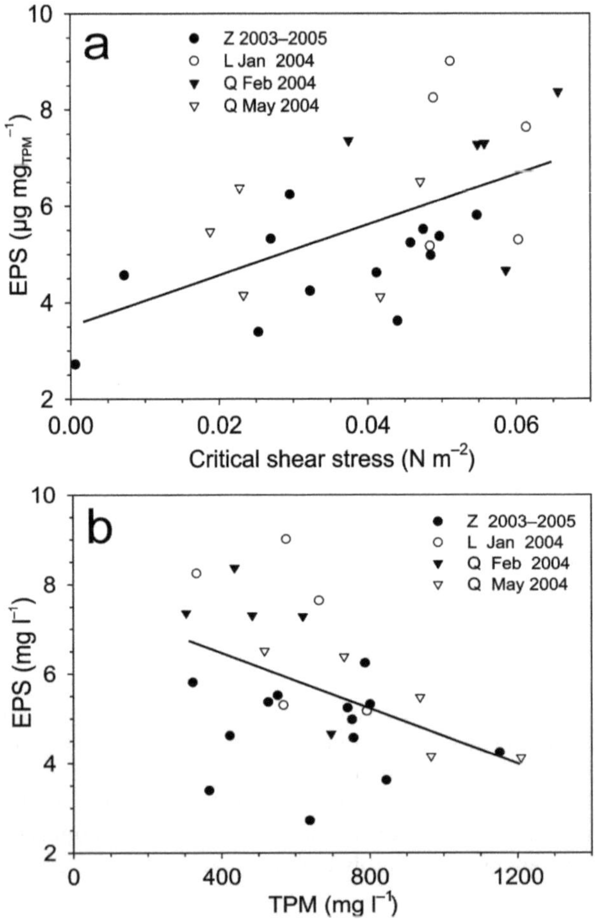

Fig. 3.23. Relationships between content of extracellular polymeric substances (*EPS*) of eroded particles and critical shear stress ($f(x) = 50.3x + 3.6$, $r = 0.537$) (**a**), and between EPS concentration and dry weight of resuspended particles (*TPM:* total particulate matter; $f(x) = 0.003x + 7.74$, $r = 0.458$) (**b**) in sediments investigated at different sites in the groyne field of the river Elbe (*Z:* central station during the investigation period 2003–2005; *L, Q:* stations of the transects *L* and *Q* sampled in January, February, and May 2004, respectively). The *straight lines* indicate the linear regression curves

The critical shear stress and the EPS content of the sediments were significantly correlated; the higher the EPS content, the higher the sediment stability (Fig. 3.23a). This relation was supported by the significant inverse correlation between the EPS concentration and the concentration of the resuspended sediments (TPM; Fig. 3.23b). From these observations it was concluded that increasing EPS contents enhanced the critical shear stress and therefore decreased the erosion risk in sediments of the groyne field of the river Elbe.

The source of EPS production in the sediments still remains unclear. Benthic microalgae, which are known to be the main EPS producers in shallow aquatic sediments (e.g., de Brouwer et al. 2005), can be discounted because of the low light intensity (2–3 $\mu E\ m^{-2}\ s^{-1}$). According to Ploug et al. (2002), light intensities of >7 $\mu E\ m^{-2}\ s^{-1}$ are necessary to support positive photosynthesis. It could be proposed that planktonic microalgae deposited in the groyne field contribute to the EPS content of the sediments. This is supported by the significant correlation between contents of chlorophyll a and EPS (colloidal fraction, Fig. 3.24a). Whereas meio- and macrofaua can destabilize sediments through their movement, they may also contribute to the stabilization of

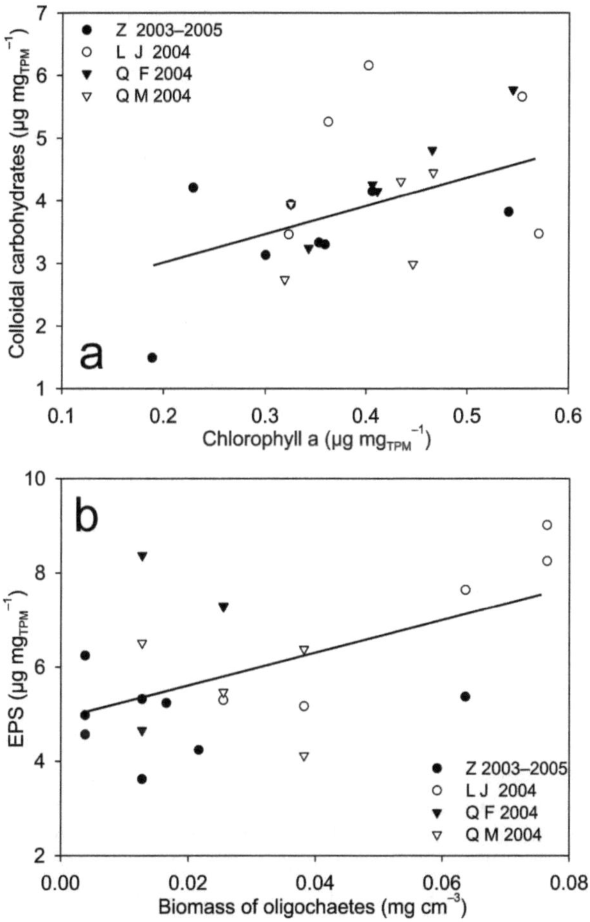

Fig. 3.24. Relationships between content of chlorophyll a and content of extracellular polymeric substances (EPS) (($f(x) = 4.6x + 2.1$, $r = 0.457$) of eroded particles (**a**), and between EPS content and biomass of oligochaetes (expressed as mg wet weight per one cm^3 of sediment) (($f(x) = 33.9x + 5.0$, $r = 0.515$) (**b**) in sediments sampled at different sites in the groyne field of the river Elbe (*Z:* central station during the investigation period 2003–2005; *L, Q:* stations of the transects *L* and *Q* sampled in January, February, and May 2004, respectively). The *straight lines* indicate the linear regression curves

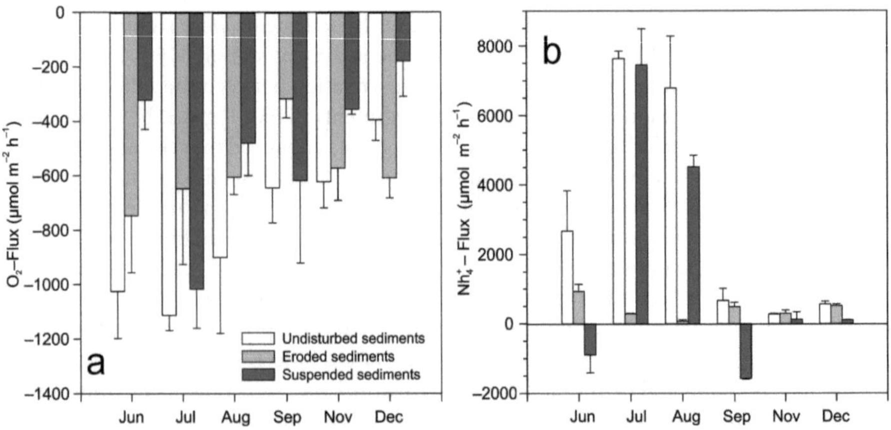

Fig. 3.25. Measurements of oxygen consumption and ammonia fluxes (dark incubation) of undisturbed, eroded, and suspended sediments sampled from the station Z during the period from June to December 2004. The *bars* represent mean values ±SD ($n = 3$)

sediments by their biogenic structures (burrows, tubes). Sediments of the groyne field were colonized by the oligochaetes *Tubifex tubifex* which live in tubes coated with mucilagenous material. A significant direct correlation was found between the biomass of the worms and the contents of EPS (Fig. 3.24b). This especially applies to the sandy mud sediments at the edges of the groyne field which were characterized by high worm biomass, high contents of EPS, and low ratios of organic carbon to nitrogen, characteristic for EPS.

EPS are especially known for their high adsorption capacity for metals (Schlekat et al. 1998). With increasing erosion of particles and EPS, the release of metals from the sediments increased (Fengler et al. 2006). The comparable patterns of mobilization of cadmium, lead, mercury, zinc, and copper pointed to the unspecific binding of the metals to EPS.

Erosion of sediments caused a stimulation of the activities of microorganisms associated with the resuspended particles. Measurements of the respiration revealed that the sum of the oxygen consumption of the eroded and the resuspended sediments clearly exceeded the oxygen consumption of the undisturbed sediments (Fig. 3.25a). In summer microbial activities associated with resuspended sediments caused a considerable release of ammonia (Fig. 3.25b). Oxygen consumption and liberation of ammonia represent major processes causing the eutrophication of aquatic environments.

3.5.5 Conclusions

The stability of the sediments of the river Elbe was reflected in sediment-specific erosion curves that varied seasonally and spatially due to variations in the chemical properties and microbial colonization of the eroded sediment particles. Exposed sandy mud sediments at the edges of the groyne field showed higher critical shear stress as compared to the more muddy sediments in the center. With increasing stability, fewer particles of larger size were liberated. Based on correlations between critical shear stress and content of EPS of the particles released it is suggested that deposited planktonic microalgae and oligochaetes contributed to the EPS content and act as biostabilizers in the sediments in the groyne field.

The fine sediment particles were main contributors to the nutrient and pollution load in the water column. Fractionation of particles according to their sedimentation speed showed that the fine fraction was characterized by enhanced contents of carbon, nitrogen and phosphorus as well as enhanced microbial colonization and activity. Particularly during the summer, bacteria associated with resuspended fine particles contributed to the eutrophication of water by high consumption of oxygen and liberation of ammonia.

Based on our observations we conclude that the sediment stability is a key parameter that predicts the erosion risk of riverine sediments. Sediments with a low stability (low critical shear stress) act as sources of nutrients and pollutants, whereas sediments with a high stability (high critical shear stress) function as sinks which have a high potential to bind nutrients and pollutants. To minimize the erosion risk of fine-grained cohesive sediments in riverine environments enhanced knowledge on the impact of sediment properties and (micro)benthic colonization on sediment stability is required for a successful application of sediment management strategies.

Acknowledgments

This study was part of the interdisciplinary research project "Sedimentdynamik und Schadstoffmobilität in Fließgewässern" (sediment dynamics and pollutant mobility in rivers (SEDYMO)) supported by the German Federal Ministry of Education and Research (BMBF). The authors are grateful to Lars Kreuzer and Ingrid Kreuzer for technical assistance. The port authority, Hamburg, and the Wasser- und Schifffahrtsamt, Lauenburg, is thanked for the help during sampling.

References

de Brouwer JFC, Bjelic S, de Deckere EMGT, Stal LJ (2000) Interplay between biology and sedimentology in a mudflat (Biezelingse Ham, Westerschelde, The Netherlands). Continental Shelf Res 20:1159–1177

de Brouwer JFC, Stal LJ (2001) Short-term dynamics in microphytobenthos distribution and associated extracellular carbohydrates in surface sediments of an intertidal mudflat. Mar Ecol Prog Ser 218:33–44

de Brouwer JFC, Wolfstein K, Ruddy GK, Jones TER, Stal LJ (2005) Biogenic stabilization of intertidal sediments: the importance of extracellular polymeric substances produced by benthic diatoms. Microb Ecol 49:501–512

Decho AW (1990) Microbial exopolymer secretion in ocean environments: their role(s) in webs and marine processes. Oceanography and Marine Biology: An Annual Review 28:73–153

Decho AW, Kawaguchi T, Allison MA, Louchard EM, Reid RP, Stephens FC, Voss KJ, Wheatcroft RA, Taylor BB (2003) Sediment properties influencing upwelling spectral reflectance signatures: The "biofilm gel effect". Limnol Oceanogr 48:431–443

Fengler G, Köster M, Meyer-Reil LA (2006) Mikrobielle Stoffumsätze an resuspendierten Sedimenten. Final report of the interdisciplinary BMBF-project: Sediment Dynamics and Pollutant Mobility in Rivers (SEDYMO)

Findlay RH, King GM, Watling L (1989) Efficiacy of phospholipids analysis in determining microbial biomass in sediments. Appl Environ Microbiol 55:2888–2893

Flemming HC, Wingender J (2001) Relevance of microbial extracellular polymeric substances (EPSs). Part II: Technical aspects. Wat Sci Technol 43:9–16
Graf G, Rosenberg R (1997) Bioresuspension and biodeposition: A review. J Mar Systems 11:269–278
Grant J, Gust G (1987) Prediction of coastal sediment stability from photopigment content of mats of purple bacteria. Nature 330:244–246
HELCOM, Helsinki Commision (1988) Guidelines for the Baltic monitoring programme for the third stage. Baltic Sea Environ Proc 27D: biological determinants. Helsinki Commission, Helsinki, pp 1–60
Köster M, Dahlke S, Meyer-Reil LA (1997) Microbiological studies along a gradient of eutrophication in a shallow coastal inlet in the southern Baltic Sea (Nordrügensche Bodden). Mar Ecol Prog Ser 152:27–39
Köster M, Meyer-Reil LA (2002) Ecology of marine microbial biofilms. In: Bitton G (ed) The encyclopedia of environmental microbiology. John Wiley & Sons, Inc., New York, pp 1081–1091
Meyercordt J, Meyer-Reil LA (1999) Primary production of benthic microalgae in two shallow coastal lagoons of different trophic status in the southern Baltic Sea. Mar Ecol Prog Ser 178:179–191
Meyer-Reil LA (1983) Benthic response to sedimentation events during autumn to spring at a shallow-water station in the western Kiel Bight. II. Analysis of benthic bacterial populations. Mar Biol 77:247–256
Meyer-Reil LA (2006) Mikrobiologie des Meeres. Eine Einführung. Facultas UTB, Stuttgart
Nichols CM, Ladiére SG, Bowman JP, Nichols PD, Gibson JAE, Guézennec J (2005) Chemical characterization of exopolysaccharides from Antarctic marine bacteria. Microb Ecol 49:578–589
Paterson DM, Tolhurst TJ, Black KS, Shayler SA, Mather S, Black I (1999) Measuring the in situ erosion shear stress of intertidal sediments with the cohesive strength meter (CSM). Est Coast Shelf Sci 49:281–294
Petersen W, Hong J, Williamoski C, Wallmann K (1996) Release of trace contaminants during reoxidation of anoxic sediment slurries in oxic water. Arch Hydrobiol Spec Issues Advanc Limnol 47:295–305
Ploug H, Zimmermann-Timm H, Schweitzer B (2002) Microbial communities and respiration on aggregates in the Elbe Estuary, Germany. Aquat Microb Ecol 27:241–248
Rieling T (2000) Remineralisation organischen Materials in Boddengewässern Mecklenburg-Vorpommerns unter besonderer Berücksichtigung der Bedeutung von Partikeln und Aggregaten. Ph.D. thesis, University Greifswald
Roast SD, Widdows J, Pope N, Jones MB (2004) Sediment-biota interactions: mysid feeding activity enhances water turbidity and sediment erodability. Mar Ecol Prog Ser 281:145–154
Schlekat CE, Decho AW, Chandler GT (1998) Sorption of cadmium to bacterial extracellular polymeric sediment coatings under estuarine conditions. Environ Toxicol Chem 17:1867–1874
Shanks AL, Edmondson EW (1989) Laboratory-made artificial marine snow: A biological model of the real thing. Mar Biol 101:463–470
Shimeta J, Amos CL, Beaulieu SE, Katz SL (2004) Resuspension of benthic protists at subtidal coastal sites with differing sediment composition. Mar Ecol Prog Ser 259:103–115
Sutherland TF, Amos CL, Grant J (1998) The effect of buoyant biofilms on the erodibility of sublittoral sediments of a temperate microtidal estuary. Limnol Oceanogr 43:225–235
Tsai CH, Lick W (1986) A portable device for measuring sediment resuspension. J Great Lakes Res 12:314–321
Westrich B, Förstner U (2005) Sediment dynamics and pollutant mobility in rivers (SEDYMO) assessing catchment-wide emission-immission relationships from sediment studies. BMBF coordinated research project SEDYMO (2002–2006). J Soils Sediments 5:197–200
Widdows J, Brinsley MD, Bowley N, Barrett C (1998) A benthic annular flume for in situ measurement of suspension feeding/deposition rates and erosion potential of intertidal cohesive sediments. Est Coast Shelf Sci 46:27–38
Widdows J, Brinsley MD, Salkeld PN, Lucas CH (2000) Influence of biota on spatial and temporal variation in sediment erodability and material flux on a tidal flat (Westerschelde, The Netherlands). Mar Ecol Prog Ser 194:23–37
Widdows J, Brinsley MD, Pope ND, Staff FJ, Bolam SG, Somerfield PJ (2006) Changes in biota and sediment erodability following the placement of fine dredged material on upper intertidal shores of estuaries. Mar Ecol Prog Ser 319:27–41
Ziervogel K (2003) Aggregation and transport behaviour of sediment surface particles in Mecklenburg Bight, south-western Baltic Sea, affected by biogenic stickiness. Ph.D. thesis, University Rostock.

Chapter 4

Transport Modeling

Mark Markofsky · Bernhard Westrich

This section deals with numerical and experimental investigation into the transport behavior of fine reactive sediment fractions interacting with dissolved contaminants in the river environment. The contributions focus on modeling transport processes including sedimentation, erosion, mixing, flocculation, sorption and degradation. In Sect. 4.1, a detailed description of a two-dimensional numerical module for dissolved and particulate contaminant transport in rivers is presented with special emphasis on the interaction of the dissolved and particulate phase and exchange processes between the water column and the uppermost active sediment layer. There is a significant influence of the mixing process at the water-sediment interface on the longitudinal dispersion of a contaminant cloud, caused for example by a flood event. The predictive two-dimensional flow and transport model TELEMAC-SUBIEF/CTM (by EdF) is extended and considered a powerful tool for predicting the dispersion and deposition of hazardous substances disposed after dredging to the fluvial environment. The case study in Sect. 4.2 is an application of the above-mentioned multifractional numerical model for the description of the disposal of highly contaminated dredged material in a regulated river showing the spatial and temporal distribution of the concentration of contaminants in the plume. The results illustrate the trapping effect of the near-bank groyne fields and the river's dead arms, where the contaminant are partly deposited and accumulate. In addition, the numerical investigation also addresses the uncertainty of the model results due to insufficient data about the concentration and fluxes of the considered contaminant. The unsteady flow conditions in the river stretch, the fluctuating input of material by the suction dredger and the interruption of the dredging operation could not be captured sufficiently by the field measurements. Therefore, the model could not be calibrated on data set with high temporal and spatial resolution. A detailed description of a three-dimensional model for calculating the fate of multiple fractions of suspended sediment in an estuarine environment with emphasis on flocculation is given in Sect. 4.3. The counteracting processes of flocculation creating larger floc size and break-up due to turbulent shear forces are described. The sediment is divided into several fractions and the interaction between these fractions is simulated. The numerical model is compared with the coagulation theory and laboratory experiments. A practical application was carried out to find structural measures for the mitigation of sedimentation in a tidal harbor within the Hamburg Harbor system in Germany. It was shown that the model can be used to predict how a current deflection wall can reduce sediment deposition.

Section 4.4 presents a technique for differentiating between various transfer fluxes of fine particles. It is stated that there are serious shortcomings in the modeling options currently available to account consistently for wash-load particle fluxes. The proposed model is based on vortex-drag concepts as an option to predict resistance-to-flow in alluvial streams displaying a dynamic bed morphology evolution. The author considers this a new perspective for simulating the transport of SPM in open channels by acknowledging the role of suction-vortices in actively maintaining the resuspension process. With this concept, the author shows the advantage of significantly lowering the number of model parameters to be calibrated. Application is made to the Scheldt Estuarine System by Antwerp.

George K. Jacoub · Bernhard Westrich

4.1 Two-Dimensional Numerical Module for Contaminant Transport in Rivers

4.1.1 Introduction

Contaminants such as heavy metals or organic pollutants are adsorbed to fine sediment particles which are transported through the river system and deposited in the regions of low flow velocities. This results in potential sources of contaminants called "hot spots" which can be eroded by floods causing deterioration of the river water quality. Therefore, the erosion, transport and deposition of contaminated sediments play a significant role in water resources engineering and management. It is a challenging task to model and predict the pathway and fate of contaminated sediments with emphasis on their spatial and temporal distribution in surface waters.

There has been intensive research trying to address the contaminant transport in river systems based on assumptions and simplifications according to each study case. This study however, presents the development of a 2-D numerical module which deals with an entire physical concept of contaminant transport in rivers. The development is based on the TELEMAC-System which consists of several modules (EDF 2002). One of them is TELEMAC-2D which solves the classical hydrodynamic Saint-Venant equations and the other one is SUBIEF-2D which handles the suspended sediment transport. Both modules are programmed in Fortran 90 and based on the Finite Element formulations with unstructured grid. The new developments are focused on the SUBIEF-2D module and named as CTM-SUBIEF-2D which stands for Contaminant Transport Module-SUBIEF-2D. The CTM-SUBIEF-2D module describes the transport of dissolved and adsorbed substances in the water column and the river bed with emphasis on the interaction between dissolved and adsorbed contaminants using the first order sorption kinetics "k_d concept". Physical processes such as sedimentation, erosion, diffusion and degradation are taken into account. The physical based model concept and the numerical implementation of the new developments are presented in the following sections.

Fig. 4.1.
Illustration of two dimensions inflow/outflow of discharge and concentration

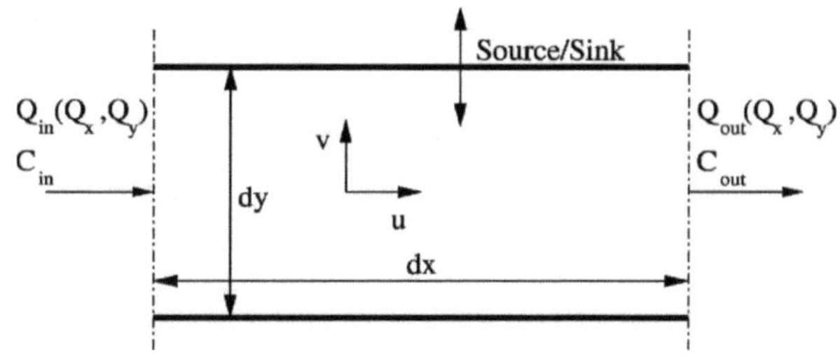

4.1.2 Module Concept and Assumption

The numerical code used in this study is the TELEMAC-System code which consists of several modules. One of them is the transport module, SUBIEF-2D, which deals with the transport of one or several tracers in a two-dimensional free surface flow. SUBEF-2D has been designed on the basis of the finite element method and the results are integrated over the water depth according to Eq. 4.1 and the concept of the mass balance of a solute which is derived for a river section (horizontal plane) with a length of dx and a width of dy as shown in Fig. 4.1.

The mass conservation equation is written in terms of a depth averaged concentration and consists of local derivative, advection, diffusion and source or sink terms as defined in Eq. 4.1.

$$\frac{\partial C}{\partial t}+u\frac{\partial C}{\partial x}+v\frac{\partial C}{\partial y}=\underline{\underline{D}}\left[\frac{\partial^2 C}{\partial x^2}+\frac{\partial^2 C}{\partial y^2}\right]+\frac{S_o}{h} \tag{4.1}$$

in which C is the substance mass per unit volume (kg m^{-3}), t is the time (s), u and v are the depth averaged velocity components in x,y-directions (m s^{-1}), $\underline{\underline{D}}$ is the dispersion tensor (m^2 s^{-1}), S_o is the source or sink (kg m^{-2} s^{-1}) and $\bar{h}$ is the average water depth (m).

As contaminant transport modeling in river systems represents a great challenge to numerically model and predict the fate of contaminated sediments in fluvial systems, a two-dimensional contaminant transport module has been developed based on SUBIEF-2D module. The developed module, CTM-SUBIEF-2D, deals with the mass conservation for five variables as shown in (Fig. 4.2): suspended sediment (SS), dissolved contaminant (C_D) and adsorbed contaminant (C_A) in the water column as well as in the river bed (G_D and G_A). The partitioning between the two variables is described by a kinetic sorption mechanism of first order in both the water column and the pore water of a river bed (Chapra 1997). The exchange between the river bed and the water column for the dissolved contaminants is driven by the concentration gradient in the water/sediment interface in a diffusive manner and for the adsorbed contaminants the exchange is controlled by sedimentation and erosion processes. The diffusion coefficient, D_S, is taken constant both in space and time and can be determined according to the experimental studies of Haag 2003. During the transport, contaminants are subject to degradation. In this research, a first-order decay is assumed for all chemical and biological reactions, i.e. the rate loss of contaminants is proportional to the concentration at any time, and it can be defined as, $-\lambda_d C$, (Thomann and Mueller 1987 and James 1993).

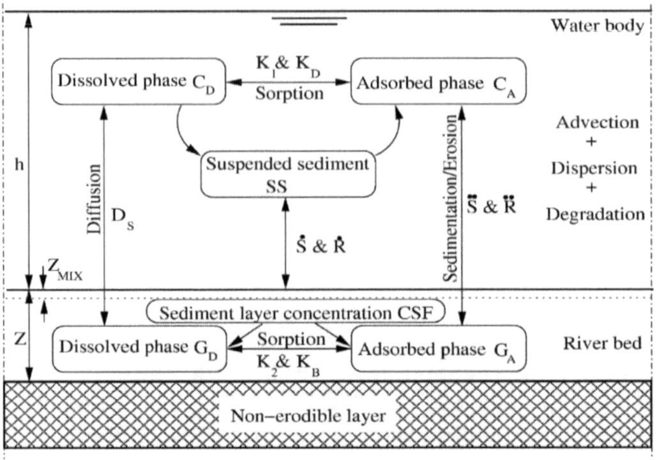

Fig. 4.2. Physical concept of the developed module CTM-SUBIEF-2D

The transport governing equations of the five variables can be given in the form of:

1. Mass conservation for suspended sediment in the water column, SS:

$$\frac{\partial \mathrm{SS}}{\partial t} + \underline{u}\,\mathrm{grad\,SS} = div\left(\underline{\underline{D}}\,\mathrm{grad\,SS}\right) - \frac{S^{\bullet} - R^{\bullet}}{\bar{h}} + S_{\mathrm{SS}}$$
$$\text{where} \quad S^{\bullet} = V_{\mathrm{S}}\mathrm{SS}\left[1 - \frac{\tau_{\mathrm{b}}}{\tau_{\mathrm{cd}}}\right], \quad R^{\bullet} = M\left[\frac{\tau_{\mathrm{b}}}{\tau_{\mathrm{ce}}} - 1\right] \tag{4.2}$$

2. Mass conservation for the dissolved contaminant in the water column, C_{D}:

$$\frac{\partial C_{\mathrm{D}}}{\partial t} + \underline{u}\,\mathrm{grad}C_{\mathrm{D}} = div\left(\underline{\underline{D}}\,\mathrm{grad}C_{\mathrm{D}}\right) - K_1\left(K_{\mathrm{D}}\mathrm{SS}C_{\mathrm{D}} - C_{\mathrm{A}}\right)$$
$$-\frac{D_{\mathrm{S}}}{Z_{\mathrm{AB}}}\frac{\left(C_{\mathrm{D}} - G_{\mathrm{D}}\right)}{\bar{h}} - \lambda_{\mathrm{d}}C_{\mathrm{D}} + S_{\mathrm{CD}} \tag{4.3}$$

3. Mass conservation for the adsorbed contaminant in the water column, C_{A}:

$$\frac{\partial C_{\mathrm{A}}}{\partial t} + \underline{u}\,\mathrm{grad}C_{\mathrm{A}} = div\left(\underline{\underline{D}}\,\mathrm{grad}C_{\mathrm{A}}\right) + K_1\left(K_{\mathrm{D}}\mathrm{SS}C_{\mathrm{D}} - C_{\mathrm{A}}\right)$$
$$-\frac{S^{\bullet\bullet} - R^{\bullet\bullet}}{\bar{h}} - \lambda_{\mathrm{d}}C_{\mathrm{A}} + S_{\mathrm{CA}} \tag{4.4}$$

4. Mass conservation for the dissolved contaminant in the river bed (pores media), G_{D}:

$$\frac{\partial G_{\mathrm{D}}}{\partial t} = \frac{D_{\mathrm{S}}}{Z_{\mathrm{AB}}}\frac{\left(C_{\mathrm{D}} - G_{\mathrm{D}}\right)}{Z_{\mathrm{MIX}}} - K_2\left(K_{\mathrm{B}}\mathrm{CSF}G_{\mathrm{D}} - G_{\mathrm{A}}\right) - \lambda_{\mathrm{d}}G_{\mathrm{D}} \tag{4.5}$$

5. Mass conservation for the adsorbed contaminant in the river bed (sediments), G_A:

$$\frac{\partial G_A}{\partial t} = K_2(K_B CSFG_D - G_A) + \frac{S^{\bullet\bullet} - R^{\bullet\bullet}}{Z_{MIX}} - \lambda_d G_A$$

$$\text{where} \quad S^{\bullet\bullet} = \frac{C_A}{SS} S^{\bullet} \ , \ R^{\bullet\bullet} = \frac{G_A}{CSF} R^{\bullet} \tag{4.6}$$

in which $S^{\bullet}$, $R^{\bullet}$, $S^{\bullet\bullet}$ and $R^{\bullet\bullet}$ are the sedimentation and erosion rates for the suspended sediment and adsorbed contaminant in both the water body and the river bed (kg m^{-2} s^{-1}), τ_b is the bed shear stress (N m^{-2}), τ_{cd} is the critical bed shear stress at which deposition begins (N m^{-2}), τ_{ce} is the critical bed shear stress for erosion (N m^{-2}), V_S is the fall velocity of suspended sediments (m s^{-1}), M is erosion coefficient (kg m^{-2} s^{-1}), Z_{AB} is the attributed thickness where the diffusion takes places into the water/sediment interface at the river bed (m), D_S is the diffusion coefficient to the river bed (m^2 s^{-1}), Z_{MIX} is the mixing layer thickness (m), λ_d is the specific decay constant coefficient (s^{-1}), K_1 is the sorption kinetics in the water body (s^{-1}), K_2 is the sorption kinetics in the river bed (s^{-1}), K_D is the equilibrium sorption coefficient in the water body (m^3 kg^{-1}), K_B is the equilibrium sorption coefficient in the river bed (m^3 kg^{-1}), CSF is the deposited or eroded sediment layer concentration (kg m^{-3}), S_{SS}, S_{CD} and S_{CA} are the source/sink terms for the suspended sediment, dissolved and adsorbed contaminants in the water body respectively, $\underline{\underline{D}}$ is dispersion tensor (m^2 s^{-1}) and $\bar{h}$ is the average water depth (m).

In the CTM-SUBIEF-2D module, a single active sediment layer of thickness Z is considered where sedimentation, erosion and mixing take place. This concept of having a single layer is supported as an option for reasonable results as long as short term simulations are performed (Di Silvio 1991 and Sieben 1996). If only sedimentation of pure and contaminated sediments occurs, then the concentration of contaminants is averagely integrated over the deposited layer thickness ΔZ. If the active layer Z is contaminated and erosion occurs, the concentration of contaminant rest is determined over the thickness $Z - \Delta Z$, Fig. 4.3.

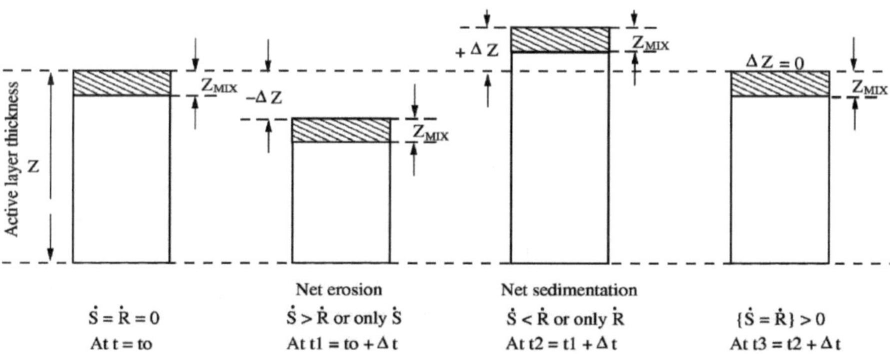

Fig. 4.3. Simplified concept of the mixing layer used in the developed module CTM-SUBIEF-2D

Mixing becomes more important and dominates the river bed contaminant concentrations if sedimentation and erosion take place simultaneously at about the same rate. The mixing process is assumed to occur at a very thin top layer of deposited sediments. In general, the thickness of this layer, Z_{MIX}, is a key parameter depending on bed shear intensity and sediment properties which change in space and time. It is difficult to determine or measure this thickness because of lack of data. The Z_{MIX} thickness should be neither too thin so that it remains the same during one numerical time-step simulation nor too large in order to reproduce effective changes of the bed composition during the simulation (Belleudy and Sogreah 2000 and 2001). Therefore, it is assumed that the minimum thickness should equal the smallest sediment fraction in order to allow mixing.

The exchange rates between the dissolved and adsorbed contaminants in both the water body and the river bed are controlled by sorption processes. These processes can be mathematically presented in terms of equilibrium distribution coefficients and sorption kinetics which are calculated considering steady-state hydraulic conditions and defined as (Santschi and Honeyman 1989 and Carrol and Harms 1999):

$$\mathrm{SSK_D} = \lim_{t\to\infty}\left[\frac{C_\mathrm{A}}{C_\mathrm{D}}\right] \quad \text{and} \quad \mathrm{CSFK_B} = \lim_{t\to\infty}\left[\frac{G_\mathrm{A}}{G_\mathrm{D}}\right] \tag{4.7}$$

In the Eqs. 4.5 and 4.6, the dissolved and adsorbed concentrations are calculated from the mass rate of contaminants deposited or eroded. The right hand side of these equations is obtained as follows:

$$\begin{aligned} \frac{\partial m_\mathrm{D}}{\partial t} &= \frac{\partial(G_\mathrm{D} Z_\mathrm{MIX})}{\partial t} = Z_\mathrm{MIX}\frac{\partial G_D}{\partial t} + G_\mathrm{D}\frac{\partial Z_\mathrm{MIX}}{\partial t} \\ \frac{\partial m_\mathrm{A}}{\partial t} &= \frac{\partial(G_\mathrm{A} Z_\mathrm{MIX})}{\partial t} = Z_\mathrm{MIX}\frac{\partial G_\mathrm{A}}{\partial t} + G_\mathrm{A}\frac{\partial Z_\mathrm{MIX}}{\partial t} \end{aligned} \tag{4.8}$$

Assuming that the Z_{MIX} is independent of time, which is acceptable when short term simulations are performed, then the Eq. 4.8 can be rewritten as:

$$\frac{\partial m_\mathrm{D}}{\partial t} = Z_\mathrm{MIX}\frac{\partial G_\mathrm{D}}{\partial t} \quad \text{and} \quad \frac{\partial m_\mathrm{A}}{\partial t} = Z_\mathrm{MIX}\frac{\partial G_\mathrm{A}}{\partial t} \tag{4.9}$$

in which m_D and m_A are the dissolved and adsorbed masses per unit area of river bed in pores media and sediments, respectively (kg m^{-2}).

In the Eqs. 4.5 and 4.6, the advection terms do not appear because the advective transport velocity in the river bed is negligible when compared to the flow velocity in the water column. Therefore, the module accounts for the vertical exchange of contaminants between the water column and the river bed. Then, these equations are treated using the Finite Difference method.

4.1.3 Numerical Implementation

The governing equations of the CTM-SUBIEF-2D module are characterized by the advection-diffusion terms and hence, they are classified as parabolic-hyperbolic equations. Two numerical methods have been implemented to treat the transport equations as follows:

1. The method of operator splitting.
2. The method of Streamline Upwind Petrov-Galerkin "SUPG".

Operator Splitting Method

The solution of the transport equation procedure has two steps, e.g., an advection step (first step) and a diffusive step (second step) as shown in Eq. 4.10.

$$\frac{\partial C}{\partial t} \approx \frac{\Delta C}{\Delta t} = \underbrace{\frac{\tilde{C} - C^n}{\Delta t}}_{\text{First step}} + \underbrace{\frac{C^{n+1} - \tilde{C}}{\Delta t}}_{\text{Second step}} \tag{4.10}$$

in which C is the substance concentration (kg m^{-3}), $\tilde{C}$ is the pre-solution concentration of C (kg m^{-3}), C^n and C^{n+1} are the concentration of C at $t = n$ and $t = n + 1$, respectively (kg m^{-3}).

In the first step, the advection term is treated by the method of characteristics "MOC" using the Runge-Kutta scheme of the first order and the interpolation of characteristics within the elements conforms to the finite elements method. According to the principles of Runge-Kutta scheme and assuming that the velocity vector at a node of an element is constant during the time step, the characteristics of each variable can be determined as follows:

$$\begin{aligned}
&\frac{\tilde{\mathrm{SS}} - \mathrm{SS}^n}{\Delta t} + \underline{u}\,\mathrm{grad}\,\mathrm{SS} = 0 \quad \Leftrightarrow \quad \frac{\mathrm{dSS}}{\mathrm{d}t} = 0 \\
&\frac{\tilde{C}_\mathrm{D} - C_\mathrm{D}^n}{\Delta t} + \underline{u}\,\mathrm{grad}\,C_\mathrm{D} = 0 \quad \Leftrightarrow \quad \frac{\mathrm{d}C_\mathrm{D}}{\mathrm{d}t} = 0 \\
&\frac{\tilde{C}_\mathrm{A} - C_\mathrm{A}^n}{\Delta t} + \underline{u}\,\mathrm{grad}\,C_\mathrm{A} = 0 \quad \Leftrightarrow \quad \frac{\mathrm{d}C_\mathrm{A}}{\mathrm{d}t} = 0
\end{aligned} \tag{4.11}$$

In the second step, Finite Difference and Finite Element methods are used for time and space discretizations, respectively. In the finite difference method, the concentration C at time step $n + 1$ is calculated from Eq. 4.12 where $\tilde{C}$ is obtained from the advection step (first step). The unknown concentration C on the right hand side of Eq. 4.12 can be replaced by the Eq. 4.13.

$$\frac{C^{n+1} - \tilde{C}}{\Delta t} = div\left(\underline{\underline{D}}\,\mathrm{grad}\,C\right) + \mathrm{Source/Sink} + \mathrm{other\ terms} \tag{4.12}$$

$$C = \theta C^{n+1} + (1-\theta)C^{n} \tag{4.13}$$

in which θ is the implicit parameter varying from 0 (fully explicit) to 1 (fully implicit). It is recommended to use θ between 0.5 and 0.6 for good and stable results. If θ is between 0 and 0.5, then the stability of the results is not ensured (Hinkelmann 2002).

Other spatial terms of the transport equation can be treated with the Finite Element method. The concentration variable C can be replaced by an approximated value C resulting from summing up the products of interpolation functions N_i and the node value C_i as shown in Eq. 4.14:

$$C^{n} = \widehat{C}^{n} = \sum_{i=1}^{i=3} N_i \widehat{C}_i^{n} = N_1\widehat{C}_1^{n} + N_2\widehat{C}_2^{n} + N_3\widehat{C}_3^{n} \tag{4.14}$$

in which i is the number of nodes per element. Three nodes per element are commonly used in the TELEMAC-System.

When Eq. 4.14 is substituted into Eqs. 4.2–4.4 (setting up all terms on the right side), they do not equal to zero any longer but have a residual error ε, Eq. 4.15. In order to eliminate this error, the method of weighted residuals is used which can be applied by multiplying the residual error by a weighting function N_j and integrating this quantity over the computational domain Ω, Eq. 4.15.

$$\frac{\partial \widehat{C}}{\partial t} - \underline{\underline{D}}\frac{\partial^2 \widehat{C}}{\partial x^2} - \underline{\underline{D}}\frac{\partial^2 \widehat{C}}{\partial y^2} - \text{Source/Sink} = \varepsilon \;, \qquad \int_{\Omega} N_j \varepsilon \,\mathrm{d}\Omega = 0 \tag{4.15}$$

Using the Standard-Galerkin method, the $N_j = N_i$, the following form can be obtained:

$$\int_{\Omega} N_j \left(\frac{\partial \widehat{C}}{\partial t} - \underline{\underline{D}}\frac{\partial^2 \widehat{C}}{\partial x^2} - \underline{\underline{D}}\frac{\partial^2 \widehat{C}}{\partial y^2} - \text{Source/Sink} \right) \mathrm{d}\Omega = 0 \tag{4.16}$$

Equation 4.16 includes turbulent diffusion and source/sink terms that can be treated using the Green-Gauss theory according to the form:

$$\int_{\Omega} div(\alpha \underline{v})\mathrm{d}\Omega = \int_{\Omega} \alpha\, div\, \underline{v}\mathrm{d}\Omega + \int_{\Omega} \mathrm{grad}\,\alpha \underline{v}\mathrm{d}\Omega = \int_{\Gamma} (\alpha \underline{v})\underline{n}\mathrm{d}\Gamma \tag{4.17}$$

in which Γ is the boundary of the computation domain, α is a scalar value, $\underline{v}$ is a vector and $\underline{n}$ is the norm vector perpendicular on the boundary line.

Applying this concept to the transport equation and assuming that $\alpha = N_j$ and $\underline{v} = \mathrm{grad}\,\widehat{C}$, the following equation can be obtained:

$$-\int_{\Omega} N_j div\left(\underline{\underline{D}}\mathrm{grad}\,\widehat{C}\right)\mathrm{d}\Omega = \underbrace{-\int_{\Gamma} N_j \underline{\underline{D}}\mathrm{grad}\,\widehat{C}\mathrm{d}\Gamma}_{=\text{zero}} + \int_{\Gamma} \mathrm{grad}\, N_j \underline{\underline{D}}\mathrm{grad}\,\widehat{C}\mathrm{d}\Omega$$

$$\mathrm{grad}\widehat{C} = \left[\theta \widehat{C}^{n+1} + (1-\theta)\widehat{C}^{n}\right] \tag{4.18}$$

The integration over the boundary of the computational domain is zero which is represented by the first term of the right side of the Eq. 4.18. Applying this equation to the transport equations, the following set of terms of the transport equation is obtained as:

$$\text{Temporal term} = \frac{\widehat{C}^{n+1} - \widehat{C}^{\text{advection}}}{\Delta t} \underbrace{\int_{\Omega} N_i N_j \mathrm{d}\Omega}_{\text{Mass Matrix}}$$

$$\text{Dispersion term} = \left[\theta \widehat{C}^{n+1} + (1-\theta)\widehat{C}^{n}\right] \underbrace{\int_{\Omega} \operatorname{grad} N_j \underline{\underline{D}} \operatorname{grad} N_i \mathrm{d}\Omega}_{\text{Dispersion Matrix}}$$

$$\begin{aligned} \text{Sorption term} = & -\left[\theta \widehat{C}_{\mathrm{D}}^{n+1} + (1-\theta)\widehat{C}_{\mathrm{D}}^{n}\right] K_1 K_{\mathrm{D}} \mathrm{SS} \underbrace{\int_{\Omega} N_i N_j \mathrm{d}\Omega}_{\text{Mass Matrix}} \\ & + \left[\theta \widehat{C}_{\mathrm{A}}^{n+1} + (1-\theta)\widehat{C}_{\mathrm{A}}^{n}\right] K_1 \underbrace{\int_{\Omega} N_i N_j \mathrm{d}\Omega}_{\text{Mass Matrix}} \end{aligned} \tag{4.19}$$

$$\text{Degradation term} = \left[\theta \widehat{C}_{\mathrm{D}}^{n+1} + (1-\theta)\widehat{C}_{\mathrm{D}}^{n}\right] \lambda_{\mathrm{d}} \underbrace{\int_{\Omega} N_i N_j \mathrm{d}\Omega}_{\text{Mass Matrix}}$$

$$\text{Source/Sink term} = S_{\mathrm{o}} \underbrace{\int_{\Omega} N_i N_j \mathrm{d}\Omega}_{\text{Mass Matrix}}$$

in which C represents the concentrations of suspended sediment (SS), dissolved contaminant (C_{D}) and adsorbed contaminant (C_{A}).

There is a great number of references in literature which handle the treatment and the implementation of the mass and dispersion matrixes of Eq. 4.19. These matrixes have already been numerically treated in the TELEMAC-System (EDF 2002). The following steps are carried out to rearrange and implement the transport equations:

1. applying the procedure mentioned in Eq. 4.19 to the five transport equations;
2. replacing $\mathrm{d}\Omega$ by $\mathrm{d}x\mathrm{d}y$ for a two dimensions model;
3. setting the unknown variables on the left-hand side of the equations and all known variables with coupling terms on the right-hand side;
4. using the Jacobi matrix to transform the derivatives of the interpolation functions after considering the Cartesian coordinates.

Herewith, a representative rearranged transport equation of the adsorbed contaminant in the water column, C_{A}, is shown in Eq. 4.20. The other transport equations of SS and C_{D} have the same structure.

$$C_A^{n+1}\left(\left[\frac{1}{\Delta t}+\theta K_1+\theta\lambda_d\right]\iint\limits_{x,y} N_iN_j\mathrm{d}x\mathrm{d}y\right)-\theta C_A^{n+1}\iint\limits_{x,y}\operatorname{grad}N_j\underline{\underline{D}}\operatorname{grad}N_i\mathrm{d}x\mathrm{d}y$$

$$=\left(\begin{array}{l}\left[\begin{array}{l}\dfrac{\tilde{C}_A^{\text{advection}}}{\Delta t}-(1-\theta)K_1\Delta tC_A^n+\underline{\underline{\theta K_1K_D\Delta tSSC_D^{n+1}}}+(1-\theta)K_1K_D\Delta tSSC_D^n-\\(1-\theta)\lambda_d+\ddot{R}-\ddot{S}\pm S_o\end{array}\right]\\\iint\limits_{x,y}N_iN_j\mathrm{d}x\mathrm{d}y\end{array}\right) \tag{4.20}$$

$$+(1-\theta)C_A^n\iint\limits_{x,y}\operatorname{grad}N_j\underline{\underline{D}}\operatorname{grad}N_i dxdy$$

According to the section above, the transport equation of dissolved and adsorbed contaminants in the river bed (G_D and G_A, respectively) are determined from:

$$\begin{aligned}&G_D^{n+1}\left[1+\theta\frac{D_S}{Z_{MIX}Z_{AB}}\Delta t+\theta K_2K_B\Delta t\right]\\&=G_D^n\left[1-(1-\theta)\frac{D_S}{Z_{MIX}Z_{AB}}\Delta t-(1-\theta)K_2K_B\Delta t\right]\\&\quad+\frac{D_S}{Z_{MIX}Z_{AB}}\Delta t\left[\theta C_D^{n+1}+(1-\theta)C_D^n\right]+K_2\Delta t\left[\underline{\underline{\theta G_A^{n+1}}}+(1-\theta)G_A^n\right]\quad\text{and,}\end{aligned} \tag{4.21}$$

$$\begin{aligned}G_A^{n+1}\left[1+\theta K_2\Delta t\right]&=G_A^n\left[1-(1-\theta)K_2\Delta t\right]+K_2\Delta t\left[\underline{\underline{\theta K_BG_D^{n+1}}}+(1-\theta)K_BG_D^n\right]\\&\quad+\left[\frac{S^{\bullet\bullet}-R^{\bullet\bullet}}{Z_{MIX}}\right]\Delta t\end{aligned} \tag{4.22}$$

Streamline Upwind/Petrov-Galerkin "SUPG"

The method focuses on how to treat the advection term using the central difference and optimal upwind scheme. The last is developed on the basis of an artificial diffusion $\tilde{k}$ (Kelly et al. 1980). The scalar value of the artificial diffusion can be calculated from Eq. 4.23 as it is implemented in the TELEMAC-System.

$$\begin{aligned}&\tilde{k}=\frac{\delta\Delta X}{2},\quad\text{where}\\&\delta=0\Rightarrow\text{Normal central scheme}\\&\delta=1\Rightarrow\text{Upwind/Petrov}-\text{Galerkin scheme}\\&\delta=C_r\end{aligned} \tag{4.23}$$

in which δ is the upwind scheme coefficient (–), Cr is the Courant number = $(u\Delta t)/(\Delta X)$, Δt is the time step (s) and ΔX is the grid distance of the element (m).

If the Standard Galerkin Weighted Residual method is used, the weighting function is considered to be continuous across inter-boundaries but it may have discontinuity at the elements nodes. The SUPG method formulation considers this effect and requires a discontinuous weighting function as shown below:

$$\varpi = \omega + p \tag{4.24}$$

in which ϖ is the weighting function according to the SUPG, ω is the continuous weighting function and p is the discontinuous weighting function of lower degree (Brooks and Huges 1982).

They found that the p function equals to the upwind artificial diffusion tensor $\tilde{k}_{ij}$ with some modifications to consider the effect of the local mean velocity at each node u.

$$\varpi = \omega + \frac{\tilde{k}_{ij} u_j \underline{\omega}}{\|u\|} \tag{4.25}$$

in which $\underline{\omega}$ is a certain function that depends on the advection term.

The weighting function $\underline{\omega}$ has to be applied to all terms of the transport equation. However, Brooks and Huges 1982 have shown that the method SUPG can not be used to treat the diffusion and the source/sink terms. Those terms can be then treated using the finite element method as mentioned above. After applying the SUPG basic functions as explained above to only the advection term in the transport equations, the following is obtained:

$$\begin{gathered}
\int_\Omega \frac{\tilde{\mathrm{SS}} - \mathrm{SS}^n}{\Delta t} \mathrm{d}\Omega + \int_\Omega \omega \underline{u} \,\mathrm{grad}\, \mathrm{SS} \mathrm{d}\Omega + \int_\Omega \frac{\tilde{k}_{ij} u_j \underline{\omega}}{\|u\|} \underline{u} \,\mathrm{grad}\, \mathrm{SS} \mathrm{d}\Omega = 0 \\
\int_\Omega \frac{\tilde{C}_\mathrm{D} - C_\mathrm{D}^\mathrm{n}}{\Delta t} \mathrm{d}\Omega + \int_\Omega \omega \underline{u} \,\mathrm{grad}\, C_\mathrm{D} \mathrm{d}\Omega + \int_\Omega \frac{\tilde{k}_{ij} u_j \underline{\omega}}{\|u\|} \underline{u} \,\mathrm{grad}\, C_\mathrm{D} \mathrm{d}\Omega = 0 \\
\int_\Omega \frac{\tilde{C}_\mathrm{A} - C_\mathrm{A}^\mathrm{n}}{\Delta t} \mathrm{d}\Omega + \int_\Omega \omega \underline{u} \,\mathrm{grad}\, C_\mathrm{A} \mathrm{d}\Omega + \int_\Omega \frac{\tilde{k}_{ij} u_j \underline{\omega}}{\|u\|} \underline{u} \,\mathrm{grad}\, C_\mathrm{A} \mathrm{d}\Omega = 0
\end{gathered} \tag{4.26}$$

Nonlinearity

The discretization methods mentioned above are implicit and this leads to difficulties when solving a system of equations which has a large number of unknowns. The TELEMAC-System provides different types of solvers; only those used in this research are mentioned here. They are the Conjugate Gradient method, the Conjugate Gradient on Normal Equation method and the Generalized Minimal Residual method (GMRES). In addition to these solvers, the CTM-SUBIEF-2D module deals with five transport equations which are coupled through the interaction terms (underlined in Eqs. 4.20–4.22). This coupling is weak in transport equations (Hinkelmann 2002) and therefore, a linear solver using the Picard method is implemented. The number of sub-iterations for this method is limited to 100; however, in some cases less than 10 iterations are needed.

Boundary Conditions

The TELEMAC-System treats three types of boundary conditions which are Dirichlet, Neumann and no boundary condition (degree of freedom). The first boundary condition prescribes the concentration of substances at the inflow or outflow boundaries. The second boundary condition deals with defined diffusive fluxes over the boundaries. Normally these fluxes are prescribed using the concentration gradient. In most cases, the diffusive fluxes over the boundaries are set to zero. The third type of boundary condition is the easiest to deal with since there is no change in the corresponding transport equation.

Module Limitations

The module deals so far with short term simulations with relatively small time steps on the order of magnitude of seconds. In addition to this, the following items are not considered:

1. Physical processes such as flocculation and consolidation.
2. Complete sediment mixing layers (multi-layer model). However, the simplified concept used has given reasonable results as compared to field measurements (Westrich, this vol.).
3. Interaction between different contaminants and different sediment fractions.
4. Chemical and biological reactions.

To show the calibration, validation or even comparison with other numerical models is beyond the scope of this section. However, a real case application for a headwater section of the Upper Rhine where data on sediment properties (erosion), sediment depth profile, contaminated fraction and suspended sediment concentration were available is shown (Westrich, this vol.). The numerical results show a good agreement with field data.

Readers who wish to know the applicability of the module in more detail, for various real cases, are referred in the first instance to Jacoub 2004, Jacoub and Westrich 2004, 2006 which show comparisons of the module results with physical model results for a flood retention reservoir and with field measurements for the river Elbe.

4.1.4 Conclusions

This research presents the development of a two dimensional numerical module for contaminant transport in rivers. The new module, named CTM-SUBIEF-2D, has been developed on the basis of the TELEMAC-System. It describes the transport of dissolved and particulate contaminants in the water column and river bed with emphasis on first order sorption kinetics. By solving five partial differential equations, the depth averaged of suspended sediment and dissolved and particulate contaminants in the water column and the river bed are calculated. Sedimentation, erosion, mixing of sediments and degradation of contaminants processes are taken into account. The module handles the interaction between one contaminant substance with one sediment fraction. The module handles the time-dependent variables and parameters such as particle settling velocity, concentration or flux of sediments, and depth dependent variables in sediments.

Numerical schemes such as Operator Splitting method and Stream Upwind/Petrov-Galerkin method (SUPG) for treating the transport equations are implemented. The five transport equations are coupled through interaction terms. Because of the weak coupling of the five transport equations, the Picard method for linearization is used. Westrich, this volume, shows an application of the developed module for a headwater section of the Upper Rhine and its numerical results with comparison to field measurements.

References

Beulleudy Ph, SOGREAH (2000) Numerical simulation of sediment mixture deposition Part 1: analysis of a flume experiment. Journal of Hydraulic Research 38(6)

Beulleudy Ph, SOGREAH (2001) Numerical simulation of sediment mixture deposition Part 2: analysis of a flume experiment. Journal of Hydraulic Research 39(1)

Brooks AN, Hughes TJR (1982) Streamline Upwind Petrov Galerkin formulations for convection dominated flows with particular emphasis on the Naiver-Stokes equations. Journal of Computer Methods in Applied Mechanics and Engineering 32:199–259

Carroll J, Harms IH (1999) Uncertainty analysis of partition coefficients in a radionuclide transport model. Journal of Water Research 33:2617–2626

Chapra SC (1997) Surface water-quality modelling. Mc Graw-Hill, New York

Di Silvio G (1991) Sediment exchange between stream and bottom: a four-layer model. Proceeding of the international Grain Sorting Seminar, Mitteilung der Versuchanstalt für Wasserbau, Hydrologie und Glaziologie, ETH, vol. 117, pp 191–196

EDF (2002) TELEMAC Modelling System, 2D-Hydrodynamics software, Principles manual-validation documents. Direction des Etudes et Recherches, Distributed by SOGREAH consultants, edn 5.0

Haag I (2003) Der Sauerstoffhaushalt staugeregelter Flüsse am Beispiel des Neckars – Analysen, Experimente, Simulationen. Mitteilung, Institute of Hydraulic Engineering (IWS), University of Stuttgart, No. 122

Hinkelmann R (2002) Efficient Numerical Methods and Information-Processing Techniques in Environment Water. Mitteilung, Institute of Hydraulic Engineering (IWS), University of Stuttgart, No. 117

Jacoub G (2004) Development of a 2-D numerical module for particulate contaminant transport in flood retention reservoirs and impounded rivers. Doctoral thesis, Institute of Hydraulic Engineering (IWS), University of Stuttgart, No. 133

Jacoub G, Westrich B (2004) 2-D numerical code to simulate the transport and deposition of dissolved and particulate contaminants in a flood retention reservoir. International Conference on Hydro-Science and Engineering, ICHE 2004, vol. 6, pp 272–274

Jacoub G, Westrich B (2006) Effect of river groyne structures on flow, sedimentation and erosion dynamics in rivers (Case study: the river Elbe). The International General Assembly Conference EGU-2006, Vienna-Austria

James A (1993) An Introduction to Water Quality Modeling. Second edition

Kelly DW, Nakazawa S, Zienkiewict OC, Heinrich JC (1980) A note of upwinding and anisotropic balancing dissipation in finite element approximations to convective diffusion problem. International Journal of Numerical Methods Engineering 15:1705–1711

Santschi PH, Honeyman BD (1989) Radionuclides in aquatic environments. Journal of Radiation Physics and Chemistry 34(2):213–240

Sieben A (1996) One dimensional models for mountain-river morphology. Communications on Hydraulic and Geotechnical Engineering, Delft University of Technology, Report no. 96-2

Thomann RV, Mueller JA (1987) Principles of Surface Water Quality Modelling and Control. Harper International edition

Joachim Karnahl · Bernhard Westrich

4.2 Two-Dimensional Numerical Modeling of Fine Sediment Transport Behavior in Regulated Rivers

4.2.1 Introduction

In most cases low flow velocities in the headwater of weirs and dams cause sedimentation of fine suspended sediments resulting in a reduction of the flow cross-section. In waterways, where the navigability is restricted and the safety of the embankments is affected, the deposited sediments must be removed. Since many river sediments are polluted by various substances, the dredging and disposal of contaminated sediments requires a feasibility study focusing on the environmental impact. A numerical transport model can provide basic information about the temporal and spatial distribution of the disposed sediments in the river channel, and their potential deposition in near bank river training structures such as groyne fields, or in harbours and cut-off meander.

A two dimensional depth averaged flow and transport model was developed to describe the transport, dispersion, sedimentation and erosion of different suspended sediment fractions as well as adsorbed and dissolved pollutants.

Contaminant transport modeling needs detailed information about suspended sediment concentration, fractional sediment fall velocity and sorption parameters. Field measurements of suspended sediment concentration, flow velocity and water level usually provide only local or cross-section averaged data which must be analyzed and evaluated for model calibration. In the field, remobilization and release of adsorbed substances to the aqueous phase is often not detectable because of chemical analysis limitations.

4.2.2 Objective

The objective of this study was to build a mathematical model to describe the processes of multifractional suspended sediment transport with adsorbed and dissolved contaminants for applications in surface waters. A depth averaged 2-D model was chosen in order to model longer time periods and larger domains. This model was applied to investigate the temporal and spatial distribution of three suspended sediment fractions, as well as the adsorbed and dissolved toxic substance *Hexachlorobenzoene* (HCB) in the plume of dredged material disposed in the tailwater of a hydro-power station at the Upper Rhine River. The model generates the concentration field and the sedimentation patterns and was used to perform a sediment mass balance. The uncertainties of the field data and the model results are discussed.

4.2.3 The Numerical Model

The hydrodynamic part of the simulation was performed by using the 2-D Finite-Element model TELEMAC 2D which solves the depth-averaged Saint-Venant equations and generates the flow field for the transport and reaction model SUBIEF 2D (EDF 2004).

Both models are independent from another, i.e. they are solved sequentially and not simultaneously. Different approaches for the bed roughness are implemented, similarly different turbulence models (constant viscosity, k-ε, Elder model), numerical schemes and solvers are available. The following section describes the basic mathematical equations for the transport model. The model domain and the model parameters are described in Sect. 4.2.4 and in Table 4.2.

Model Development

To simulate the transport of contaminated particles the SUBIEF 2D code was extended to model multifractional sorption processes, including the sedimentation and erosion processes of the particles with adsorbed substances. The model allows the implementation of chemical and biological reactions like transformation and degradation, as well as diffusive flux of the dissolved substances at the water-sediment interface. The basic model concept is shown in Fig. 4.4.

The 2-D transport equations for suspended sediments, dissolved and adsorbed substances are generally given in the following form:

$$\frac{\partial C_i}{\partial t} + \underline{u}\,\mathrm{grad}\,C_i = \frac{1}{h} div\left(h\underline{\underline{D}}\,\mathrm{grad}\,C_i\right) \pm \text{source/sink} \tag{4.27}$$

with C_i = concentration (kg m^{-3}) (suspended sediment, dissolved or adsorbed substances), t = time (s), u = velocity (m s^{-1}), $\underline{\underline{D}}$ = dispersion tensor (m^2 s^{-1}).

In the following case study the formulation of Elder (1959) is used for the dispersion coefficient:

$$D_{l,t} = u_* h k_{l,t} \quad \left(\frac{\mathrm{m}^2}{\mathrm{s}}\right) \tag{4.28}$$

with u_* = friction velocity (m s^{-1}), h = water depth (m), $k_{l,t}$ = calibration constants in longitudinal and transversal flow directions.

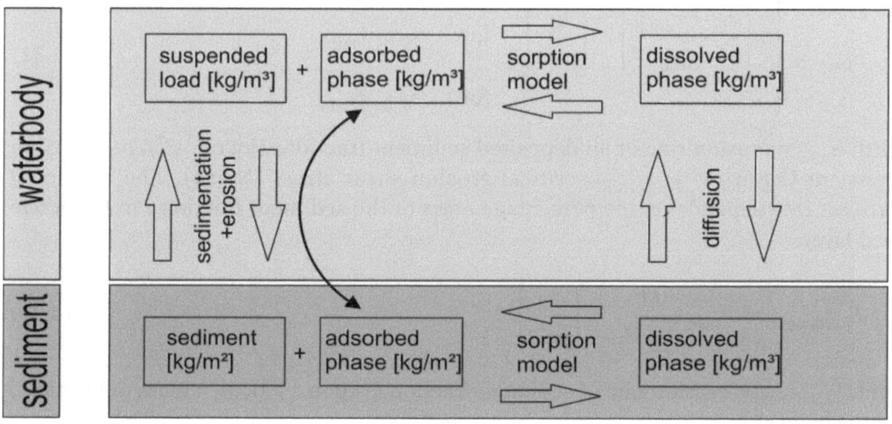

Fig. 4.4. Basic model concept

Erosion and deposition are the main source and sink terms in the suspended sediment transport equation. For the deposition of cohesive sediment the following formulation (Krone 1962) is used:

$$S_{i,\text{sediment}} = \begin{cases} w_{s,i}\, C_i \left(1 - \dfrac{\tau_0}{\tau_{\text{crit,Si}}}\right) & \left(\dfrac{\text{kg}}{\text{m}^2\text{s}}\right) \quad \text{for} \quad \tau_0 < \tau_{\text{crit,Si}} \\ 0 & \qquad\qquad \text{for} \quad \tau_0 \geq \tau_{\text{crit,Si}} \end{cases} \tag{4.29}$$

with $S_{i,\text{sediment}}$ = sedimentation rate of fraction i (kg m^{-2} s^{-1}), $w_{s,i}$ = respective fall velocity of fraction i (m s^{-1}), τ_0 = bed shear stress (N m^{-2}), $\tau_{\text{crit,Si}}$ = critical deposition shear stress of fraction i (N m^{-2}).

The sedimentation rate of the adsorbed substances $S_{i,\text{adsorbed}}$ associated with the fraction i depends on the ratio between the concentration of the adsorbed substance $C_{i,\text{adsorbed}}$ and the concentration of the suspended sediment $C_{i,\text{sediment}}$:

$$S_{i,\text{adsorbed}} = S_{i,\text{sediment}} \frac{C_{i,\text{adsorbed}}}{C_{i,\text{sediment}}} \quad \left(\frac{\text{kg}}{\text{m}^2\text{s}}\right) \tag{4.30}$$

The critical deposition shear stress can be approximately assumed constant or optionally described by the energy approach after Westrich and Juraschek (1985) by adaptation to multiple fractions as described by Dreher (2005):

$$\tau_{\text{crit,Si}} = \frac{(\rho_s - \rho_w) g h \sum w_{s,i}\, C_{i,\text{sediment}}}{k u \rho_s} \quad \left(\frac{\text{N}}{\text{m}^2}\right) \tag{4.31}$$

with g = gravity constant (m s^{-2}), k = bed form and grain size specific constant defined as required energy for suspension over available dissipative energy from the flow (–), ρ_s = sediment density (kg m^{-3}).

For the erosion of (cohesive) sediments the approach by Kuijper et al. (1989) based on Partheniades (1965) is used with adaptation to multiple fractions as follows:

$$E_{\text{total}} = \begin{cases} m \left(\dfrac{\tau_0 - \tau_{\text{crit,E}}}{\tau_{\text{crit,E}}}\right) & \left(\dfrac{\text{kg}}{\text{m}^2\text{s}}\right) \quad \text{for} \quad \tau_0 > \tau_{\text{crit,E}} \\ 0 & \qquad\qquad \text{for} \quad \tau_0 \leq \tau_{\text{crit,E}} \end{cases} \tag{4.32}$$

with E_{total} = erosion rate of all deposited sediment fractions (kg m^{-2} s^{-1}), m = erosion constant (kg m^{-2} s^{-1}), $\tau_{\text{crit,E}}$ = critical erosion shear stress (N m^{-2}). The fractional erosion rate depends on the percentage mass of the sediment fraction i in the active bed layer:

$$E_{i,\text{sediment}} = E_{\text{total}} \frac{M_i}{\sum M_i} \quad \left(\frac{\text{kg}}{\text{m}^2\text{s}}\right) \tag{4.33}$$

with $E_{i,\text{sediment}}$ = erosion rate of sediment fraction i (kg m^{-2} s^{-1}), M_i = mass of fraction i in the bed (kg).

The erosion rate of the adsorbed substances $E_{i,\text{adsorbed}}$ associated with the fraction i depends on the ratio between the mass of the adsorbed substance in the sediment $M_{i,\text{adsorbed}}$ and the mass of the sediment $M_{i,\text{sediment}}$:

$$E_{i,\text{adsorbed}} = E_{i,\text{sediment}} \frac{C_{i,\text{adsorbed}}}{C_{i,\text{sediment}}} \quad \left(\frac{\text{kg}}{\text{m}^2\text{s}}\right) \tag{4.34}$$

Chemical, biological and degradation processes are treated as user defined exchange terms for each substance. The Freundlich or Langmiur isotherms (Stumm and Morgan 1996) are usually used for sorption processes. In the following case study, the partitioning coefficient K_d, based on the Freundlich isotherm, with first order kinetics is implemented as follows:

$$J_{\text{sorption},i} = k_t\left(K_d\, C_{S,i}\, C_d - C_{A,i}\right) \quad \left(\frac{\text{kg}}{\text{m}^3\text{s}}\right) \tag{4.35}$$

with $J_{\text{sorption},i}$ = mass flux due to sorption (kg m^{-3} s^{-1}), k_t = kinetic sorption factor (s^{-1}), K_d = partitioning coefficient (l kg^{-1}), $C_{S,i}$ = sediment concentration of fraction i (kg m^{-3}), C_d = dissolved concentration of a substance (kg m^{-3}), $C_{A,i}$ = concentration of a substance adsorbed onto fraction i (kg m^{-3}).

4.2.4 Case Study: Disposal of Dredged Material

Site Description

In the headwater of the Iffezheim weir which is the lowest downstream hydropower station in the Upper Rhine River, fine cohesive sediments had to be removed to reestablish the original flow cross-section and to ensure the freeboard required for the safety of the embankment against flooding. The material was dredged by a suction dredger and disposed through a 3.5 km long pipeline to the tailwater during low water levels with a corresponding river discharge ranging from 700 to 1 500 m^3 s^{-1} to avoid sediment deposition on the flood plains. Measurements were performed before, during and after the dredging in the main channel, in typical groyne fields and harbor areas over a distance of 100 km. These included local flow velocities, suspended sediment concentration, grain size distribution and turbidity. Sediment traps were also deployed at different locations.

The numerical investigation is focused on the upper 22 km long river reach immediately downstream of the Iffezheim weir between river km (rkm) 334.0–356.2. The confluence of the Murg tributary is located 8 km downstream of the weir. Three gauging stations are inside the model area (Fig. 4.5): Iffezheim (rkm 336.6), Plittersdorf (rkm 340.3) and Lauterbourg (rkm 349.3). The measured discharge at Plittersdorf is used as the inflow boundary condition. Figure 4.5 shows a typical grain size distribution of the suspended sediment and the model area with the locations of the cross-sections at which measurements and gauging took place.

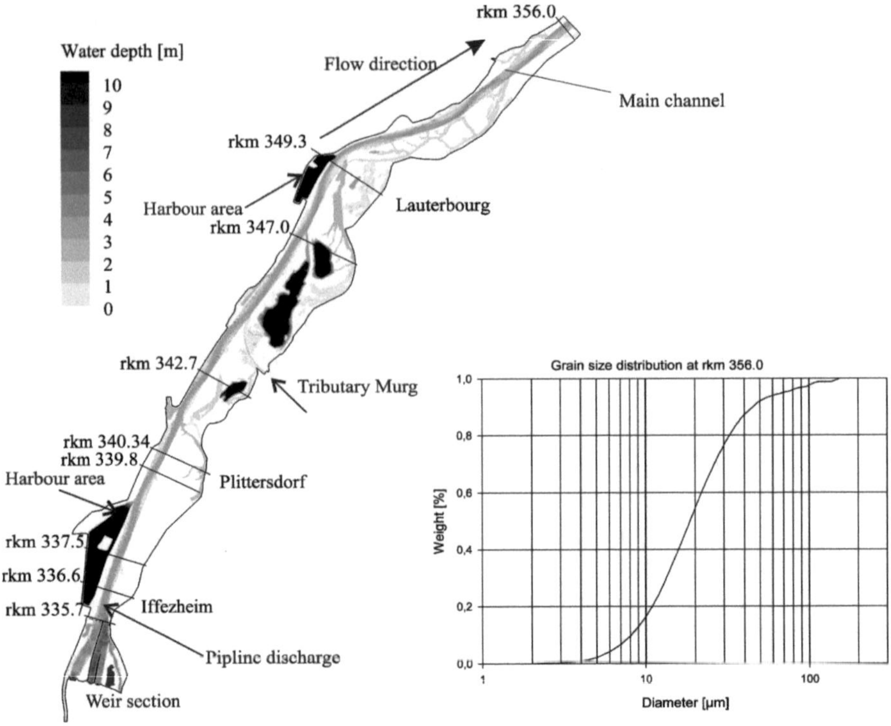

Fig. 4.5. General layout of tailwater section

Model Input Data from Field Measurements

During the dredging and disposal period, which lasted from Jan. 18 to Feb. 17, 2005, about 150 000 m^3 of sediment was discharged through the slurry pipeline. Figure 4.6 shows the discharge, the dredged material mass flux and the HCB concentration of the dredged material during the disposal period. During the dredging operation there was a flood in the Murg tributary from January 20–22, 2005 and a small flood in both the river Rhine and Murg from February 11–17, 2005. The variation of the mass discharge pumped by the dredger is caused by the temporary interruption of the suction dredger. Sediment samples were taken at the suction dredger to investigate grain size distribution, HCB concentrations and other chemical parameters of the dredged material. The percentage of sediment particles with a diameter greater than 200 μm was less than 5% by weight. Hence three sediment fractions 0–20, 20–60, 60–200 μm were modeled to consider different adsorbed HCB concentrations and different sedimentation behavior. Adsorbed HCB was found in all three sediment fractions. To overcome the heterogeneous grain size distribution and therefore also the adsorbed HCB concentration, cumulative samples were taken, indicated by the step like dashed line in Fig. 4.6, showing the HCB concentration. However the cumulative samples of the 3 fractions still show fluctuations in the fractional percentage, e.g., for the 20–60 μm fraction a range from 18.4 to 24.8% (see Table 4.1).

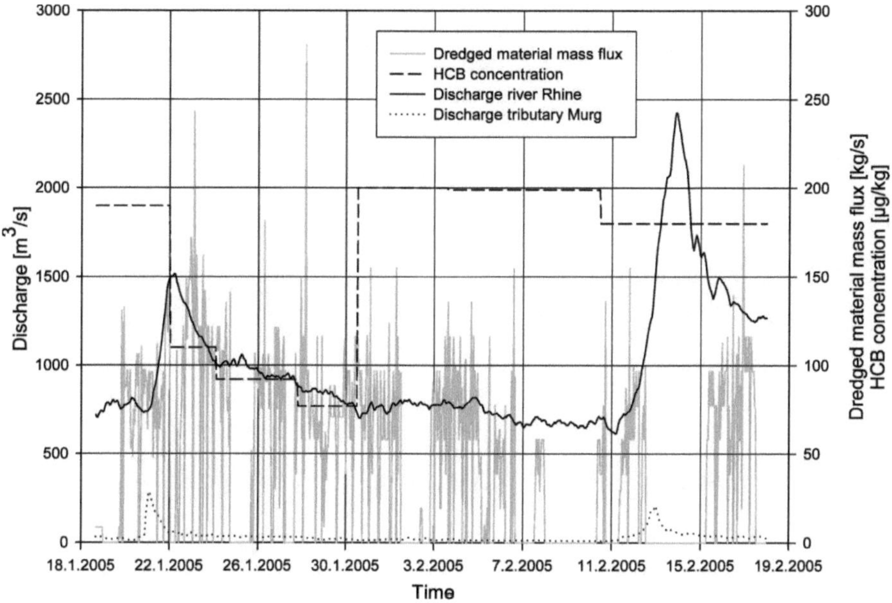

Fig. 4.6. Discharge, dredged material mass flux and HCB concentration of the dredged material

Table 4.1. Sediment distribution and HCB concentration

	Input data	0–20 µm fraction (%)	20–60 µm fraction (%)	60–200 µm fraction (%)	HCB concentration ($\mu g\ kg^{-1}$)
Background	9 – 40 $mg\ l^{-1}$	42.0	46.5	11.5	85
Flushed sediment	0 – 280.6 $kg\ s^{-1}$	48.0 – 71.1	18.4 – 24.8	6.8 – 25.7	92 – 200

The background suspended sediment concentration in the Rhine was nearly constant at 9 $mg\ l^{-1}$, except for 7 days from February 11–18, 2005 with an increased background concentration up to 40 $mg\ l^{-1}$ due to a high discharge of 2 425 $m^3\ s^{-1}$. The adsorbed HCB background concentration has been monitored as 85 $\mu g\ kg^{-1}$ and this value was used in the model calculations. Since no measurements in the Murg tributary exist, the background suspended sediment concentration is also assumed to be 9 $mg\ l^{-1}$. Since the dissolved HCB is below the chemical detection limit the background concentration was assumed to be zero.

Model Domain Description and Setup

The model domain area is ~29.3 km^2, and discretizised by 88 910 nodes with 175 870 elements. The smallest node distance is 1.6 m, the largest node distance is 63.6 m (mean ~15 m) the element areas range from 1.6 m^2 up to 1 274 m^2. The typical node distance in the main channel and the groyne fields is around 20 m, the groyne structures have a finer resolution with node distances from 1.6 to 10 m. The mesh resolution on the

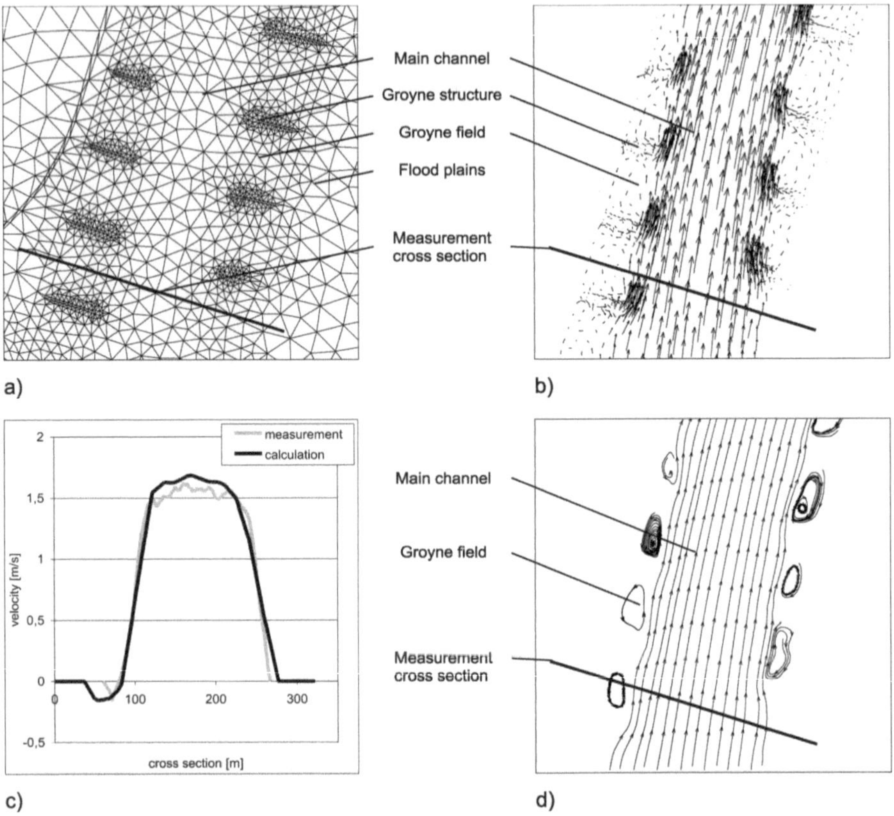

Fig. 4.7. a Mesh resolution, **b** flow field, **c** flow velocities, **d** streamlines

flood plains is larger than in the main channel, except the flow paths from small side streams. Figure 4.7a shows a mesh-section including the measurement cross-section at Iffezheim.

In a first step the hydrodynamic model for the flow field was calibrated with a steady state discharge and then applied to the 30 day unsteady flushing period with a time step of 5 seconds. The model run for the flow field took around 12 h computation time using 32 parallel processors.

In a second step the transport model should be calibrated but due to high fluctuation of the flushing mass and the available measured data a calibration cannot be performed, so different scenarios were calculated. A transport model run with 32 parallel processors took around 2.5 days computation time with a time step of 2 seconds.

The Flow Field

The tailwater at the weir is regularly fed with gravel to avoid bed erosion. Therefore, downstream of the weir there is some bed elevation change due to both the dumping of gravel material and bed load transport. The channel bathymetry was measured in the fall of 2001. Several water level measurements were conducted. For the hydraulic model calibration the water level measurement of November 6, 2001 at a constant low flow discharge of 810 $m^3 s^{-1}$ was used, to minimize the effect of the sediment feeding on the bed elevation. The difference between calculated and measured water level is in the range of ±5 cm.

The unsteady flow model calculation for the 30 days flushing period (Discharge see Fig. 4.6) was compared with the data of the three gauging stations Iffezheim, Plitterdorf and Lauterbourg. At Iffezheim the calculated water levels differs between −49 cm to +30 cm compared with the measured values and have a root mean square error (RMSE) of 8.4 cm. At Plittersdorf the difference ranges from −31 to +29 cm (RMSE 6.8 cm) and at Lauterbourg from −21 to +32 cm (RMSE 6.0 cm).

The largest water level difference was observed during the rising phase of the small flood events. The discrepancy is mainly caused by the fact, that the data from Plittersdorf, which is 6 km downstream of the weir are used as upper model boundary condition although there is a time lag of about 1–2 h. By accounting for this time shift better model fit was achieved, e.g., at Iffezheim the water level differences will be reduced to −29 to +6 cm (less than 10% of water depth). Additionally it is evident that the sediment feeding during the 4 years period between the flow model calibration and application has changed the bed morphology and therefore has an impact on flow resistance and water level. A significant change of the river bed conditions can be detected immediately downstream of the sediment feeding area. However, this effect is decreasing with distance to the weir. Nevertheless, the comparison between the measured flow velocities (BfG 2005) and the calculated values shows a good agreement, the maximum difference is about 13%. Figure 4.7b show the flow field for a low flow situation of 780 $m^3 s^{-1}$, where the groynes are not overtopped, with high flow velocities in the main channel, eddies in the groyne fields and no flow on the flood plains. This can also be seen in Fig. 4.7d where the streamlines are shown. Figure 4.7c shows the measured and calculated velocities. The difference in the flow velocities in the groyne field results from the measurement method (ADCP), which enables only good measurements for higher flow depth. Finally, the model was calibrated for a steady state low flow situation (810 $m^3 s^{-1}$) and not for unsteady flow situations.

The Transport Model

The transport model describes the transport behavior of the three suspended sediment fractions in the main channel, in adjacent groyne fields and harbor areas. The model does not properly resolve the near-field mixing in the vicinity of the pipeline discharge because the resolution of the model in the vicinity would need to be considerably high. Similarly buoyancy effects and reducing mixing due to the highly concentrated suspensions discharged from the pipe were also not represented in the model. The simulation of the flow field has some uncertainties, which must be keep in mind for the interpretation of the transport model results. Flocculation processes where not taken into account, due to the fresh water conditions and the low concentrations in the far field of the dredged sediment.

Several simulations with different physical input parameters were done. An estimation of the sensitivity to the input parameters is thus possible.

Table 4.2. Physical parameters for the transport model variations

Parameter (Reference)	Unit	V 1	V 2	V 3	V 4	V 5	V 6
Fall velocity w_s (Zanke)	m s^{-1}	0.00009 (for 0– 20 µm) 0.00142 (for 20– 60 µm) 0.01390 (for 60–200 µm)					
Erosion constant m (Metha)	kg m^{-2} s^{-1}	0.00005					
Critical erosion shear stress $\tau_{crit,E}$	N m^{-2}	0.5	2	100	0.5	2	100
Critical deposition shear stress $\tau_{crit,S}$ (van Rijn, Westrich)	N m^{-2}	0.07	0.07	0.07	0.07	Eq. 4.31	Eq. 4.31
Calibration constant k_l (Elder)	–	6.0					
Calibration constant k_t (Elder)	–	0.6					
Partitioning coefficient k_d (Boguslavsky, WSA)	l kg^{-1}	0	0	0	13 000	0	0
Kinetic factor k_t (WSA)	1/s	0	0	0	1/86 400	0	0

The physical parameters were taken from literature values, all physical parameters used in each of the six simulations (V 1–V 6) are shown in Table 4.2. Simulation V 5 was done to evaluate the effect of sorption processes. The fractional distribution is described by the fall velocity and is calculated for the mean value of the diameter range (e.g., 10 µm for the 0–20 µm range).

In V 1 to V 3 the critical erosion shear stress is varied and was assumed to be the mean value between freshly deposited sediment and softly consolidated sediments in the river reservoirs (0.5 N m^{-2}) and normal consolidated sediments (2 N m^{-2}), or set to 100 N m^{-2} to simulate the maximum possible sedimentation rate. The critical deposition shear stress was assumed to be the mean value of the experimental data of van Rijn (1993), or taken from Eq. 4.32 in V 5 and V 6. The conservative assumption that no HCB sorption processes occur was chosen as a first approximation for the model variation V 1–V 3 and for V 5 + V 6. In the variation V 4 a mean value of k_d (13 000 l kg^{-1}) chosen from literature (Boguslavsky 2000 and WSA 2002) and a value of k_t (1/86 400 s^{-1}) chosen from literature (WSA 2002) were assumed. The dissolved HCB in the bottom has a minor influence on whole system, therefore diffusion at the sediment–water interface and sorption processes within deposited sediments were neglected.

Discussion of the Model Results

The transport rates in terms of kg s^{-1} at different cross-sections of V 1 were compared with the measured transport rates of the BfG (2005) and shown in Fig. 4.8. The concentration measurement were in the form of several turbidity measurements in a cross-

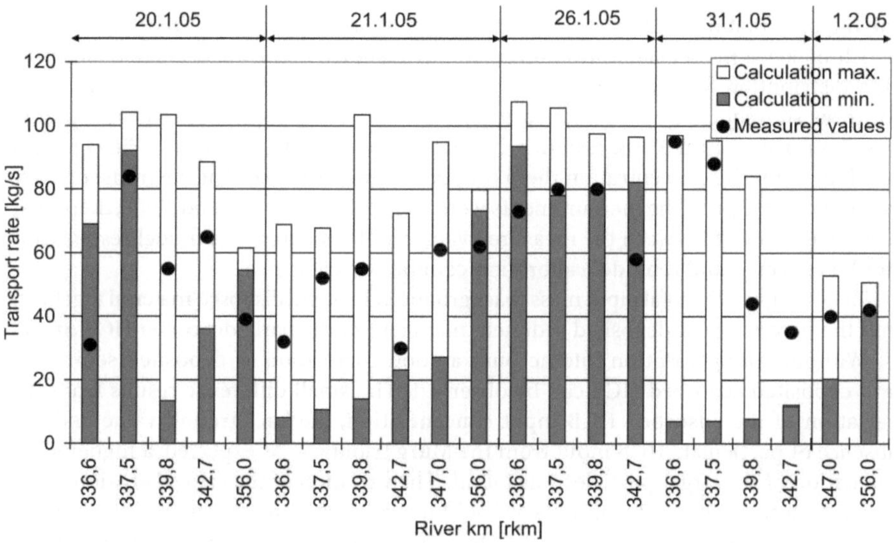

Fig. 4.8. Measured and calculated suspended sediment transport rates

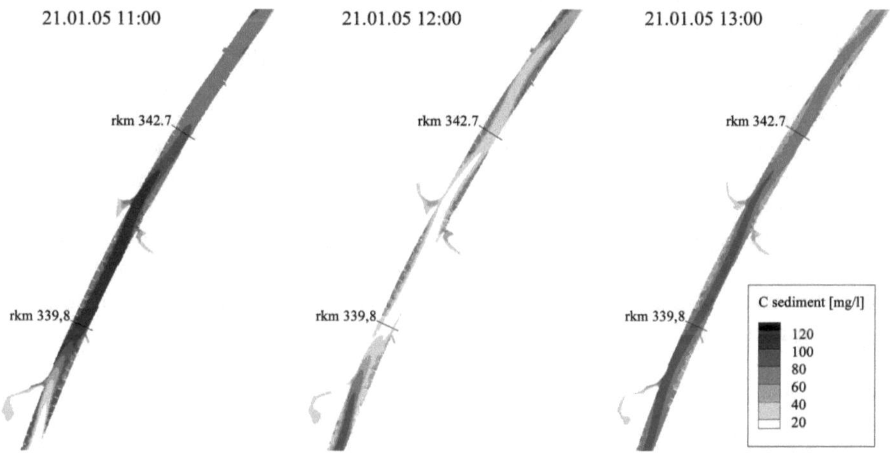

Fig. 4.9. Concentration fields

section. The turbidity sensor was calibrated at each cross-section with the suspended sediment concentrations determined from water samples. The transport rate is calculated by integration of the measured concentrations multiplied with the local flow velocity at the measurement point. During the survey the flow velocities and the concentrations substantially change in a given cross-section and therefore, the measurements do not represent an instantaneous concentration field and transport rate.

The calculated transport rates are instantaneous values and the minimum and maximum value in Fig. 4.8 represent the variation of the transport rate during the measurement period. The high variation of the calculated transport rates is caused by the high fluctuation of the dredged material input and intermittent dredging activity, which makes a model calibration difficult. For instance Fig. 4.9 shows the concentration distribution from model results between the cross-section rkm 339.8 and 342.7 for three subsequent hours on January 21, 2005. At 11:00 a highly concentrated plume passing through rkm 339.8, resulting in a transport rate of 103.5 kg s^{-1} (Fig. 4.8), but one hour later the concentration there dropped dramatically so that the transport rate was only 14.1 kg s^{-1}. The measurements were taken between 11:00 and 12:00, therefore, a good agreement between the measured values and instantaneous model results cannot be expected and a model calibration cannot be performed.

Table 4.3 gives the total input mass (background values and disposed material together) and the percentage of deposited sediment mass and particularly deposited HCB mass.

Without taking sorption into account a good correlation of deposited sediments and deposited adsorbed HCB can be observed. The small difference results from the variation of the adsorbed HCB input concentration, the background value and the absence of particulate HCB input from the Murg tributary. As expected, a higher sedimentation of the larger particles was found. The variation of the critical erosion shear stress between V 1–V 3 (especially V 3 without any erosion) shows this effect more clearly. For fraction 3 an increase of sedimentation between V 1 and V 3 from 13.4 to 32.73% can be seen. Sedimentation mainly occurs near to the harbor entrance and in the groyne fields. Although the numerical mesh size in the side arms on the flood plains is very course, some deposition is still predicted.

By assuming sorption in V 4 a significant decrease of the deposited adsorbed HCB can be predicted. In the main channel a gradual increase of dissolved HCB (background value is zero) from the pipeline outlet to the end of the model domain due to desorption is calculated. But with the assumed sorption kinetic factor $k_t = 1/86\,400$ (s^{-1}) the residence time of 3–4 h in the main channel is not sufficient to reach the equilibrium state. Furthermore, a gradual dispersive release of dissolved HCB from groyne fields and harbor areas into the main channel during dredging breaks can be predicted. Analogously suspended sediments in the groyne fields are partly released in the main channel and partly deposited in the groyne field during dredging breaks.

Table 4.3. Total mass input and percentage of deposited sediment and HCB mass

	Fraction 1 (0–20 µm)	Fraction 2 (20–60 µm)	Fraction3 (60–200 µm)	HCB on Fraction 1	HCB on Fraction 2	HCB on Fraction 3
Total mass input	70 237 t	25 456 t	18 127 t	10.406 kg	3.821 kg	2.885 kg
Sed. V 1	2.11%	7.23%	13.40%	2.24%	7.12%	13.14%
Sed. V 2	2.26%	8.50%	18.19%	2.40%	8.51%	18.33%
Sed. V 3	2.86%	12.95%	32.73%	3.03%	13.33%	33.62%
Sed. V 4	2.08%	7.05%	13.09%	1.41%	5.90%	11.34%
Sed. V 5	3.32%	19.71%	43.28%	3.39%	19.07%	42.07%
Sed. V 6[a]	7.83%	38.97%	73.57%	8.24%	40.50%	74.58%

Comparing V 5 with V 2, which are only different by using an energy approach for the critical deposition shear stress (Eq. 4.31), shows a strong increase of sediment deposition in the groyne fields. This approach leads also to a strong deposition of suspended sediments in the main channel at the outlet of the flushing pipeline, which was observed in the nature but not found in V 1–V 4. The energy approach simulates the descent of the plume from the pipe onto the bed better than a constant critical shear stress. In V 5 and V 6 which use an energy approach, sedimentation is the dominating factor for all three fractions, whereas in V 1–V 3 the choice of the critical erosion shear stress has a stronger influence for the larger particles.

4.2.5 Conclusions

The 2-D transport model has proven to be a useful tool for a supplementary investigation and sensitivity analysis on the transport and sedimentation behavior of a contaminated suspended sediment plume in a navigational channel with groyne fields. The numerical study provides a deeper insight into the multifractional suspended sediment transport and physico-chemical processes as related to remobilization of adsorbed contaminants. Furthermore, it allows a calculation of the spatial and temporal contaminant distribution in the river system which could not be covered by the field data.

The present knowledge in modeling different sediment fractions is relatively poor, especially with respect to the erosion of separate fractions and need further research work. Nevertheless, the physical based numerical model is an engineering approach which enables a description of the most relevant physical processes, including sorption and degradation. However, more detailed and frequent field measurements are necessary to minimize the uncertainties of the physical and chemical parameters, especially the suspended particle settling velocity, the suspended sediment concentration and the pollutant specific sorption parameters. High spatially and temporally resolved data would also allow detailed calibration and validation of numerical models for predicting fine sediment transport in rivers.

Acknowledgments

We would like to thank the German Waterways and Shipping Office Freiburg (WSA) the German Federal Institute of Hydrology (BfG) and the German Federal Waterways Engineering and Research Institute (BAW) for their helpful cooperation. The investigation presented here was part of the project SEDYMO (*Se*diment *Dy*namic and *Mo*bility), funded by the German Federal Ministry of Education and Research (BMBF).

References

BfG (2005) Ergebnisse aus dem begleitenden Untersuchungsprogramm für die Umlagerung von Baggergut in die fließende Welle unterhalb der Staustufe Iffezheim/Rhein. BfG report 1474, Koblenz, in German

Boguslavsky S (2000) Orcanigc Sorption and Cation Exchange Capacity of Blacial Sand, Long Island. State University of New York, online published: http://pbisotopes.ess.sunysb.edu/reports/boguslavsky/

Dreher T(2005) Selektive Sedimentation von Feinstschwebstoffen in Wechselwirkung mit wandnahen turbulenten Strömungsbedingungen. Online published: http://elib.uni-stuttgart.de/opus/volltexte/2005/2263/; urn:nbn:de:bsz:93-opus-22633, in German

Elder JW (1959) The dispersion of marked fluid in turbulent shear flow. Journal of Fluid Mechanics 5(4):544–580

Electricite de France (EDF) (2004) TELEMAC modelling system. Distributed by SOGREAH consultants

Gualtieri C (2004) Interaction Between Hydrodynamics and Mass-Transfer at the Sediment-Water Interface. iEMSs: Manno, Switzerland, 2004. ISBN 88-900787-1-5

Krone RB (1962) Flume studies of the transport of sediment in estuarial shoaling processes. Hydr. Eng. Lab. and Sanit. Eng. Res. Lab., Univ. of California, Berkeley

Partheniades E (1965) Erosion and deposition of cohesive soils. Amer. Soc. Civ. Eng., J. Hydraulics Division, HY 1:105–139

Kuijper C, Cornelisse JM, Winterwerp JC (1989) Research on erosive properties of cohesive sediments. J. Geophysical Research 94(C10):14341–14350

Metha, AJ (1988) Laboratory Studies on Cohesive Sediment Deposition and Erosion. In: van Leussen W (ed) Physical Processes in Estuaries. Springer-Verlag Berlin Heidelberg New York, pp 427–445

Stumm W, Morgan JJ (1996) Aquatic Chemistry. Wiley-InterScience, New York Chichester Brisbane Toronto Singapore

van Rijn L (1993) Principles of sediment transport in rivers, estuaries and coastal seas. Aqua Publications Amsterdam Oldemarkt

Westrich B, Juraschek M (1985) Flow transport capacity for suspended sediment. International association for hydraulic research 21st Congress, Melbourne, vol. 3, pp 590–594

Westrich B (1988) Fluvialer Feststofftransport – Auswirkung auf die Morphologie und Bedeutung für die Gewässergüte. Oldenburg Verlag, München Wien, in German

WSA Freiburg (2002) Schlußbericht der Arbeitsgruppe "Baggerungen". Distributed by WSA Freiburg, in German

Zanke U (1982) Grundlagen der Sedimentbewegung. Springer-Verlag Berlin Heidelberg New York, in German

Dirk Ditschke · Mark Markofsky

4.3 A Non-Equilibrium, Multi-Class Flocculation Model

4.3.1 Introduction

Cohesive sediments build flocs which have settling velocities which are some orders of magnitude higher than those of the single particles. The floc size and thus the settling velocity depend on the turbulence intensity and the sediment concentration and properties. Generally it is assumed that a specific floc size is associated with a given flow regime or that at least the variance in the floc size distribution is small. In most cases only one sediment-class is used in models of suspended sediment transport in river or estuarine flow. In-situ and laboratory measurements show that this assumption may not be correct (Bornhold 1992). The floc size can vary an order of magnitude above or below the mean diameter. Therefore, it can be important to consider different sediment classes in order to refine and improve both deposition and suspended sediment concentration predictions. A special case where the consideration of different sediment classes is essential is the transport of polluted sediments since many pollutants are transported with the finest fractions.

The challenge of modeling multiple fractions of suspended sediment is that the flocs change their size due to flocculation and break-up. It is not sufficient to divide sediment into several fractions and compute the transport of each fraction. For flocculation and break-up processes it is necessary to consider an interaction between these fractions.

Although the flocculation time can play an important role in the prediction of sediment concentrations and deposition, the floc-size may be limited by the settling time (Winterwerp 1999). But also if the flow conditions change quickly, for example if sediment laden water flows from the rapidly flowing river into a harbor, the time which is needed to build up a floc-size equilibrium may be important for the deposition pattern in the harbor.

4.3.2 The Fractionated Flocculation Model

Interaction between Fractions

To meet this challenge the numerical model TELEMAC-3D (Hervouet 1991) was extended to treat different sediment classes. For each sediment class the three-dimensional transport equation is solved. Deposition and erosion are simulated for each class and the total bed evolution computed. Each sediment class is described by a settling velocity, w_s and a mean-particle diameter D. If only one parameter is available, the other is computed using Eq. 4.36 after Winterwerp (1998):

$$w_s = \frac{\alpha}{18\beta}\frac{(\rho_s - \rho_w)g}{\mu}D_p^{3-n_f}\frac{D^{n_f - 1}}{1 + 0.15\mathrm{Re}^{0.687}} \tag{4.36}$$

The particles are assumed to be spherical ($\alpha = \beta = 1$), with a fractal dimension of $n_f = 2$. The primary particle diameter D_p is derived from measurements as the D_{50} of the non-flocculated sediment.

This equation is designed for cohesive sediments but works also for non-cohesive sediments by assuming spherical particles ($\alpha = \beta = 1$), with a fractal dimension $n_f = 3$ and a Stokes-regime ($Re \ll 1$). The settling velocity is the main parameter for the transport computation. The mean-diameter D of the sediment fraction is used for the exchange with the bed-morphology model SediMorph (Malcherek 2005).

With a multi-class sediment model it is possible to simulate a non-cohesive suspended load with a wide-spread size distribution or to follow specific sediment fractions. When combined with the bed-morphology model, the deposition of polluted sediment fractions can be simulated.

The second step is the simulation of cohesive sediment behavior through an interaction between the sediment classes. The flocculation process is simulated as an exchange from one sediment class into another with a higher settling velocity.

It is assumed that larger flocs always have a higher settling velocity and within each class all particles have the same diameter and settling velocity. In the other direction, break-up is defined as a transfer from one sediment class into another with a lower settling velocity and smaller diameter.

Of the different flocculation and break-up mechanisms in estuarine waters, turbulence induced flocculation and break-up is dominant (Winterwerp 1999). After a con-

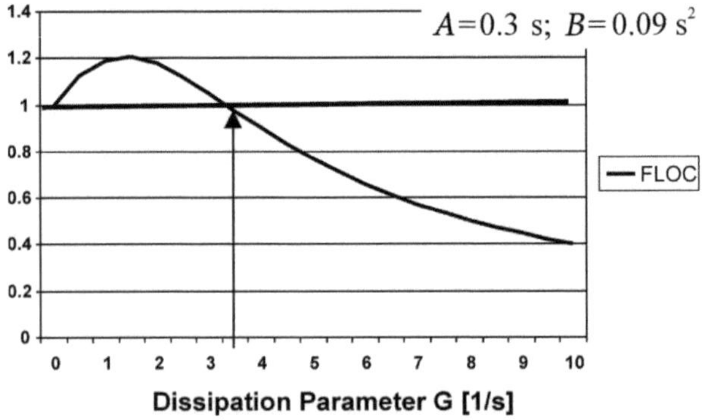

Fig. 4.10. Variation of the FLOC-parameter for increasing values of the dissipation parameter *G*

cept of Dyer (1989) flocculation increases with increasing turbulence due to the higher collision probability of flocs. With increasing turbulence, the shear stresses in the water rise and more and more flocs are destroyed. After a given turbulence intensity, break-up dominates flocculation. Based on the work of Argaman and Kaufman (1970), Van Leussen (1994) formulated this concept using Eq. 4.37 for the settling velocity w_s of cohesive sediments:

$$w_s = w_0 \frac{1 + AG}{1 + BG^2} \tag{4.37}$$

The dissipation parameter $G\,(\mathrm{s}^{-1}) = (\varepsilon/\nu)^{1/2}$ is used to represent the turbulence intensity and can be computed with a k-ε-model. The parameters A and B are empirical values for flocculation and breakup; w_0 = settling velocity of a particular sediment class in still water.

The parameter FLOC (Eq. 4.38) is introduced to determine whether flocculation or break-up is dominant at a given turbulence intensity. At low G, FLOC becomes greater than 1 which leads to an increase in the settling velocity and indicates flocculation. At higher G FLOC becomes less than 1. The settling velocity decreases as break-up dominates.

$$\frac{1 + AG}{1 + BG^2} \equiv \mathrm{FLOC} \rightarrow \mathrm{FLOC} = \begin{cases} > 1 \rightarrow \text{Flocculation} \\ < 1 \rightarrow \text{Breakup} \end{cases} \tag{4.38}$$

Figure 4.10 shows the variation of the FLOC-parameter for increasing G. The values for the empirical parameter $A = 0.3$ s and $B = 0.09$ s^2 were determined through a calibration of the Weser Estuary (Malcherek et al. 1995) and do not necessarily apply to the general case.

Modeling of the Time Dependency

The flocculation process of cohesive sediment in nature at low concentrations is quite slow and can last several hours (Lick et al. 1992). The flocculation-time depends on the collision probability and the rate of collisions leading to growth of sediment flocs. Based on the coagulation-theory (Smoluchofski 1917) the diameter of the floc has no major influence on the speed of the flocculation process. The collision probability is mainly dependent on the turbulence intensity and the sediment concentration, while the probability that particles stick together depends primarily on the sediment physical and biological properties. These properties are lumped in the concept of "stickiness" in which the Extracellular Polymer Substance (EPS) is supposed to have the greatest influence for mud flocs (De Brouwer et al. 2002; Fengler et al. 2004). Unfortunately, a coherent relationship between the EPS-concentration and the stickiness of the flocs does not yet exist.

Break-up of flocs is generally much faster than flocculation. The primary factors are the forces generated by turbulent shear and the resistance of the flocs against these forces. The resistance is also associated with "stickiness". Contrary to the flocculation process, the resistance against the external forces is dependent on the floc diameter.

The time-dependency is realized in the numerical model by transferring in each time step only a small portion of a sediment fraction into the next larger class. An effectiveness-factor for flocculation $\varepsilon_{\text{floc}}$ and one for break-up $\varepsilon_{\text{break}}$ are introduced. For a sediment fraction i which has a smaller fraction j and a larger fraction k, the change of the concentration due to flocculation is computed as follows:

$$\frac{\partial c_i}{\partial t} = \left(\varepsilon_{\text{floc},j}\, c_j - \varepsilon_{\text{floc},i}\, c_i\right) / \partial t \tag{4.39}$$

The formulation for break-up is similar:

$$\frac{\partial c_i}{\partial t} = \left(\varepsilon_{\text{break},k}\, c_k - \varepsilon_{\text{break},i}\, c_i\right) / \partial t \tag{4.40}$$

Within each time step the change of a concentration c_i in the case of flocculation is computed as the transfer of the smaller fraction j into the next larger fraction k. If break-up is dominant, (FLOC < 1) the change comes from the break-up of the larger fraction k into the smaller class j. As the sediment is only shifted from one class into another the conservation of mass is ensured.

The effectiveness-factor ε parameterizes the factors influencing flocculation or break-up. For flocculation $\varepsilon_{\text{floc}}$ includes the effects of the turbulence intensity, the concentration and stickiness. The turbulence computed as G and the concentration c have theoretically a linear influence on the flocculation velocity (Smoluchofski 1917). As Smoluchofski does not consider break-up this assumption has to be modified. Even if flocculation is dominant, break-up also takes place. The effect of break-up is considered in the FLOC-parameter, which is used to represent the influence of turbulence in the fractionated flocculation model. Flocculation and break-up are equal when FLOC $= 1$. The stickiness of the flocs is parameterized with K_1 and is used to calibrate the model. Thus the flocculation effectiveness, $\varepsilon_{\text{floc}}$ can be given by:

$$\varepsilon_{\text{floc}} = K_1(\text{FLOC} - 1)c \tag{4.41}$$

The break-up effectiveness, $\varepsilon_{\text{break}}$ includes the turbulence intensity, stickiness, the floc diameter and a calibration parameter K_2:

$$\varepsilon_{\text{break}} = K_2\left(\frac{1}{\text{FLOC}} - 1\right)D \tag{4.42}$$

In nature the floc-size is limited by the Kolmogorov-length (Van Leussen 1997). The minimum size distribution is that of single particles. The Kolmogorov-length λ_0 can be computed with a k-ε-model from $\lambda_0 = (\nu^3/\varepsilon)^{1/4}$. If this length becomes smaller than the mean diameter of a sediment fraction, flocculation of smaller fractions is less than the break-up.

The determination of a lower limit for the floc size is more complicated. Due to the concept of settling velocity classes in the model there is no distinction between a single particle which cannot be broken by turbulent shear stresses and a small floc with the same settling velocity. It is not possible to define a minimum concentration for each fraction from an analysis of the single particle size distribution. For example, the concentration of the smallest fraction can decrease to zero because all small particles are bound in flocs and thus associated with a larger fraction. For this reason, the characteristic parameters of a size distribution D_{10}, D_{50} and D_{90} are used to ensure a proper minimum size distribution. The D_{10} defines the diameter which is larger than that of 10% of the particles, D_{50} and D_{90} are defined accordingly.

This is realized in the model as follows: if the actual D_{90} is smaller or equal to the minimum D_{90}, then break-up of flocs with a diameter larger than the minimum D_{90} is stopped. The same happens for flocs with a diameter between D_{50} and D_{90} at the D_{50}-border, and at the D_{10}-border for flocs between D_{10} and D_{50}.

Comparison with the Coagulation Theory

A first estimation of the quality of the flocculation model is found by comparing the numerical results with the theoretical values of the coagulation theory (Smoluchofski 1917). Therefore, a short overview of the theoretical solution of the coagulation theory is given here. Under the assumption that flocculation takes place between two flocculation classes, the variation of a floc class k over time can be written as:

$$\left[\frac{\partial c_k}{\partial t}\right]_{\text{FLOC}} = \frac{1}{2}\sum_{i+j=1} \beta_{ij}\bar{A}_{ij}c_i c_j - c_k\sum_{i=1}^{\infty}\beta_{ik}\bar{A}_{ik}c_i \tag{4.43}$$

The first term on the right hand side of Eq. 4.43 is floc growth due to the flocculation of flocs belonging to two smaller floc-classes i and j; the second term describes the reduction of class k due to flocculation into a larger fraction. The function β is the probability that the particles i and j collide and $\bar{A}$ the probability that a new floc is formed due to the collision.

It is possible to convert this general flocculation equation into Eq. 4.44 for the time-dependent concentration c_t of a floc-class with the concentration c_0 at $t = 0$ (Malcherek 1995):

$$c_t = \frac{c_0}{\left(1 + \left(\frac{\beta\bar{A}}{2}\right)\frac{6c_0}{\pi D^3\rho_s}t\right)} \tag{4.44}$$

Fig. 4.11.
Schematic representation of the rectilinear (*left*) and curvilinear models (*right*) (after Han and Lawler 1992)

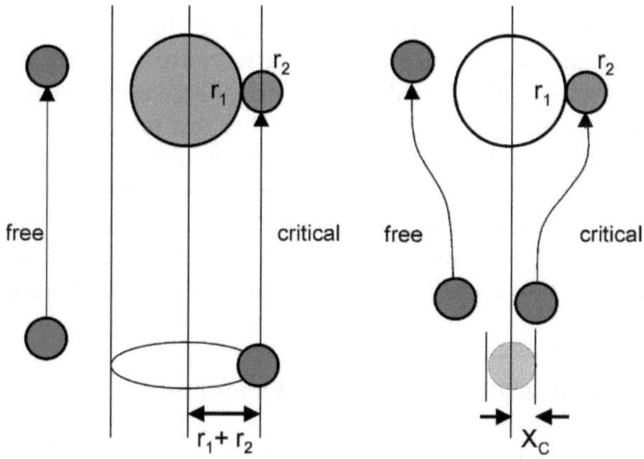

Fig. 4.12.
Comparison of the fractionated flocculation model (—) with the coagulation theory: rectilinear model (□); curvilinear model (◀)

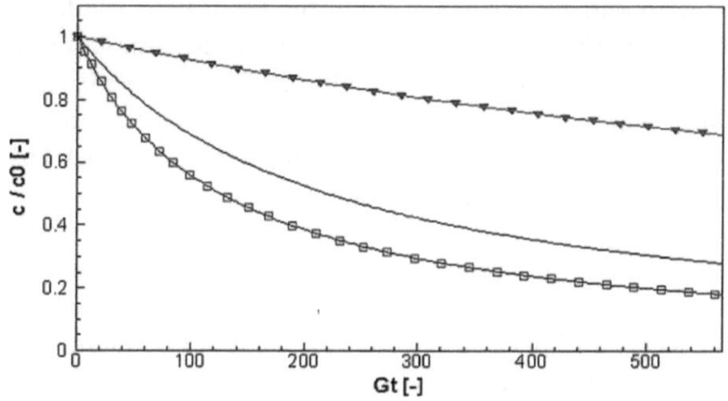

For the collision probability β, different formulations have to be set depending on the collision mechanism. As we neglect the influence of Brownian motion and differential settling in estuaries, only the collision probability due to turbulence is considered:

$$\beta_{\text{Turb}} = \frac{G}{6}\left(D_i + D_j\right)^3 \tag{4.45}$$

It should be noticed that Eq. 4.45 overestimates the flocculation intensity and the settling velocity, because near-field effects are neglected. Laboratory experiments have shown that in the case of differential settling, the trajectories of small particles are deflected by larger particles. This leads to a major decrease of the collision probability (Stolzenbach and Elimelich 1994). Han and Lawler (1992) argued that this near-field effect is present even if two particles come close due to turbulent fluctuations. They proposed a "curvilinear model" with a correction-factor e_{cor} (Eq. 4.46) as an extension to the "rectilinear model" presented above:

$$e_{\text{cor}} = \exp\left[-3.4 + 0.62\log(\gamma) + \psi(3.5 - 1.2\log(\gamma))\right] \tag{4.46}$$

with $\psi = D_j/D_i$, $\gamma = 8H_A/(3\pi w_s D_i^2)$ and H_A is the Hamaker-constant, typically given as 3.9×10^{-20} for particles in water. Figure 4.11 demonstrates the different approaches for the rectilinear and the curvilinear models.

Additional investigations have shown that the rectilinear model overestimates the flocculation rate since near-field effects are neglected, but the curvilinear model underestimates the flocculation intensity because the permeability of the flocs is not taken into account (Li and Logan 1997). For this reason the results of the fractionated flocculation simulation introduced here should lie in between these two models.

Figure 4.12 shows a comparison between the fractionated flocculation model and both the rectilinear and the curvilinear coagulation theory models. For clarity, only the decrease in concentration of a floc class due to flocculation into a larger class is shown. The y-axis is normalized using the concentration at $t = 0$.

4.3.3 Analysis of Laboratory Experiments

The fractionated flocculation model was used to analyze laboratory experiments. The experiments were carried out at the Institute of Hydromechanics of the University of Karlsruhe (Kühn and Jirka 2006) within the "SEDYMO" research group (Fine *Se*diment *Dy*namics and Pollutants *Mo*bility in Rivers) (Förstner 2004). With a differential settling column (Fig. 4.13) it is possible to generate a controlled turbulence field similar to that of a natural river without having any advective transport.

The size distribution of kaolin flocs was measured with an inline microscope. By generating different turbulence profiles a regime with mainly flocculation was established as a flocculation test. Afterwards the turbulence was increased so that break-up was dominant in the whole water column. For a detailed description of the experiment see Kühn and Jirka (this volume). It should be mentioned, that the laboratory experiments where conducted at low concentrations, with low settling velocities when compared with natural flocs.

The set up of the numerical model differs from the laboratory test. Instead of a small tank without advection, a straight open channel with a constant velocity profile is used. The channel is 500 m long and 100 m wide to avoid boundary effects. Cyclic boundary conditions were prescribed in order to have the sediment concentration

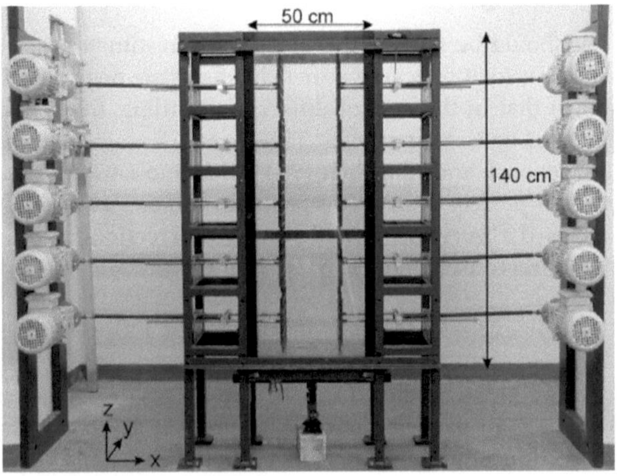

Fig. 4.13.
A differential settling column (Kühn and Jirka 2006)

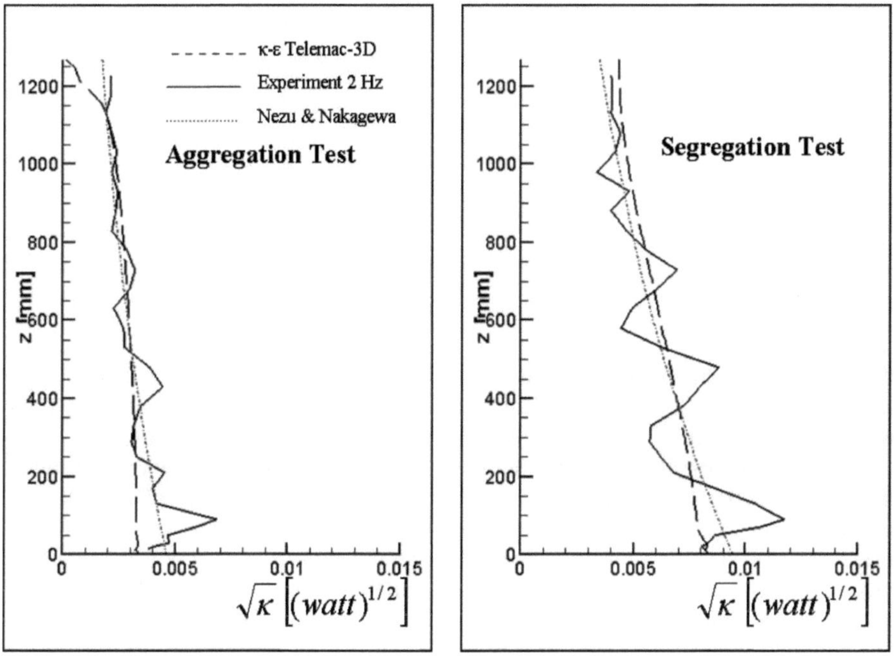

Fig. 4.14. Turbulence profiles of the numerical model, the experiment and theoretical values of Nezu and Nakagawa 1993 (*left:* aggregation; *right:* segregation)

independent of the advection. The same sediment mass flow which leaves the domain is added to the inflow as a boundary condition. The flow velocity is calibrated to meet the turbulence profile in the laboratory. Figure 4.14 shows the turbulence profiles calculated with the numerical model as well as the experimental data and the theoretical values of Nezu and Nakagawa (1993). In both the aggregation and the segregation test cases the flow velocities are quite low. In the numerical model, mean velocities of 5 cm s^{-1} for the flocculation case and 10 cm s^{-1} for the break-up case was chosen.

Aggregation

The laboratory experiments show that in the lower 40% of the water column break-up exceeds flocculation; at about 70% maximum flocculation is observed. With the numerical model the break-up is dominant for $G = 0.17\ s^{-1}$; the maximum flocculation at $G = 0.09\ s^{-1}$. The best fit of the flocculation parameters yielded $A = 3$ s and $B = 17\ s^2$. This shows that Kaolin-flocs are very unstable. For example, the boundary between flocculation and break-up in the Weser-Estuary was found to be at $G = 3.5\ s^{-1}$. Other laboratory flocculation experiments with artificial flocs use values for G of several hundred s^{-1} (Lick et al. 1992; Flesch et al. 1999).

An increase in the floc size was observed in the experiments after 4 hours. The floc-size distribution and the mean floc diameter were measured. In the numerical model,

Table 4.4. Discretisation of floc fractions

D (µm)	%	C (mg l^{-1})	w_s (mm s^{-1})
5	2	0.010	1.80×10^{-2}
10	9	0.035	3.60×10^{-2}
15	32	0.115	5.39×10^{-2}
20	60	0.140	7.19×10^{-2}
30	90	0.150	1.08×10^{-1}
40	100	0.050	1.44×10^{-1}
60	100	0.010	2.16×10^{-1}

the floc-size distribution at the beginning of the experiment was discretized into 7 floc-size fractions (Table 4.4). The associated concentrations and settling velocities (Eq. 4.36) were computed with $\alpha = \beta = 1$, $n_f = 2$, $\gamma_s = 2.650$ kg m^{-3} and $D_p = 18.2$ µm.

The floc-size distribution after 4 hours in the model was compared with the experimental data at the location of maximum flocculation, i.e. at $z/h = 0.68$. The stickiness parameter for flocculation K_1 was adjusted to match the numerical results with the measured floc-size distribution (Fig. 4.15 and Table 4.5). Best results were achieved using a floc-size dependent stickiness. This seemed to indicate that smaller flocs have in this case a higher floc-building probability (see "Evaluation").

$$K_1 = \frac{0.7 \cdot 10^{-8}}{D^{1.6}} \tag{4.47}$$

Segregation

G is greater than 0.17 s^{-1} in the entire water column for the segregation test case, so break-up is dominant in the whole domain. In the experiment a steady state floc-size distribution was reached after 5 minutes. Since this distribution did not change after the turbulence intensity was increased, it can be concluded that all the flocs were destroyed. The numerical and measured floc-size distributions are in good agreement (Fig. 4.16 and Table 4.6). For break-up, the stickiness-parameter can be described by Eq. 4.48:

$$K_2 = 500 D^{1.25} \tag{4.48}$$

Evaluation

The numerical model can be used to analyze the results of the laboratory experiments. A relationship between the floc-size and the effectiveness of the flocculation and break-up processes was found.

The somewhat unexpected experimental result, that smaller flocs had a higher building capacity than larger ones is attributed to the nearly total lack of biological influences and the low concentrations and small particle and floc diameters in the

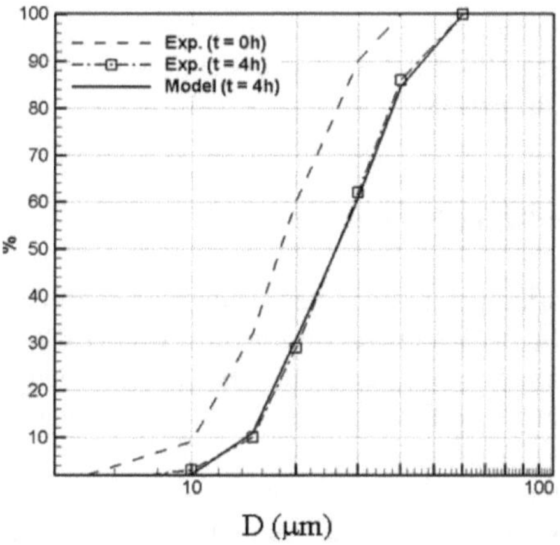

Fig. 4.15. The floc-size distribution after 4 hours in the flocculation experiment and in the numerical model

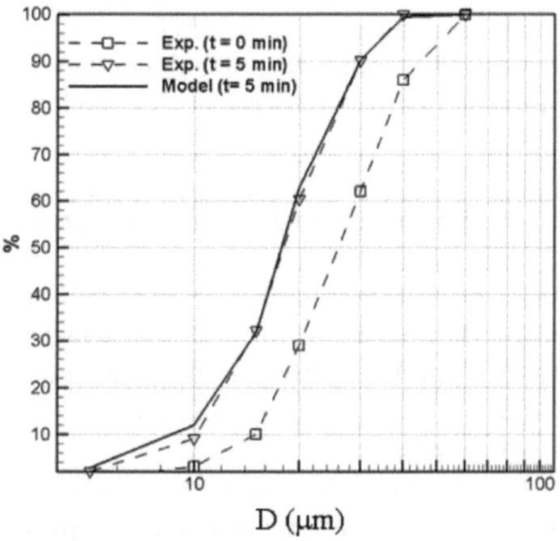

Fig. 4.16. The floc-size distribution after 5 minutes in the break-up experiment and in the numerical model

Table 4.5. Comparison of numerical and experimental results for flocculation

D (μm)	Size dist. (%) Exp. $t = 0$	Size dist. (%) Exp. $t = 4$ h	Size dist. (%) Model $t = 4$ h
5	2.0%	0.5%	0.1%
10	9.0%	3.0%	2.1%
15	32.2%	10.0%	10.9%
20	60.3%	29.0%	30.8%
30	90.1%	62.0%	60.8%
40	100.0%	86.0%	84.5%
60	100.0%	100.0%	100.0%
D_{50}	18.2 μm	26.4 μm	26.4 μm

Table 4.6. Comparison of numerical and experimental results for break-up

D (μm)	Size dist. (%) Exp. $t = 0$	Size dist. (%) Exp. $t = 5$ min	Size dist. (%) Model $t = 5$ min
5	0.1%	2.0%	2.7%
10	2.1%	9.0%	12.0%
15	10.9%	32.2%	31.8%
20	30.8%	60.3%	62.9%
30	60.8%	90.1%	90.3%
40	84.5%	100.0%	99.5%
60	100.0%	100.0%	100.0%
D_{50}	26.4 μm	18.2 μm	17.9 μm

experiments. The main influence for the stickiness is then not the biology but electrochemical forces, which are more effective for smaller particles.

Nevertheless, further investigations are necessary due to given uncertainties. Uncertainties can originate from the measuring technique since single flocs which appear in a small frame are counted and the probability of catching one of the bigger flocs is quite low. As such, the largest fractions may be underestimated. The choice of the discretization of the fractions in the flocculation model may influence the empirical parameters.

The flocculation model does not simulate a pure flocculation process. Using Eq. 4.38 for the FLOC-parameter always computes a kind of balance between flocculation and break-up. Flocculation is therefore always dependent on break-up to a certain degree. It is possible that a floc-size dependency of the flocculation process can be deduced from the floc-size dependency of the break-up process.

4.3.4 Application to Sedimentation in an Estuarine Harbor

As an example for the simulation of sedimentation in an estuarine harbor, the Koehlfleet in the Hamburg Harbor area was selected. In 1990, a current deflection wall (CDW) was built at the harbor mouth to reduce sedimentation. The Hamburg Port Authority claims an average annual reduction of the harbor sedimentation of about 35% over a 6-year period.

In order to calculate the effect of such a wall in other harbor basins, a research project was started in 2001 to simulate this process with a numerical model (Ditschke and Markofsky 2003). The results showed a good representation of the flow fields with and without the CDW and of the water exchange between the estuary and the harbor; but the comparison between the computed and measured sedimentation showed a need for improvement. Therefore, the flocculation model described above was applied to the harbor model.

The data base in this area is quite good for flow and sediment. But floc size distributions were available for only one day (Bornhold 1992). From these measurements and the known average conditions for sediment concentrations and properties, initial values for the computation were extrapolated. The measured floc size distribution is simulated with 6 fractions. The largest fraction is set at the value of the Kolmogorov length computed with the turbulence model. The tabulated concentrations are average concentrations for flood and ebb tide (Table 4.7). The settling velocity for each fraction is computed from Eq. 4.36.

Initial numerical tests with a straight channel were conducted to roughly calibrate the parameters of the multi-class flocculation model. Reasonable results were found for:

$$\frac{1+0.5G}{1+0.07G^2} = \text{FLOC} \tag{4.49}$$

Table 4.7.
The settling velocity for each fraction after Winterwerp (1999)

D (µm)	w_s (mm s^{-1})	Conc. Ebb (mg l^{-1})	Conc. Flood (mg l^{-1})
5	0.08	7	10.5
15	0.24	8.2	12.3
50	0.80	8.2	12.3
150	2.27	16.4	24.5
180	3.55	52.6	78.8
500	6.10	12.9	19.3
1 000	9.68	0.01	0.01

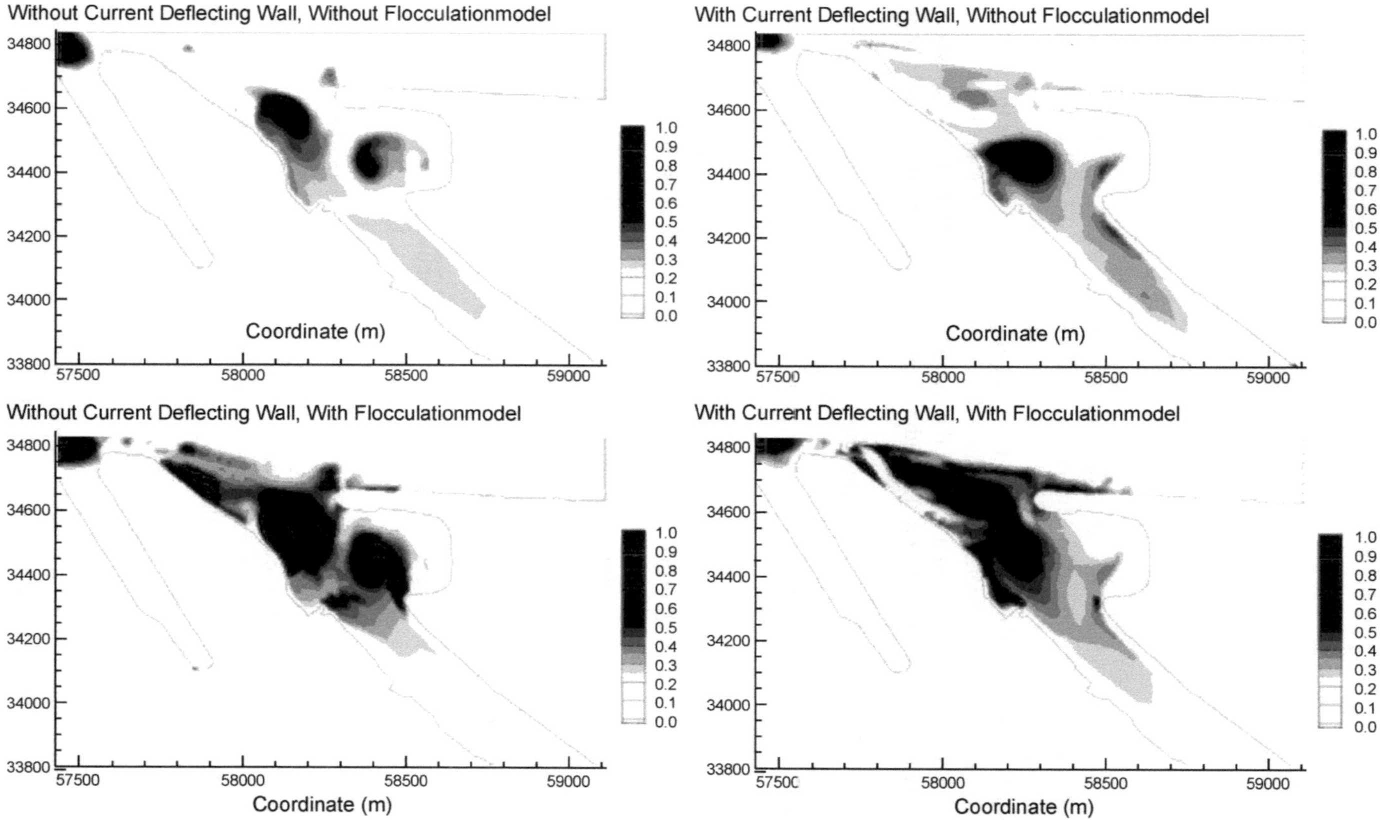

Fig. 4.17. Normalized deposition patterns without (*above*) and with (*below*) a multi-class flocculation model and without (*left*) and with (*right*) a current deflection wall (CDW)

Table 4.8.
Deposition (m^3) with and without a multi-class flocculation model and with and without a current deflection wall (CDW)

	1-Class flocculation	Multi-Class flocculation
Without CDW	1 690	2 710
With CDW	1 840	2 300
Difference	150	–410
%	+9%	–15%

The values for $A = 0.5$ and $B = 0.07$ are considered to be realistic because they are much closer to the values found by Malcherek et al. (1995) ($A = 0.3$ S; $B = 0.09$ S^2, see Fig. 4.10) than the values for the artificial sediment mentioned above.

The sedimentation during a single tide was computed with and without the multi-class flocculation model and with and without the current deflection wall. The results with this model show a reduction of 24% in the calculated deposition (Table 4.8) when compared with the 1-class simulation. The sedimentation patterns are also significantly different (Fig. 4.17).

The deposition is much higher in the computation with segmented flocculation model because the fraction with the largest settling velocity is not explicitly represented in a 1-class simulation but rather created by the flocculation model in areas with low turbulence (Kolmogorov length > 1 mm) i.e. within the harbor. Naturally a high settling velocity leads to increased sedimentation.

The calculated 15% reduction in sedimentation with the CDW during a single tide is of the same order of magnitude as the mean 6 year annual decrease of 35% stated by the authorities. Although one can not expect to hit the long term annual average with the computation of one single tide, this result indicates that the reproduction of the sedimentation processes in the harbor entrance is much better when one uses a segmented flocculation model.

To further understand the reasons for the improvement it is worth taking a closer look at the turbulence distribution near the harbor bottom and at the harbor entrance.

Figure 4.18 shows the dissipation parameter G at 5 cm above the bottom for both runs with and without a deflecting wall. At places where G is larger than (approximately) 7 Hz, break-up dominates flocculation. But even more important in this context is the Kolmogorov length (λ_0). If $G > 1$ Hz, λ_0 is <1 mm, and the largest flocs break. If $G > 4$ Hz, 0.5 mm flocs will also break. With a deflection wall, these values were exceeded over extensive bottom reaches and also in the middle of the water column. Values of about 3 Hz can be seen in the deflection channel and values for $G > 7$ in the vortex region of the wall.

4.3.5 Conclusions

At present three-dimensional numerical models generally use either one suspended sediment class with an advanced settling velocity formulation or use a multi-class

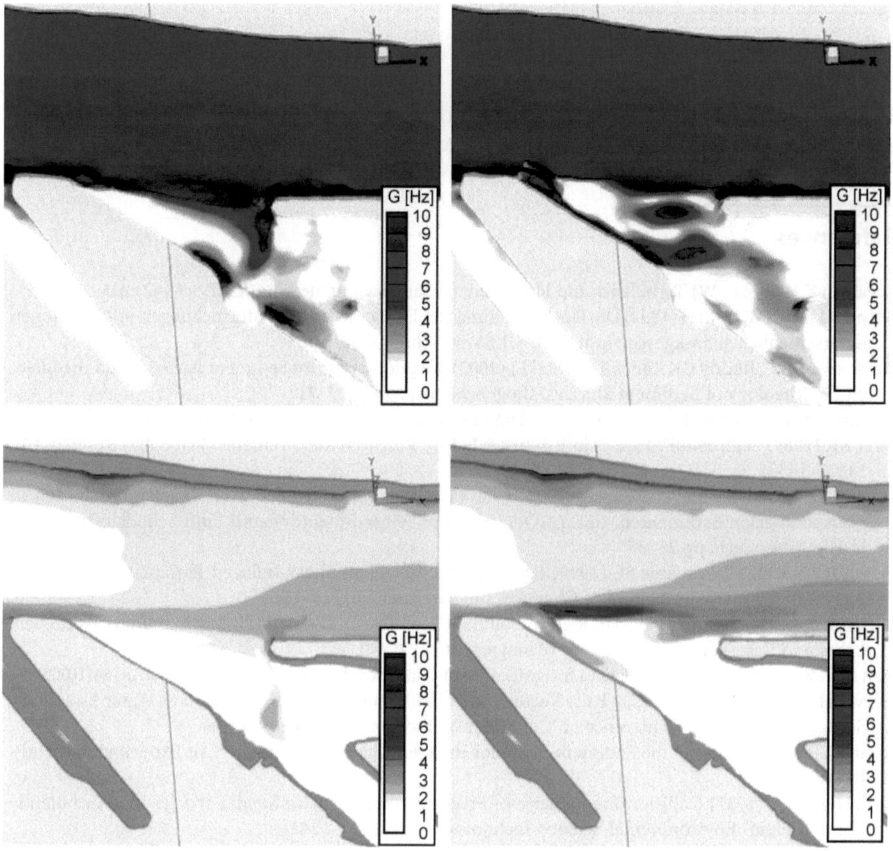

Fig. 4.18. The dissipation parameter G 5 cm above the bottom (*above*) and in 8 m water depth (*below*) without (*left*) and with (*right*) a CDW

approach for the different fractions of suspended sediment. The restriction in such models is that they either do not have the capability of dealing with a wide spread floc- or particle-size distribution or they cannot react to changing hydrodynamic conditions which generate a strong change in the floc-size distribution.

In order to refine and to improve suspended sediment concentration and deposition predictions, the numerical model Telemac-3D was extended to treat multiple classes of suspended sediment. The model presented here simulates the particle size distribution. For example, it is now possible to be consistent with a fractionated bed morphology model and to consider flocculation processes by calculating the generation of larger floc classes out of the primary particles. The time-dependence of the flocculation and breakup processes was implemented using an empirical effectiveness-parameter.

The calculations were found to be in good agreement with theoretical values and laboratory data and significantly improved predictions for harbor deposition.

Acknowledgments

The investigations presented here were part of the project SEDYMO (*Se*diment *Dy*namic and *Mo*bility) funded by the German Federal Ministry of Education and Research (BMBF).

References

Argaman Y, Kaufman WJ Turbulence and Floculationt. J Sanitary Engineering ASCE 96(SA2):223–241
Bornhold J, Puls W, Kühl H (1992) Die Flockenbildung von Elbeschwebstoff: Untersuchungen mit Fraktionen unterschiedlicher Sinkgeschwindigkeit. GKSS-Forschungszentrum
De Brouwer JFC, Ruddy GK, Jones TER, Stal LJ (2002) Sorption of EPS to Sediment Particles and the Effect on the Rheology of Sediment Slurries. Biogeochemistry(61), 57–71
Ditschke D, Markofsky M (2006 in Print) A Time Dependent Flocculation Model Intercoh 2005
Dyer KR (1989) Sediment Processes in Estuaries: Future Research Requirements. J Geophys Res 94(C10): 14327–14332
Fengler G, Köster M, Meyer-Reil LA (2004) Charakterisierung mikrobieller Lebensgemeinschaften in resuspendierten Sedimenten. SEDYMO-Workshop: Feinsedimentdynamik und Schadstoffmobilität in Fließgewässern, pp 25–26
Flesch JC, Spicer PT, Pratsinis SE (1999) Laminar and Turbulent Shear-Induced Flocculation of Fractal Aggregates. American Insitute of Chemical Engineers 45(5):1114–1124
Förstner U (2004) Sediment Dynamics and Pollutant Mobility in Rivers: An Interdisciplinary Approach. Lakes and Reservoirs: Research and Management (9):25–40
Han M, Lawler DF (1992) The (Relative) Insignificance of G in Flocculation. J Am Water Works Ass 84(10):79–91
Hervouet JM (1991) TELEMAC, a Fully Vectorized Finite Element Software for Shallow Water Equations. Second International Conference on Computer Methods and Water Resources
Kühn G, Jirka GH (2006) Fine Sediment Behavior in Open Channel Turbulence: an Experimental Study. Intercoh 2005
Li X, Logan BE (1997) Collision Frequencies of Fractal Aggregates with Small Particles in a Turbulently Sheared Fluid. Environmental Science Technology 31(4):1237–1242
Lick W, Lick J, Ziegler K (1992) Flocculation and its effect on the vertical transport of fine-grained sediments. Hydrobiologia (235/236):1–16
Malcherek A (2005) Mathematical Module SediMorph – Validation Document Version 1.1. In Technical Report. Bundesanstalt für Wasserbau
Malcherek A, Markofsky M, Zielke W (1995) Numerical Modelling of Settling Velocity Variations in Estuaries. In: Arch. Hydrobiol. Spec. Issues Advanc. Limnil 47:353–362
Malcherek A (1995) Mathematische Modellierung von Strömungen und Stofftransportprozessen in Ästuaren, Institut für Strömungsmechanik und elektronisches Rechnen im Bauwesen, Universität Hannover, Bericht Nr. 44/1995, Hannover
Nezu I, Nakagawa H (1993) Turbulence in Open Channel Flow. IAHR/AIRH Monograph Series Balkema Publishers, Rotterdam
Smoluchofski M (1917) Versuch einer Mathematischen Theorie der Koagulationskinetik Kolloider Lösungen. Zeitschrift für Physikalische Chemie 92:129–168
Stolzenbach KD, Elimelich M (1994) The Effect of Density on Collision Between Sinking Particles: Implication for Particle Aggregation in the Ocean. Journal of Deep Sea Research I 41(3):469–483
Van Leussen W (1994) Estuarine Macroflocs and their Role in Fine-Grained Sediment Transport. Dissertation, University of Utrecht
Van Leussen W (1997) The Kolmogorov Microscale as a Limiting Value for the Floc Size of Suspended Fine-Grained Sediments in Estuaries. Cohesive Sediments, pp 45–62
Winterwerp JC (1998) A Simple Model for Turbulence Induced Flocculation of Cohesive Sediments. Journal of Hydraulic Engineering Research 36(3):309–326
Winterwerp JC (1999) On the Dynamics of High-Concentrated Mud Suspensions. Dissertation, TU Delft

Michel Verbanck · Aurélie Larcy · Nicolas Huybrechts · Jean-Pierre Vanderborght

4.4 Suction-Vortex Resuspension Dynamics Applied to the Computation of Fine-Particle River Fluxes

4.4.1 Addressing the Fine-Particle Issue

River and estuarine management on the basin scale is receiving renewed attention in the agenda of EU member states, mainly under the prescriptions of the 2000/60 Water Framework Directive. One of the important components of the 'ecological status' to be reached in surface water obviously derives from the interaction processes between the dissolved and the particulate phases. In terms of pollutant transfer, particles with the finest sizes are those that cause the greatest environmental concern (Kausch and Michaelis 1996; Owens et al. 2005). To explore the various management options, environmentally oriented numerical river models should therefore be able to reproduce the transfer fluxes of these fine particles (Förstner 2004).

In this context, a first problem is that a part of this suspended particulate matter (SPM) can be transported as wash-load (WL) or very fine particle load, notably in a state of autosuspension. According to Bagnold (1962), autosuspended particles have a settling velocity w_s not exceeding the value US (i.e. the product of mean stream velocity U and slope of the energy grade line S). Among environmentalists, it is widely recognized that these fine (slow-falling) particles are key vectors of pollutants in river basins because of their specific surface and adsorption capacity. In the following it will be shown that there are serious shortcomings in the modeling options currently available (Rendon-Herrero 1974; Wang 1979) for accounting consistently for wash-load particle fluxes. Distinguishing, in a given SPM record, what can be designated as WL is one thing; being able to predict the WL flux as a function of the hydrological sollicitations developing in the course of years (or decades) in the river watershed is another, certainly more difficult, proposition.

A second difficulty in this area is that a small number of storm events of very high intensity, lasting only a few days, can deliver the largest fraction of the annual particle flux at a considered river cross-section (Salomons and Brils 2004). In these circumstances, the whole riverbed moves rheologically, a key feature not adequately accounted for in traditional hydraulic models (Yen 2000). In more general terms, there is no sufficient consideration of the dynamical equilibrium between alluvial channel resistance, bedform development and bed-material load discharge, and for the role turbulence plays in these feedback interactions (Cao and Carling 2002; Verbanck 2004a; Huybrechts and Verbanck 2006). To address the issue, a research action is underway in the framework of the EU Integrated project Aquaterra under FP6. Model development is based on vortex-drag concepts recently proposed by Verbanck (2006) as a viable option for predicting resistance-to-flow in alluvial streams displaying a dynamic bed morphology evolution. This opens new perspectives for simulating the transport of SPM in open-channels, acknowledging the role of suction vortices (Dinkelacker 1982) in actively maintaining the resuspension process. The concept is physically based and has, up to now, been applied with some success to a number of well documented fluvial systems. It has the advantage of significantly lowering the number of model parameters to be calibrated (see Sect. 4.4.3).

4.4.2 Approach

What Is Meant by 'Wash-Load' in This Study

Historically, the transport of bed material, either over the bed or in suspension, was categorized as 'bed-material load' while the very fine suspended sediments not present in appreciable quantities in the bed were designated as wash-load (Einstein 1950). When suspended load (SL) is distinguished from bed-load (BL), the classification is conceptual; it is the mode of transport which is used to make the distinction. Unfortunately, the definition of washload is considerably more ambiguous. On one side, the classification is based on the origin of the particles (coming from the watershed or from upstream river stretches, as opposed to the local riverbed). On the other side, 'not present in appreciable quantities' often means considering all grains of a size less than, for example, the d_{10} of the riverbed material to be wash-load (Einstein 1950). The problem is evidently that the two criteria do not necessarily match. An alternative perspective on very fine SPM transport is to identify the fall-velocity threshold under which particles are likely to be maintained in the water column by the autosuspension process (Bagnold 1962; Wang 1979). This is an interesting approach to the problem, as it more clearly introduces a dynamic threshold, less static (as a systematic use of the d_{10} criterion mentioned above). Along a similar line of thought, we choose in this study to have a purely empirical definition of wash-load (section below). The originality of the approach is that it starts from the dynamics of what is observed in suspension (irrespective of the grain size characteristics of the local riverbed). Adopting a dynamical threshold between wash-load and suspended load evidently implies that exchanges between the two components occur on a regular basis, both in the temporal and spatial frame. This is a well-known difficulty of the problem, especially in the estuarine environment, where a large part of the very fine suspensions are cohesive and have a tendency to form flocs, directly influencing their settling properties (ten Brink 1997; Chen et al. 2005; Desmit et al. 2005). All mentions of 'wash-load' in this contribution thus correspond to the working definition given as Eq. 4.50 (meaning i.e. that wash-load and suspended load are mutually exclusive).

Identifying 'Wash-Load' Contributions in Experimental Records

The coarse particles moving in permanent contact with the riverbed (q_{BL}), characterized by high settling velocities in the turbulent fluid medium ($w_s / u_* \geq 3$), are known to be predominantly inert (granular) and are therefore not taken into account explicitly here:

$$\text{observed SPM flux} = q_{WL} + q_{SL} \tag{4.50}$$

The total flux of suspended particle material observed in the water column at a given river location is thus perceived as the sum of two fluxes which, in a 1-D (low-concentration) perspective as adopted here, can be seen as the product of flow discharge Q ($m^3 s^{-1}$), volumetric concentration (either C_{WL} or C_{SL}) (–) and particle

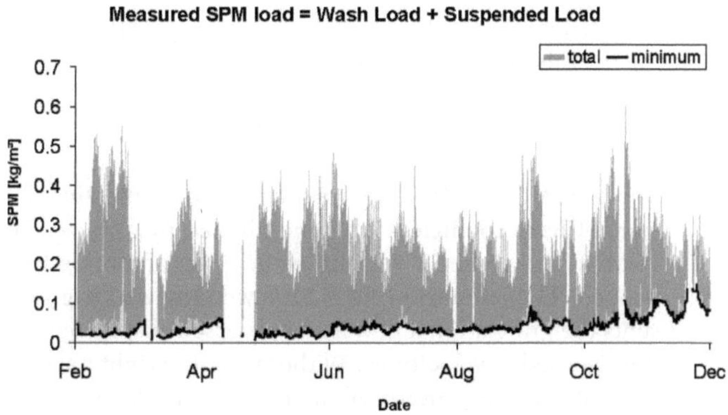

Fig. 4.19. Eleven months of continuous turbidity records in the Scheldt Estuary

density ρ_s (kg m^{-3}). A typical distribution between $\rho_s\ C_{WL}$ and $\rho_s\ C_{SL}$ from an experimental survey of the tidal river Scheldt (Antwerp site, St. Anna landing stage) is illustrated on Fig. 4.19. Wash-load contributions will be the highest when high discharges from the upstream watershed (November and December in this case) have been able to mobilize pollutant sources distributed on the surfaces exposed to runoff. However, as highlighted by Symader et al. (this volume), the runoff remobilization process can present strong local features, with the most intense events not necessarily associated with the highest wash-load fluxes.

In this estuarine case, a technique has been found to overcome the difficulty of having q_{WL} and q_{SL} unresolved in the SPM records. As the stream velocity reverses twice per tide, the minimum observed SPM concentration during a tidal cycle corresponds to a slack with velocity dropping to zero. The minimum concentrations (joined by the thick black line in Fig. 4.19), which correspond to particles which do not deposit in the low-velocity slack periods occurring four times a day, are considered in this study to reflect the long-term variation of the wash-load contribution (according to the terminological restrictions of WL as provided above). In this way, a new signal can be generated, corresponding more closely to the suspended-load contribution q_{SL}, namely particles which have the local riverbed as their transient source and sink. Progress in solving the problem raised by Eq. 4.50 is therefore possible if a robust sediment transport model, accounting for the dynamic behavior of the alluvial bed and the associated resuspension process, can be shown to reproduce the q_{SL} component in an acceptable way. An attempt to do this is presented below.

Adaptation of the Bagnold Suspension Model

In view of the complexity of the problem, stationary sediment transport and transport capacity approaches are, at this stage, retained as convenient working hypotheses. Implementation of the model itself is currently performed only locally (zero-D approach) with notably no account for the longitudinal distribution of riverbed deposits. As will be seen, there is ample scope for further model elaboration, as these hypotheses correspond to a gross over-simplification of particle transport processes (Burt et al. 1997; Winterwerp and van Kesteren 2004) occurring in natural, estuarine systems such as the one addressed here.

Based on Bagnold's stream-power concepts (1966), Verbanck (1995) suggested studying the value of $\eta_{SL} C_*$ in the following power balance equation (W m^{-2} of riverbed area):

$$(\rho_s - \rho_l) g w_s C_{SL} R_{Hy} = \rho_l \eta_{SL} C_* u_* (u_*^2 - u_{*c}^2) \tag{4.51}$$

in which R_{Hy} stands for hydraulic radius, w_s effective settling velocity, C_* non-dimensional Chezy coefficient (the one divided by $g^{1/2}$), η_{SL} suspension efficiency, u_* shear velocity, and u_{*c} shear velocity at the onset of motion for sediments forming the channel bed. The left part of Eq. 4.51 is well-known (Bagnold 1966; Yalin 1977) and represents the gravitational power of immersed suspended sediments on their way towards the riverbed at velocity w_s. Without η_{SL}, the right part of Eq. 4.51 is the stream power, calculated as the product of one-dimensional velocity and shear stress in excess of that required to put the deposited particles into motion. The power balance simply expresses that the gravitational power of the transported material uses a certain fraction (η_{SL}) of the available stream power to remain suspended in the water column. This assimilates a river flow to a certain sediment-transport engine, characterized by a given efficiency. Comparatively to previous versions of the suspension model (Verbanck 1996; Ma and Verbanck 2003), Eq. 4.51 also specifies, that under a certain threshold power, the concentration of bed-material particles transported in the water column drops to zero.

In Bagnold's (1966) approach, particles denser than water are maintained suspended in the column through the upward components of the burst cycle, a non-isotropic turbulent process inherent to high-speed flow close to a wall boundary (the riverbed in this case). It is therefore logical to try to relate the sediment transport efficiency to the burst intensity (Leeder 1983; Verbanck 2004b). Resuspension by burst activity implies that the flux of turbulent fluid momentum away from the boundary is exceeding, at least by a small amount, the one returning towards it. With obviously a plane bed in mind, Bagnold examined what are the consequences of these turbulent features developing over a unit area of riverbed. As a result of the net transfer of *upward* momentum, there is an equivalent pressure disturbance created at the boundary, the instantaneous magnitude of which is difficult to quantify. However, averaged over a sufficient number of burst cycles, the overall excess pressure is denoted Δp. Expressed per unit of bed area it has units (N m^{-2}). The suspension efficiency derived by Bagnold actually relies on a model for the prediction of the maximum possible value of Δp (denoted $\Delta p^\wedge$ as a reminder of the resuspension effect) as a function of stream power conditions. Making $\Delta p^\wedge$ directly proportional to the shear stress distributed over the unit bed area (see Table 4.9) he derived for $\eta_{SL} C_*$ the constant value $\kappa^{1.5}$ (= 0.266). He was satisfied that the choice of this constant value, produced semi-theoretically for a plane bed, reasonably predicted the transport capacity condition for suspended sediments in alluvial rivers. There is thus in this approach no explicit consideration for the bed morphology condition and its consequences in terms of resistance to flow.

Table 4.9. Full magnitude of pressure change created at the boundary by burst activity, $\Delta p^\wedge$ (N m^{-2})

Bagnold 1966	$\Delta p^\wedge = \kappa \tau_0$	Semi-theoretical (boundary considered as a plane)
This study (in which, by definition, $m \geq 1$)	$\Delta p^\wedge = \left(\frac{2\pi}{m}\right)^{10/13} \kappa^2 \tau_0$	Deduced from actual alluvial channels (with bedforms)

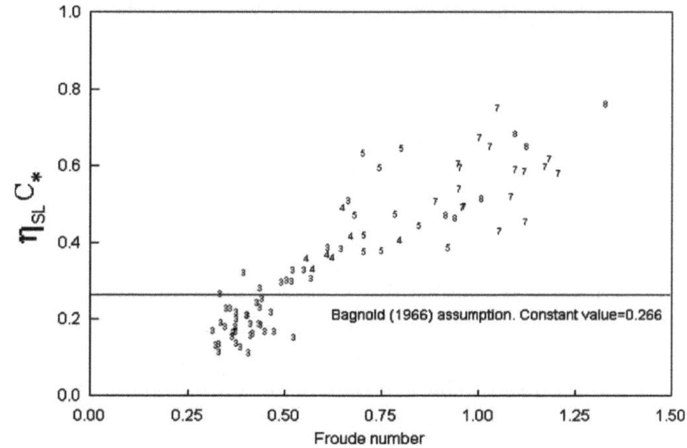

Fig. 4.20.
Questioning the validity of Bagnold's assumption, based on flume runs by Guy et al. 1966 ($d_{50} = 0.19$ mm, 0.27 mm, 0.28 mm), and Stein 1965 ($d_{50} = 0.40$ mm). Observed sediment transport rates q_{SL} are used, by application of Eq. 4.51, to deduce the values taken by $\eta_{SL} C_*$ when dune bedforms are progressively washed out and replaced by the bedform configurations typical of the upper alluvial regime

In more recent developments, interpretation of flume and field data performed by Bennett (1973) suggested that considering the product $\eta_{SL} C_*$ as invariant was not a sufficient approach to the problem, given the extended particle-size and stream-power ranges experienced in natural alluvial stream dynamics. Accordingly, Celik and Rodi (1984, 1991) proposed improving the prediction of $\eta_{SL} C_*$ by making it dependent on the value of the relative roughness. The relevance of this criterion for particles as fine as those of interest here was questioned, however, by Westrich and Juraschek (1985) who conducted flume studies of non-depositing flows loaden with quartz particles as fine as 26 μm. By comparing the $\eta_{SL} C_*$ values observed, over erodible and non-erodible boundaries, they were able to demonstrate the significant influence of the separation effect produced by bedform development. A further investigation of the role of the morphological adaptation of the movable bed in controlling the efficiency of suspended sediment transport was recommended (Yalin and Da Silva 2001).

Considering only non-cohesive, single-grain particles under capacity transport in a flume, the value of $\eta_{SL} C_*$ is plotted in Fig. 4.20 as a function of the alluvial regime, reflected both by the ambient Froude number $U/(gD)^{1/2}$ (abscissa) and by the observed bedform configuration phase (datapointer numbered from BF1 to BF8, following the bedform nomenclature suggested by Simons and Richardson 1966).

Figure 4.20 suggests that it is only in a limited alluvial regime range (BF3, corresponding to the fully developed dune configuration) that field sediment transport capacities would be acceptably predicted using Bagnold's assumption. Although Bagnold commented that the $\kappa^{1.5}$ constant would primarily apply to 'higher stages of flow' and upper regimes (bedform BF5 to BF8), this is not substantiated by the general trend depicted in Fig. 4.20. In fair retrospect, it is possible that the fully developed dune configuration, as observed in actual alluvial rivers, has constituted the bulk of experimental evidence examined in the light of Bagnold's suspension model, contributing to its undeniable success among practitioners (Peters 1976). In this study, we retain our confidence in Bagnold's rational leading to the general power balance. However, in Eq. 4.51, we introduce a non-constant value of $\eta_{SL} C_*$, under the form of a dependence with Rossiter modes as recently introduced in alluvial hydraulics studies (Verbanck 2006; Huybrechts and Verbanck 2006).

Inferring Resuspension from Vortex-Generated Wall-Pressure Signals

Rather than skin-friction drag or form-drag, we rely on vortex maintenance to explain how rivers flowing over alluvial bedforms with a characteristic wavelength λ_{BF} (Fig. 4.21) dissipate mean motion energy. The effect of topographic forcing, basically a non-turbulent process, will be addressed below. Shear layer vortices are generated behind the dune crest, where flow separates, and impinge on the back of the next dune at the distance x_R, where streamlines reattach. Experimental evidence collected in large sandbed alluvial channels (Venditti and Bauer 2005; Best 2005) showed that secondary eddies, generated at the point of impingement, could be strong enough to be perceptible at the water surface, displaying characteristic periodicity. Levi (1983a) studied similar oscillatory flow processes with frequency f_{vortex}, that arise when a restrained fluid body, of characteristic length dimension L, interacts with a free flow of mean velocity U. Prediction of the frequency of the periodic perturbations was possible with a Strouhal law of the form:

$$f_{\text{vortex}} = \frac{1}{2\pi}\frac{U}{L/m} \tag{4.52}$$

in which $m = 1$ corresponds to the fundamental frequency ($m = 2, 3$, etc. being the higher harmonics, Levi 1983b). The role of control factor m in Eq. 4.52 is the exact counterpart, in aerodynamics, of the dynamical mode number intervening in the Rossiter equation: as soon as it takes an integer value, resonance is active and allows the self-sustained oscillations to be maintained, the most efficient resonance effect taking place at $m = 1$ (Rossiter 1962). Using Eq. 4.52 as a way to predict the periodicity of the vortices in the separation region implies that $m/2\pi$ is the characteristic Strouhal number: a first application of Eq. 4.52 is thus the prediction of the frequency of oscillation. Reversely, spatially periodic structures engendered by the oscillation are controlled by

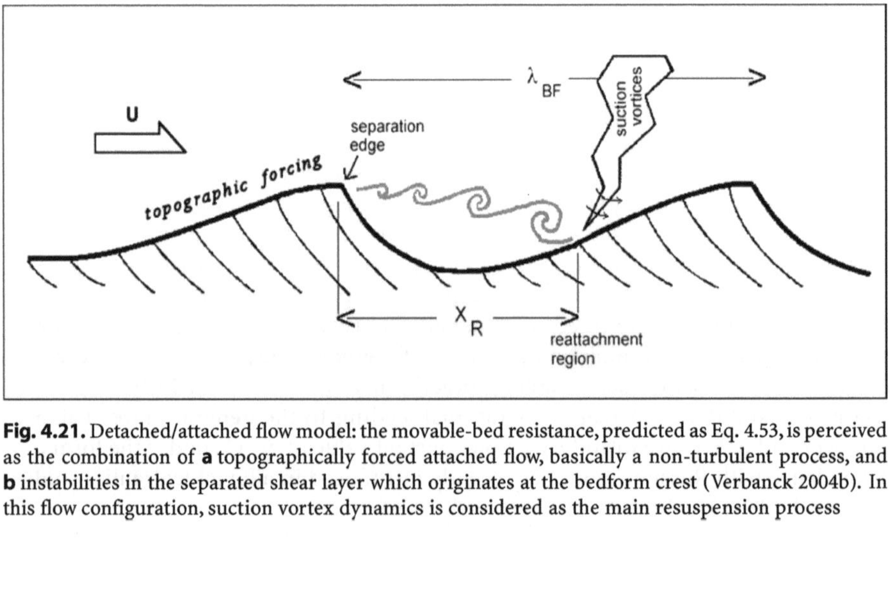

Fig. 4.21. Detached/attached flow model: the movable-bed resistance, predicted as Eq. 4.53, is perceived as the combination of **a** topographically forced attached flow, basically a non-turbulent process, and **b** instabilities in the separated shear layer which originates at the bedform crest (Verbanck 2004b). In this flow configuration, suction vortex dynamics is considered as the main resuspension process

the value of m, allowing m to be placed at the center of a revised bedform typology. In this respect as well, $m = 1$ is to be seen as the condition of the most effective water-transport engine.

On this basis, the attached-detached flow model developed by Verbanck (2004b) provides a way to predict the value of S, the slope of the energy grade line, for a quasi-2-D turbulent open-channel flow over mobile bedforms:

$$S = \left[\frac{\text{gravity forces}}{\text{stationary inertial forces}}\right]^{-5/3} \left[\frac{\text{stationary inertial forces}}{\text{nonstationary inertial forces}}\right]^{-10/3} \tag{4.53}$$

This formulation in two components (based on the Froude number and Strouhal numbers) allows the identification of the modes of energy transfer. In this formulation, bursting activity is undoubtedly reflected by the second term on the right side. It is the one that we want to interpret in light of the pressure change $\Delta p^\wedge$ created at the boundary by the periodic vortex impingement at the reattachment point shown in Fig. 4.21, and the possible link with secondary turbulent structures active in resuspending particles. Here again, an analogy can be made with what is known in experimental aerodynamics. Detailed examination of pressure patterns created by turbulent flow along a wall has led Dinkelacker (1982) to conclude that an important part of pressure disturbances can be attributed to a special form of burst activity. With a flow model he showed that 'tornado-like' vortices developing over the wall zone could be responsible for significant momentum transfer across the boundary layer, while explaining the pressure observations at the wall. Turbulent mixing achieved in this way would be very efficient and we see no reason why a similar process and its resuspension consequences should not apply in hydrodynamics (Baud and Hager 2000) and, in the present instance, to the near-bed zone in a river. Accordingly, 'suction-vortices' is the designation retained here to represent the nearly-vertical turbulent structures (Fig. 4.21) which, at their base, literally suck out the particles deposited on the back of the next bedform (Ha and Chough 2003) and then propel turbid clouds upwards. These clouds often take the form of characteristic twirling boils (Gyr and Hoyer 2006), as represented in the sketch above.

When these boils are powerful enough to be observed at the water surface, study of their periodicity (as performed experimentally by, for example, Jackson 1976; Kostaschuk 2000) has shown that it is reasonably described by a Strouhal law prediction. This is the reason why we place the Strouhal number (as occurring in Eq. 4.53) in the center of the resuspension modeling effort. In a flow configuration such as the one illustrated in Fig. 4.21, the topographical forcing repeated from λ_{BF} to λ_{BF} will generate a gravity wave tending to reconstitute a planar, free surface. At the considered disturbance scale, surface tension effects can be neglected. The speed of propagation of the gravity wave is thus simply predicted as (Airy's law):

$$c = \sqrt{\frac{g}{2\pi/\lambda_{BF}} \tanh \frac{2\pi D}{\lambda_{BF}}} \tag{4.54}$$

with D being the total depth and g the acceleration of gravity. The ratio of average stream velocity U to celerity c is called the generalized Froude number (Fr_g). It is this general form which is retained in the logic of Eq. 4.53 (where it actually occurs as Fr_g to the power 10/3). Equation 4.54 shows that it is only for high values of the ratio λ_{BF}/D that Fr_g converges to the traditional Froude number $U/(gD)^{1/2}$ (the one used, e.g., in Fig. 4.20).

The evidence collected within the Aquaterra project suggests that the pressure change $\Delta p^\wedge$ responsible for sediment resuspension actually involves the Strouhal number in a way which presents a striking similarity with Eq. 4.53:

$$\left[\frac{\text{stationary inertial forces}}{\text{nonstationary inertial forces}}\right]^{-10/3} = \left[\frac{\Delta p^\wedge}{\kappa^2 \tau_0}\right]^{-13/3} \tag{4.55}$$

It is possible that Eq. 4.55, and the value of the exponents (which are retained as such because of the strong standing of Eq. 4.53 so far, Luong and Verbanck 2007) will in the future be further interpreted in the light of turbulent wall-pressure patterns. It is indeed interesting to note that a similar minus ten-third dependency law has been observed to fit reasonably turbulent wall-pressure signatures as expressed in the frequency spectrum form (Black 1966; Hinze 1975).

Equation 4.55 can also be put into the form of Table 4.9, which clarifies how we would now predict the value of $\Delta p^\wedge$ (compared with Bagnold's suggestion for a riverbed treated essentially as a plane boundary, with no explicit consideration of bedform influence).

In the special condition of in-phase waves ($m = 1$), the magnitude of $\Delta p^\wedge$ reaches a maximum value for a given mean distributed bed stress. The so-called 'antidune standing wave' condition thus constitutes the most efficient bed configuration to produce extreme sediment flux for a given τ_0. This, however, also needs to be appreciated in the light of Eq. 4.53, which implies that, for a given Froude number, $m = 1$ will create the least shear stress at the bed (headloss varying as $m^{10/3}$). The transport capacity formula emerging from this analysis should therefore reflect the dual intervention of the control factor m on particle transport processes, once as an enhancing factor, and once as an inhibition term. This opposition between mixing performance and fluid-flow efficiency is well known in other fluid-mechanics problems, such as turbomixer design and chemical reactor engineering.

The combination of Eq. 4.51 with Eq. 4.55 gives an expression for the volumetric concentration of suspended load (Eq. 4.56). Compared with existing transport capacity approaches (Yalin 1977; Molinas and Wu 2001; Yang 2005 and references therein), this is, to our knowledge, the first sediment transport equation which explicitly takes into account the bedform configuration phase, through the control factor m. It is therefore particularly suited to the interpretation of observations in natural alluvial streams, such as the turbidity records obtained in the Scheldt Estuary in Antwerp.

$$C_{SL} = \kappa^3 \left(\frac{2\pi}{m}\right)^{15/13} \frac{u_* S}{\Delta w_s} \left(\frac{\theta - \theta_c}{\theta}\right) \tag{4.56}$$

The prefactor κ^3 is a simple numerical coefficient (0.071) from the data analysis, and the κ^3 notation was retained because it was the same as the square of the $\kappa^{1.5}$ initially suggested – on theoretical grounds – by Bagnold (1966). θ is the non-dimensional shear stress and θ_c the corresponding threshold value from the Shields diagram. Δ stands for the relative excess density. By the mere logic of its derivation, Eq. 4.56 only applies to the suspended-load component. It is not supposed to reflect bed load

q_{BL} (w_s/u_* ratios significantly higher than 3). For volumetric concentrations higher than 1%, concentration modifiers also need to be introduced, notably to account for the influence on the bulk suspension density and for hindering effects on particle settling behavior at very high suspension concentrations (Wan and Wang 1994).

In the follow-up to the discussion on the magnitude of the sediment-transport efficiency, it is noteworthy that suggesting Eq. 4.56 as a transport capacity formula is equivalent to predicting $\eta_{SL} C_*$ as a function of m raised to the power –15/13. This provides a reasonable explanation for the general shape of Fig. 4.20: the grouping $\eta_{SL} C_*$ takes its highest values when ambient Froude number conditions allow m to get sufficiently close to its minimum possible value ($m = 1$, as expression of Rossiter fundamental mode). In the upper alluvial regime, this only occurs in the so-called 'in-phase waves' bedform configuration (Verbanck 2006). This would explain why datapoints in the right part of Fig. 4.20 are systematically associated with the highest sediment-transport efficiency values. River flows over two-dimensional dunes (that we generally associate with the second possible harmonics $m = 2$) would, in this respect, represent significantly less efficient sediment-transport machines. This corroborates field observations collected from natural alluvial streamflows (Toffaleti 1968). A closer, more appropriate, examination of the influence of bedforms of various extent would be possible through the use of generalized Froude number Fr_g, provided that experimental λ_{BF} values are made available as part of the riverflow characterization effort (see Eq. 4.53).

4.4.3 Comparison of the Suspension Model with Field Data

Automated turbidity records have been obtained to characterize Scheldt estuarine waters in Antwerp since 1998. Samples are taken every 30 min, about 1.5 m below the water surface, from a floating landing stage (Sint Anna site). The device is fit with an internal fouling control, guaranteeing good precision even at low SPM values. Absorbance is measured at 660 nm wavelength. Kaolinite standards are used to convert the

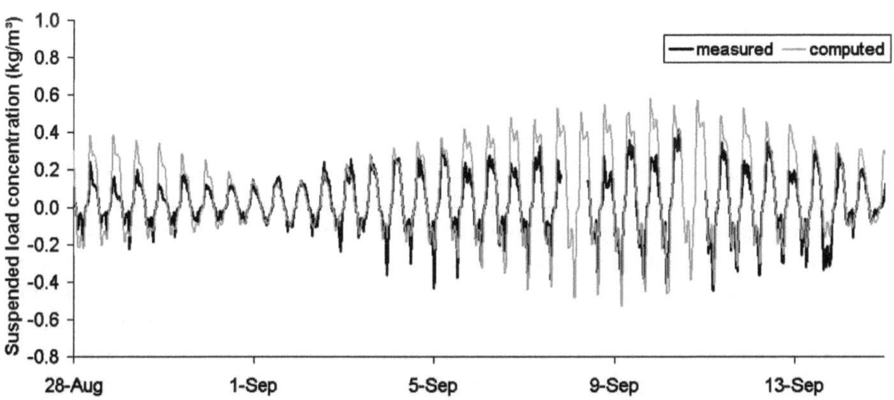

Fig. 4.22. Comparison between experimental suspended solids concentrations (*thick blackline*) and suction-vortex resuspension model output (*light gray line*) for the Scheldt at Antwerp

absorbance signal into SPM concentrations. Figure 4.19 provided an example of a record lasting a few months, illustrating how the 'wash-load' contribution could be taken out of the overall signal. Zooming on a much shorter period, presenting only the suspended-load contribution, Fig. 4.22 shows how local deposited sediments are responsive to stream power increases corresponding to the regular tidal inversions.

Stream velocities in Sint Anna (unfortunately not measured locally) were obtained by a one-dimensional hydrodynamic model proposed by Regnier (1997) for that part of the estuary. The value of the effective settling velocity in the turbulent medium, as occurring in Eq. 4.56, is extracted from the locally observed suspended sediment profiles along the vertical ($w_s = 0.5$ mm s^{-1}) following the method proposed by Verbanck et al. (2002). On this basis, results of the suction-vortex resuspension model, as applied to the eighteen-day continuous sequence, are plotted in Fig. 4.22. To account for the specific estuarine behavior (actually a strong, non-stationary sediment transport system), an invariant time-lag correction of minus 35 min (delay of observed SPM values compared to what they would be in the case of instantaneous response to the hydrodynamic changes) was imposed. To obtain the result shown in Fig. 4.22, the bedform wavelength λ_{BF} (cf. celerity of the gravity wave in generalized Froude number Fr_g, Eq. 4.53) was tuned to 24 m, a value not incompatible with the range of bedform wavelengths (5 m $\leq \lambda_{BF} \leq$ 70 m) usually observed at the site (Francken et al. 2004). Figure 4.22 suggests that, under neap-tide conditions, suspension capacity transport can be observed, fitting very nicely the predictions of the suction-vortex resuspension model. By contrast, this is not true during the largest part of spring tides. It is very likely that, when stream power conditions are the highest, the local source of easily erodible deposits is completely exhausted, exposing old layers characterized by very low erodibility (and very low mud content). If confirmed, this observation in itself could present some operational interest, notably in terms of the sampling strategy to be adopted in further estuarine environmental surveys.

Concluding Remarks

Regarding suspended load predictions, model results proposed so far for the highly dynamical Scheldt sediment system are encouraging, although, surprisingly, they were obtained without special consideration for the cohesive nature of the deposited material. If suspended load can be predicted along the lines suggested in this study, this could mean for the future that the computed q_{SL} for a given river (even a non-tidal one) could be substracted from the experimental SPM record collected during a sufficiently long time. With Eq. 4.50, this would allow the generation of a new, relatively clear q_{WL} experimental signal (no longer obscured by the dynamics of local river deposits). Although obtained in an indirect way, the reconstituted q_{WL} record could serve as a basis for the development of a new generation of wash-load prediction models (irrespective of any terminology arguments about what these WL contributions may be). Assessment and management of non-point sources indeed constitute one of the major challenges of future environmental protection of rivers. The achieved progress is noticeable regarding the control of (municipal and industrial) pollution point-source discharges.

In another respect, consideration of the differences highlighted by the model output between neap- and spring-tide conditions could be exploited from a straightforward, operational point of view. For further environmental surveys, a model assisted sampling strategy could be considered aiming to optimize, in the temporal and spatial frames, the collection of sediment deposits in highly dynamic estuarine and riverine systems. Sampling and monitoring costs are indeed only justified if there is a sufficient guarantee of sample representativeness. Related concerns are increasing among the EU member states, in their effort to launch the surveillance schemes imposed by the 2000/60 Water Framework Directive.

Acknowledgments

The automated water-quality monitoring station along the Scheldt was operated with the financial help of the Belgian federal state (Management Unit of the Mathematical Model of the North Sea, MUMM). Dr. Pierre Wollast and Mr. Didier Bajura were very instrumental in doing the design and field operation. The first author is much indebted to the discussions shared with Prof. Albrecht Dinkelacker, Max-Planck-Institut für Strömungsforschung, Göttingen. The efficient help of Mrs. Arielle Cornette, operational assistant, is also gratefully acknowledged. Two anonymous reviewers made very constructive comments allowing a previous version of the manuscript to be substantially improved. The study contributes to the AquaTerra Project 'Integrated Modelling of the River-Sediment-Soil-Groundwater System' funded by the European 6th Framework Programme, research priority 1.1.6.3 Global Change and Ecosystems (European Commission, Contract no. 505428-GOCE). It is part of Flux3 'Input/Output Mass Balances in River Basin: Dissolved and Solid Matter Load', a sub-component of the AquaTerra Integrated Project.

References

Bagnold RA (1962) Auto-suspension of transported sediment; Turbidity currents. Proceedings of the Royal Society of London. Series A, Mathematical and Physical Sciences, vol. 265, no. 1322, pp 315–319
Bagnold RA (1966) An approach to the sediment transport problem from general physics, US Geological Survey Professional Paper 422-I, U.S. Department of the Interior
Baud O, Hager WH (2000) Tornado vortices in settling tanks. J Env Eng, 126(2):189–191
Bennett JP (1973) An investigation of the suspended load transport efficiency in the Bagnold equation. IAHR Intl Symp River Mechanics, Bangkok, pp 455–463
Best J (2005) Kinematics, topology and significance of dune-related macroturbulence: some observations from the laboratory and field. Spec Publs Int Ass Sediment 35:41–60
Black TJ (1966) Some practical applications of a new theory of wall turbulence. Proc. Heat Transfer and Fluid Mech. Inst., pp 366–386
ten Brinke WBM (1997) Temporal variability in aggregate size and settling velocity in the Oosterschelde, The Netherlands. In: N Burt, WR Parker, Watts J (eds) Cohesive Sediments. John Wiley & Sons, Chichester
Burt N, Parker WR, Watts J (Eds) (1997) Cohesive Sediments. John Wiley & Sons, Chichester
Cao Z, Carling PA (2002) Mathematical modelling of alluvial rivers: reality and myth. Proc Inst Civil Engrs, Water and Maritime Eng, Part 1 Sept 02 (3), pp 207–219; Part 2 Dec 02 (4), pp 297–307
Celik I, Rodi W (1984) A deposition – entrainment model for suspended sediment transport. Sonderforschungsbereich 210, Universität Karlsruhe
Celik I, Rodi W (1991) Suspended sediment-transport capacity for open channel flows. ASCE J Hydr Eng, 117(2):191–204

Chen MS, Wartel S, VanEck B, Van Maldegem D (2005) Suspended matter in the Scheldt Estuary. Hydrobiologia 540:79–104
Desmit X, Vanderborght J-P, Regnier P, Wollast R (2005) Control of phytoplankton production by physical forcing in a strongly tidal, well-mixed estuary. Biogeosciences 2:205–218
Dinkelacker A (1982) Do tornado-like vortices play a role in turbulent mixing processes? In: Structure of turbulence in heat and mass transfer (Zaric ZP, Editor) Washington DC, Hemisphere Publishing Corp., pp 59–72
Einstein HA (1950) The bed-load function for sediment transportation in open channel flows, USDA Soil Conservation Service (Washington DC), Technical Bulletin no. 1026
Francken FD, Wartel SD, Parker RD, Taverniers ED (2004) Factors influencing subaqueous dunes in the Scheldt Estuary. Geo-Marine Letters 24(1):14–21
Förstner U (2004) Sediment dynamics and pollutant mobility in rivers: An interdisciplinary approach. Lakes and Reservoirs: Research and Management 9:25–40
Guy HP, Simons DB, Richardson EV (1966) Summary of alluvial channel data from flume experiments, 1956-61. USGS Professional Paper 462-I, 96 p
Gyr A, Hoyer K (2006) Sediment transport, a geophysical phenomenon. Fluid Mechanics and its Applications Vol 82, Springer, Dordrecht, 279 p
Ha HK, Chough SK (2003) Intermittent turbulent events over sandy current ripples: a motion-picture analysis of flume experiments. Sedimentary Geology 161(3):295–308
Hinze JO (1975) Turbulence. McGraw-Hill, New York, 2nd ed, 790 p
Huybrechts N, Verbanck MA (2006) Fully-coupled 1D model of mobile-bed alluvial hydraulics with a closure drawn from Rossiter modes resonance concepts. 7th Belgian National Congress on Theoretical and Applied Mechanics, Mons
Kausch H, Michaelis W (eds) (1996) Suspended particulate matter in rivers and estuaries, Advances in Limnology, 47, Stuttgart, 573 p
Kostaschuk R (2000) A field study of turbulence and sediment dynamics over subaqueous dunes with flow separation. Sedimentology 47:519–531
Leeder MR (1983) On the dynamics of sediment suspension by residual Reynolds stresses–Confirmation of Bagnold's theory. Sedimentology 30:485–491
Levi E (1983a) A universal Strouhal law. J Eng Mech, 109(3):718–727
Levi E (1983b) Oscillatory model for wall-bounded turbulence. J Eng Mech 109(3): 728–740
Luong GV, Verbanck MA (2007) Froude number conditions associated with full development of 2D bedforms in flumes. Proc 5th IAHR Symp River, Coastal & Estuarine Morphodynamics, Enschede (NL), acceptedMa TL, Verbanck MA (2003) The minimum slope for preventing accumulation of solids in newly designed sewers. Tribune de l'Eau, no. 624/4:50–59
Molinas A, Wu B (2001) Transport of sediment in large sand-bed rivers. J Hydr Res 39(2):135–146
Owens PN, Batalla RJ, Collins AJ, Gomez B, Hicks DM, Horowitz AJ, Kondolf GM, Marden M, Page MJ, Peacock DH, Petticrew EL, Salomons W, Trustrum NA (2005) Fine grained sediment in river sediments: Environmental significance and Management issues. River Research and Applications 21:693–717
Peters JJ (1976) Sediment transport phenomena in the Zaire River. In: Nihoul JCJ (ed) Bottom turbulence. Elsevier Oceanography series 19, pp 221–236
Regnier P (1997) Long-term fluxes of reactive species in strong tidal estuaries: Model development and application to the Western Scheldt Estuary. Ph.D. thesis, Chemical Oceanography Lab, ULB
Rendon-Herrero O (1974) Estimation of washload produced on certain small watersheds. ASCE J Hyd Eng 100(HY7):835–848
Rossiter JE (1962) The effect of cavities on the buffeting of aircraft. Royal Aircraft Establishment, Tech. Memo. 754:1962
Salomons W, Brils J (eds) (2004) Contaminated sediments in European River Basins. SedNet final summary report, 80 p
Simons DB, Richardson EV (1966) Resistance to flow in alluvial channels. USGS Prof Paper, 422-J
Stein RA (1965) Laboratory studies of total load and apparent bed load. J Geophys Res 70(8):1831–1842
Toffaleti FB (1968) A Procedure for Computation of the Total River Sand Discharge and Detailed Distribution, Bed to Surface, Technical Report no. 5, Committee of Channel Stabilization, Corps of Engineers, U.S. Army, November

Venditti JG, Bauer BO (2005) Turbulent flow over a dune: Green River, Colorado. Earth Surf. Process. Landforms 30:289–304

Verbanck MA (1995) Transferts de la charge particulaire dans l'égout principal de la ville de Bruxelles. Ph.D. thesis, ULB Dept Water Pollution Control, 193 p

Verbanck MA (1996) Assessment of sediment behaviour in a cunette-shaped sewer section. Water Science and Technology 33(9):49–60

Verbanck MA (2004a) Sediment-laden flows over fully-developed bedforms: first and second harmonics in a shallow, pseudo-2D turbulence environment. In: Jirka GH, Uijttewaal WSJ (eds) Shallow Flows. AA Balkema, pp 231–236

Verbanck MA (2004b) Sand transport at high stream power: towards a new generation of 1D river models? Proc 9th International Symposium on River Sedimentation, Yichang, China, Invited paper, pp 307–318

Verbanck MA (2006) How fast can a river flow over alluvium? J Hydraulic Research, in press

Verbanck MA, Laaji A, Niyonzima A (2002) Computing River Suspended Load over Bedforms in the Lower, Transition and Upper Hydraulic Regime. In: Bousmar D, Zech Y (eds) River Flow. Sweet and Zeitlinger, Lisse, pp 625–632

Wang SY (1979) A review of the effective power of sediment suspension in open channel flow – On the criterion of distinguishing bed material load from wash load. Bulletin of Science 24(9):410–413 (in Chinese)

Wan ZH, Wang ZY (1994) Hyperconcentrated flow. IAHR Monograph, AA Balkema, Rotterdam, 289 p

Westrich B, Juraschek M (1985) Flow transport capacity of suspended sediments. XXIst IAHR congress, Melbourne, vol. 3, pp 590–594

Winterwerp JC, van Kesteren WGM (2004) Introduction to the physics of cohesive sediments in the marine environment. Developments in Sedimentology 56, Elsevier, 559 p

Yalin MS (1977) Mechanics of sediment transport, 2nd ed. Pergamon Press, Oxford

Yalin MS, Ferreira Da Silva AM (2001) Fluvial processes. IAHR Monograph, International Association of Hydraulic Engineering and Research, Delft, The Netherlands

Yang SQ (2005) Sediment transport capacity in rivers. J Hydr Res 42(3):131–138

Yen BC (2000) From modeling the Yellow River to river modeling. In: Soong D, Yen BC (eds) First Sino-U.S. Joint Workshop on Sediment Transport and Sediment Induced Disasters, Beijing 15–17 March 1999, Post-Workshop Summary, NSF, 70 p

Chapter 5

Catchment Modeling

Ulrich Kern

This chapter addresses different monitoring and modeling approaches to investigate and quantify sources and pathways of sediments, nutrients and pollutants within river catchments.

Within four sections, the authors apply a variety of methods focusing on issues of different temporal and spatial scales ranging from event-driven sediment transport in a small mountainous catchment to annual load and mass balance in large river basins. The goal of this chapter is to improve the knowledge in assessing source-related load balances in river catchments.

Starting point of the first chapter is the European Water Framework Directive (WFD) which takes a combined approach of emission limit values and environmental quality standards to pollution control. WFD requirements are identified and present concepts to calculate substance specific load and mass balance in river catchments are discussed with respect to their capacity and limitation in assessing point source and non-point-source pollution.

Section 5.2 presents a field study investigating the effects of land cover changes on sediment transport in an alpine watershed by means of a coupled, spatially distributed event-based rainfall-runoff-erosion model. The choice of the sediment transport formula may significantly affect the calculated magnitude of erosion and deposition. Hence, a better understanding of the driving mechanism of erosion in relation to overland flow is required.

Section 5.3 evaluates the flood dependent transport processes of suspended sediment and pollutants in the Middle Elbe River. Event-driven matter transport is due to the complex interaction of: inflow from tributaries with different pollutant load; the role of groyne fields acting either as a source or as a sink for particulates; and the deposition on floodplain areas. Transport of the trace metals depends on their sediment-water partitioning and the grain-size of the suspended sediment.

Section 5.4 aims at mean annual loads of phosphorus from diffuse and point sources in large river basins. The empirical emission model presented is based upon so-called phosphotopes which are regarded as homogeneous types of sub-areas representing discontinuous source areas for non-point phosphate inputs. As shown for the Ems and Rhine catchments this phosphotope modeling concept may identify hot spot areas of pollution and hence, may support options and plans for a river basin management and remediation measures according to WFD. Further experience in monitoring and modeling of pollutant emission and transport in river catchments over a large range of spatial and temporal scales is still necessary for the implementation. The following four papers show the broad spectrum of work and give inspiration for future research in the field of catchment-related pollution control and management.

Ulrich Kern · Frank Wendland · Ekkehard Christoffels

5.1 Catchment Modeling of Emissions from the Perspective of WFD Implementation

5.1.1 Introduction

The Water Framework Directive (WFD) sets ambitious objectives for the protection of European water resources (EC 2000). For priority substances and other pollutants environmental quality standards have been defined for surface water. An understanding of the sources and pathways of these substances within river catchments is crucial to establish effective monitoring programs and to develop emission control strategies as a part of cost-effective programs of measures. In this context, source-related emission modeling offers the potential to support WFD implementation, since these catchment models provide annual load balances and highlight the relevance of various pollutant sources and pathways (Fig. 5.1).

The objective of this chapter is to discuss the value of catchment-related emission modeling from the viewpoint of WFD implementation. Since this contribution is related to the SEDYMO research program (Westrich and Förstner 2005), emphasis is given to fluvial systems. Trace metals are chosen as a substance group of environmental concern which is particle-associated to a large extent. Instream processes such as erosion, sedimentation or biogeochemical transformation are in the focus of other con-

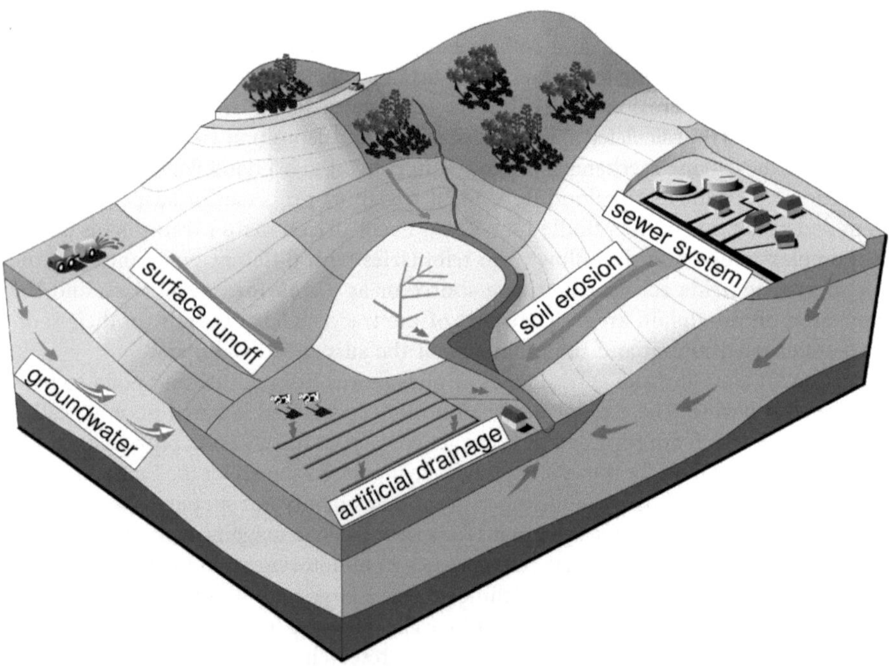

Fig. 5.1. Sources and pathways of pollutants in river basins

tributors and hence, are not considered here. The results presented are from the research project SAFE (Becker et al. 2005) and are related to the 1 800 km^2 catchment of the Erft River, a left tributary of the lower Rhine located in the western part of Germany.

In a first step, the WFD is briefly presented and its requirements for catchment modeling are described. Next, pathways of pollutants are analyzed with respect to data availability and their temporal process behavior. The capabilities of present modeling concepts which calculate substance load budgets in river catchments are investigated. The deficits, obstacles and perspectives of catchment models are considered.

5.1.2 The WFD and Its Requirements for Catchment Modeling

On 23 October 2000, the "Directive 2000/60/EC of the European Parliament and of the Council establishing a framework for the Community action in the field of water policy" (EC 2000), commonly known as the European Water Framework Directive (WFD), was adopted. The key elements and objectives of the Directive are as follows (Blöch 2004):

- protection of all waters aiming at good status, as a rule, at the latest by 2015, linked to a non-deterioration principle
- comprehensive monitoring systems for all waters
- coherent water management based on river basins
- combined approach of emission limit values and quality standards
- economic instruments to support environmental objectives
- mandatory public participation
- streamlining legislation, and ensuring a single, coherent managerial frame.

The Directive foresees a timetable for implementation using a stepwise approach consisting of: initial river basin analysis (finalized 2005), the implementation of monitoring programs by 2007, the establishment of programs of measures and river basin district management programs until 2009, the implementation of measures until 2012 and the achievement of good status of water resources until 2015.

Within the first river basin analysis, in Germany published in spring 2005, 60% of all surface water bodies were deemed to be at risk, 26% possibly at risk and 16% not at risk of failing the WFD objectives (BMU 2005). One major reason for waters being at risk was found to be deficits in hydromorphology, including river continuity. Another is related to discharges from point and diffuse sources affecting the water quality of surface waters, which is the focal point of this study.

The Combined Approach

The WFD takes a combined approach to pollution control first by setting emission controls to limit pollution at the source (e.g., waste water, agricultural fertilizers) and secondly, by establishing water quality objectives for bodies of water. In every case, the more stringent of the two will apply (De Toffol et al. 2005). Thus Member States will have to set down in their programs of measures both the limit values to control emissions from individual point sources and environmental quality standards (EQS) to limit the cumulative impact of such emissions as well as those from diffuse sources of pollution. For surface waters, EQS have been defined for priority substances and certain other pollutants considering the annual average (AA-EQS) and the maximum allowable concentration (MAC-EQS) (EC 2006).

A meaningful combined approach is strongly connected to a sound understanding of the origin, pathways and depots of pollutants within river catchments. Surface waters may receive loadings of pollutants from various point and diffuse sources. Knowledge about the pollutant origin and especially about the dominant input pathways is prerequisite to develop successful strategies for emission avoidance and emission control.

The relevance of external sources in comparison to the background values for the apparent instream pollution has to be investigated. In the past, especially particle-bound pollutants such as most of the heavy metals and hydrophobic organic micropollutants may have accumulated in aquatic sediments due to historical pollution. In this case, pollution of river bed has to be considered and may even far exceed the impact of external sources. These 'areas of concern' have to be identified within a river basin and assessed with respect to their ecological damage potential using a 'weight of evidence' approach (Heise et al. 2004; Heise et al. 2005).

The WFD Scales

It is noteworthy to recognize that the WFD's water quality targets are connected to different spatial scales. Here, three levels may be distinguished, which have to be embedded within a management framework (Fig. 5.2).

The protection of the marine environment is a major issue in Europe's water policy. This is manifested within international agreements such as HELCOM for the Baltic Sea (EC 1994a,b) and OSPAR for the North-East Altantic (EC 1997), which aim at a preservation of biodiversity of the marine ecosystem, protection from eutrophication, hazardous and radioactive substances and the effects of offshore exploration of oil and gas. From the marine viewpoint only the entire load of nutrients and pollutants from a river basin is important without considering its precise origin.

However, knowledge about the 'areas of concern' as well as the 'input pathways of concern' is required to set efficient management priorities on the catchment scale. Additionally, legislative and financial instruments rather than technical measures will be used at this scale to support the achievement of the environmental goals.

The other end of the WFD scale is defined by the single water bodies for which, as a rule, the good status has to be achieved by 2015. For these river sections, operational monitoring and cost-effective measures have to be developed. Hence, when considering water quality targets, a detailed understanding of the pollutant emission pathways is needed on the local scale.

On the regional scale, management units may be defined by groups of water bodies with similar conditions or impacts.

With respect to temporal discretization, substance load balances are needed at least on an annual basis. Firstly, this is to identify trends within the WFD reporting cycle, as the Directive demands river basin management plans to be established and updated every six years. Secondly, the annual basis allows to compare modeling results with the AA-EQS for priority substances and pollutants in surface waters.

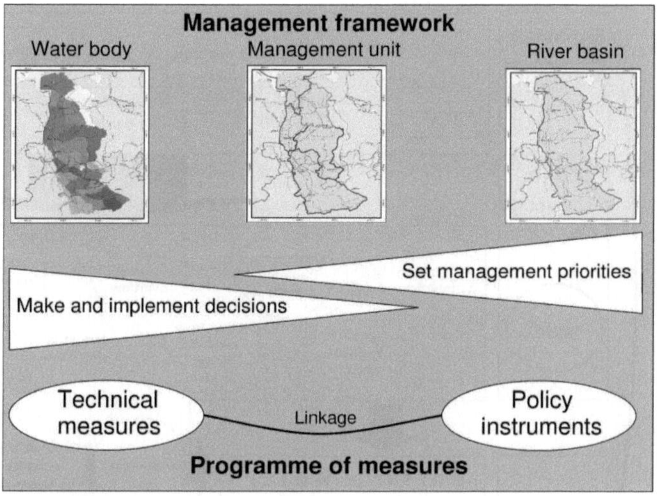

Fig. 5.2.
WFD management framework

The Contribution of Emission-Related Catchment Models

Modeling load balances of substance emissions within river basins supports for WDF implementation by:

1. Contributing to development of monitoring strategies
2. Supplementing available monitoring data
3. Facilitating data analysis of emission and instream pollution
4. Helping to verify complience with environmental quality standards
5. Identifying and assessing the relevance of pollutant pathways
6. Enabling development of instruments and measures for pollution control (e.g., by scenario analyses)
7. Providing a basis for implementation of the 'polluter pays principle'

5.1.3 Calculating Emission Balances in River Systems

Emission Pathways

To calculate the substance loadings which a water body receives is a challenging task since a network of numerous environmental pathways has to be considered (Fig. 5.3). Each pathway may be defined by three parts: mobilization from a specific source, a transport route, and the input to the water body. For example, the roofs of buildings may consist of zinc which is mobilized by rainfall and transported via roof runoff into a combined sewer system. Zinc may be discharged into a river in this example either via the outflow of the waste water treatment plant or, in case of overloading, via combined stormwater overflow. At this point, the question arises if it is advantageous or even necessary to describe the entire pathway (including mobilization, transport route and input) or if it would be sufficient or favorable to consider only the input. Generally,

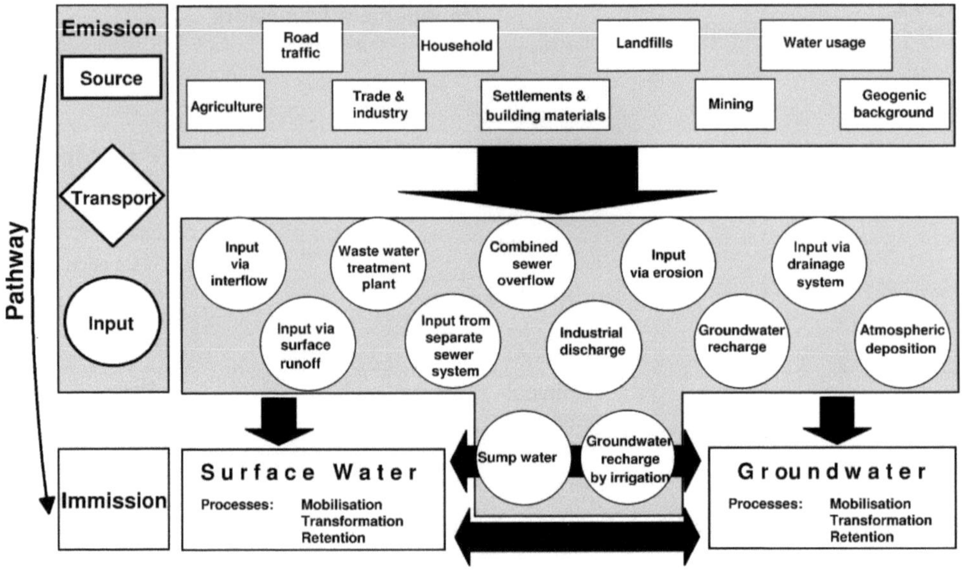

Fig. 5.3. Environmental pathways of pollutants in river basins

a clear answer cannot be given to this question since the choice of the model concept depends on both the objectives of the investigation and the availability of data. However, the recommendation can be given that a modeling approach which is limited to the input is sufficient if either the input is well known or if it is fed from heterogeneous sources which are not easily identifiable. A highly sophisticated process – based model describing the whole pathway seems to be less appropriate to be applied in large river basins than a conceptual model which is developed to set management priorities. On the other hand the same conceptual model developed for the large scale may be too coarse to support decisions on a water body scale.

Contaminant sources are usually classified into two general forms, point sources and non-point sources, each of which poses specific problems regarding monitoring, quantification and management. Point sources of pollution are those originating from a single, well defined location. As such, they are often readily identified and generally easily controlled and monitored. Examples of point sources of pollution relevant to water and sediment management in river basins include industrial discharge, effluent from waste water treatment plants (WWTPs) and combined sewer overflow (CSO). Non-point (diffuse) sources of contaminants are those originating from a wide area within the catchment. As their location of discharge into a water body is not well defined the identification, and in particular the control, of these sources presents a challenge to water pollution management. Examples of diffuse sources of pollution include atmospheric deposition and the input of contaminants via surface runoff, interflow, groundwater runoff and erosion.

It is important to realize that some of the input pathways are continuous processes whereas others are event-driven. In view of the fact that the WFD requires emission load balances on an annual basis, on the one hand, those pollutant inputs which are defined by continuous, almost uniform graphs of substance loadings are the easiest to quantify and to control. Runoff from WWTP, groundwater recharge or pumping of sump water in mining areas are examples of this kind of processes. On the other hand, the description and simulation of pollutant inputs defined by discontinuous pathways which are steered by low frequency, high impact (LFHI) events is most difficult. Input via soil erosion, surface runoff and CSO belong to this group. As their fate is strongly affected by erosion, particulate nutrients and pollutants are difficult to handle.

Model Validation

Validation of emission load balances is an issue of outstanding relevance. Since only few input pathways are monitored by measuring the discharged loadings, the overall load balances can only be checked by means of environmental monitoring data from the receiving water bodies. Here, two problems have to be envisaged: First, the water quality of a water body may be influenced or even controlled by historical sediment pollution rather than by external sources (see Sect. 5.1.2). Second, instream processes such as erosion, sedimentation and biogeochemical transformation processes may be important for particle-bound and/or reactive compounds. Today, most of the catchment models do not account for these internal riverine processes in detail but use constant or discharge-dependent retention factors instead. Therefore, presented comparisons assuming that the sum of emissions minus retention is equal to monitored loading may be inadequate due to the deficiencies in process-description for the water bodies themselves.

5.1.4 Obstacles and Strategies in Catchment Modeling

In this section different approaches to quantify emissions are discussed for selected pathways. As an example of a continuous point source, effluent of waste water treatment plants (WWTP) is considered. Soil erosion is highlighted as discontinuous pathway from non-point source which is controlled by LFHI events.

Emissions from Continuous Point Source: Input Via WWTP Effluent

In the following presentation, different possibilities of estimating annual effluent loads from WWTP are investigated and discussed for the substance group of trace metals.

Figure 5.4 provides a schematic view of a conventional WWTP receiving its inflow from the sewage system. Purification is accomplished through fixation of a portion of the inflowing trace metals in the sewage sludge. The fluvial system is affected by the WWTP effluent. Hence, the effluent seems to be the ideal location to monitor and calculate the pollutant loading.

Calculation, or better estimation of the mean annual pollutant load is given by the product of the annual WWTP runoff and the mean annual concentration of the regarded pollutant in the WWTP effluent. In practice, a surveillance program is either continuous or discontinuous at certain monitoring intervals providing e.g., hourly or daily values. In case of lacking monitoring data annual WWTP discharge can be esti-

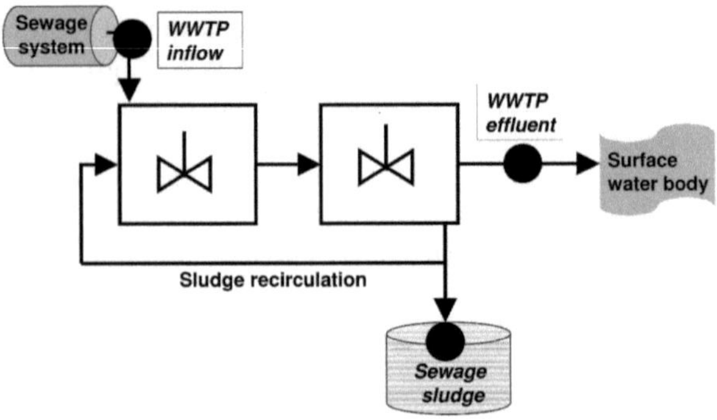

Fig. 5.4.
Points to quantify emissions from waste water treatment plants

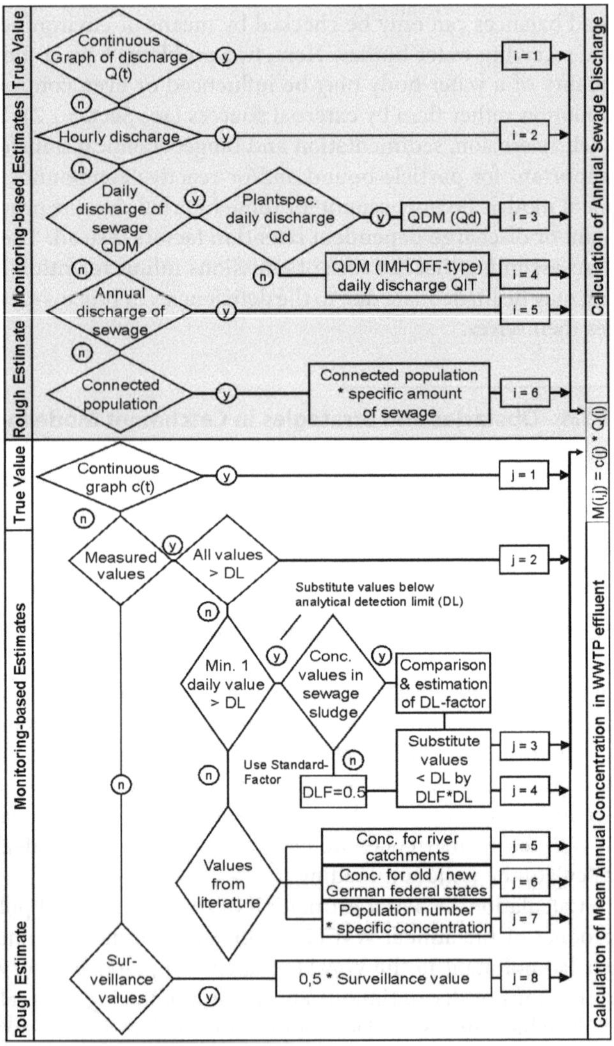

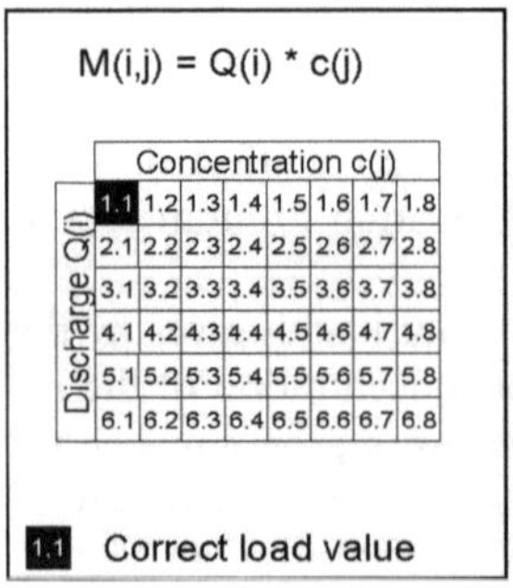

Fig. 5.5.
Stepwise approach to calculate annual WWTP effluent loads

mated by the design capacity of the plant or connected inhabitants (Zessner and Lindtner 2005), and annual concentration is usually derived either from literature values or from the surveillance value of the plant's operating permit, which can only provide a rough estimate of the true concentration. For calculation of both discharge and concentration, a decision tree can be set up which goes from greatest to poorest data availability (Fig. 5.5).

When combining the different estimates for the annual means of discharge and concentration, an $(n \times m)$-matrix can be set up where every matrix element refers to a specific load estimate. The upper left matrix element (1,1) refers to the correct load value, as it is determined by continuous monitoring, whereas the lower right one, e.g., matrix element (n, m), reflects the load estimate which is expected to be the least accurate, since it is not based on any monitoring data. Straightforward calculation of monitoring data generates a certain matrix element e.g., a specific annual load estimate. Depending on the given monitoring frequency this load estimate will be more or less accurate. From the matrix approach it becomes apparent that monitoring strategy has to be taken into account to ensure reliable emission data.

Another problem in practice is caused by concentration values below the analytical detection limit (DL). For trace metals, conventional analytical methods such as optic emission spectrometry (OES) might not be sensitive enough to detect trace metals in WWTP effluents. In this case a common procedure is to substitute values below DL to avoid data gaps in order to perform load calculations. The half detection limit value is most widely used to substitute analytical values below DL. For log-normally distributed analytical values, Fig. 5.6 depicts that this approach is only suitable if the minority of monitoring data is lower than DL. Otherwise, especially if only 10% or less of the

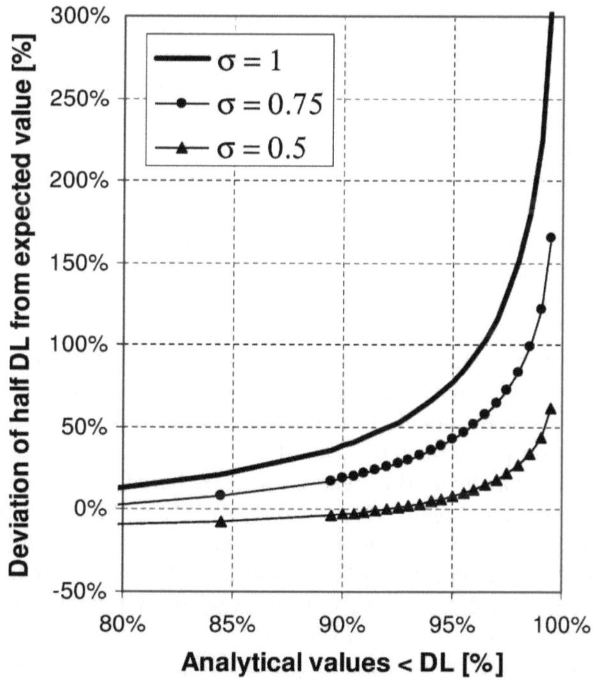

Fig. 5.6. The detection limit problem: deviation of the half detection limit value from the expected true value as a function of the percentage of values below detection limit. Results are given for different standard deviations σ assuming log-normally distributed analytical values

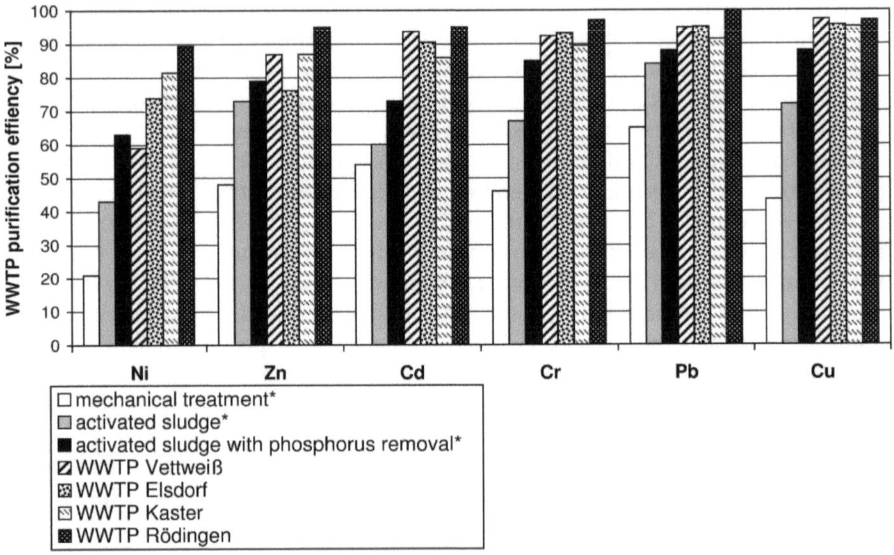

Fig. 5.7. The purification efficiency problem: trace metal retention efficiencies of WWTPs with different design characteristics. Mean literature values according to Fuchs et al. (2002) are indicated with * and compared with median values of three WWTPs (Vettweiss, Elsdorf and Kaster) using activated sludge with phosphorus removal and the membrane activated sludge WWTP at Rödingen

monitored data are detectable, the true concentration might be tremendously overestimated. In this case monitoring data are insufficient and do not form a basis for annual load estimates. For example, the WWTP surveillance program of the state of North Rhine-Westphalia, Germany, using conventional OES in combination with inductive coupled plasma (ICP-OES), revealed that 99% of concentration values for lead and cadmium were below DL (MUNLV 2006).

The solution to this difficulty is either to use more sensitive analytical methods such as mass spectrometry (ICP-MS) or to employ enrichment techniques prior to the trace metal detection. Since increasing analytical effort raises the costs of any surveillance program, it should be determined if there are further alternatives in calculating the pollutant loading of WWTP effluent.

DL problem may be avoided by considering the WWTP inflow or, even more promisingly, by using the sewage sludge where pollutants become enriched throughout the treatment process (Fig. 5.4). With either alternative, WWTP emissions can only be derived if the purification efficiency of the WWTP is well known. Thus, plant-specific data on pollutant retention efficiency have to be gathered. As a fall-back, literature values may be used instead if these are based upon the same design characteristic of the treatment plant. However, for WWTPs within the Erft River catchment, retention efficiencies for trace metals measured by Becker et al. (2005) appear to be considerably higher than reported by Fuchs et al. (2002) for WWTP in the Rhine catchment (Fig. 5.7).

Figure 5.8 depicts a comparison of mean annual loads of trace metals for 46 municipal WWTP plants in the Erft River catchment. The overall load of the examined trace metals derived from sewage sludge is at an average of approximately 5.0 t yr^{-1},

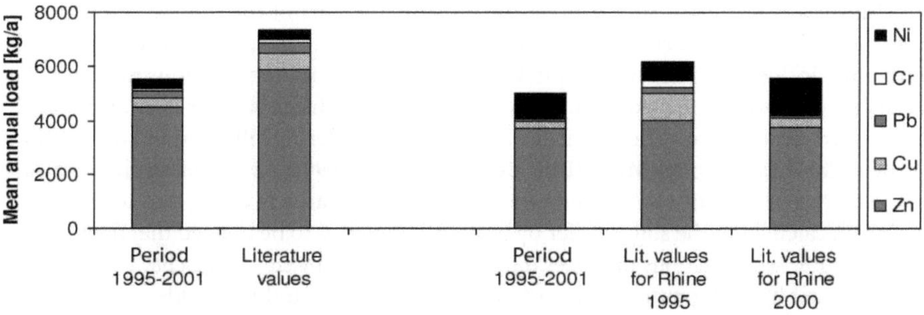

Fig. 5.8. Emissions from 46 WWTPs in the Erft River catchment. Comparison between the WWTP effluent method (*left*) and the sewage sludge approach (*right*) for trace metals

which is only slightly smaller than the load of 5.5 t yr^{-1} in the effluent. Based on this result, the sewage sludge method can be recommended as an alternative to the effluent method, despite significant deviations in value for individual trace metals such as nickel or lead. For both methods, literature values overestimate the values based on monitoring.

Emissions from Discontinuous Non-Point Source: Soil Erosion

As an example of a discontinuous non-point source, soil erosion is highlighted in this section. This pathway is significant for nutrients, especially phosphorus, and particle-bound pollutants and may form a large portion of the entire emissions as known for large river basins (Böhm et al. 2000).

Soil erosion depends on the characteristics of the catchment (hill slope, soil properties, vegetation) and those of the precipitation events (intensity, amount and course of rainfall). Erosion monitoring data is sparse since the events are unpredictable.

Due to the lack of appropriate measured data, either conceptual models or physically based models are employed to quantify soil erosion. The conceptual approach is generally based on the Universal Soil Loss Equation (USLE, Auerswald 1998) and provides mean annual values. Process oriented erosion models such as AGNPS (USDA 2006), EROSION 3D (Schmidt 1996; von Werner 2002), EUROSEM (Morgan et al. 2006) and WEPP (USDA 2006) simulate the physical processes of soil mobilization and soil transfer for discrete precipitation events. In the following both approaches are described and results are compared for the example of the upper Rotbach, a sub-basin of the Erft River, Germany.

Conceptual approach of erosion modeling. The USLE adapted to German conditions by Schwertmann et al. (1990), is employed to calculate the mean annual loss of soil (Eq. 5.1):

$$A = R \cdot K \cdot L \cdot S \cdot C \cdot P \tag{5.1}$$

A = Mean annual soil loss (t $(\text{ha a})^{-1}$), R = rain erosivity factor (N h^{-1}), K = soil erodibility factor (t h $(\text{ha N})^{-1}$), L = slope length factor (–), S = slope steepness factor (–), C = crop and management factor (–), P = protection measures factor (–).

Using Eq. 5.1 to estimate the potential erosion, the part of the mobile soil which is transported into the surface waters can be calculated if delivery and transfer of soil under the influence of heavy precipitation are taken into account. Various methods are available to determine the sediment delivery ratio, *(a)* accounting only for steep areas which are located close to rivers (Behrendt et al. 1999); *(b)* considering hill slope and the portion of farm land (see model MONERIS, Behrendt et al. 1999); or *(c)* classifying the hill slope by the probability with which the abutting, slopy area is connected to a rivulet.

When calculating the soil transfer by erosion it is always to be noticed that the portion of small grain-size fractions increases during the transport by surface discharge. This sorting is caused by preferential settling of larger soil particles and is considered by the enrichment ratio (ER, see Sect. 5.4.3) employing one of the following procedures: *(a)* constant substance-specific ER (Fuchs et al. 2002); *(b)* ER depending on the long term mean of the annual sediment input (Auerswald 1998); *(c)* ER accounting for the correlation between sorption surface and grain size of soil particles when simulating the graded transport of several grain size fractions; *(d)* ER given by the ratio of measured substance concentrations in river bottom sediment to the mean monitored soil contamination (Böhm et al. 2001).

The substance load by soil erosion, which is transported into the river during heavy precipitation, is calculated by Eq. 5.2 from the mean annual soil loss (Eq. 5.1), the sediment delivery ratio, the enrichment ratio and the substance concentration in the soil.

$$\mathrm{SE} = E\frac{\mathrm{SDR}}{100}\frac{\mathrm{ER}}{100}c_{\mathrm{soil}}\frac{A}{1000}R \tag{5.2}$$

SE = substance input via erosion (kg), E = potential erosion (t ha^{-1}), R = rainfall erosivity factor (%), SDR = sediment delivery ratio (%), ER = enrichment ratio (%), c_{soil} = substance concentration in the soil (mg kg^{-1}), A = area of farm land (ha).

Process oriented approach of erosion modeling. These models simulate the soil erosion in small catchments for single precipitation events. The model EROSION 3D (Schmidt 1996; von Werner 2002) which is employed here, is applicable in catchments of up to approximately 400 km^2. Spatial resolution of grid cells is between 5 and 20 m, temporal discretization is by time steps of 10 min. The model EROSION 3D consists of two main components, a GIS-module and the program core. The GIS-module is used for the digital relief analysis. The program core of the model takes into account, among others, the following processes of soil erosion: infiltration of precipitation according to Green and Ampt (Schmidt 1996); generation of runoff (excess of infiltration and retention in troughs); unsoldering of soil particles from the soil surface; transport of particles and deposition depending on the transport capacity of the surface runoff; enrichment of fine particles along the transport path.

Input parameters of deterministic erosion models are numerous and can be subdivided into three groups. Digital terrain model; soil parameters (distribution of grain sizes, storage density, content of organic carbon, initial water content, resistance against erosion, coefficient of roughness, degree of coverage, correction factor, associated precipitation situation); precipitation parameters (intensity, interval of discretization, duration of precipitation).

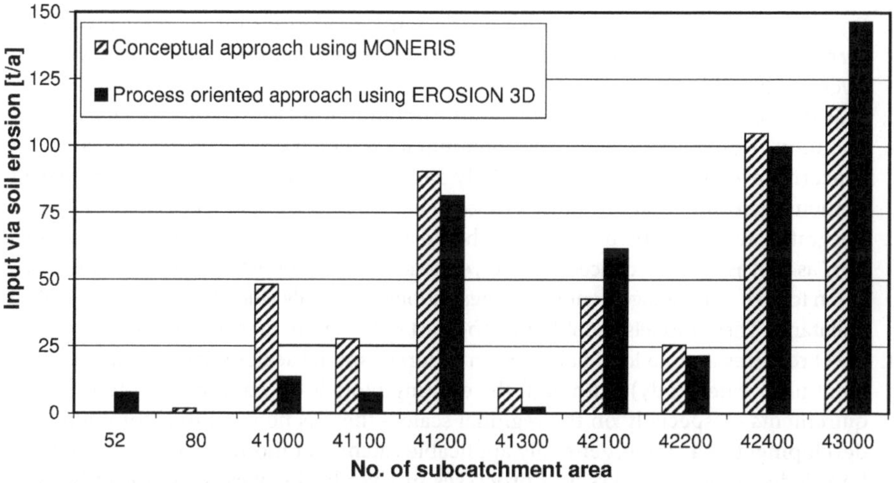

Fig. 5.9. Sediment input via soil erosion in different areas of the upper Rotbach catchment

Model application. For the year 2000, the annual sum of soil erosion and deposition in the upper Rotbach catchment, a river system in the slopy setting of the northern Eifel mountains. Figure 5.9 shows the results of the conceptual method following the approach of MONERIS (Behrendt et al. 1999) and those of the process oriented EROSION 3D model assuming that all the farmland is covered with winter wheat.

The comparison of the model results for the sub-catchments of the upper Rotbach shows remarkable resemblances. This applies to the spatial allocation of the erosion as well as for the absolute quantities.

Despite of these promising results, it must be stressed that matter input via soil erosion can only be assessed with large uncertainties. Most seriously, model validation by means of instream data is affected by sparse data and riverine sediment transport which may blot out the effect of soil erosion (see Sect. 5.1.3).

From the viewpoint of WFD both approaches are supplementary and thus, their combination is most promising. The conceptual approach may serve to assess the significance of the soil erosion pathway in comparison to other human pressures and impacts on the river basin scale. The process oriented approach is to investigate the event-driven effects of soil erosion and to evaluate the significance of measures, e.g., for soil conservation, on the local and regional scale.

5.1.5 Conclusions

From the perspective of WFD implementation, there is a substantial need for catchment modeling and in particular for substance load balances. These models contribute to the combined approach. They may support decisions to set up monitoring programs, to develop cost-effective instruments and measures for pollution control and to implement the 'polluter pays principle'. Application and further development of catchment models should address the following considerations:

1. Emission-related catchment models are designed to identify 'input pathways of concern' e.g., they aim at identifying the relevant external sources and impacts within a river catchment.
2. This type of models disregards instream contamination due to historical sediment pollution. Hot spots of sediment contamination which are referred to as 'areas of concern' have to be identified separately and require a risk assessment with respect to their environmental relevance (Heise et al. 2004; Heise et al. 2005).
3. Integrated water quality management should account for both, 'input pathways of concern' as well as 'areas of concern'. Hence, emission-related models cannot be used in isolation to arrive at management recommendations, especially when looking at particulate pollutants. These models should rather be part of the entire managerial framework.
4. WFD requires annual load balances on different spatial scales (river basin, management unit, water body). Although the existing models do not meet all the WFD requirements – especially on the regional scale – there is no need to recommend the development of a single, generally applicable catchment model. For one, the level of knowledge about transfers and processes in river basins varies significantly from one water body or management unit to another. For another, the availability of model input parameters is very different. Therefore, applying catchment models should not be reduced to the automatic use of a specific, straightforward algorithm. Instead, the regional applicability of a certain model should first be checked against the objective and the scale of application, the regional data availability and the level of knowledge about pollutant sources and pathways. A variety of emission-related models which offer data-dependent ways of calculating pollutant discharges are thus needed. Robustness checks and sensitivity analysis are strongly recommended to increase reliability of the calculated load balances.
5. At present, catchment models do not consider the fate of pollutants within rivers in an appropriate way. Especially for particle-bound and reactive compounds, either these models have to be further improved, or a coupling of emission-related catchment models with surface water quality models has to be envisaged to overcome the existing deficits. Hence, better collaboration between "catchment modelers" and "instream modelers" is recommended.

Acknowledgments

The authors acknowledge the German Ministry of Education and Research (BMBF) for funding of the SAFE project (no. 0330063/0330063A).

References

Auerswald K (1998) Bodenerosion durch Wasser. In: Richter G (ed) Bodenerosion – Analyse und Bilanz eines Umweltproblems. Wissenschaftliche Buchgesellschaft, Darmstadt, pp 33-42 (in German)

Becker A, Christoffels E, Großkinsky B, Hiller A, Kern U, Krump R, Thormann D, Firchhof W, Palm N, Tiedemann K (2005) Verbundvorhaben Stoff- und Datenmanagement in Flusseinzugsgebieten am Beispiel von Schwermetallen in der Erft (SAFE). Institut für Siedlungswasserwirtschaft, RWTH Aachen und Erftverband, Bergheim (in German)

Behrendt H, Bach M, Kunkel R, Opitz D, Pagenkopf WG, Dannowski R, Deumlich D (1999) Nährstoffbilanzierung der Flussgebiete Deutschlands, UBA Texte 75/99, Umweltbundesamt Berlin (in German)
Blöch H (2004) European water policy and the water framework directive: an overview. JEEPL 3:170–178
BMU (ed 2005) Die Wasserrahmenrichtlinie – Ergebnisse der Bastandsaufnahme 2004 in Deutschland. Bundesministerium für Umwelt, Naturschutz und Reaktorsicherheit, Berlin (in German)
Böhm E, Hillebrand T, Marscheider-Weidemann F, Schrempp C, Fuchs S, Scherer U, Lüttgert M (2000) Emissionsinventar Wasser für die Bundesrepublik Deutschland. UBA Texte 53/00, Umweltbundesamt Berlin (in German)
Böhm E, Hillenbrand T, Marscheider-Weidemann F, Schempp C, Fuchs S, Scherer U (2001) Bilanzierung des Eintrags prioritärer Schwermetalle in Gewässer. UBA Texte 29/01, Umweltbundesamt Berlin (in German)
De Toffol S, Achtleitner S, Engelhard C, Rauch W (2005) Challenges in the implementation of the Water Framework Directive: case study of the alpine river Drau, Austria. Wat Sci Technol 52(9):243–250
EC (1994a) Council Decision 94/156/EC on the accession of the Community to the Convention on the protection of the marine environment of the Baltic Sea Area 1974 (Helsinki Convention), OJ L73, 16.03.1994
EC (1994b) Council Decision 94/157/EC on the conclusion, on behalf of the Community, of the Convention on the protection of the marine environment of the Baltic Sea Area (Helsinki Convention as revised in 1992), OJ L73, 16.03.1994
EC (1997) Council Decision 98/249/EC on the conclusion of the Convention for the protection of the marine environment of the North-East Atlantic (Paris Convention), OJ L104, 03.04.1998
EC (2000) Directive 2000/60/EC of the European Parliament and the Council establishing a framework for Community action in the field of water policy, OJ L327, 22.12.2001
EC (2006) Proposal for a Directive of the European Parliament and of the Council on environmental quality standards and pollution control in the field of water policy and amending the Directive 2000/60/EC. COM 2006(397) final, 17.07.2006, Internet at http://ec.europa.eu/environment/water/water-dangersub/surface_water.htm
Fuchs S, Scherer U, Hillebrand T, Marschner-Weidemann F, Behrendt H, Opitz D (2002) Schwermetalleinträge in die Oberflächengewässer Deutschlands. UBA-Texte 54/02, Umweltbundesamt Berlin (in German)
Heise S, Förstner U, Westrich B, Jancke T, Karnahl J, Salomons W, Schönberger H (2004) Inventory of historical contaminated sediment in Rhine Basin and its tributaries. Report on behalf of the Port of Rotterdam. Internet at http://www.zu-harburg.de/ut/bis/Projects.htm
Heise S, Claus E, Heininger P, Krämer Th, Krüger F, Schwartz R, Förstner U (2005) Studie zur Schadstoffbelastung der Sedimente im Elbeeinzugsgebiet – Ursachen und Trends. Report on behalf of the Hamburg Port Authority (in German)
Morgan RPC, Quinton JN, Rickson RJ (2006) European Soil Erosion Model. EUROSEM Web Site. Internet at http://www.silsoe.cranfield.ac.uk/nsri/research/erosion/eurosem.htm
MUNLV (ed) (2006) Entwicklung und Stand der Abwasserbeseitigung in Nordrhein-Westfalen. Ministerium für Umwelt und Naturschutz, Landwirtschaft und Verbraucherschutz des Landes Nordrhein-Westfalen, Düsseldorf (in German)
USDA (2006) *A*gricultural *N*on-*P*oint *S*ource Pollution Model. AGNPS Web Site. Internet at http://www.ars.usda.gov/Research/docs.htm?docid=5233
USDA (2006) Water Erosion Prediction Project. WEPP Web Site. Internet at http://topsoil.nserl.purdue.edu/fpadmin/weppmain
Schmidt J, von Werner M, Michael A (1996) Entwicklung und Anwendung eines physikalisch begründeten Simulationsmodells für die Erosion geneigter landwirtschaftlicher Nutzflächen, Berliner Geogr. Abhandlungen 61, Eigenverlag, Berlin (in German)
Schwertmann U, Vogl W, Kainz M (1990) Bodenerosion durch Wasser – Vorhersage des Abtrags und Bewertung von Gegenmaßnahmen. Ulmer Verlag, Stuttgart, 2nd edn (in German)
von Werner M (2002) EROSION 3D Benutzerhandbuch Version 3.0. GeoGnostics Software, Berlin (in German)
Westrich B, Förstner U (2005) Sediment dynamics and pollutant mobility in rivers (SEDYMO). JSS 5(4):197–200
Zessner M, Lindtner S (2005) Estimations of municipal point source pollution in the context of river basin management. Wat Sci Technol 52(9):175–182

Jörg Kirsch · Paolo Burlando · Peter Molnar · Elke Hinz

5.2 Modeling the Effects of Land Cover Changes on Sediment Transport in the Vogelbach Basin, Switzerland

5.2.1 Introduction

Land use changes are often considered to be the reason for increased flood frequency and magnitude as well as enhanced erosion. In small mountain basins, rapid changes in vegetation land cover may occur due to anthropogenic or natural causes, with pronounced effects on hydrological processes and the local ecosystem. With deforestation or windstorm damage, an increased proportion of the land surface may be exposed to more intensive overland flow processes, providing an enhanced sediment supply to the fluvial system and increasing the risk for downstream areas during severe flood events. It is generally accepted that the adaptation of appropriate land use strategies will lead to a more effective prevention of runoff and sediment production. In addition, careful management of the vegetation cover may lead to the ecosystem being less exposed to windstorm damages and more effective in regenerating after disturbances. A sustainable forest management which minimizes the risks of floods and erosion and strengthens the ecosystem has to be based on a proper understanding and evaluation of basin response to changes in vegetation cover.

Unfortunately, direct observations of land use change effects are very limited, in particular in combination with adequate hydrological, sediment transport and erosion measurements. Therefore, the quantification of sustainable forest management practices often relies on mathematical modeling and scenario analysis. Scenarios play an important role because they use expert knowledge to identify realistic catchment changes, which are subsequently used to assess the relevant hydrologic response by means of model simulations. Thus, distributed rainfall-runoff models, coupled with sediment transport and erosion/deposition models and used together with land cover change scenarios, can help in a quantitative assessment and evaluation of erosion patterns resulting from catchment response to scenarios of different land uses.

As part of a larger study (Kirsch and Burlando 2005) the present investigation focuses on the development and application of a coupled spatially distributed and GIS-integrated event-based rainfall-runoff-erosion model, and on its subsequent application to the analysis of catchment response to land cover change scenarios. Specifically this paper presents the investigation of the effects on overland and channel potential erosion and transport rates in a small mountain basin in Switzerland. The paper reports the results of a preliminary study which aims at showing the potential of modeling as a replacement of extensive, demanding and often unfeasible 1:1 investigations, whereas the study is designed as an investigation framework for ungauged basins, especially targeting the study of basin changes. It does not address and resolve open scientific issues concerning erosion and sediment transport issues which are still debated in the literature. Therefore, assumptions which have been proposed and extensively used in the literature are neither refuted nor proved, but studied from the prospective of modeling relevant changes within the basin over time. Conversely, while showing the power of mathematical modeling, the study expresses a word of caution on the (direct application) of equations available from the literature without reflecting, even in a qualitative fashion, on the physical processes.

The analyzed land cover change scenarios consist of hypothetical forest damages caused by windstorms similar to historically observed storms in Switzerland (BUWAL 2002). The results focus on the changes in hillslope and channel erosion rates, on the modification of the duration of the erosion event and of the total sediment load. The analyses were carried out by means of two different hillslope sediment transport approaches, to assess their sensitivity and their ability to predict erosion in space and time as influenced by land cover change.

5.2.2 The Case Study

The study catchment Vogelbach is located in the Alptal, central Switzerland (Fig. 5.10). Covering an area of 1.55 km^2, its elevation ranges between 1 060 m at the outlet and 1 545 m at the highest location. Average annual precipitation measured at the catchment weather station is around 2 100 mm yr^{-1} with highest precipitation in the summer. The average annual runoff from the catchment area is about 1 590 mm, resulting in an annual runoff coefficient of about 0.74. Based on station measurements, the rainfall DDF curves give for the return period 2.33 and 100 years the rainfall amount of 28 mm and 69 mm (1 hour duration), and 82 mm and 156 mm (24 hour duration). The mean runoff is about 0.078 $m^3 s^{-1}$, while the highest recorded peak was about 6 $m^3 s^{-1}$. The estimate of the centennial flood is approximately 11.5 $m^3 s^{-1}$ (BWG, 2003).

Land cover consists mainly of forest (63%), pasture and meadow (31%). Hillslope steepness ranges between 20 and 40% with an average gradient of 37%. Most common soils in the basin are Gleysol and Regosol. The Swiss digital soil map (Geostat 1997) and specific literature (Burch 1994) indicate shallow soils (30–40 cm) with a large clay content causing a generally low infiltration capacity. Forested areas have a slightly higher infiltration capacitiy (compared to unforested areas) which leads to a lower degree of water logging. In general, geomorphologic properties and vegetation/land cover distribution within the Vogelbach catchment represent well the boundaries of hydrologically similar reacting areas. The erosion patterns and erosion intensities are also strongly related to hillslope morphology and vegetation cover; in par-

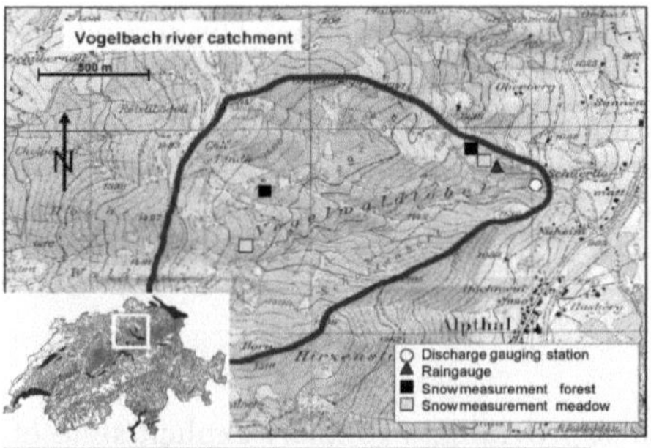

Runoff measurements	since 1967
Time resolution	10 min

Drainage Area	155 ha
Exposition	ESE
Altitude Range	1060 m – 1545 m
av. hillslope gradient	37 %
av. main channel gradient	18 %
Drainage density	ca. 8 km/km²
Forested area	63 %
Wetted area	25 %
Pasture area	12 %
Geology	Flysch

Fig. 5.10. Location, characteristics and instrumentation of the Vogelbach Basin

ticular steep hillslopes connected to the channels are susceptible to erosion, soil creep and shallow landslides. The main channel is incised into the hillslopes and has an average steepness of 18% and an average width of 5.5 m (Milzow et al. 2006). The longitudinal profile surveyed by Milzow et al. (2006) shows a step-pool morphology with occasional cascading sections and exposed bedrock. The median bed sediment particle size is around $d_{50} = 120$ mm, most bed material ranges between 10 and 1 000 mm in size. The catchment is instrumented with an automatic gauging station measuring discharge with a time resolution of 10 minutes. Precipitation is recorded with the same time resolution at a weather station located within the basin.

Sediment transport measurements are carried out by hydrophones, automatically recording the impulses of bed load particles larger than 10 mm at the basin outlet. Suspended load monitoring is not provided within the current equipment of the catchment. To estimate the total sediment load, measurements and relations of the neighboring catchment (Erlenbach) are used. There, calibrated hydrophones and a sediment retention basin allow the monitoring of total sediment and bed load sediment transport (Rickenmann 1997).

Spatial data used within this study are commonly available data within Switzerland. The spatially distributed land cover data for the base scenario ("present state") are derived from the raster map "Arealstatistik72" (100×100 m) of the Geostat Database. This map is the basis for developing hypothetical land use change scenarios for forest damage and management. In addition, these data are expanded by information and data of Burch (1994) and Walthert et al. (2003). Several floods between 1986 and 1990 were used to calibrate and validate the hydrological model (see the following section), and simulate the impact on erosion induced by land cover changes. The analysis focuses here particularly on a flood event of 1986 with a peak of 4.7 $m^3 s^{-1}$, corresponding approximately to a 8 year return period flood. The total event precipitation was around 46 mm with a duration of 3 hours and a maximum intensity of 7.5 mm in 10 min. The runoff coefficient for this event was around 0.45. Sediment yield was derived using observation data for the same event in the Erlenbach catchment and is estimated to be within the range 1 100–1 900 m^3 of sediment. This range was estimated using two different approaches, both based on the relation between erosion-effective runoff volume and sediment volume. Rickenmann (1997) defined a linear relationship between these two parameters. The similarity of the two catchments is used for a simple transfer of the sediment volume by the erosion-effective runoff volume. The second approach assumes a different regression between erosion-effective runoff volume and sediment for the two catchments and uses an area-related transfer by the area-normalized erosion rate. The above mentioned range is defined by the different sediment volumes resulting from the two approaches.

5.2.3 The Modeling Framework

The setup of the modeling framework aims at providing a spatially explicit suite of models on the basis of which impacts of land cover changes on the spatial and temporal distribution of runoff and erosion can be analyzed. An event-based, spatially distributed hydrological model, coupled to an event-based, spatially distributed erosion model is developed for this purpose.

The Hydrological Model (Fest98mod)

The hydrological model used to simulate the catchment response is based on an early version of the FEST98RS model extensively presented in Montaldo et al. (2004; the reader is referred to this work for further details about the hydrological model). The model provides a convenient framework because of its event-based approach and its spatially explicit distributed nature and land cover parameterization. The spatially distributed nature of the model requires the availability of a digital elevation model (DEM), and of raster based thematic maps of land cover and soil characteristics, as well as spatially distributed precipitation input. The fundamental model components are the surface runoff production and concentration modules. Subsurface flow processes are modeled by a single or multiple linear reservoirs arranged in sequence.

Surface runoff production is determined by a modified SCS-CN method (SCS 1972) and routed through hillslope and channel cells by the variable parameter Muskingum-Cunge method (vMCM, Cunge 1969; Ponce and Yevjevich 1978). Both methods are implemented on a raster basis.

The CN method uses soil and land cover information to the derive soil storage capacity. Kuntner (2002) modified the method to allow accounting for intermittent precipitation and optimized the CN values for Swiss catchments by extensive calibration and validation. Excess rainfall is routed by the vMCM, differentiating between overland and channel flow by the area-threshold method. A fraction-factor approach separates the infiltrated rainfall into deeper percolating water and subsurface flow, which is routed to the outlet by a linear reservoir scheme. This connects the storage of the subsurface horizons linearly to the output Q ($m^3 s^{-1}$) at the outlet. The linearity is described by the storage coefficient k (s), which defines the decay of the storage S (m^3) over time $Q = (1/k)S$.

The vMCM is a nonlinear coefficient routing method representing an approximation of the diffusive wave. Its parameterization (essentially the wave celerity and the hydraulic diffusivity) is variable in space and time thus accounting for effective physical hillslope and channel properties (slope, esp. slope/channel width) and for the space-time variability of the inflow rate. Ponce (1983) and Brunner (1989) compared it against fully unsteady flow equations, and stated a good comparison over a wide range of conditions. To calculate overland flow, hillslope width is considered here equal to the DEM cell size (25 m in this case). Channels are conversely assigned a geometry in the form of a relation between depth and width, which is derived by direct survey or by scaling relationships dependent on the drainage area. Details about the equations can be found in Ponce and Yevjevich (1978), Ponce (1989), Ponce and Chaganti (1994), whereas a description of their implementation in rainfall-runoff modeling is found in Montaldo et al. (2004).

The parameters that could not be derived by direct knowledge of the basin characteristics were derived from literature and adapted to the Vogelbach by calibration and subsequent validation. The calibration of the model is thus based on Kuntner (2002) who carried out extensive calibrations over different scales, including the catchment considered in this study. Main task of Kuntner (2002) was the calibration of the CN-Numbers for Switzerland and the preparation of a Swiss-wide SCS-soil type map.

Relevant hydrological parameters for the rainfall excess were the CN-value, the Soil-type (derived) and the initial loss. All parameters of the flow routing module were in a first step derived from literature and maps or from previous field studies as in the case of the depth-width relation of flow area and the roughness for channel and overland flow, which were estimated on the basis of Chow (1973), Chaudhry (1993) and Engmann (1986). The storage coefficient for the subsurface routing was initially derived from the recession limb of the measured hydrographs. All parameters were altered in their physically meaningful range to obtain the best fit for the calibration events of Kuntner (2002). In this paper, one event is considered for calibration of the hydrological model, three further events for validation. All of the considered events were tested against several criteria, among them the Nash-Sutcliff efficiency (with values between 0.95 and 0.98).

The Sediment Transport and Erosion Model

The coupled erosion model determines event-based sediment transport and erosion/deposition rates both on hillslopes and in channels assuming transport limited conditions. Because of the likely very limited contribution of the suspended sediment fraction due to the coarse nature of the large part of the sediments, the study focuses only on bedload transport in the channel. In general, erosion/deposition is computed from the sediment mass balance for each cell: $dq_S / dx = D$, where q_S represents the sediment mass flow (kg $m^{-1} s^{-1}$), x the flow distance (m) and D the net erosion/deposition rate (kg $m^{-2} s^{-1}$). Following Molnar et al. (2006) and assuming transport limited conditions, sediment fluxes can be set equal to the sediment transport capacity T_c, that is $q_S \approx T_c$. Transport limited conditions give a maximum potential for sediment transport and erosion/deposition and allow us to identify the most sensitive areas and the overall erosion potential of the basin.

Channel (bedload) sediment transport is implemented by the Schoklitsch equation for bedload in steep mountain streams, in which the critical flow q_c ($m^2 s^{-1}$) is estimated by an empirical relation based on channel gradient (S) and particle diameter d_{16} developed by Bathurst et al. (1987):

$$q_S = \frac{2.5}{\rho_S / \rho} S^{2/3} (q - q_c) \quad \text{with} \quad q_c = 0.21\sqrt{(gS^{-1.12})} d_{16}^{1.5} \tag{5.3}$$

Here, q ($m^2 s^{-1}$) is the specific water discharge rate, S (–) the slope, ρ (kg m^{-3}) the density of water and ρ_S (kg m^{-3}) the density of bed material.

Many sediment transport equations exist for hillslope overland flow, depending on the prevalent flow hydraulics and sediment characteristics (e.g., Julien and Simons 1985; Prosser and Rustomji 2000). In this paper, two commonly applied formulae were selected to investigate the effects of land use changes over time.

The first relation studied is based on *shear stress* (e.g., Foster and Meyer 1972; Mitas and Mitasova 1998) and gives the specific volumetric discharge as:

$$q_S = \frac{K}{\rho_S} \cdot (\tau_0 - \tau_c)^{\mu} \quad \text{with} \quad \tau_0 = g \rho r_{hy} s \tag{5.4}$$

where K is the coefficient of effective transport capacity (s), τ_0 the bed shear stress (N m^{-2}), τ_c the critical shear stress (N m^{-2}), μ an exponent (–) and r_{hy} the hydraulic radius (m). The second relation is based on *runoff*, assuming laminar flow over hillslope surfaces. Referring to Julien and Simons (1985), Prosser and Rustomji (2000) and Kilinc (1972) the general formula is given by

$$q_S = \frac{\alpha}{\rho_S} S^{\beta} q^{\gamma} i^{\delta} (1 - \tau_c / \tau_0)^{\varepsilon} \tag{5.5}$$

with q the specific water discharge (m^2 s^{-1}), i the rainfall intensity (m), ρ the density of water (kg m^{-3}), β, γ, δ, ε exponents (–) and α a transport capacity factor (–). Transport limited condition implies $\tau_0 \gg \tau_c$. Julien and Simons (1985) discussed the characteristics of overland flow and noted that overland flow can occur under turbulent and laminar conditions. They stated that the relation of viscous to inertia forces in combination with small flow depth forces the flow to be laminar up to a certain Reynolds number *Re*, which is indicated by Shen and Li (1973) to be approximately $Re = 2\,000$, after which smooth turbulent flow starts. The assumption of laminar flow is supported by Hessel (2002), where in field measurements on plot sizes with similar (or even steeper) slopes, land cover and velocities, the Reynolds number seldom exceeded the critical value.

The assumption of transport limited (conditions) implies that as much material is supplied as can be transported by the runoff. Raindrop impact – one of the most important factors for sediment detachment – plays a minor role for the transport limited case. Raindrop impact may influence under certain circumstances the flow conditions and therefore the sediment transport rate indirectly. However, in this study we assume the raindrop impact on the soil surface to be negligible due to the predominantly large sediment size and because large part of the catchment area is forested. For this latter reason there is no evidence of mud- or debris-flow development on the steep slopes. In addition, the interaction of erosion and sediment transport with the vegetation coverage and the overland flow has been approached conceptually by defining a characteristic roughness for each cover, rather than modeling with more sophisticated techniques such as those discussed by Nepf (1999). This seems to be indeed reasonable because of the consistency with the modeling scale which is based on a 25×25 m raster, which is rather coarse for the representation of small scales erosion features.

Land Cover Scenarios

Several hypothetical scenarios of land cover changes were developed on the basis of an extensive literature review. They all originated from the actual land cover derived from the raster maps of "Arealstatistik72". In a comprehensive study (Kirsch and Burlando 2005) two main driving scenarios (windstorm, climate change) were developed to study their possible effects on land cover and in turn on runoff, sediment yield, and erosion/deposition patterns. Here, only the windstorm scenario is illustrated.

After the Vivian (1990) and Lothar (1999) wind storms, the Swiss Federal Office for the Environment carried out several studies which analyzed the behavior and coping strategies of ecosystems (natural vs. man influenced) after heavy windstorms, with a focus on the re-establishment and growth of vegetation (forest in particular) after

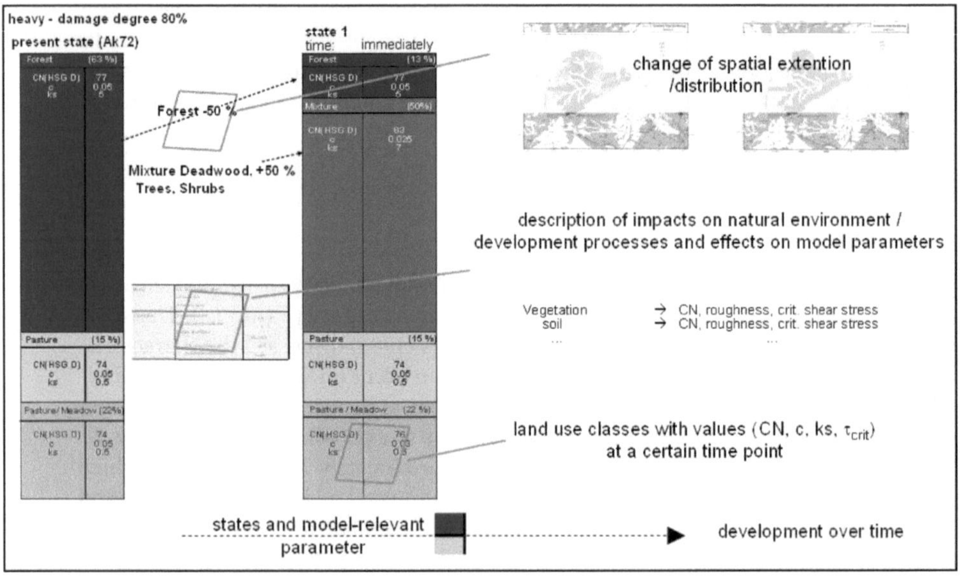

Fig. 5.11. Conceptualization of the time-space scenario evolution and the relevant model parameters and example of transition between phases (**a**) and (**b**) (adapted from Kirsch and Burlando 2005)

damaging events (e.g., Lässig and Schönberger 1997; Lässig and Motschalow 2000). On this basis, the developed windstorm scenario was divided into two subscenarios (removal vs. no removal of woody debris), of which only the latter is reported here. We developed a hypothetical, but qualitatively plausible evolution of vegetation cover in time and space after a damaging windstorm event, to describe the natural recovery of the forest. The land use change scenarios were translated into spatial and temporal changes of the model-relevant parameters, thus resulting in appropriate parameter sets for each condition within the scenario-development states. The scenarios were not only defined in terms of spatial and temporal changes but also in terms of different magnitudes of impact. Scenarios with a damage degree of 80% and 40% of the forest cover were considered at three different states in time, i.e. *(a)* immediately after the event and *(b)* 2 and *(c)* 7 years later. Figure 5.11 summarizes the construction and description of the parameter sets and their development over time – i.e. phases *(a)*, *(b)*, *(c)* – and space.

5.2.4 Results

A sensitivity analysis was carried out to illustrate the importance of input parameters and a baseline simulation was carried out to compare forest damage scenarios (for more details see Molnar et al. 2006; Kirsch and Burlando 2005; Hinz 2004). Table 5.1 lists the parameter values for the different scenarios and approaches considered in this paper. The investigated scenarios overall show similar impacts on runoff production, overland and channel flow velocity and depth (see Fig. 5.12). In the damaged areas, stronger runoff generation can be observed, associated with higher velocities due to smaller roughness.

Table 5.1. Parameter values for the base scenario simulation (see also Fig. 5.11)

Rainfall-runoff model			Erosion model		
$CN_{(forest/meadow/pasture)}$	77/74/76	(–)	$d_{16\,(channel)}$	22	mm
$k_{h\,(forest/meadow/pasture)}$	0.5/5/5	$m^{1/3}\,s^{-1}$	$\rho_{s\,(channel/hillslope)}$	2000/800	$kg\,m^{-3}$
k_{ch}	25	$m^{1/3}\,s^{-1}$	$\tau_{c\,(forest/meadow)}$	60–100/15	Pa
λ	0.05	(–)	α (Eq. 5.5)	6×10^4–2×10^5	
			β (Eq. 5.5)	1.66	
			χ (Eq. 5.5)	2.035	
			K (Eq. 5.4)	5×10^{-5}–1×10^{-4}	
			μ (Eq. 5.4)	1–1.5	

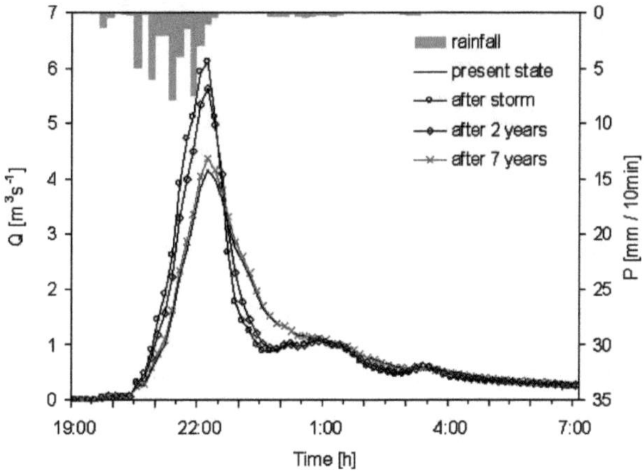

Fig. 5.12. Hydrograph comparison for the base and windstorm scenarios

This leads to faster flow concentration and higher peak flows as well as a faster flood wave rise, implying a potential increase of erosion in the channel. After 2 and 7 years, as the vegetation partially re-establishes, the basin response converges to that of the base scenario. This behavior reflects the changes of the model parameters corresponding to the land cover modifications.

Channel bedload sediment transport responds to the resulting flow scenarios accordingly. The 80% damage scenario (illustrated by Fig. 5.13a) produces stronger impacts on the hydrograph and sediment transport than the 20% one. Conversely, the simulation of overland flow erosion in response to scenarios shows a different behavior depending on the sediment transport formula used. Figure 5.13b shows that the shear stress and the runoff approach produce similar sedimentographs, but instantaneous values and volumes differ by a scale factor of about 0.5. The shear-stress approach was calibrated by Hinz (2004), and therefore assumed to be in the right order of magnitude. However, this suggests that a crucial point for any catchment scale quantitative modeling of overland sediment transport capacity is connected with the application and choice of sediment transport formulas. Their dependence

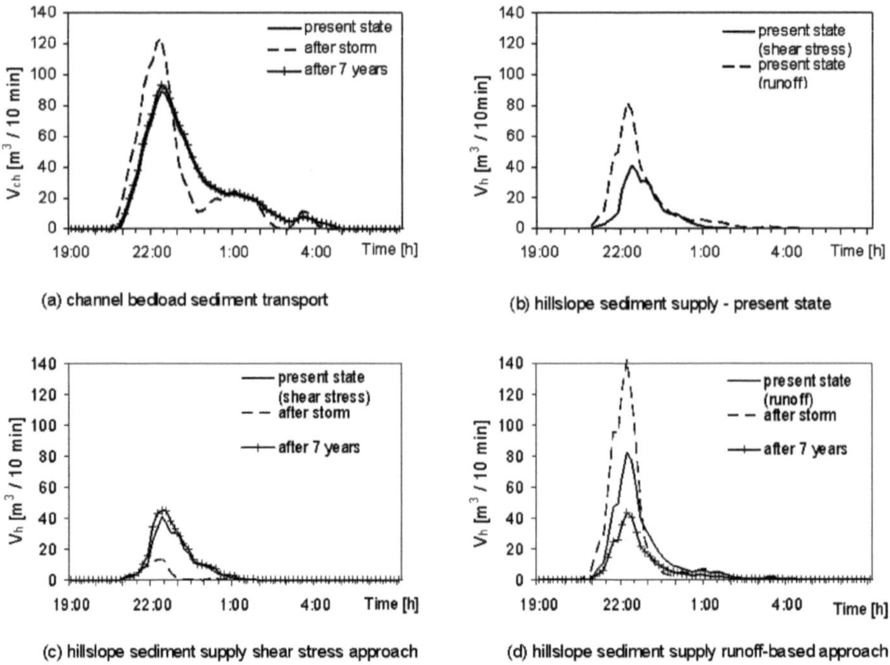

Fig. 5.13. Comparison of model results for different modeling approaches and different scenarios

on flow velocities, depths or flow rates reflects the changes of the surface hydraulics with a direct transfer to modeled erosion rates and transport capacities. This is well illustrated by Fig. 5.13c and d, showing, for a damage degree of 80%, a 28% decrease in hillslope sediment supply following the event when the shear stress based equation is used. Lower overland flow depths are the result of the balance between a drop in surface resistance and an increase in surface runoff following the event. On the other hand the runoff based equation predicts an increase in hillslope sediment supply due to the increase in surface runoff. As vegetation re-establishes in time, the volume of overland flow returns almost to the original conditions. The transport capacity volume still increases up to 118% because of the non-linear effect due to $\gamma \neq 1$ in Eq. 5.5.

5.2.5 Concluding Remarks

This study illustrates how scenarios and models may be developed and used to investigate in a spatially and temporally explicit fashion the impacts of land cover change on runoff, sediment yield and erosion/deposition patterns at the catchment scale. A distributed hydrological model was applied to several events with good results. The coupled hillslope erosion model that relates the transport rate to shear stress and flow depth, shows that especially hillslope sediment transport has to be tackled carefully.

Indeed, the choice of the sediment transport formula may significantly affect the magnitude of the erosion/deposition in relation to its functional dependence on the hydraulics of overland flow. In this sense, the modeler can influence the results by the way in which the equations are chosen and the model parameters are adjusted to reflect land cover change scenarios. Equations 5.4 and 5.5 use different hydraulic variables to determine the erosion rates. In the case of the shear stress approach, changes in surface roughness on hillslopes in the scenarios affect flow depth, which may decrease despite an increase in surface runoff following forest damage. The runoff approach directly reflects the increase in surface runoff due to the land use change. Both approaches can be used after calibration to match observed volumes in time-static modeling, meaning modeling based on the current state of the catchment. The differences shown in the resulting sedimentographs as well as in the predicted response to land cover changes (time-variable modeling) point out the importance of understanding the driving mechanism of erosion in relation to overland flow processes. The decision which of the two approaches is more appropriate depends on the laminar or turbulent flow characteristics and on soil erodibility. It is apparent from the literature that this question requires further experimental investigations across different scales to clarify open issues. To address these connections additional field studies have to be carried out in which shallow overland flow on different slopes and land cover is observed and analyzed together with sediment transport. Among these, erosion studies on natural hillslopes carried out by means of silt fences can be mentioned, as well as plot-size studies to evaluate man-made land cover changes, to study local effects for different slopes, soils and land cover combinations. This is necessary to overcome the limitations of insufficient or weak calibration and validation of existing models due to limited or lacking field data. In particular this targets conceptual models improving their predictive capabilities and acceptance.

Acknowledgments

This study was partially funded by the Grant SN20/02 of the Division "Schutzwald und Naturgefahren"/BAFU. Digital elevation data were provided by Swisstopo, land-use and soils data by Geostat and streamflow, sediment transport and precipitation data by WSL Birmensdorf.

References

Bathurst JC, Graf WH and Cao HH (1987) Bed load discharge equations for steep mountain rivers. In: Thorne CR, Bathurst JC, Hey RD (eds) Sediment transport in gravel-bed rivers. pp 453–491, Wiley & Sons Ltd.

Brunner GW (1989) Muskingum-Cunge Channel Routing. Lecture Notes, Hydrological Engineering Center, U.S. Army Corps of Engineers, Davis, CA

Burch H (1994) Ein Rückblick auf die hydrologische Forschung der WSL im Apltal. In: Hydrologie kleiner Einzugsgebiete, Beiträge zur Hydrologie der Schweiz, no. 35, pp 18–33, SGHL, WSL Birmensdorf, Switzerland (in German)

BUWAL (2002) (ed) Lothar Zwischenbericht, Materielle und Finanzielle Bilanz 2001. BUWAL, Ittigen

BWG (2003) Hochwasserabschätzung in schweizer Einzugsgebieten – Praxishilfe. Berichte des BWG, Nr. 4, Bern

Chaudhry MH (1993) Open-Channel Flow. Prentice Hall, Englewood Cliffs

Chow VT (1973) Open-Channel Hydraulics. Mc Graw-Hill Book Company

Cunge JA (1969) On the subject of a flood propagation computation method. J Hydraul Res 7(2):205–230

Engmann (1986) Roughness coefficients for routing surface runoff. J Irrig Drain Eng ASCE 112(1):39–53

Foster GR, Meyer LD (1972) A closed-form soil erosion equation for upland areas. In: Shen HW (ed) Sedimentation – Symposium to honor Prof. H.A. Einstein, 12.1–12.19, Colorado State University, Ft. Collins, USA

GEOSTAT (1997) User's manual of GEOSTAT database. Fed. Office of Statistics, Bern, Switzerland

Hessel R (2002) Modelling soil erosion in a small catchment on the Chinese Loess Plateau. Ph.D. Dissertation – Proefschrift, NBC: 38.09: fysische geografie. Universiteit Utrecht, Netherland

Hinz E (2004) Auswirkungen von Landnutzungsänderungen auf das Abfluss- und Erosionsverhalten. ETH Diploma Thesis, ETH Zürich, Switzerland, http://e-collection.ethbib.ethz.ch (in German)

Julien PY, Simons DB (1985) Sediment transport capacity of overland flow. Trans. ASAE 28(3):755–762

Kilinc MY (1972) Mechanics of soil erosion from overland flow generated by simulated rainfall. Ph.D. Dissertation, Colorado State University, Ft. Collins, USA

Kirsch J, Burlando P (2005) The Influence of Different Land Uses and Catchment Training on Runoff Production and Mitigation of Water Related Hazards in Small Mountainous Basins. Final Report of BUWAL contract SN 20/02, Chair of Hydrology and Water Resour. Manag., ETH Zürich, Switzerland (in German)

Kuntner R (2002) A methodological framework towards the formulation of flood runoff generation models suitable in alpine and prealpine regions. Ph.D. Dissertation, no. 14699, ETH Zürich, Switzerland, http://e-collection.ethbib.ethz.ch

Lässig R, Schönenberger W (1997) Was passiert, wenn man die Natur sich selber überlässt? – Ergebnisse der Sukzessionsforschung auf Windwurfflächen. Aus: Laufener Seminarbeiträge 1/1997, pp 76–74

Lässig R, Motschalow SA (2000) Wiederbewaldung nach Lothar (4). Vielfältige Strukturen nach Windwurf in Naturwäldern. Wald und Holz 81/12:39–43

Mitas L, Mitasova H (1998) Distributed erosion modelling for effective erosion prevention. Water Resour Res 34:505–516

Montaldo N, Mancini M, Rosso R (2004) Flood hydrograph attenuation induced by a reservoir system: analysis with a distributed rainfall-runoff model. Hydrol Process 18:545–563

Milzow C, Molnar P, McArdell BW, Burlando P (2006) Spatial organization in the step-pool structure of a steep mountain stream (Vogelbach, Switzerland). Water Resour Res 42, W04418, doi:10.1029/2004WR003870

Molnar P, Burlando P, Kirsch J, Hinz E (2006) Model investigations of the effects of land-use changes and forest damages on erosion in mountainous environments. In: Rowan JS, et al. (eds) Sediment Dynamics and the Hydromorphology of Fluvial Systems. IAHS Publ. 306, pp 589–600

Nepf HM (1999) Drag, turbulence, and diffusion in flow through emergent vegetation. Water Resour Res 35(2):479–489

Ponce VM (1983) Development of Physically Based Coefficients for the Diffusion Method of Flow Routing, Final Report to the USDA, Soil Conservation Service. Lanham, MD

Ponce VM, Yevjevich V (1978) Muskingum-Cunge method with variable parameters. ASCE Hydraul Div HY12:1663–1667

Prosser I, Rustomij P (2000) Sediment transport capacity relations for overland flow. Prog. Phys. Geogr. 24(2):179–193

Rickenmann D (1997) Sediment transport in Swiss torrents. Earth Surf. Proces. Landf. 22:937–951

SCS – Soil Conservation Service (1972) National Engineering Handbook, Section 4, Hydrology. U.S. Department of Agriculture, Washington D.C., U.S.A.

Shen HW, Li RM (1973) Rainfall effects on sheet flow over smooth surface. Journal of Hydraulics Divisions, ASCE:771–792

Walthert L, Blaser P, Lüscher P, Luster J, Zimmermann D (2003) Langfristige Waldöokosystem-Forschung LWF in der Schweiz. Kernprojekt Bodenmatrix, Ergebnisse der ersten Erhebung 1994–1999, ETH Zürich, Switzerland, http://e-collection.ethbib.ethz.ch (in German)

Martina Baborowski · Frank Krüger · Olaf Büttner · Peter Morgenstern
Ingo Lobe · Wolf von Tümpling · Holger Rupp · Helmut Guhr

5.3 Transport and Fate of Dissolved and Suspended Particulate Matter in the Middle Elbe Region during Flood Events

5.3.1 Introduction

Although flood events are natural events, they may have an impact on the condition of a river system, i.e. when polluted deposits are involved in the process. The quality of the transported matter is mainly affected by the origin of the flooding water as well as the remobilization of deposits within the catchment area (e.g., groyne fields, lock-and-weir systems, mining and industrial areas, sewage plants). Transport and fate of a contaminant in the water body are significantly influenced by the ratio of the concentrations in the dissolved and the particulate phase. Depending on their morphology, floodplains act as sink for suspended matter (Engelhardt et al. 1999; Friese et al. 2000; Hanisch et al. 2005; Costa et al. 2006).

In the Middle Elbe groyne fields are the characteristic morphological feature. The resuspension of deposited, not yet consolidated groyne field sediments can lead to the first SPM (suspended particulate matter) peak in the course of a flood, a long time before the flood crest has arrived (Spott and Guhr 1996; Baborowski et al. 2004). All further peaks are the result of matter flow and inputs further upstream of the location, e.g., tributary inflows. Therefore the remobilization processes in groyne fields upstream of a sampling location play the most important role for the right starting time of a flood sampling campaign (Baborowski et al. 2005).

The assessment of sediment dynamic and pollutant mobility in rivers requires the integration of various experimental and modeling techniques (Förstner et al. 2000; Förstner 2004). They have to include aspects of erosion (Gust and Müller 1997; McNeal et al. 1996; Haag et al. 2001), transport (Brunk et al. 1997) and deposition (Asselmann and Middelkoop 1995; Kronvang et al. 2002; Krüger et al. 2006). In the paper results of investigations on transport and deposition in the middle part of the river Elbe are presented.

5.3.2 Aim

Former flood investigations along the Elbe River are not consistent with respect to sampling locations, begin of sampling, sampling times and methods. Furthermore they often do not consider the travel time of matter transport within the river (Baborowski et al. 2005). To overcome these deficits, investigations were performed considering both, local specific erosion values of the discharge as well as the travel time of the flood wave along a river stretch. The results provide a basis for a better understanding of the flood dependent transport processes in the Middle Elbe.

To assess the risk potential of the polluted river SPM for the floodplains, a well described floodplain (Büttner et al. 2006) was chosen to show exemplarily the potential sink function of floodplains for SPM transported with the flood wave.

5.3.3 Study Site

The investigations were carried out in the middle part of the river Elbe at river km 318, left bank near Magdeburg, and at river km 454, left bank in Wittenberge (German mileage). The considered stretch covers 136 km of the total length of the German part of the Elbe, which altogether amounts to 730 km.

There are approximately 1 950 typical groynes located between the two sampling sites. Along the whole course of the river Elbe in Germany there are approximately 6 900 groyne fields. The geographical position of the sampling sites is shown in Fig. 5.14.

The Sampling Site Magdeburg is part of the monitoring program of the International Commission for the Protection of the Elbe (IKSE/MKOL). The water quality at this site depends on inputs of stretches of the upper Elbe (Czech part, Dresden industrial region) as well as inputs from the polluted tributaries Mulde and Saale. Their confluences are 59 km and 27 km upstream of the location respectively. Therefore under normal discharge conditions the water quality at this point represents the pollution situation of the Middle Elbe.

Between sampling site Magdeburg and the Sampling Site Wittenberge no further significant pollution input occurs. The total retention area between both sites is around 19 700 ha (IKSE/MKOL 2001). The Floodplain Schönberg (upstream of Wittenberge) covers an area of ~200 ha.

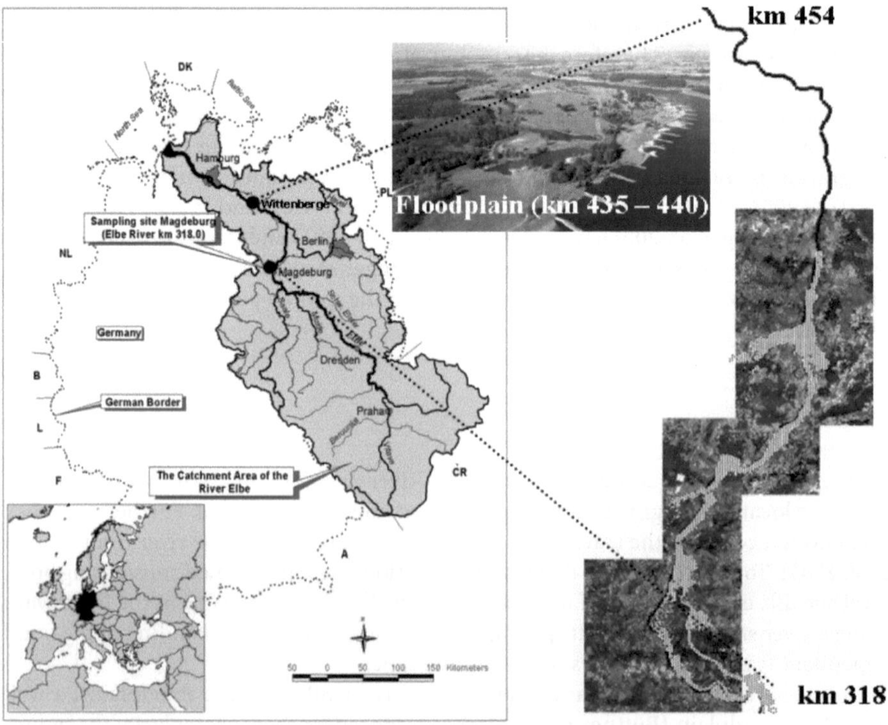

Fig. 5.14. Catchment area of the river Elbe with position of the sampling sites

5.3.4 Methodology

Investigations of two different flood waves were performed during a spring flood in February and March 2005, to get a better understanding of the main pollution potentials and pollution pathways in the Middle Elbe catchment area.

Sampling Strategy

Daily water samples were taken at the sampling sites in Magdeburg and Wittenberge considering local specific discharge threshold values as starting point for the measurement campaign.

For the sampling site Magdeburg at river km 318 the discharge threshold is about 800 $m^3 s^{-1}$ (Baborowski et al. 2004; Spott and Guhr 1996). For Wittenberge a threshold value of 1 080 $m^3 s^{-1}$ was chosen. At this critical value the Elbe water begins to flow into the floodplain.

The investigations were supplemented by monitoring deposited matter in the floodplain Schönberg using artificial lawns as sediment traps (Friese et al. 2000; Krüger et al. 2006). The artificial lawns aim at the simulation of the natural soil coverage (pasture or mown grassland). With respect to the natural topography, the traps were arranged at sampling points which cover a typical spectrum of morphological units in riparian areas like bayou of flood channel, not drained depression and plateau positions.

Main Parameter Water Samples

Dry weight of SPM was measured according to German Industrial Standards (DIN 38409 part H2, filtration onto Whatman GF/F glass fiber filter). The particle size distribution of SPM was counted in the range of 2 to 200 µm using an optical instrument, based on single particle evaluation (Baborowski 2002).

Heavy metals and arsenic (As) were analyzed in filtered (<0.45 µm) and unfiltered samples. The filtered samples were acidified with HNO_3, the unfiltered digested with HNO_3/H_2O_2 in a microwave equipment.

Aluminum (Al), iron (Fe), manganese (Mn) and zinc (Zn) were determined by optical emission spectrometry with inductively coupled plasma (ICP-OES), while arsenic, cadmium (Cd), chromium (Cr), copper (Cu), lead (Pb), nickel (Ni), uranium (U) and zinc (Zn) were measured using mass spectrometry also with inductively coupled plasma (ICP-MS).

Floodplain Sediments

To estimate the sediment deposition during the flood artificial lawns which act as sediment traps were arranged at 7 positions (5 replicates) on the floodplain.

After drying, the collected sediments were analyzed by means of energy dispersive X-ray fluorescence (EDXRF) with regard to their contents of heavy metals and arsenic.

Total organic carbon of the sediments was analyzed in triplicate with a C(H)NS analyzer (Elementar vario EL) after drying at 105 °C and grinding of the samples.

Load Calculation

The deposition (D) between Magdeburg and Wittenberge (taking into account a constant travel time of 2 days) was calculated as:

$$D = \sum_{i} \left[F_{\mathrm{M}}(t_i) - F_{\mathrm{W}}(t_{i+2}) \right] \;, \quad i = 1,\ldots,n \tag{5.6}$$

with
- $F_{\mathrm{M}}(t_i) = Q_{\mathrm{M}}(t_i)C_{\mathrm{M}}(t_i)$ and $F_{\mathrm{W}}(t_i) = Q_{\mathrm{W}}(t_i)C_{\mathrm{W}}(t_i)$
- F_{M}, F_{W}: contaminant load at Magdeburg and Wittenberge, respectively
- Q_{M}, Q_{W}: discharge at Magdeburg and Wittenberge, respectively
- C_{M}, C_{W}: contaminant concentration at Magdeburg and Wittenberge, respectively
- t_i: i^{th} date

5.3.5 Results and Discussion

The particle size distribution of the transported SPM exemplarily is given in Fig. 5.15 for the Magdeburg sampling site. In agreement with the results of prior studies (Baborowski et al. 2004; Baborowski 2002) changes of the particle number concentrations in the range between 2 and 20 µm dominated during both flood waves. Particles < 20 µm constitute the largest portion of the total particle number (Fig. 5.16). This fraction is especially important for the transport of pollutants which tend to sorption. Their portion of mass of the SPM is small.

From Fig. 5.15 the remobilization of sediments at the early stage of the flood can be seen. Thereby the peaks of particle number concentrations of both grain size fractions as well as the concentration of SPM occurred at the same time (Fig. 5.16). While the concentrations of SPM and of the particle fraction > 20 µm decreased continuously after the maximum is reached, the particle concentrations < 20 µm remained elevated for longer, especially during the second wave. This indicates the supply of fine grained sediments from the tributaries Mulde and Saale. Owing to the storage capacities in its lower reach (Mulde reservoir, Saale lock-and-weir systems), a shift towards fines rather than coarse particles has been demonstrated in prior investigations (Baborowski et al. 2004). Hence, the flood dependent matter transport in the middle part of the river Elbe

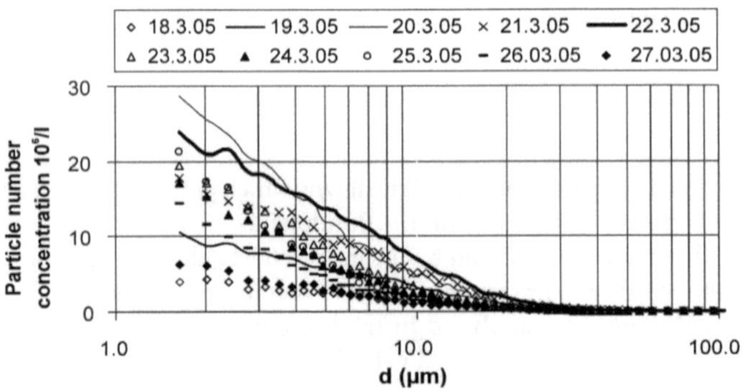

Fig. 5.15. Particle size distribution of the SPM at river km 318

Fig. 5.16.
Particle number concentration (*N*) and dry substance (*DS*) of the SPM

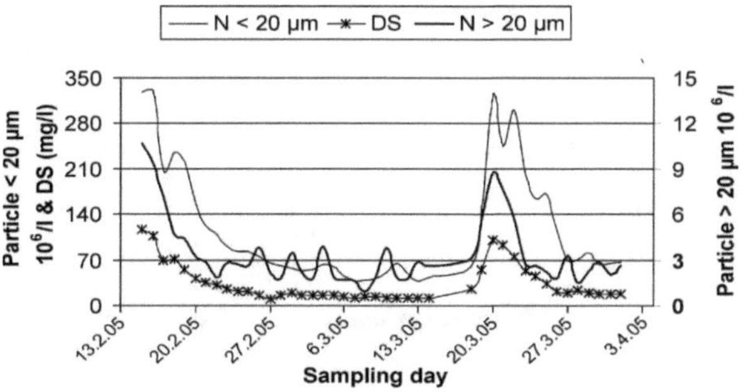

Fig. 5.17.
Discharge (*Q*) and dry substance (*DS*) of SPM at river km 318 and river km 454

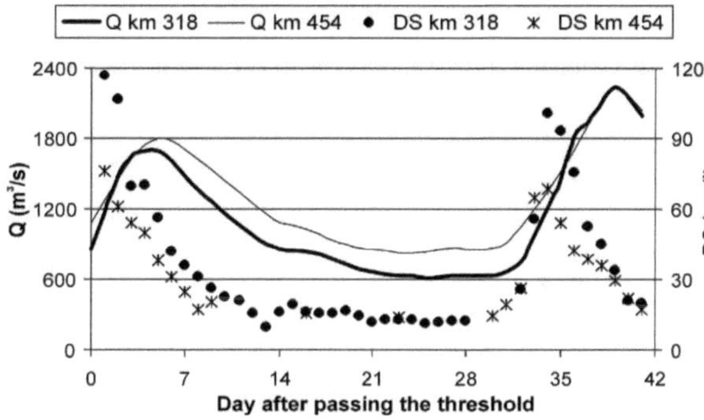

seems to be rather limited by transport capacity of the groyne field sediments than by supply of sediments from the tributaries Mulde and Saale.

The travel time of the Elbe River between Magdeburg and Wittenberge was about 2 days. Therefore, if flow data of river km 318 are compared with flow data of river km 454, the corresponding time series are shifted by two days to account for the time the water takes to flow along this river stretch. Consequently the data on the *x*-axis are normalized by means of the time since the threshold has been exceeded for the first time.

As expected the highest SPM concentrations at both sampling sites were measured directly after passing the discharge threshold due to remobilization of groyne field sediments (Fig. 5.17).

This result is important for transport modeling. To determine initial conditions as well as boundary conditions in numerical pollution transport modeling, suspended sediment concentrations are required. However, if the number of available data is insufficient, it is necessary for reliable model calculations to provide the typical graph of SPM concentration based on the sparse data available. Therefore, local knowledge of the typical course of SPM during a flood event is essential. E.g. it has to be taken into account during all steps of model processing that the SPM maximum occurs prior to the maximum discharge (Lawler et al. 2006). Uncertainties in forecasting flood extend and depositions are often caused by uncertainties in boundary conditions rather than uncertainties in model parameters (Lane 1998).

Moreover the results, shown in Fig. 5.17, indicate decreasing SPM values along the river stretch between river km 318 and 454. Corresponding to the SPM, the concentrations of heavy metals and As decreased as well (Fig. 5.18).

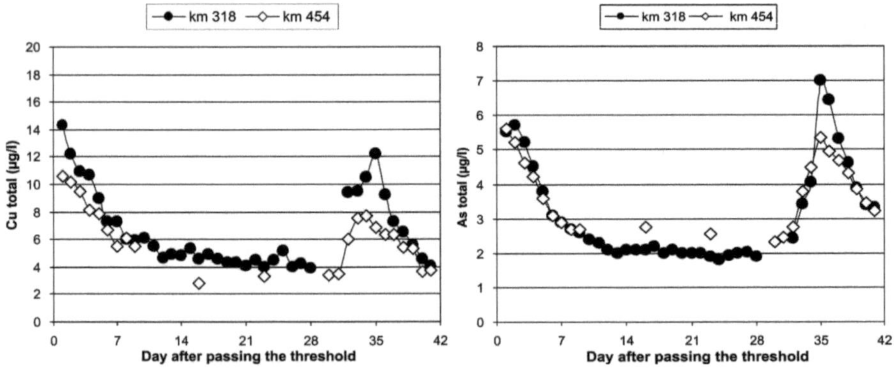

Fig. 5.18. Graph of Cu (*left*) and As (*right*) at river km 318 and river km 454

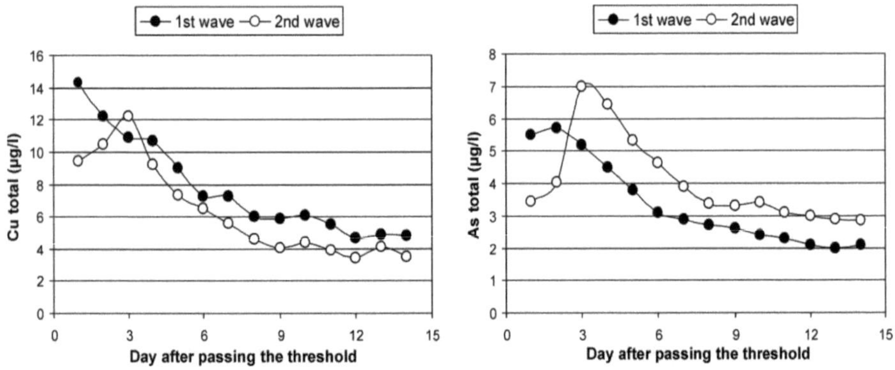

Fig. 5.19. Comparison of both waves for Cu (*left*) and As (*right*) at river km 318

Graphs of concentrations during the first and the second wave show a different course. To give an example, the comparison of both waves is presented for the sampling site Magdeburg in Fig. 5.19. The maximum concentrations of Cu and As do not show the same values in both waves and also occur at different times after passing the discharge threshold, indicating the influence of different sources of pollution.

The ore and companion elements Cu and As are typical pollutants originating from former mining activities in the catchment areas of the Elbe tributaries Mulde and Saale. Despite of a significant decrease of pollution in the river Elbe over the last fifteen years, contaminations with heavy metals, e.g., in sediments, are still present in the river basin (Heise et al. 2005). During flood the pollutants of these secondary deposits are remobilized by erosion and hence, appear in the water body. Due to regional geographic peculiarities in the Middle Elbe as well as results of prior studies, the element As can be considered as indicator for pollution inputs of the river Mulde, whereas Cu acts as indicator for inputs of the river Saale (former copper shale mining). Therefore, together with the evaluation of the discharge development in the Elbe and its tributaries (data not shown) during the flood, the dominating influence of the Saale on the first wave and of the Mulde on the second wave is reflected in Fig. 5.19.

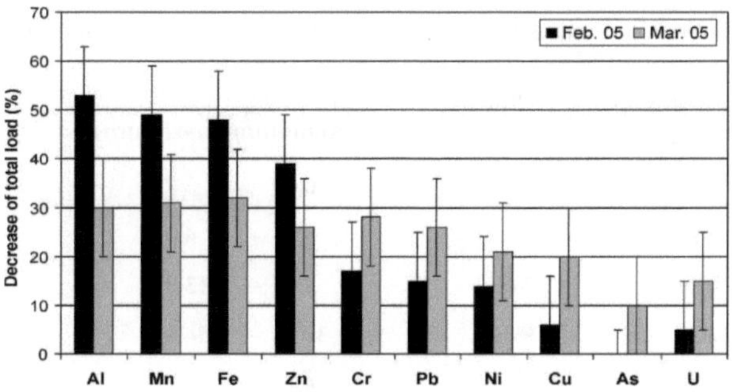

Fig. 5.20.
Decrease of total element load between river km 318 and river km 454

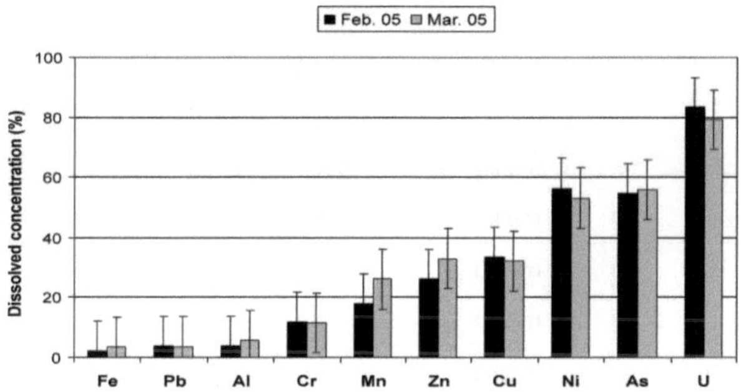

Fig. 5.21.
Percentage of transport in dissolved phase

An overview of the longitudinal matter transport is given in Fig. 5.20. It shows that over the entire flood event, pollutants are retained in the river reach under consideration.

The trace metal loads decreased by up to 50% between the two sampling sites. Thereby, the decrease of element loads varied significantly between the pollutants. It was highest for clay reference compounds (Al, Fe, Mn). The trapping of the clay fraction by sedimentation is evident and much larger than observed for the other, also mainly in particulate form transported elements Cr, Pb and Zn (Fig. 5.21). On this occasion a grain-size effect seems to be the reason for the different settling behavior of the pollutants. In contrast, the distribution between dissolved and particulate phase is more decisive for the observed different transport behavior of the elements Ni, Cu, As and U. Especially Ni, As and U were mainly transported in the dissolved fraction (55 to 80%, Fig. 5.21), responsible for the transport to the sea.

Moreover, the decrease of element loads for Al, Fe, Mn and Zn was higher during the first wave than observed for the second wave (Fig. 5.20). This can be interpreted as indication that these elements are originate to a larger extent from the groyne fields and exhibit, in contrast to the other elements, a hysteresis behavior. Cyclically decreasing SPM and element concentrations in the case of Elbe flood waves following each other in quick succession have been described in detail by Spott and Guhr (1996). However, this result together with the findings discussed for Fig. 5.16 underlines the hypothesis that the transport in the Middle Elbe is rather limited by transport capacity than by supply of sediments.

In summary, the reduction of loads can be explained by high retention due to sedimentation in floodplains, old arms and slack water zones. For example, the SPM retention between river km 318 and km 454 was calculated to be about 19 000 t for the first

Table 5.2. Composition of the sediments deposited on the floodplain Schönberg

Parameter	Unit	1st wave, $n = 7$ Minimum–maximum	2nd wave, $n = 7$ Minimum–maximum
Deposition	$t\ ha^{-1}$	0.09 – 1.8	1.0 – 4.0
Si/Al	–	3.1 – 4.0	3.6 – 4.6
Al_2O_3	%	7.8 – 13.8	10.4 – 11.8
SiO_2	%	39.4 – 54.9	45.8 – 53.9
Fe_2O_3	%	4.4 – 6.7	5.0 – 6.1
C	$g\ kg^{-1}$	71 – 134	80 – 115
N	$g\ kg^{-1}$	6.7 – 14.1	8.2 – 12
S	$g\ kg^{-1}$	2.2 – 4.7	2.5 – 3.3
Mn	$mg\ kg^{-1}$	492 – 2556	1775 – 4593
Zn	$mg\ kg^{-1}$	700 – 3350	943 – 1182
As	$mg\ kg^{-1}$	26 – 56	47 – 63
Cr	$mg\ kg^{-1}$	83 – 128	106 – 123
Cu	$mg\ kg^{-1}$	104 – 173	96 – 128
Ni	$mg\ kg^{-1}$	42 – 67	49 – 84
Pb	$mg\ kg^{-1}$	158 – 1138	141 – 163
U	$mg\ kg^{-1}$	0 – 4	2 – 4

flood wave. This is about 25% of the total SPM load passing km 318. Balancing these findings with the total floodplain area of 19 700 ha (MKOL/IKSE 2001), a deposition of nearly 1 t ha^{-1} can be estimated for the first flood wave.

The calculations of the deposition in the sediment traps of the floodplain Schönberg underline these findings (Table 5.2).

The average SPM deposition was 0.98 t ha^{-1} for the first wave and 2.02 t ha^{-1} for the second wave. The overall retention between km 318 and km 454, basing on the sediment trap investigations of both waves for the whole period of floodplain inundation can be estimated for Zn: 68 t, Pb: 13 t, Ni: 3 t, Cu: 7 t, As: 3.3 t and U: 0.15 t.

This demonstrates a potentially risk for the land use resulting from the function of floodplains as a sink for polluted SPM transported to the sea. Nevertheless, the determined deposition rates and contamination loads indicate that single floods in the area of the Lower Middle Elbe can hardly influence the quality of the soils in the central floodplain whose pollutant loads are the result of a pollutant input for decades and centuries. On the one hand, still today the contamination of the recent flood sediments corresponds to a large extent to the quality of the topsoils (Krüger et al. 2005). On the other hand, 0.09–4.0 t ha^{-1} sediment input (Table 5.2) with an assumed bulk density of 1 g cm^{-3} (after Krüger et al. 2005) solely amounts to a maximum of 0.6% of the 0–10 cm soil depth interval that is used as standardized soil sampling. This is not sufficient to alter the pollution state of this soil depth interval significantly.

5.3.6 Conclusions

Discharge related investigations allow a better understanding of the flood dependent matter transport. In view of both, local specific erosion thresholds as starting times for the measurement at a sampling site as well as the travel time along the regarded river stretch, different pollution pathways in the river system can be detected. Furthermore a more detailed description of the floodplains sink function for sediments and pollutants becomes possible, if instream and floodplain investigations are combined.

Acknowledgments

The authors would like to thank the staffs of the UFZ departments River Ecology, Soil Physics and Analytics for assisting in field and laboratory work.

Parts of the work were supported by the European Union FP6 Integrated Project AquaTerra (Project no. 505428) under the thematic priority sustainable development, global change and ecosystems.

References

Asselmann NEM, Middelkoop H (1995) Floodplain sedimentation: quantities, patterns and processes. Earth Surf. Process. Landforms 17:687–697

Baborowski M, Claus E, Friese K, Pelzer J, von der Kammer F, Kasimir P, Heininger P (2005) Comparison of different monitoring programs of the 2002 summer flood in the river Elbe. Acta Hydrochim Hydrobiol 22(2005):404–417

Baborowski M, von Tümpling W, Friese K (2004) Behaviour of suspended particulate matter (SPM) and selected trace metals during the 2002 summer flood in the river Elbe (Germany). Hydrol Earth Syst Sc 8(2):135–150

Baborowski M (2002) Characterisation of Suspended Particulate Matter (SPM) in the river Elbe (Germany) by survey of the particle size distribution. Dresden, vol. II Matter and particle transport in surface and subsurface flow. In: Water Resources and Environment Research. Proceedings of ICWRER 2002, pp 23–27

Büttner O, Otte-Witte K, Krüger F, Meon G, Rode M (2006) Numerical modeling of floodplain hydraulics and suspended sediment transport and deposition at the event scale in the middle river Elbe, Germany. Acta Hydrochim Hydrobiol 34(3):265–278

Brunk BK, Weber-Shirk M, Jensen-Lavan A, Jirka GH, Lion LW (1996) Modeling natural hydrodynamic systems with a differential-turbulence column. J Hydraul Eng 122:373–380

Costa AT, Arias Nalini Jr. H, de Tarso Amorim Castro P, Carvalho de Lena J, Morgenstern P, Friese K (2006) Sediment contamination in floodplains and alluvial terraces as an historical record of gold exploitation in the Carmo River basin, Southeast Quadrilatero Ferrifero, Minas Gerais, Brazil. Acta Hydrochim Hydrobiol 34(3):245–256

Engelhardt C, Krüger A, Karrasch B, Baborowski M (1999) Input-output balances of nutrients and plankton in a flooded area of the lower Odra. Acta Hydrochim Hydrobiol 27:325–330

Förstner U (2004) Sediment dynamics and pollutant mobility in rivers: An interdisciplinary approach. Lakes and Reservoirs: Research and Management 9:25–40

Förstner U, Jirka GH, Lang C, et al. (2000) Significance of sediments in aquatic ecosystems – interdisciplinary process studies on fine sediment dynamics and pollutant mobility in flowing waters. In: Assessment of chemicals – Concepts for Sediments and Marine Ecosytems (in German). pp 75–109, 8th BUA-Colloquium, GDCh Monograph 17. 11 January 1999, Frankfurt,Germany

Friese K, Witter B, Brack W, Büttner O, Krüger F, Kunert M, Rupp H, Miehlich G, Gröngröft G, Schwartz R, van der Veen A, Zachmann DR (2000) Distribution and fate of organic and inorganic contaminants in a river floodplain – results of a case study on the river Elbe, Germany. In: Wise DL, Trantolo D, Cichon EJ, Inyang HI, Stottmeister U (eds) Remediation Engineering of Contaminated Soils. Marcel Dekker, New York, Basel, pp 375–428

Gust G, Müller V (1997) Interfacial hydrodynamics and entrainment functions of currently used erosion devices. In: Burt N, Parker R, Watts J (eds) Cohesive Sediments. pp 149–174, John Wiley & Sons, Chichester

Haag I, Kern T, Westrich B (2001) Erosion investigation and sediment quality mesurements for a comprehensive risk assessment of contaminated aquatic sediments. Sci Total Environ 266:249–257

Hanisch C, Zerling L, Junge FW, Czega W (2005) Verlagerung, Verdünnung und Austrag von schwermetallbelasteten Flusssedimenten im Einzugsgebiet der Saale. Abhandlungen der Sächsischen Akademie der Wissenschaften zu Leipzig. Mathematisch-naturwissenschaftliche Klasse 64(1), 164 pp

Heise S, Claus E, Heininger P, Krämer Th, Krüger F, Schwartz R, Förstner U (2005) Studie zur Schadstoffbelastung der Sedimente im Elbeeinzugsgebiet, Ursachen und Trends. Im Auftrag der Hamburg Port Authority

IKSE/MKOL (2001) Bestandsaufnahme des vorhandenen Hochwasserschutzniveaus im Einzugsgebiet der Elbe. Magdeburg

Kronvang B, Falkum O, Svendsen LM, Laubel A (2002) Deposition of sediment and phosphorus during overbank flooding. Verh Int Verein Limnol 18:1289–1293

Krüger F, Schwartz R, Kuhnert M, Friese K (2006) Methods to calculate sedimentation rates of floodplain soils in the middle region of the Elbe River. Acta Hydrochim Hydrobiol 34(3):175–187

Krüger F, Meissner R, Gröngröft A, Grunewald K (2005) Flood induced heavy metal and arsenic contamination of Elbe River floodplain soils. Acta Hydrochim Hydrobiol 33(5):455–465

Lawler DM, Petts GE, Foster IDL, Harper S (2006) Turbidity dynamics during spring storm events in an urban headwater river system: The Upper Tame, West Midlands, UK. Sci Total Environ 360:109–126

Lane SN (1998) Hydraulic modelling in hydrology and geomorphology: a review of high resolution approaches. Hydrological Processes 12:1131–1150

McNeal J, Taylor C, Lick W (1996) Measurements of erosion of undisturbed bottom sediments with depth. J Hydraul Eng 122:316–324

Spott D, Guhr H (1996) The dynamics of suspended solids in the tidally unaffected area of the river Elbe as function of flow and shipping. Arch Hydrobiol Spec Issues Advanc Limnol 47:127–133

Björn Tetzlaff · Frank Wendland

5.4 Modeling P-Fluxes from Diffuse and Point Sources in Heterogeneous Macroscale River Basins Using MEPhos

5.4.1 Introduction

As the inventory of the EU-water framework directive (EU-WFD) revealed in 2004, high nutrient inputs are a major concern for most inland and coastal waters in Germany. Consequently the "good status" in water quality will not be achieved for a high percentage of water bodies. To improve water quality until 2015 the EU-WFD demands the drawing-up of detailed river basin management plans and programmes of measures until 2009. This provokes a demand for suitable instruments and models resp. to be applied for the area- and pathway-differentiated quantification of nutrient inputs. Model applications should aim at the localization of critical source areas within river basins as a basis for proposing measures of emission control which are adapted to site properties.

In the following the new empirical emission model MEPhos is presented, which enables a systemic quantification of P-loads in large heterogeneous river basins (Tetzlaff 2006). MEPhos differentiates between diffuse and point sources as well as between eight pathways.

The model is applied for the macroscale basins of the river Ems (12 900 km²) and parts of the river Rhine (12 200 km²). The river Ems basin is to be found in the north-western German lowlands and is dominated by sandy and boggy soils under intensive agricultural use. The investigation area Rhine is characterized by upland conditions with forest and grassland use. Furthermore the catchment comprises of sub-regions with high industrial and population densities (partly up to 2 900 citizens per km²).

5.4.2 MEPhos Model Description

The new model MEPhos consists of following features (Tetzlaff 2006, Fig. 5.22):

- Area-differentiated modeling approach for the quantification of mean annual P-inputs from point and diffuse sources to surface waters
- Differentiation between the eight pathways artificial drainage, groundwater-borne runoff, erosion, wash-off, rainwater sewers, combined sewer overflows, municipal waste water treatment plants, industrial effluents
- Consideration of P-retention in running and standing waters

The area-differentiated modeling approach to quantify diffuse P-inputs via artificial drainage, groundwater-borne runoff, erosion and wash-off is based on phosphotopes (Fig. 5.22). Phosphotopes are regarded as homogeneous types of sub-areas representing discontinuous source areas for diffuse phosphate inputs. Analogous to hydrotopes,

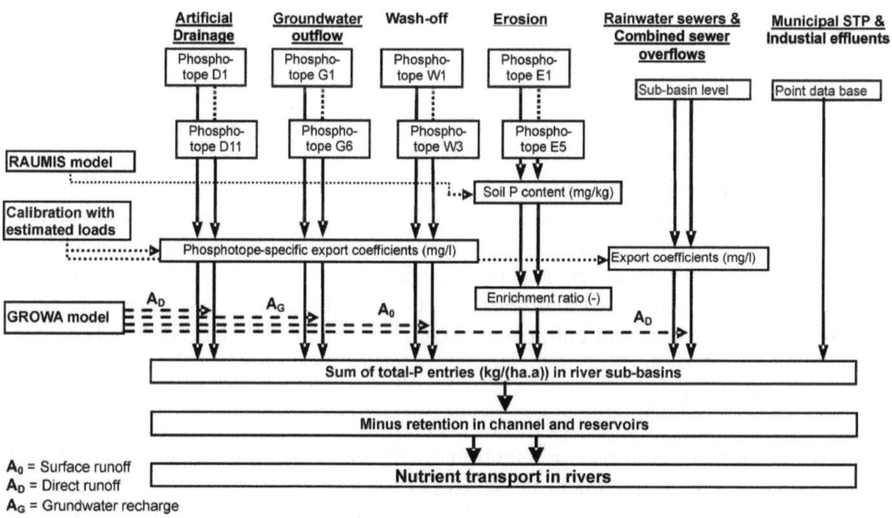

Fig. 5.22. Overview of the MEPhos model

phosphotopes include a set of different parameters, which control P-emissions. These parameters like soil types, land use and hydraulic connectivity to surface waters can be quantified for macroscale investigation areas by already available data sets from federal and state authorities as well as by results from hydrologic and agro-economic models. Phosphotopes are processed in GIS by classification and intersection of data sets. Details are given in Tetzlaff et al. (2007).

After modeling mean annual P-inputs via all eight pathways the emissions are summed up for river sub-basins related to water quality gauges (Fig. 5.22). Then the mean annual load of an upstream sub-basin is added and the P-retention subtracted (Eq. 5.7). This enables a validation of modeled P-loads by comparison with mean annual P-loads estimated from measured water quality and discharge data.

$$F_{\mathrm{EZG}} = \left[\left(\sum F_{\mathrm{Dr}} + \sum F_{\mathrm{Gw}} + \sum F_{\mathrm{Abschw}} + \sum F_{\mathrm{Eros}} + \sum F_{\mathrm{Tk}}\right) + \sum F_{\mathrm{KA}} + \sum F_{\mathrm{ID}} + \sum F_{\mathrm{Mw}}\right] + F_{\mathrm{OL}} - R_{\mathrm{F}} - R_{\mathrm{S}} \quad (5.7)$$

F_{EZG} = estimated load for river sub-basin, F_{Dr} = P-inputs via artificial drainage, F_{Gw} = P-inputs via groundwater-borne runoff, F_{Abschw} = P-inputs via wash-off, F_{Eros} = P-inputs via erosion, F_{Tk} = P-inputs via rainwater sewers, F_{KA} = P-inputs from municipal waste water treatment plants, F_{ID} = P-inputs from industrial effluents, F_{Mw} = P-inputs via combined sewer overflows, F_{OL} = P-load of an upstream river sub-basin, R_{F} = P-retention in running waters, R_{S} = P-retention in standing waters.

The retention of P due to settling and biogeochemical transformation within surface waters is modeled separately for running and standing waters. Behrendt and Opitz (2000) have developed an approach to describe the retention in rivers integratively, which is applied in the MEPhos model (Eq. 5.8). The coefficients a and b account for the influence of the catchment size and were quantified by Behrendt and Opitz (2000).

$$F = \frac{1}{1 + R_{\mathrm{F}}} E \quad \text{and} \quad R_{\mathrm{F}} = a q^{b} \quad (5.8)$$

F = modeled load for running waters (t a^{-1}), R_{F} = factor to describe P-retention in running waters (–), E = sum of modeled P-emissions (t a^{-1}), q = discharge per unit area (l s^{-1} km^{-2}), a, b = coefficients (–).

Equation 5.8 cannot record the increased P-retention occurring e.g. in reservoirs due to reduced flow velocity, increased travel time as well as higher sedimentation rate (Alexander et al. 2002; Kirchner and Dillon 1975). As a consequence, the particulate P-load transported as suspended matter decreases. According to Molot and Dillon (1993) P-retention in reservoirs and enbarraged lakes can be calculated as follows (Eq. 5.9):

$$R_{\mathrm{S}} = \prod \exp(-k_{\mathrm{r}} q^{-1}) \quad (5.9)$$

R_{S} = P-retention in reservoirs, k_{r} = retention coefficient ("settling velocity") (m a^{-1}), q^{-1} = reciprocal areal hydraulic load of reservoirs (a m^{-1}).

The following section focuses on modeling P-inputs to surface waters via erosion and presents an example of MEPhos results for erosion-related inputs.

5.4.3 Modeling Sediment and P-Inputs to Surface Waters Via Erosion

P-inputs to surface waters via erosion are controlled by soil loss from arable land within river basins, sediment delivery ratio, P-content in the top soil as well as by P-enrichment during the erosion process (Frede and Dabbert 1999; Auerswald 1989). Therefore diffuse P-inputs to surface waters via erosion are modeled after Eq. 5.10, based on the universal soil loss equation adapted to German conditions (Schwertmann et al. 1990):

$$E = (R \cdot K \cdot LS \cdot C) \cdot S \cdot PG \cdot ER \tag{5.10}$$

E = mean annual P-inputs via erosion (kg ha^{-1} a^{-1}); R = rain erosivity factor (N ha^{-1} a^{-1}); K = soil erodibility factor (t h ha^{-1} N); LS = combined slope length and steepness factor (–); C = crop and management factor (–); S = sediment delivery ratio (%/100); PG = content of total-P in the top soil (mg kg^{-1}); ER = enrichment ratio (–).

Because of the high sensitivity of the factors L and S (Auerswald 1989) they are not quantified by standard literature values, but calculated as LS-factor using algorithms of Moore and Wilson (1992). The slope shape is another important relief feature, because it controls the concentration of surface runoff (Auerswald et al. 1988). For modeling erosion with MEPhos the slope shape is taken into account by correction factors after Prasuhn and Grünig (2001) for the modification of the LS-factor.

With regard to proposals for efficient eutrophication reduction measures (Sect. 5.4.5), modeling of diffuse P-inputs by MEPhos has to be performed on the basis of phosphotopes, i.e. area-differentiated. Phosphotopes for modeling erosion inputs represent source areas for release of sediment and particulate-P and are made up by erodible arable land which is hydraulically connected to the network of rivers and flow paths. Morphological flow paths are modeled using a highly-resolved digital terrain model and the algorithm D Infinity (D∞) (Tarboton 1997). The parameters needed for the application of D∞ are calibrated with flow paths mapped in a test area in order to guarantee a representative and reliable simulation of flow paths. After selecting those flow paths, which are connected to the river network, the remaining paths are buffered with stripes of 2×30 m width according to findings of Sommer and Murschel (1999) and Fried et al. (2000) resp. Only those sub-areas contained in the 60 m strips are regarded as hydraulically connected. Then phosphotopes for modeling P-entries via erosion are derived by intersecting the data sets of erodible arable land with the buffered network of flow paths and rivers. Depending on their erosion potential five different phosphotopes are distinguished. By performing this process of disaggregating the eroding and hydraulically connected arable land, sediment deliveries to surface waters are also determined. MEPhos sediment deliveries vary between 3 and 29% with a mean of 11%. Werner et al. (1991) estimated the mean sediment delivery of former Western Germany at 8%.

According to Eq. 5.10 the level of particulate P-inputs results not only from the sediment delivered to surface waters, but also from the P-content of the top soil and the enrichment ratio. The calculation of the total-P content in the top soil is based on P-surpluses and clay content following Behrendt et al. (1999). P-surplus is modeled on a county level employing the agro-economic model RAUMIS developed and run by the German Federal Agricultural Research Centre (IAP and FAL 1996), the clay content is given by soil maps on a scale of 1:50 000 (Fig. 5.22).

For considering the selective effect of erosion induced by water an enrichment ratio has to be determined. For this a method is employed in the MEPhos model requiring measured water quality, above all about the P-content of suspended matter (Behrendt et al. 1999). According to Eq. 5.11 the enrichment ratio for river sub-basins is given by:

$$ER = \frac{P_S}{P_{OA}} \tag{5.11}$$

ER = enrichment ratio (–), P_S = P-content of suspended matter given at discharges above a "critical" level (mg kg^{-1}), P_{OA} = P-content in the top soil of sediment delivering arable land (mg kg^{-1}).

Then the mean annual P-entry (1995–1999) via erosion is modeled using Eq. 5.10. An example of the erosion modeling results is shown for eastern parts of the river Rhine basin in Fig. 5.23.

A check for validity of MEPhos modeling results can be performed by a comparison with mean annual P-loads estimated from measured data on both water quality and discharge. However, this requires quantification of the total sum of diffuse and point P-inputs via all eight pathways (see following chapter). A validation of modeling results for one single pathway is not feasible.

5.4.4 Total P-Inputs from Diffuse and Point Sources and Validation of Model Results

Diffuse P-inputs via the pathways artificial drainage, groundwater-borne runoff and wash-off are also modeled with MEPhos and are added to the erosion-related P-inputs. As shown in Fig. 5.23, mean annual P-inputs vary between <0.1 and >15 kg P ha^{-1} a^{-1} in both investigation areas. The river Ems basin is characterized by larger and coherent source areas. P-inputs between 0.5 and 3 kg P ha^{-1} a^{-1} are typical for the northern part and can be traced back to phosphotopes of the category "deep-ploughed cultivated raised bog soils under agricultural use" emitting P via artificial drainage. This pathway accounts also for the highest diffuse P-emissions of more than 10 kg P ha^{-1} a^{-1}, which result from raised bog soils under grassland use.

In the sub-basins of the river Rhine low P-inputs of <0.1 kg P ha^{-1} a^{-1} dominate, which result from geogenic background, represented by phosphotopes of the groundwater pathway. Medium and high P-inputs appear as small spots and are a consequence of erosion on arable land (Sect. 5.4.3, Fig. 5.23). P-inputs via erosion vary between less than 0.1 and more than 15 kg ha^{-1} a^{-1}. Erosion-related P-inputs are characterized by a small spatial extent of sediment delivery areas with highly differing input levels (Fig. 5.23, enlarged section). The small spatial extent of the phosphotopes results from the disaggregation of erodible arable land, its intersection with buffered flow paths and the application of highly-resolved data sets, above all the digital terrain model with a resolution of 10×10 m^2. High erosion-related P-inputs are to be found mainly in sub-regions with steep slopes and widespread arable land use (Fig. 5.23). For the Ems and Rhine investigation areas mean annual P-inputs of 2.94 kg km^{-2} a^{-1} and 14.8 kg km^{-2} a^{-1} resp. are modeled. The difference reflects the contrasting natural and land use conditions between the two areas (Sect. 5.4.1).

Fig. 5.23.
Mean annual P-inputs (1995–1999) from diffuse sources in the investigation areas Ems and Rhine (enlarged: spatial patterns of erosion induced P-inputs in eastern parts of the river Rhine sub-basin)

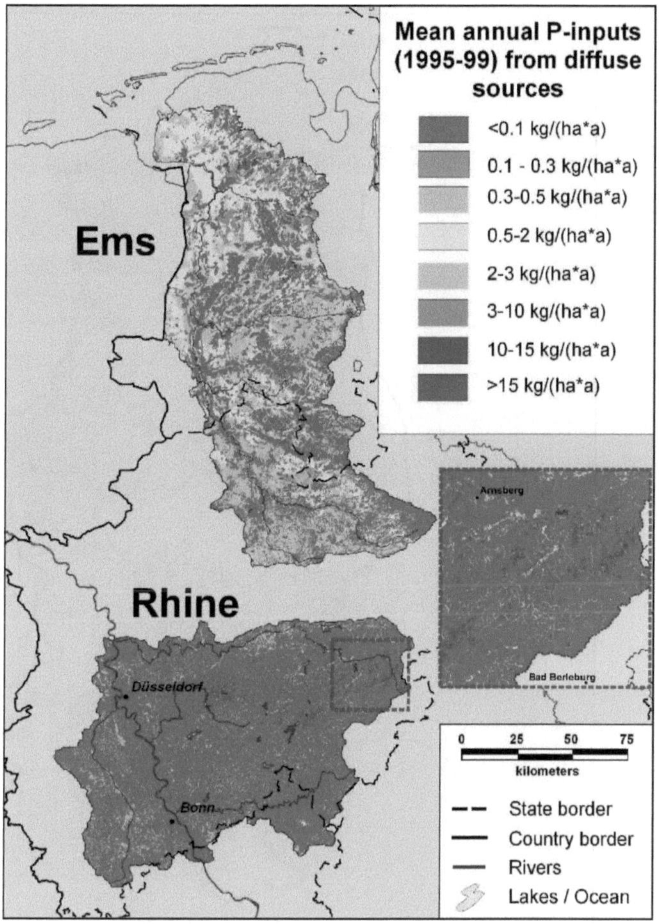

After modeling mean annual P-inputs via all other pathways diffuse and point source emissions can be summed up. This leads to mean annual total loads for the period 1995–1999, which amount to 1 666 t a^{-1} in the river Ems basin and to 1 574 t a^{-1} for the river Rhine sub-basin. Figures 5.24 and 5.25 show the relevance of all eight pathways for the mean P-input.

In the entire river Ems basin diffuse P-emissions account for 87% of all P-inputs to surface waters. It can be stated for nearly all of the 56 sub-basins that diffuse inputs dominate and that the pathway artificial drainage is of highest importance in this lowland river basin. The percentage of inputs via artificial drainage differs between 16 and 89% in total, it is above 50% in most cases (Fig. 5.24). The highest loads from non-point sources are modeled for the lower reaches of the river Ems and its tributaries, which corresponds with the decreasing population density and the increasing intensity of agricultural activities from south to north. Due to the origin of these high loads from artificially drained agricultural land, i.e. as mainly soluble reactive and therefore highly algae-available P, the agricultural activities on raised bog soils result in a tre-

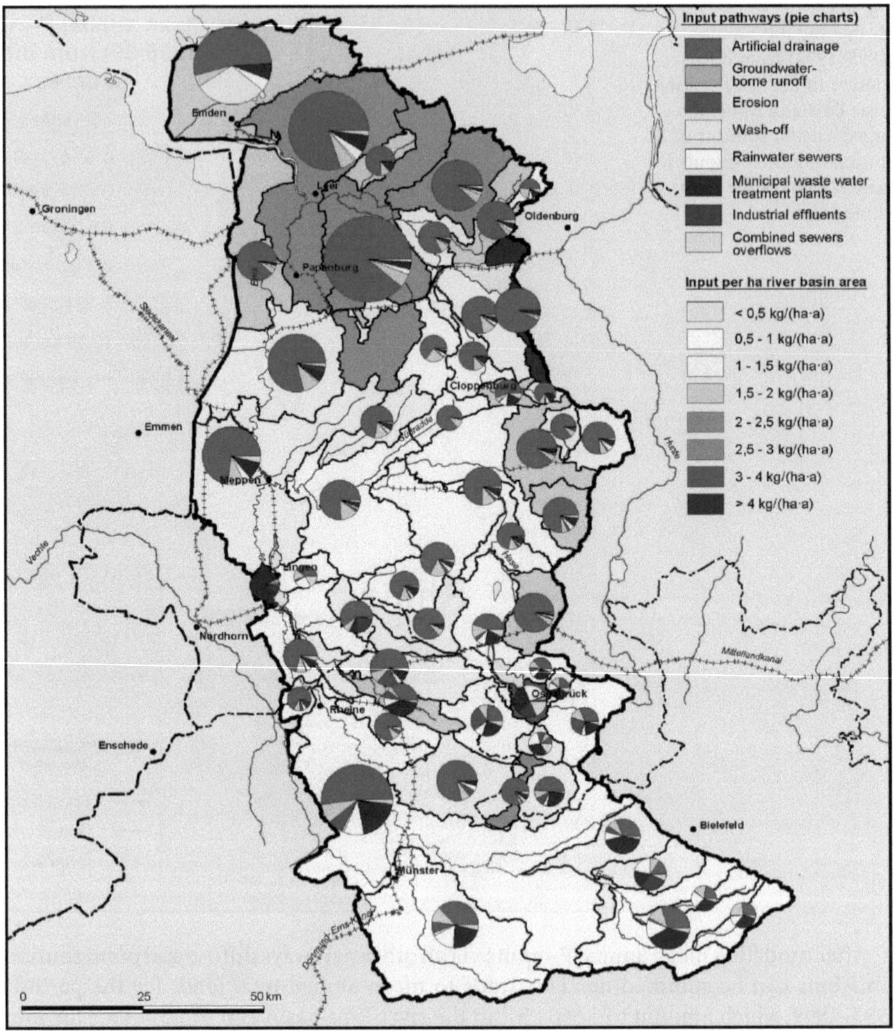

Fig. 5.24. Mean annual P-inputs (1995–1999) of sub-basins and relevance of pathways for the river Ems basin

mendous local and regional eutrophication potential within surface waters. Furthermore the P-retention during fluvial transport is relatively low because of the short distance between the source areas and the mouth of the river Ems. As a consequence, also the wadden sea receives a high phosphate input from diffuse sources.

While the mean total P-inputs (1995–1999) in the investigation area Rhine equal those in the Ems area, the mean relation between diffuse and point source inputs of 32:68 differs significantly. The main reasons are the high densities of population and industry, above all at the river Rhine and along the lower reaches of the river Ruhr.

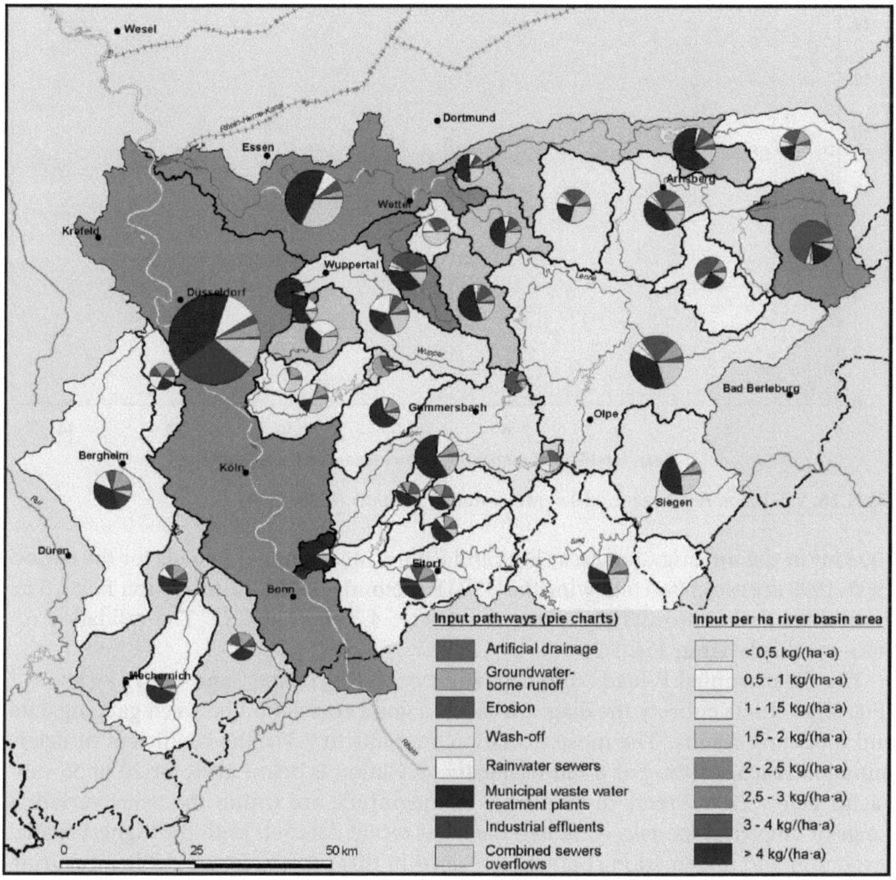

Fig. 5.25. Mean annual P-inputs (1995–1999) of sub-basins and relevance of pathways for the river Rhine sub-basin

High emissions via combined sewer overflows play an important role, too. When the MEPhos modeling results are examined for river sub-basins, areas where diffuse emissions are a major concern are revealed also for the investigation area Rhine (Fig. 5.25). They are located in the upper reaches of the rivers Erft and Ruhr, i.e. in the south-west and east of the investigation area. As a result of arable land use on steeper slopes, partly on loess soils, soil erosion is responsible for these increased diffuse P-loads.

The validity of the MEPhos modeling results is tested by a comparison with mean annual P-loads estimated from measurements of daily discharge and monthly total-P-concentration during the time period 1995–1999. When selecting the gauging stations attention was paid to achieve a great variability with respect to basin size, natural conditions and population density. Furthermore the extent of gaps in the time series of the measured data should be as small as possible. These requirements are fulfilled for 32 resp. 24 gauge-related sub-basins with sizes between approximately 50 and

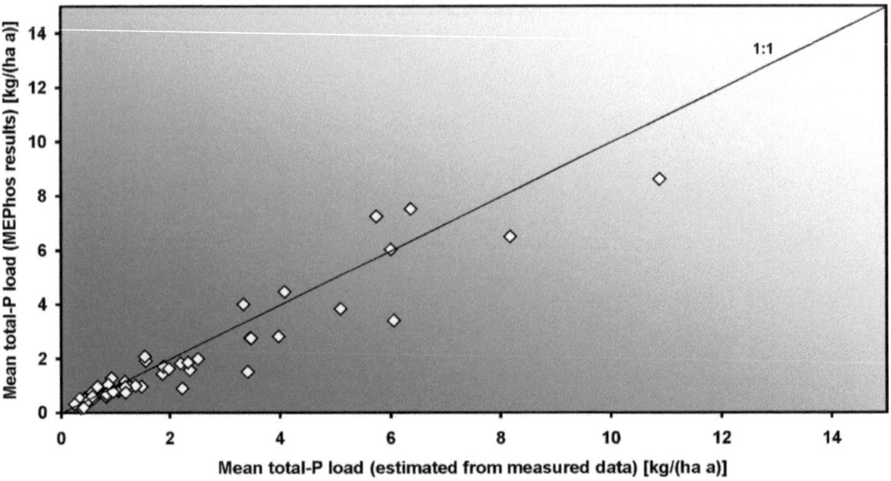

Fig. 5.26. Validation results of modeled mean annual P-inputs (1995–1999)

500 km^2 in the investigation areas Ems and Rhine. Mean annual P-loads for the period 1995–1999 are estimated following the OSPAR method (OSPAR Commission 1998). The general validation process is explained in Sect. 5.4.2 and Fig. 5.22. The validation results are displayed in Fig. 5.26.

The mean annual P-loads differ between ca. 0.1 kg ha^{-1} a^{-1} and ca. 11 kg ha^{-1} a^{-1} (Fig. 5.26). In its entirety the diagram shows a good correlation between gauging data and modeling results. The mean deviation amounts to 7.3%, the coefficient of determination reaches 89%. For 8 sub-basins the deviation is below 10%, for 20 of 56 sub-basins below 20%. Errors in this order of magnitude are within the usual variation range of empirical models. Although the most recent data sets with the highest spatial resolution and information content were used in this project, unavoidable measuring and interpolation errors are undoubtedly involved. Systematic errors causing the deviations could not be identified. Because of the restricted data availability in the field of urban drainage and sanitation, the model routines for the quantification of mean annual P-inputs via rainwater sewers and combined sewer overflows have to get by on a limited number of parameters. With respect to these restrictions the model results can be regarded as valid.

5.4.5 Conclusions and Management Options for Tackling Eutrophication

The MEPhos model in its existing form is suitable for the quantification of mean annual P-loads from diffuse and point sources in large river basins. Based on valid model results management options for the reduction of eutrophication can be proposed. When aiming at reasonable cost-effectiveness-relations the focus of reduction measures has to be on "hot spots", i.e. critical source areas with small spatial extent and high emission. In the river Ems basin the phosphotope "drained raised bog soils under grassland use" takes up less than 4% of the basin area and emits about 30% of all diffuse P-inputs.

Land use changes with respect to nature conservation could be one possibility to lower P-emissions. In the river Rhine sub-basin P-inputs via erosion make up ca. 35% of all diffuse inputs, originating from about 19% of the basin area. Erosion protection measures both on sediment delivering land and along water courses (on-site and off-site) are regarded as suitable management options. But due to the fact that in the river Ruhr basin a series of reservoirs act as sediment sink and due to the dominance of point sources, tackling diffuse sources in the river Rhine sub-basin has only little effect on the general water quality situation. A far larger improvement would be achieved by taking technical measures for small and medium sewage treatment plants, e.g., P-elimination techniques, and by extension of measures against combined sewer overflows.

Acknowledgment

This research work was financed by the German Federal Ministry for Education and Research (BMBF) in the framework of the project REGFLUD which was part of the research priority river basin management.

References

Alexander RB, Elliott AH, Shankar U, McBride GB (2002) Estimating the sources and transport of nutrients in the Waikato River Basin, New Zealand. Water Resources Res 38:1268–1291

Auerswald K (1989) Prognose des P-Eintrags durch Bodenerosion in die Oberflächengewässer der BRD. Mitt Dtsch Bodenkundl Ges 59/II: 661–664

Auerswald K, Flacke W, Neufang L (1988) Räumlich differenzierende Berechnung großmaßstäblicher Erosionsprognosekarten – Modellgrundlagen der dABAG. Z Pflanzenernähr Bodenk 151:369–373

Behrendt H, Opitz D (2000) Retention of nutrients in river systems dependence on specific runoff and hydraulic load. Hydrobiologia 410:111–122

Behrendt H, Huber P, Kornmilch M, Opitz D, Schmoll O, Scholz G and Uebe R (1999) Nutrient emissions into river basins in Germany. Umweltbundesamt, Berlin

Frede HG, Dabbert S (eds) (1999) Handbuch zum Gewässerschutz in der Landwirtschaft. 2nd ed, Landsberg

Fried JS, Brown DG, Zweifler MO, Gold MA (2000) Mapping Contributing Areas for Stormwater Discharge to Streams Using Terrain Analysis. In: Wilson JP, Gallant JC (eds) Terrain Analysis – Principles and Applications. New York, pp 183–203

IAP, FAL (eds) (1996) Entwicklung des gesamtdeutschen Agrarsektormodells RAUMIS96 – Endbericht zum Kooperationsprojekt. Bonn and Braunschweig-Völkenrode

Kirchner WB, Dillon PJ (1975) An empirical method of estimating the retention of phosphorus in lakes. Water Resources Res 11:182–183

Molot LA, Dillon PJ (1993) Nitrogen mass balances and denitrification rates in central Ontario Lakes. Biogeochem 20:195–212

Moore ID, Wilson JP (1992) Length-slope factors for the revised universal soil loss equation: Simplified method of estimation. J Soil Water Cons 49:174–180

OSPAR Commission (ed) (1998) Principles of the comprehensive study on riverine inputs and direct discharges (RID). Reference 1998–05

Prasuhn V, Grünig K (2001) Evaluation der Ökomaßnahmen – Phosphorbelastung der Oberflächengewässer durch Bodenerosion. SchrR FAL 37, Zürich-Reckenholz

Schwertmann U, Vogl W, Kainz M (1990) Bodenerosion durch Wasser – Vorhersage des Abtrags und Bewertung von Gegenmaßnahmen. 2nd ed., Stuttgart

Sommer M, Murschel B (1999) Erosion und Nährstoffabtrag. In: Dabbert S, Herrmann S, Kaule G, Sommer M (eds) Landschaftsmodellierung für die Umweltplanung. Berlin, pp 68–79

Tarboton DG (1997) A new method for the determination of flow directions and upslope areas in grid digital elevation models. Water Resources Res 33:309–319

Tetzlaff B (2006) Die Phosphatbelastung großer Flusseinzugsgebiete aus diffusen und punktuellen Quellen. Research Centre Jülich, Umwelt/Environment, 65, Jülich

Tetzlaff B, Kreins P, Kunkel R, Wendland F (2007) Area-differentiated modelling of P-fluxes in heterogeneous macroscale river basins. Water Science and Technology 55(3):123–131

Werner W, Olfs HW, Auerswald K, Isermann K (1991) Stickstoff- und Phosphateintrag in Oberflächengewässer über "diffuse Quellen". In: Hamm A (ed) Studie über Wirkungen und Qualitätsziele von Nährstoffen in Fließgewässern. Sankt Augustin, pp 665–764

Chapter 6

Sediment-Water Interactions

Ellen L. Petticrew

This chapter addresses sediment-water interactions and comprises five sections, with the papers investigating a variety of chemical, physical and biological processes that influence the behavior of sediments in natural waters. The approaches used range from small-scale laboratory studies that attempt to simulate natural conditions, to in-stream manipulation of sediments, to seasonal studies of in-stream sediments collected in a variety of environmental conditions. The goal of this type of work, as indicated in most of the sections, is to improve our knowledge of the processes regulating sediment-water interactions so as to allow the most suitable approach to managing the aquatic environment, whether it be for maintaining benthic habitats or decision-making regarding the removal of contaminated sediments.

The initial chapter addresses issues and methods for source identification of organic matter in riverine suspended sediments as it has been shown to have a significant influence on the morphology and the transport behavior of flocculated fine sediments. The values of both C:N ratios and stable isotopes of C and N allow seasonal differentiation of suspended organic matter in a northern, temperate productive stream. While this paper deals with the particulate organic matter moving as aggregated or flocculated sediments in a natural stream system, Sect. 6.2 presents a laboratory study investigating the role of various factors which regulate fine sediment aggregation and stability. Ionic concentration is the dominant factor associated with aggregation in this experimental system. Increased levels of heavy metal adsorption onto natural riverine sediment, as opposed to kaolin clay, suggests a concern regarding the storage of heavily polluted natural sediments in environments which promote aggregation and therefore sedimentation. Section 6.3 evaluates the influence of variable, natural aquatic levels of dissolved organic matter (DOM), which is also a factor that can influence sediment-water interactions, on the sorption and desorption kinetics of select organic contaminants. This work was undertaken in order to better inform knowledge of issues related to contaminant remobilization from polluted sediments. While the experimental range of DOM was not found to be significant, the hydrodynamic conditions and contaminant contact time were found to influence desorption. Section 6.4 combines meso-scale field and laboratory experiments to address resuspension and transport behavior of channel-stored riverine sediment. The results of the field experiment confirm the natural heterogeneity of riverine surficial fine sediments associated with the 'fluffy layer' on the channel bed that is not well represented in laboratory experiments. This suggests that sediment-water laboratory experiments can underestimate pollutant mobility in natural systems. Section 6.5 presents results of an experiment using a benthic chamber to assess metal mobility associated with different levels of critical shear stress, which modifies both the resuspended loads and the physical form (aggregated versus dispersed) of the sediments. The implication of eroding sediments to the depth of the anoxic layer was evaluated. It is clear from this approach to investigating the interaction of physical and chemical aspects of sediment and water that using sediments retrieved from natural environments is much more appropriate, but also more problematic due to natural variability of a number of factors including sediment mineralogy. Relatively recent recognition of the importance of managing sediments in aquatic systems suggests that continued research on sediment-water interactions is necessary over a range of scales and in a variety of natural and controlled laboratory conditions. The following five papers reflect the type of work that should help address some current management issues.

Ellen L. Petticrew · Jennifer L. McConnachie

6.1 Identifying Variable Organic Matter Sources in Riverine Suspended Sediments

6.1.1 Introduction

The settling and storage of fine-grained sediments in the interstices of fluvial gravel beds can have significant implications on both sediment conveyance in catchments and aquatic habitat quality. Given that suspended fine-grained sediment (<63 µm) moves not only as individual particles, but also as particle aggregates or flocs, there has been a relatively recent research emphasis on characterizing these structures and the conditions which enhance their growth and settling in freshwater aquatic environments (Kranck et al. 1993; Droppo and Ongley 1994; Petticrew 1996; Liss et al. 1996; de Boer 1997; Phillips and Walling 1999).

Flocs are comprised of both organic and inorganic material, bound together by a combination of physical, chemical, and biological forces (Droppo et al. 1997). The rate of flocculation depends on site-specific variables such as ionic and suspended sediment concentration, shear stress, pH, and organic source and supply (Droppo and Ongley 1994). This is evident in the comparison between marine and freshwater systems. Flocs are prolific in marine environments, where high concentrations have afforded them being termed "marine snow" (Alldredge and Silver 1988). Conversely, riverine flocculation is less apparent visually, and hydrologic conditions were preliminarily thought to be too energetic to facilitate floc-building. Flocculation is now a well-documented phenomenon in freshwater lotic systems (e.g., Droppo and Ongley 1994; Petticrew 1996; Phillips and Walling 1999), although the resulting particles are typically an order of magnitude smaller than their marine counterparts (e.g., 10^2 to 10^3 µm diameter).

The main operational difference between these riverine and marine systems is ionic concentration. Flocculation in saline environments has been attributed mainly to the high electrolyte concentration (Alldredge and Silver 1988; van Leussen 1999), while freshwater systems are characterized by much lower salinity (> 10^3 µS cm^{-1} for marine at 25 °C versus ≪10^3 µS cm^{-1} for freshwater; Kalff 2002). Van Leussen (1999) states that the role of salt flocculation is currently in question, and that organic binding agents may play an important function in the process for both environments. Droppo (2001) states that the significant biological component controlling flocculation, including both formation and destruction processes, is speculated to be organic microstructures referred to as extracellular polymeric substances (EPS). EPS appear to assist in regulating the shape and size, as well as internal complexity, of flocs by acting as bridges between inorganic particles and other components that make up floc structure. All of these morphometric characteristics inevitably influence floc density and settling velocity, which determine the nature of suspended sediment transport. EPS are typically thought of as microbial structures actively developed by microbes for several key functions. The relationship between organic matter supply and EPS is not completely understood, but the use of these structures for bacterial attachment to particles is evident (e.g., Liss et al. 1996; Droppo 2001). Dissolved and particulate organic matter must then, at the very least, have an indirect role in EPS formation as nutrients for bacterial growth and cellular activity. Both the quantity and quality of organic matter should influence the production of EPS. High quality organic matter here implies that it is readily bioavailable in that it is easily processed by microbes and metabolized efficiently. The ratio of carbon to nitrogen in organic matter has been used as means of characterizing seston quality (Bouillion et al. 2000) with lower ratios reflecting better quality, as nitrogenous products, desired by bacteria, increase. At most, organic matter could produce EPS-type molecules during decomposition that assist in particle binding in the water column. Either way, it appears that organic matter source and quality may be crucial in regulating the development of freshwater flocs. Temporal variability in floc morphology may therefore be due to changes in the sources of inorganic and organic material. Investigation of this idea requires accurate definition of organic sources and an estimate of their residence time in the aquatic system.

The use of stable isotope analysis to differentiate organic source material has increased dramatically in recent years (Griffiths 1998; Phillips and Gregg 2001). Stable isotopic tracers have been used to monitor flows of organic matter (i.e., trace trophic relations) in marine (Peterson et al. 1985; Cifuentes et al. 1988; Fry 1988; Hedges et al. 1988) and freshwater systems (Bunn et al. 1989; France 1995). Others (e.g., Kline et al. 1990; Bilby et al. 1996; Ben-David et al. 1998; McConnachie and Petticrew 2006) have utilized this technique to characterize the introduction of marine nutrients into freshwater environments. The ultimate goal of this technique is to determine the proportional contributions of multiple sources of organic matter to a mixture found in nature. Linear mixing models are used for this purpose to examine two source, single isotope, or three source, dual isotope, signatures (Phillips 2001; Phillips and Gregg 2001).

Riverine Sources of Organic Matter

Organic material introduced to river/stream ecosystems is classified as being derived from either autochthonous (generated within the stream) or allochthonous (generated from the watershed) sources. Both of these sources vary temporally and spatially. Seasonal variation of terrestrial sources is attributed to the presence of species and hydrologic regime. The composition of the riparian species changes over time, as does the proportional species contribution to streams (Johnson and Covich 1997). Stage height and precipitation also facilitate transfer of terrestrial material to streams (Koetsier et al. 1997; Tockner et al. 1999), both laterally and longitudinally. Hydrological conditions regulate the in-stream presence and distribution of biological organisms (e.g., periphyton and invertebrates) (Allan 1995), the transport and transfer of solutes (e.g., ions and nutrients) (Webster and Ehrman 1996), and the provenance and movement of organic material (Minshall et al. 1985). Water temperature, which is intimately linked to air temperature, varies both seasonally and diurnally, as well as among locations due to regional differences in climate, elevation, extent of streamside vegetation, and relative importance of groundwater inputs (Allan 1995). In-stream production of algae and periphyton is dependent on environmental factors such as insolation and temperature (Allan 1995; Sand-Jensen 1998), as well as terrestrial nutrients.

Not only do hydrological factors affect the relative proportion of various sources of organic material, but the seasonal availability of source types does as well. As floodplains are inundated during overbank flow periods, stored organic matter may enter the stream. This occurs during episodic events such as spring melt (i.e., freshet) and rain events. Rain, and associated wind, has an added impact in that allochthonous material may get blown into the stream from riparian areas. Autochthonous sources are linked to several environmental factors, including temperature, light intensity, available nutrients, and discharge, where favorable conditions are more likely to occur during low flow periods. Allochthonous sources often follow seasonal lifecycles, where leaves and needles are shed in autumn, and new growth does not occur again until spring. A third source of organic matter exists in many watersheds that are linked to marine environments. Anadromous salmon, migrating upstream to spawn, are known to introduce important marine-derived nutrients to freshwater systems (Bilby et al. 1996). Spawning female salmon create redds, or nests for eggs, by digging up gravels, simultaneously leaving a depression for deposition of eggs and sperm. By excavating the nest they clean out fine material stored in interstitial spaces that may hinder oxygen

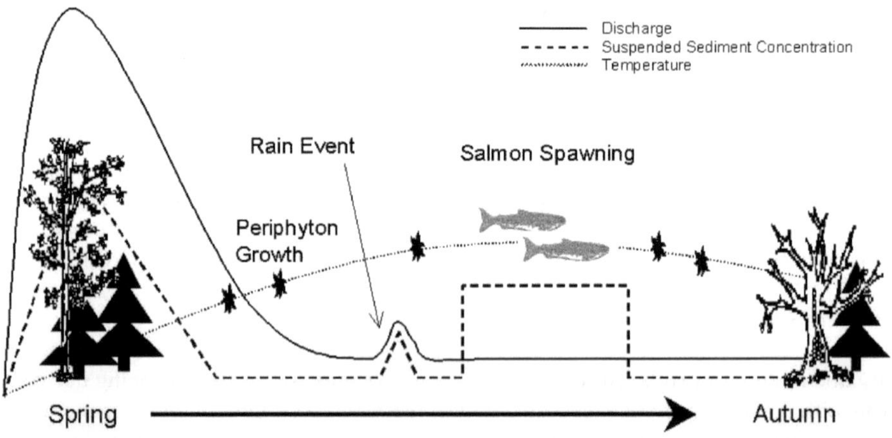

Fig. 6.1. A hypothetical depiction of seasonal patterns of suspended sediment concentration, discharge, and temperature as well as the factors which influence them in a temperate, nival dominated, fish-bearing stream. While only one rain event is shown, several typically occur during the season. Relative and absolute relationships between the variables depicted here are approximate

transfer for eggs (Soulsby et al. 2001). After spawning the female covers the eggs with excavated gravels thereby generating highly permeable sediments in the redd. Shortly thereafter (days) both the female and male die and the carcasses are left to rot in-stream providing a pulse of bioavailable, marine-derived nutrients. Figure 6.1 depicts a generalized model for these dominant sources and the timing of organic matter along with the hypothetical pattern for inorganic sediment concentrations in a typical snowmelt-dominated stream system.

Study Objective, Rationale and Approach

The objective of the work presented here was to evaluate the seasonal changes in the dominant type of organic matter incorporated in the suspended sediment of a productive fish-bearing, temperate stream system. The assumption is that in temperate forest areas such as the study region, the quality and quantity of organic material incorporated into the stream system varies over the season depending on the source types available, hydrologic regime, and conditions for biological processing. Organic matter of good quality, and/or high concentrations, may significantly increase the size of flocs, resulting in faster settling rates. This may be attributed to two factors: *(1)* flocculation is directly facilitated by the organic matter, or macromolecules comprising organic material; and/or *(2)* organic matter provides the nutrients and habitat for bacteria, which exude polymeric fibrils that bind particles together (Droppo et al. 1997).

The approach used to characterize organic matter sources was the measurement of regular carbon and nitrogen content along with stable isotopes of carbon and nitrogen in suspended sediment collected over one open-water season in a productive salmon-bearing stream. The sampling strategy was intended to capture important hydrologic and biologic events within the study system, which would influence the input and transport of various organic matter sources. Samples were collected *(1)* for springmelt and baseflow to allow for comparison between instances of minimum and maximum suspended material, *(2)* for summer rain events to incorporate resuspension of material from the bed gravels, and *(3)* for periods when different organic sources were evident (e.g., spawning salmon). Further work on the effect of the organic matter type of aggregate size was also undertaken but is not presented here.

6.1.2 Methods

Study Site

The study watershed of O'Ne-eil Creek (Fig. 6.2) is located in the Hogem Range of the Omenica Mountains in the Takla Lake region of northern British Columbia, Canada. As part of the most northern extent of the Fraser River watershed (55° N, 125°50' W), O'Ne-eil (also known as Kynoch) basin features a range in relief from 700–1 980 m above sea level. Surficial material is comprised of glacial tills and lacustrine clays at higher elevation and fine-grained glaciolacustrine sediment in the lowland areas (Ryder 1995). Encompassed within the Engelmann Spruce Sub-alpine Fir (ESSF) biogeoclimatic zone, the basin is relatively small (~75 km^2), but O'Ne-eil Creek is an

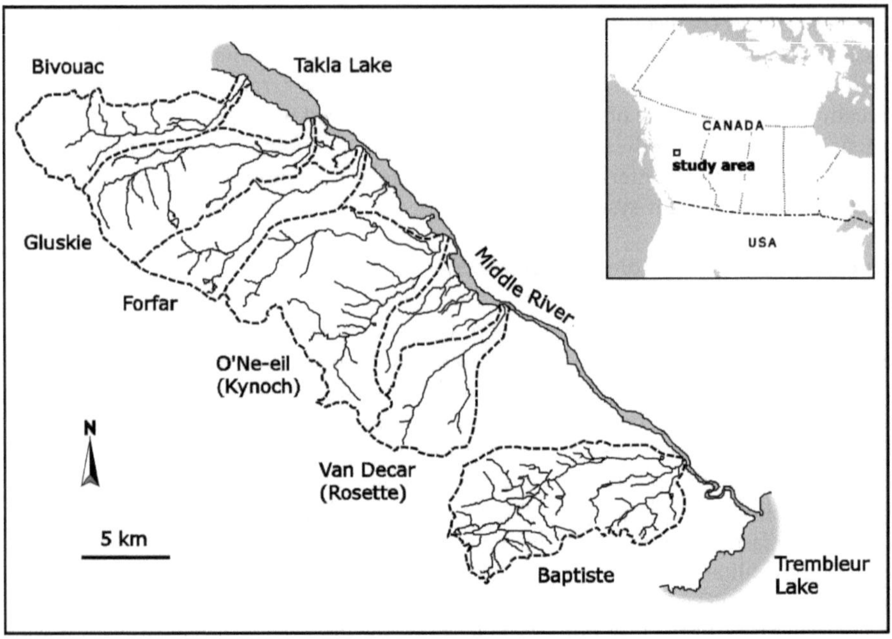

Fig. 6.2. Map of the Stuart-Takla region of northern British Columbia. Note O'Ne-eil watershed in the center

important fish-bearing stream, where annual migration of Pacific sockeye (*Oncorhynchus nerka*) salmon is well documented by Canadian Department of Fisheries and Oceans (DFO). The mainstem channel of O'Ne-eil Creek is approximately 20 km in length and 10 to 12 m wide at the mouth. The study reach exhibits favorable spawning habitat with appropriate substrate size distribution in low gradient (0.5–2%) riffles (Petticrew 1996). Little anthropogenic disturbance has occurred within this watershed specifically, however, a forest service road enables access to the lower reaches. One site in O'Ne-eil Creek, downstream of the forestry access bridge, and approximately 1 500 m upstream of the mouth, was sampled during the period of May 18 to August 21, 2001. River stage levels were monitored continuously by DFO using a pressure transducer and used in conjunction with rating curves to determine discharge. DFO also maintained precipitation gauges in the area.

Earlier quantification of organic matter inputs to the Stuart-Takla streams for 1996 by Johnston et al. (1998) indicate that riparian litter inputs for the period between July and October varied between 10 and 300 g m^{-2} in dry weight, with the majority being of deciduous rather than coniferous origin. The magnitude of vegetation introduced to the streams decreased logarithmically with distance from stream banks, while the mean areal loadings decreased as channel width increased. In-stream productivity in the form of benthic algal biomass increased in late summer in response to introduction of salmon carcass derived nutrients. Johnston et al. (1998) report that the organic inputs from the in-stream rotting of salmon carcasses exceeded those of riparian leaf litter in Stuart-Takla streams.

Sample Event Selection

This study was designed to determine what type of organic matter was associated with stream suspended sediment over one open water season in O'Ne-eil Creek. In order to evaluate the seasonal changes in fine sediment composition it is necessary to collect samples over a range of watershed events which transport or deliver inorganic and organic matter. These were partitioned into five discrete event types, including: *(1)* springmelt during the period of rising water levels; *(2)* summer pre-spawn flow conditions, but isolated from rain events; *(3)* rain events; *(4)* the period during active spawning; and *(5)* post-spawn, where no actively spawning or live salmon were present but decay of carcasses occurred. Springmelt is a period characterized by high discharge, and this is when material stored in-channel and on the floodplain is first flushed through the system resulting in elevated suspended particulate matter levels. Less suspended load is moved during lower flow levels associated with pre-spawn flows, where the source material is predominantly in-stream. Rain events are characterized by higher than baseflow discharge, where suspended sediment concentrations are expected to increase, and potentially comprise a combination of in-stream and terrestrial inputs. Five rain events were sampled during pre-spawn, salmon spawn, and post-spawn, so they reflect the combined effects of resuspension of gravel-stored material that occurs during storms and the predominant source of organic matter for the particular sampling date. The period of active salmon spawn combines the introduction of anadromous organics and biological disturbance of gravels due to the digging of redds. Decaying salmonid organic matter is expected to be delivered to the system post-spawn, but, because live fish are no longer present, disturbance of gravels is minimal. Seasonal patterns of suspended particulate matter and suspended organic matter concentration, as well as their carbon and nitrogen signatures were identified by replicate sampling (3–5 dates) within each of the above stated event types.

Suspended Sediment Collection

A mixture of sediment and water was collected using wide-mouthed 1-l Nalgene bottles from a designated site in a well-mixed portion of the stream channel. The bottles were transported to the field laboratory where values of suspended particulate matter (SPM) and suspended organic matter (SOM) concentrations were obtained through gravimetric determination. The bottles were mixed well to obtain representative samples and then a known volume of stream water was filtered through pre-combusted and pre-weighed 47 mm diameter glass-fiber filters with a nominal pore size of 0.7 µm. Triplicate samples were filtered, dried at ~60 °C and then weighed to allow the calculation of SPM concentration. The filters were then ashed in a muffle furnace at 550 °C for an hour to remove the organic material and then reweighed, allowing for determination of SOM concentration.

Stable Isotopes of Carbon and Nitrogen

Stable isotope mass spectrometry (University of British Columbia, School of Oceanography Stable Isotope Laboratory) was used to characterize seasonal sources of organic matter. The isotope ratios for both organic tissue and suspended sediment filters were measured and expressed relative to conventional standards as δ values defined as:

$$\delta X\ (‰) = (R_{sa}/R_{std} - 1) \times 1\,000 \quad (6.1)$$

where X is ^{13}C or ^{15}N, R_{sa} is the isotopic ratio of the sample (either $^{13}C/^{12}C$ or $^{15}N/^{14}N$), and R_{std} is the isotopic ratio of the standard (PeeDee Belemnite for carbon and air for nitrogen). The technique enables assessment of seasonal distribution of organic matter sources by comparing isotopic ratios from collected source material with those from suspended sediment samples. Regular carbon and nitrogen content (^{12}C and ^{14}N) was measured prior to stable isotopes in order to calculate the sample's C:N ratios. This ratio is often used to estimate sources for organic matter because autochthonous material exhibits much lower values (<15) than allochthonous material (Owen et al. 1999).

Tissue from terrestrial vegetation (e.g., spruce needles and birch, willow, and alder leaves), in-stream periphyton and algae, and salmon flesh was collected and stored in 1.2 ml centrifuge tubes and freeze-dried. Isotopes of carbon and nitrogen, as well as routine C:N ratios, were determined for each of these source materials. It should be noted that algae and periphyton are classified separately here even though the term benthic algae typically incorporates both. The algae collected for this study was free-floating and visually distinguishable from periphyton.

Duplicate samples of suspended sediment were collected, as indicated above, for isotope analysis and filtered onto pre-combusted and pre-weighed glass fiber filters, which were freeze-dried and analyzed as per the tissue samples. Seasonal trends were assessed using both carbon and nitrogen isotopes and ratios of percentage carbon and nitrogen for these filter samples.

A dual isotope (C, N), three end-member (algae, salmon, and terrestrial vegetation) mixing model based on mass balance equations (Phillips and Gregg 2001) was used to define these trends more quantitatively, and to examine relationships more conclusively. The spreadsheet used to determine source proportions, variances, standard errors, and confidence intervals can be accessed at http://www.epa.gov/wed/pages/models.htm.

6.1.3 Results

O'Ne-eil Creek discharge, local precipitation and the timing of sampling events are shown in Fig. 6.3a. Figure 6.3b indicates the concentrations of suspended particulate matter and suspended organic matter observed over the open water period. In 2001 the springmelt discharge began in late May. Five rising limb discharge dates were sampled which show elevated sediment concentrations. Flows attenuated in late June when five lower flow, lower sediment concentration, pre-spawn dates were sampled in June and early July. Five storm events were sampled over the summer season on July 7 and 9, and August 2, 3 and 21. These five events exhibited a range of suspended sediment response to the change in flows associated with precipitation. The estimated sockeye escapement was 13 893 with the fish arriving in the stream on July 20 and die-off beginning around early August. Five dates were sampled in active spawn and three in the post-spawn period.

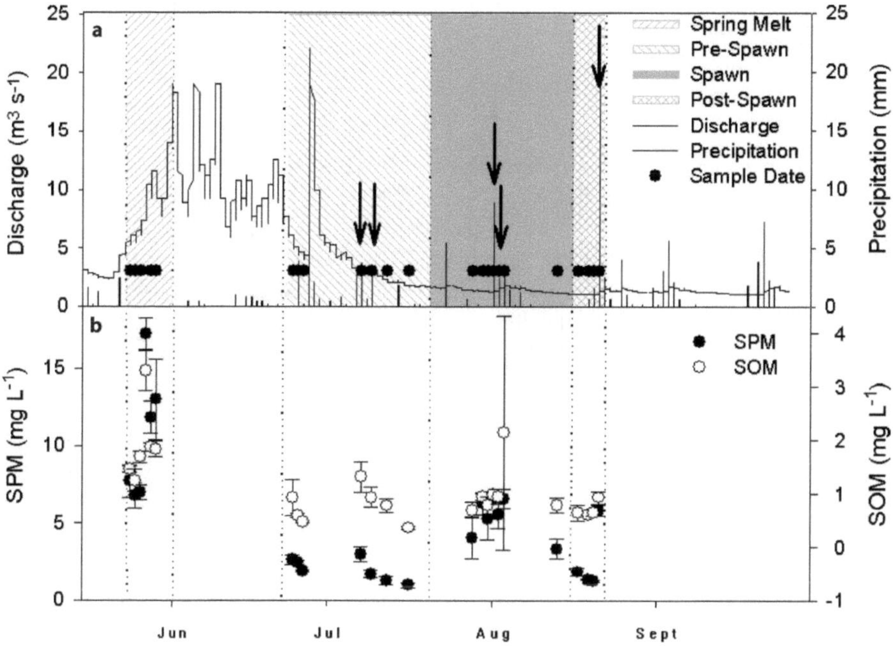

Fig. 6.3. a Schematic of sampling approach during the 2001 open water season. *Points* represent sampling times, *shaded areas* depict the four major event types of rising limb of springmelt, lower pre-spawn flow, and the active spawning period and post-spawn (i.e., no live/spawning salmon). *Arrows* designate sampled rain events. **b** Concentrations of SPM and SOM sampled over the open water season

Stable Isotopes of Carbon and Nitrogen

Initially, the proportion of carbon and nitrogen comprising the suspended sediment fraction was determined. Figure 6.4 shows the seasonal pattern of the ratio of carbon to nitrogen along with the discharge values. The pattern in the C:N ratio shows high values occurring during the spring period and remaining fairly constant until the salmon are introduced to the stream reach. The C:N ratio decreases while the salmon spawn and carcasses are present in-stream. The post-spawn samples are significantly different ($P < 0.05$) from both spring melt and pre-spawn samples. Figure 6.5 provides a breakdown of the C:N ratios for tissue types sampled as source material within and adjacent to O'Ne-eil Creek over the season. Mean values for tissue samples taken several times over the season of allochthonous, or terrestrially derived, vegetation exhibits C:N ratios greater than 15, ranging from 18.4 ±1.23% for willow leaves to 41.9 ±7.42% for spruce needles.

Autochthonous, or in-stream, organic matter is characterized here by ratios <15; the lowest values being for salmon flesh with 3.4 ±0.07%, while periphyton and algae ratios are 8.3 ±1.22 and 9.2 ±0.21%, respectively.

Stable isotope analysis of O'Ne-eil Creek source materials were undertaken on the same samples as those measured for C:N ratios that are portrayed in Figure 6.5. Statistically different carbon and nitrogen isotopic signatures were found for the three main groups of organic source material allowing for source separation techniques. Allochtho-

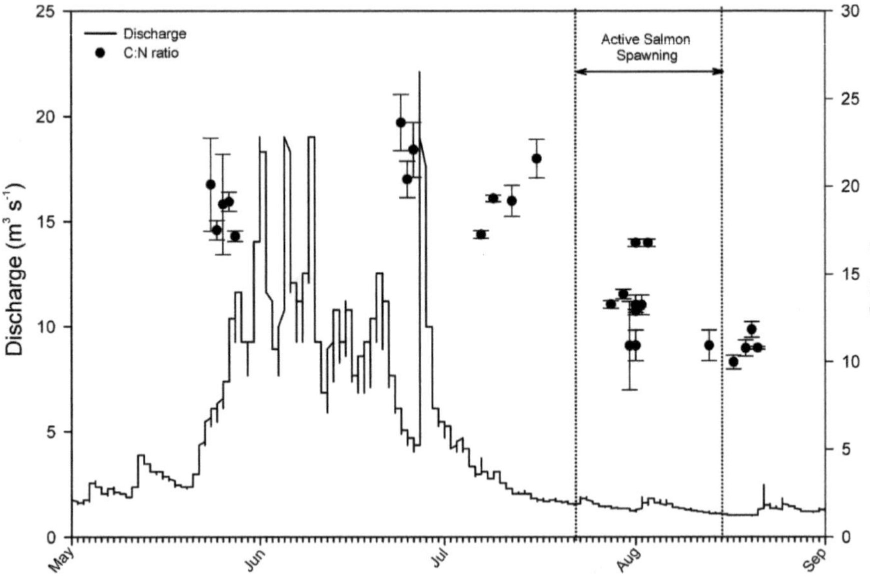

Fig. 6.4. Seasonal trend in C:N ratio of suspended sediment. *Dotted lines* indicate active spawning start (July 20) and end (August 15) dates. *Error bars* are ±1 SE

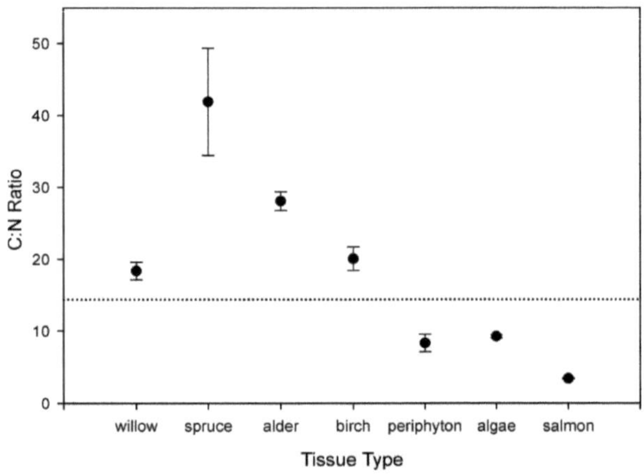

Fig. 6.5. Averaged C:N ratios for different types of organic source material sampled in or adjacent to O'Ne-eil Creek. *Dotted line* separates allochthonous (terrestrially derived) from autochthonous (produced in-stream) organic material

nous (terrestrial) material had a δ^{13}C value of 28.8 (±0.94) and a δ^{15}N value of 1.1 (±2.68). Autochthonous (algae) material had a δ^{13}C value of 21.0 (±0.50) and a δ^{15}N value of 10.8 (±0.09). The salmonid samples had a δ^{13}C value of 35.3 (±0.76) and a δ^{15}N value of 0.5 (±0.25) (Table 6.1). Seasonal analysis of the mixture of organic material in the suspended sediment reveals a trend of enrichment of the stable (heavier) isotope over the season for both carbon and nitrogen, although the latter is much more pronounced (Fig. 6.6). A slight increase in carbon isotope ratio is seen from springmelt to low flow with δ^{13}C values of −26.8 ±0.09 and −26.6 ±0.10‰, respectively. Once salmon enter the reach, the isotopic signal increases to −26.1 ±0.18‰. The

Table 6.1. Partitioning of organic matter source contributions to suspended sediment load in O'Ne-eil Creek as modeled for the five event types of springmelt, pre-spawn flow, rain events, spawn, and post-spawn using dual isotopic signatures (δ^{13}C and δ^{15}N). Note that the model used here limits the total percentage of the three sources types to be 100. This explains the presence of negative values to balance those that are >100%

	Mixture (Sediment)	Terrestrial vegetation	Salmon flesh	Algae
Springmelt				
δ^{13}C (‰) (SE)	–26.8 (0.06)	–28.8 (0.94)	–21.0 (0.50)	–35.3 (0.76)
δ^{15}N (‰) (SE)	2.0 (0.11)	1.1 (2.68)	10.8 (0.09)	0.5 (0.25)
Sample size	10	14	4	3
Source proportions (%) (SE) – calculated		110.9 (20.0)	8.7 (8.8)	–19.7 (11.6)
95% confidence limits (%)		68.4–100	0–27.8	0–4.5
Pre-spawn flow				
δ^{13}C (‰) (SE)	–26.6 (0.07)	–28.8 (0.94)	–21.0 (0.50)	–35.3 (0.76)
δ^{15}N (‰) (SE)	2.4 (0.09)	1.1 (2.68)	10.8 (0.09)	0.5 (0.25)
Sample size	10	14	4	3
Source proportions (%) (SE) – calculated		105.8 (19.0)	12.9 (8.4)	–18.6 (11.0)
95% Confidence limits (%)		65.3–100	0–31.0	0–4.4
Rain events				
δ^{13}C (‰) (SE)	–26.2 (0.11)	–28.8 (0.94)	–21.0 (0.50)	–35.3 (0.76)
δ^{15}N (‰) (SE)	3.8 (0.31)	1.1 (2.68)	10.8 (0.09)	0.5 (0.25)
Sample size	10	14	4	3
Source proportions (%) (SE) – calculated		80.8 (16.5)	27.1 (7.3)	–7.9 (9.5)
95% confidence limits (%)		46.6–100	11.8–42.4	0–11.6
Spawn				
δ^{13}C (‰) (SE)	–26.2 (0.11)	–28.8 (0.94)	–21.0 (0.50)	–35.3 (0.76)
δ^{15}N (‰) (SE)	4.3 (0.38)	1.1 (2.68)	10.8 (0.09)	0.5 (0.25)
Sample size	10	14	4	3
Source proportions (%) (SE) – calculated		68.6 (15.7)	32.8 (7.0)	–1.4 (9.0)
95% confidence limits (%)		36.2–100	18.3–47.2	0–17.1
Post-spawn				
δ^{13}C (‰) (SE)	–25.6 (0.07)	–28.8 (0.94)	–21.0 (0.50)	–35.3 (0.76)
δ^{15}N (‰) (SE)	5.6 (0.18)	1.1 (2.68)	10.8 (0.09)	0.5 (0.25)
Sample size	6	14	4	3
Source proportions (%) (SE) – calculated		46.0 (9.5)	46.8 (4.1)	7.2 (5.7)
95% confidence limits (%)		26.2–65.8	38.1–55.5	0–18.9

peak of enrichment occurs just prior to the post-spawn period, where some salmon are still alive, but earlier returns have already begun to rot in situ. After this, the isotopic ratio decreases towards a signal more similar to the period prior to salmon presence;

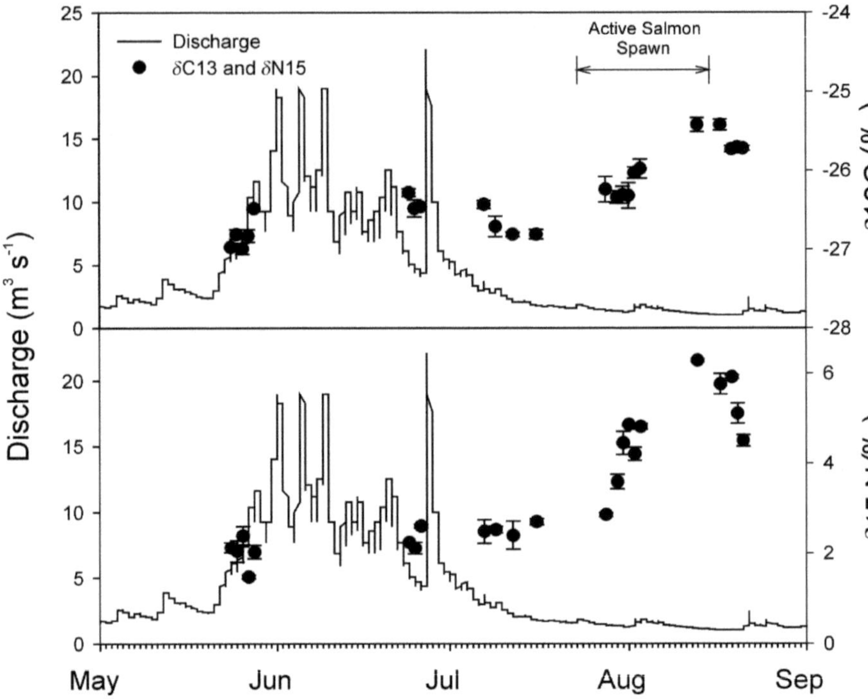

Fig. 6.6. Seasonal trends for stable isotopes of carbon and nitrogen from suspended sediment. The salmon spawning period is designated. *Error bars* are ±1 SE

however, the measured trend ends before the decline is complete, so the grouped average for the post-spawn event shows the highest enrichment as compared to all other event types (−25.6 ±0.07‰). The nitrogen isotope exhibits a similar pattern with enrichment of δ^{15}N from springmelt to pre-spawn flow of 2.02 ±0.11 to 2.42 ±0.09‰, a steep increase over the active salmon spawning period with a average of 4.27 ±0.38‰, and further enrichment after salmon die-off is complete (5.60 ±0.18‰).

The combined results of the two isotopic elements are typically shown as in Fig. 6.7, and there is clearly an increasing trend, where springmelt < low flow < rain ≈ spawn < post-spawn. Rain events were sampled over three different event types (pre-spawn flow, spawn, and post-spawn), and the average is approximately equal to that of the salmon period due to the linearity of the relationship observed in these event types (Figure 6.7). In fact, linear regression was applied to the semi-logged ungrouped data, and the relationship is significant with $r^2 = 0.663$ for $N = 46$ and $\alpha = 0.05$.

For modeling purposes, the data derived from the suspended sediment filters were compared to the tissue samples, using the dual isotope, three end-member format developed by Phillips and Gregg (2001). The results indicate that the predominant source of organic matter to the suspended sediment load changed over the season, although the 95% error bars are very large (Table 6.1). For the springmelt period, the model suggests that all organic material is derived from terrestrial vegetation, while the salmon and algal sources are contributing minimal amounts to the suspended sediment sup-

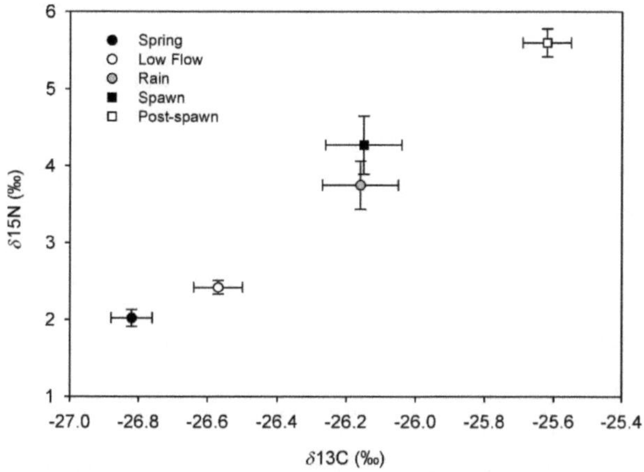

Fig. 6.7.
Stable isotopes for suspended sediment filters as grouped by event type. Bi-directional *error bars* are ±1 SE for both variables

ply. Similarly, terrestrial vegetation dominates as the organic source for the pre-spawn suspended sediments. A shift begins toward a positive salmon contribution for the rain events category, with between 66 and 95% of the proportion remaining terrestrially derived, and 21 to 34% being of salmon origin. The divergence continues for the active salmon spawn, where the suspended sediment samples were composed of anywhere from 56 to 81% of vegetative tissue, 27 to 38% of salmon flesh, and 0 to 6.1% of algae. The post-spawn period was characterized by an overlap of terrestrial and salmon inputs, 46.0% ±8.6 and 46.8 ±3.7%, respectively, and minimal algal inputs (7.2 ±5.2%). Periphyton was excluded from the model because it utilizes nutrients from the water column (Johnston et al. 1998), which complicates modeling because the periphyton isotopic signature is then a mixture of the end-members. Phillips and Gregg (2001) suggest that samples from the three source populations should be independent.

6.1.4 Discussion and Conclusion

Both the carbon to nitrogen ratios and the stable isotope analysis of suspended sediment reflect differences in organic matter composition seasonally. The changing signal of C:N from suspended sediment filters indicates the sediment load response to the seasonal change in organic matter type. Figure 6.4 displays the transitional decrease in C:N ratio caused by salmon presence within the study reach from a ratio greater than 15 to a ratio less than 15. This dilution of C:N ratios through the addition of autochthonous (i.e., in-stream) matter results in much lower values than those of allochthonous (i.e., terrestrially derived) materials that dominate earlier in the season. Figure 6.5, showing the C:N ratios for terrestrial- versus aquatic-derived organic source material for O'Ne-eil Creek, confirms this, as terrestrial sources are characterized by ratios greater than 15, whereas the in-stream supply falls below 15 as suggested by Owen et al. (1999). Thus, it appears that the suspended sediment load adopts an elemental signal related to changing type of source material. Although the ratio does not decrease as far as to match the measured ratio for salmon flesh, the presence of salmon detritus is detectable in the suspended sediment load when stable isotopes are used.

It is clear that stable isotopes provide more specific information for characterizing organic matter sources and differentiating the timing of their contribution to the stream system. Different types of organic tissue exhibit distinctly different isotopic signals, which enables differentiation between samples and even the tracing of trophic pathways through systems (Fry 1988). The data presented in Fig. 6.6 indicates little temporal variation in $\delta^{13}C$ (–27 to –26‰) and $\delta^{15}N$ (2 to 3‰) until the salmon enter the reach. At this point, a steep increase in the heavier isotopes of each element occurs; steeper for nitrogen than carbon. Then, when live salmon are no longer present in the stream, both carbon and nitrogen isotopes fall off; probably as salmon-introduced nutrients from decay processes are utilized or flushed downstream. This trend is supported by Ben-David et al. (1998), who reported that spawning Pacific salmon exhibit higher proportions of heavier carbon and nitrogen isotopes ($\delta^{13}C = -18.65 \pm 0.18‰$ and $\delta^{15}N = 13.01 \pm 0.13‰$) than terrestrial plants (means of $\delta^{13}C \approx -27‰$ and $\delta^{15}N \approx 0‰$; France 1997). Isotopic content of the event groups is clearly differentiated in Fig. 6.7 indicating the usefulness of this method in distinguishing suspended sediment organic matter types. The role of decaying salmon in introducing marine derived nutrients to the stream is supported in this data set by the elevated values during post-spawn.

Dual isotope, three end-member modeling corroborates the visual stable isotope pattern (Fig. 6.7), although a significant amount of variation in proportion of contributors exists (Table 6.1). Phillips and Gregg (2001) performed sensitivity analysis of this linear mass balance model. They found that large differences in isotopic signal between sources reduced the error (i.e., doubling the difference reduced the uncertainty by half). Further, sample size is important when dealing with source samples exhibiting similar isotopic signatures. Thus, increasing the number of samples collected, especially for terrestrial vegetation, which varies significantly in terms of species and season (France 1995), should improve the resolution of this model. As well, collecting tissue samples in a more continuous manner over the season, rather than the episodic approach presented here, and applying the model to sources collected at the same time as the mixture, may also assist investigation of natural patterns.

The stable isotope analysis substantiates earlier indications that organic source type is of significant importance for flocculation in O'Ne-eil Creek (McConnachie and Petticrew 2006). The observed contributions of organic matter to the suspended sediment follows the general pattern of terrestrial and in-stream contribution as depicted in the schematic of Fig. 6.1. Other reported results from this study area indicate larger aggregate particles were found during salmon spawn and decay (Petticrew 1996; McConnachie 2003; Petticrew and Arocena 2003; McConnachie and Petticrew 2004), a time when the dominant organic source type changes from terrestrial material to higher quality, marine-derived organic matter introduced to the freshwater system by the salmon vector. Both carbon and nitrogen signals changed considerably in terms of the heavier isotopes for the spawn compared to the pre-spawn flow period; however, samples taken up to mid-July were equal to or less than the springmelt concentrations, indicating a predominance of the terrestrial organic matter over that period.

The relationship between organic content and particle morphology, when sampled in a natural environment, is confounded by the range of variables (e.g., shear velocity, temperature) influencing aggregate size and stability, which make it difficult to isolate the influence of organic matter source material. However, laboratory evaluations indicate that the quality of the organic matter, as it relates to a suitable bacterial substrate, influences the floc development due to bacterial EPS. Therefore, characterizing the temporal and spatial changes of organic contributions is an important factor in assessing riverine sediment storage and transfers.

Acknowledgments

The opportunity to attend the Sedymo Conference in Harburg, Germany and to participate in this book was made possible by Ulrich Förstner. An initial review of the paper by Philip Owens is much appreciated. The field assistance of Antje Ullrich and Beka McConnachie is acknowledged as are sample analysis and laboratory assistance provided by Leslie Chamberlist and Tauqeer Waqar. This work was undertaken as part of the larger Stuart-Takla Fisheries-Forestry Interaction Project. Hydrologic data and field facilities were donated by the Canadian Department of Fisheries and Oceans, and Canadian Forest Products. Funding for this work was provided by Natural Science and Engineering Research Council to both authors and the Northern Land Use Institute and Fisheries Renewal British Columbia grants to ELP.

References

Allan JD (1995) Stream ecology: structure and function of running waters. Chapman and Hall, London
Alldredge AL, Silver MW (1988) Characteristics, dynamics and significance of marine snow. Progress in Oceanography 20:41–82
Ben-David M, Hanley TA, Schell DM (1998) Fertilization of terrestrial vegetation by spawning Pacific salmon: the role of flooding and predator activity. Oikos 83:47–55
Bilby RE, Fransen BR, Bisson PA (1996) Incorporation of nitrogen and carbon from spawning coho salmon into the trophic system of small streams: evidence from stable isotopes. Canadian Journal of Fisheries and Aquatic Sciences 53:164–173
Bouillon S, Mohan PC, Sreenivas N, Dehairs F (2000) Sources of suspended organic matter and selective feeding by zooplankton in an estuarine mangrove ecosystem as traced by stable isotopes. Marine Ecology Progress Series 208:79–92
Bunn SE, Barton DR, Hynes HBN, Power G, Pope MA (1989) Stable isotope analysis of carbon flow in a tundra river system. Canadian Journal of Fisheries and Aquatic Sciences 46:1769–1775
Cifuentes LA, Sharp JH, Fogel ML (1988) Stable carbon and nitrogen isotope biogeochemistry in the Delaware Estuary. Limnology and Oceanography 33:1102–1115
de Boer DH (1997) An evaluation of fractal dimensions to quantify changes in the morphology of fluvial suspended sediment particles during baseflow conditions. Hydrological Processes 11:415–426
Droppo IG (2001) Rethinking what constitutes suspended sediment. Hydrological Processes 15:1551–1564
Droppo IG, Ongley ED (1994) Flocculation of suspended sediment in rivers of southeastern Canada. Water Research 28:1799–1809
Droppo IG, Leppard GG, Flannigan DT, Liss SN (1997) The freshwater floc: a functional relationship of water and organic and inorganic floc constituents affecting suspended sediment properties. Water, Air and Soil Pollution 99:43–45
France RL (1995) Critical examination of stable isotope analysis as a means for tracing carbon pathways in stream ecosystems. Canadian Journal of Fisheries and Aquatic Sciences 52:651–656
France RL (1997) Stable carbon and nitrogen isotopic evidence for ecotonal coupling between boreal forests and fishes. Ecology of Freshwater Fish 6:78–83
Fry B (1988) Food web structure on Georges Bank from stable C, N, and S isotopic compositions. Limnology and Oceanography 33:1182–1190

Griffiths H (1998) Stable isotopes, integrations of biological, ecological and geochemical processes (Environmental Plant Biology Series). Bios, Oxford

Hedges JI, Clark WA, Cowie GL (1988) Organic matter sources to the water column and surficial sediments of a marine bay. Limnology and Oceanography 33:1116–1136

Johnson SL, Covich AP (1997) Scales of observation of riparian forests and distributions of suspended detritus in a prairie river. Freshwater Biology 37:163–175

Johnston NT, Fuchs S, Mathias KL (1998) Organic matter sources in undisturbed forested streams in the north-central interior of BC. Riparian Ecosystems Research Program Newsletter. Forest Renewal BC, Canada

Kalff J (2002) Limnology: inland water ecosystems. Prentice Hall, New Jersey

Kline TC, Goering JJ, Mathisen OA, Poe PH, Parker PL (1990) Recycling of elements transported upstream by runs of Pacific salmon: I. $\delta^{15}N$ and $\delta^{13}C$ evidence in Sashin Creek, southeastern Alaska. Canadian Journal of Fisheries and Aquatic Sciences 47:136–144

Koetsier P, McArthur JV, Leff LG (1997) Spatial and temporal response of stream bacteria to sources of dissolved organic carbon in a blackwater stream system. Freshwater Biology 37:79–89

Kranck K, Petticrew EL, Milligan TG, Droppo IG (1993) In situ particle size distributions resulting from flocculation of suspended sediment. Coastal and Estuarine Study Series 42:60–74

Liss SN, Droppo IG, Flannigan D, Leppard GG (1996) Floc architecture in wastewater and natural riverine systems. Environmental Science and Technology 30:680–686

McConnachie JL (2003) Seasonal variability of fine-grained sediment morphology in a salmon-bearing stream. M.S. Thesis, University of Northern British Columbia, Prince George

McConnachie JL, Petticrew EL (2004) Hydrological and biological event based variability in the fine-grained sediment structure of a small undisturbed catchment. In: Golosov V, Belyaev V, Walling DE (eds) Sediment transfer through the fluvial system. IAHS Pub 288. IAHS Press, Wallingford, pp 459–465

McConnachie JL, Petticrew EL (2006) Tracing organic matter sources in riverine suspended sediment: Implications for fine sediment transfers. Geomorphology 79:13–26

Minshall GW, Petersen RC, Cummins KW, Bott TL, Sedell JR, Cushing CE, Vannote RL (1985) Interbiome comparison of stream ecosystem dynamics. Ecological Monographs 53:1–25

Owen JS, Mitchell MJ, Michener RH (1999) Stable nitrogen and carbon isotopic composition of seston and sediment in two Adirondack Lakes. Canadian Journal of Fisheries and Aquatic Sciences 56:2186–2192

Peterson BJ, Howarth RW, Garritt RH (1985) Multiple stable isotopes used to trace the flow of organic matter in estuarine food webs. Science 227:1361–1363

Petticrew EL (1996) Sediment aggregation and transport in northern interior British Columbia streams. In: Walling DE, Webb BW (eds) Erosion and sediment yield: global and regional perspectives. IAHS Pub 236. IAHS Press, Wallingford, pp 313–319

Petticrew EL, Arocena JM (2003) Organic matter composition of gravel-stored sediments from salmon bearing streams. Hydrobiologia 494:17–24

Phillips DL (2001) Mixing models in analyses of diet using multiple stable isotopes: a critique. Oecologia 127:166–170

Phillips DL, Gregg JW (2001) Uncertainty in source partitioning using stable isotopes. Oecologia 127:166–170

Phillips JM, Walling DE (1999) The particle size characteristics of fine-grained channel deposits in the River Exe Basin, Devon, UK. Hydrological Processes 13:1–19

Ryder JM (1995) Stuart-Takla watersheds: terrain and sediment sources. Work Paper 03/1995. BC Ministry of Forests, Victoria

Sand-Jensen K (1998) Influence of submerged macrophytes on sediment composition and near-bed flow in lowland streams. Freshwater Biology 39:663–679

Soulsby C, Youngson AF, Moir HJ, Malcolm IA (2001) Fine sediment influence on salmonid spawning habitat in a lowland agricultural stream: a preliminary assessment. Science of the Total Environment 265:295–307

Tockner K, Pennetzdorfer D, Reiner N, Schiemer F, Ward JV (1999) Hydrological connectivity, and the exchange of organic matter and nutrients in a dynamic river-floodplain system (Danube, Austria). Freshwater Biology 41:521–535

van Leussen W (1999) The variability of settling velocities of suspended fine-grained sediment in the Ems Estuary. Journal of Sea Research 41:109–118

Webster JR, Ehrman TP (1996) Solute dynamics. In: Hauer FR, Lamberti GA (eds) Methods in stream ecology. Academic Press Inc., San Diego, 145–160

Fritz Hartmann Frimmel · Gudrun Abbt-Braun · George Metreveli
Markus Delay · Christian Heise

6.2 Aggregation and Sorption Behavior of Fine River Sediments

6.2.1 Introduction

The suspension and deposition behavior of fine particles in rivers is of high environmental relevance and depends on numerous physical-chemical factors (Lick 1982). Flow turbulence, particle size distribution, particle surface charge, ionic strength, and the concentration of organic matter (OM) are key parameters associated with these processes (Amos et al. 1992; Buffle and Leppard 1995a,b). Fine particles can interact with organic and inorganic pollutants (Grolimund et al. 1996; Roy and Dzombak 1997; Kretzschmar et al. 1999) and facilitate their transport in river systems, but they also may contribute to the immobilization of pollutants due to particle sedimentation.

In this contribution, the role of different physical-chemical parameters on the aggregation behavior of fine river sediments from the river Elbe and a model solid phase (kaolin) was investigated as well as the interaction of these solid phases with inorganic pollutants (copper, lead, zinc). Focus was put on the role of organic matter, particle surface charge, and ionic strength on the aggregation and sorption behavior of the river sediment and kaolin. The aim of the work was to assess the potential impact of particle mediated transport of contaminants on river systems.

6.2.2 Materials and Methods

Sample preparation in all experiments was carried out by using demineralized water (Milli-Q Plus, Millipore). For pH adjustment and for sediment washing HCl (suprapur, Merck) and NaOH (Merck) were used.

Solid Phases

Sediment sampling and properties. Channel bed sediment samples were collected from the river Elbe at Buhnenfeld (stream km 607.5, depth: 2 m). After sample collection, pore water was separated from the sediment. To do this, sediment samples were centrifuged at 4 000 rpm (2 150 × g) for 60 min (ROTANTA 460 RS, Andreas Hettich GmbH & Co KG). After centrifugation, the supernatant was filtered with 0.45 µm membranes and analyzed for concentration of dissolved organic carbon (DOC) (Sievers 820, Portable Total Organic Carbon Analyser, Sievers Instruments Inc., USA). The centrifuged sediment was freeze-dried (1.030 mbar, –20 °C; Christ, Alpha 2-4; Sed_0). A portion of the freeze-dried material was analyzed for mineral content using X-ray diffractometer (Siemens Diffrac 11). The results of X-ray diffractometry measurements are shown in Table 6.2.

To remove organic or inorganic (metal species) constituents, the freeze-dried sediment Sed_0 was washed with the following solutions and in the following order: solution of NaOH (0.1 mol l^{-1}), demineralized water, solution of HNO_3 (0.1 mol l^{-1}) and

Table 6.2.
Mineral content of original (freeze-dried) sediment

Minerals	Content in % (mass/mass)
Quartz	23
Feldspar	2
Mica	32
Kaolinite	37
Swelling minerals	3
Dolomite	1
Calcite	2

demineralized water. Each washing step was carried out three times and for 24 h. Finally the washed sediment was freeze-dried (Sed_w).

Kaolin. Kaolin H 3 GF (Dorfner, $d_{50} = 3\ \mu m$) was used as reference material for a mineral solid phase.

Aggregation Behavior of the Sediment

For the stability characterization of fine sediments samples of Sed_o and Sed_w were prepared with demineralized water and pore water ($n = 3$). The concentration of sediment was set to $2\ g\ l^{-1}$. In the supernatant, collected following sedimentation of the larger particles (>1 µm), pH value, electrical conductivity, particle size and zeta potential were determined. Particle size and zeta potential were measured using Zetasizer Nano ZS (Malvern Instruments, laser: 4 mW He-Ne, $\lambda = 633$ nm). For the particle size measurements dynamic light scattering technique was used (scattering angle: 173°). The electrophoretic mobility was detected by means of the laser doppler electrophoreses. The zeta potential was calculated from electrophoretic mobility using the Smoluchowski equation (Müller 1996).

Titration experiments. For a detailed characterization of the aggregation behavior of sediment particles, titration experiments were carried out by means of a Zetasizer and an Autotitrator (MPT-2, Malvern Instruments). The titration was done in a plastic beaker, which was connected through a capillary system and a peristaltic pump with a folded capillary zeta potential cell (DTS 1060, Malvern Instruments). The sediment samples (Sed_o, Sed_w, $\rho = 2\ g\ l^{-1}$ each) were prepared with demineralized water. A volume of 10 ml of the supernatant, obtained after sedimentation of the larger particles (>1 µm), was titrated with solutions of $MgSO_4$ (0.01; 0.1 and 0.5 mol l^{-1}) in the concentration range from 0 mmol l^{-1} to 50 mmol l^{-1} to study the influence of the ionic strength on the aggregation of the particels. The pH value remained constant at 7 during the titration. Furthermore, the same samples were titrated from a pH value of 7 to pH 2 with HCl (0.1 mol l^{-1} and 1 mol l^{-1}) and from a pH value of 7 to 12 with NaOH (1 mol l^{-1}). After each adjustment of pH value and $MgSO_4$ concentration, the particle size and electrophoretic mobility was measured three times. Between the measurements and during the adjustments of pH value and $MgSO_4$ concentration, the sample was circulated and stirred. The titration experiments were controlled automatically by software.

Stirring tank experiments. In cooperation with the Institute for Hydromechanics, Universität Karlsruhe (TH) (Prof. G. H. Jirka, Dipl.-Ing. G. Kühn), experiments in a stirring tank were performed in order to investigate the influence of ionic strength on the aggregation behavior of fine particles. The stirring tank had a volume of 1.2 l (inner diameter: 100 mm, height: 150 mm). A $MgSO_4$ solution (2 mol l^{-1}) was added stepwise (20 steps of 5 ml, each) to a suspension of freeze-dried river Elbe sediment (Sed_o; $\rho = 250$ mg l^{-1}). The suspension was stirred continuously (150 rpm) to prevent settling of particles.

For particle detection, an optical measurement technique based on an In-Line Microscope (Aello 7000) with a CCD camera system was used. With the instrumental set-up, particles between 4 and 500 μm could be detected.

Column Leaching Experiment and Size Exclusion Chromatography

Column leaching experiment. To examine the release of organic compounds from the river Elbe sediment, column leaching experiments were performed. A column (Alltech Grom, ECO GR-G4550Z, 450 × 50 mm) filled with 600 g of the freeze-dried sediment was sequentially leached with 950 ml of the following eluents: demineralized water, solutions of $MgSO_4$ (0.05 and 0.01 mol l^{-1}), and a solution of NaOH (0.1 mol l^{-1}).

A HPLC unit (Amersham Pharmacia, ÄKTA Explorer 100) with pump and UV detector ($\lambda = 254$ nm) was used at a flow rate of 1 ml min^{-1}. In the column outlet, UV absorption, pH value and electrical conductivity were detected online.

Size exclusion chromatography (SEC). For characterization of the organic matter in the eluates, the analytical SEC system LC-OCD with a TSK HW 50 S column and a phosphate buffer (0.028 mol l^{-1}) was used. The system was described in detail by Huber and Frimmel (1992). The system was also used to determine the DOC concentrations of the eluates. The DOC concentrations were calculated from bypass DOC peak areas by an external calibration of the DOC detector with potassium hydrogen phthalate as standard. Samples with a DOC above 10 mg l^{-1} were diluted with buffer prior to measurement.

Sorption of Heavy Metals

Sorption kinetics of heavy metals (Cu, Pb and Zn) on Sed_w and kaolin particles in the presence and in the absence of dissolved organic matter (DOM) were investigated in batch experiments. The metal stock solutions for the sorption experiments were prepared by using: $CuCl_2 \cdot 2\,H_2O$ (Merck), $PbCl_2$ (Merck), and $ZnCl_2$ (Merck). These salts were dissolved in a solution of HCl (10 mmol l^{-1}). The concentration of metals was set to 10 mmol l^{-1}. Freeze-dried sediment was dispersed and extracted with NaOH solution (0.1 mol l^{-1}) (solid to liquid ratio: 50 g/0.5 l) and shaken by a vertically spinning mixer. The extract was used as solution to study the influence of dissolved organic matter (DOM) in the batch experiments. The concentration of the solid phases (sediment and kaolin) was set to 5 g l^{-1}. The heavy metal concentration (50 μmol l^{-1}) and DOM concentration (40 mg l^{-1}) were adjusted by using metal stock solutions and sediment extract respectively. 40 ml samples were shaken in plastic beakers (Greiner Bio-one) using a vertically spinning mixer (REAX 20, Heidolph).

The sorption experiments were carried out at pH values of 5 and 7. After mixing of the components three aliquot samples (3.5 ml each) were sub sampled between 1 and 500 min, transferred to the centrifuge tubes and centrifuged in an ultracentrifuge (Optima TLX, Beckman Coulter) at 100 000 rpm (417 000 × g) for 20 min. After centrifugation, 1.5 ml of supernatant from each tube was obtained. The supernatants were collected in one tube (4.5 ml), stabilized with 1% HNO_3 (suprapur, Merck) and analyzed for metal concentration with inductively coupled plasma optical emission spectrometer (ICP-OES, Vista-Pro CCD Simultaneous, Varian).

6.2.3 Results and Discussion

Aggregation Behavior of the Sediment

Both original and washed river sediment (Sed_o and Sed_w) particles dispersed in demineralized water showed a negative zeta potential between −19 mV and −21 mV (see Fig. 6.8). Due to the high ionic strength, the original and washed sediment dispersed in pore water showed a slight decrease of the negative zeta potential (−16 mV). This trend is associated with a significant increase of the particle size from 500 nm to 1 500 nm. This is due to the compression of the electrical double layer by the adsorption of cations from the pore water onto the particle surface. As a consequence of the decreasing repulsive electrostatic forces, the particles aggregate rapidly. The pH value remained constant between 7 and 8. It is interesting to note that washing of sediment particles showed no significant influence on their aggregation behavior.

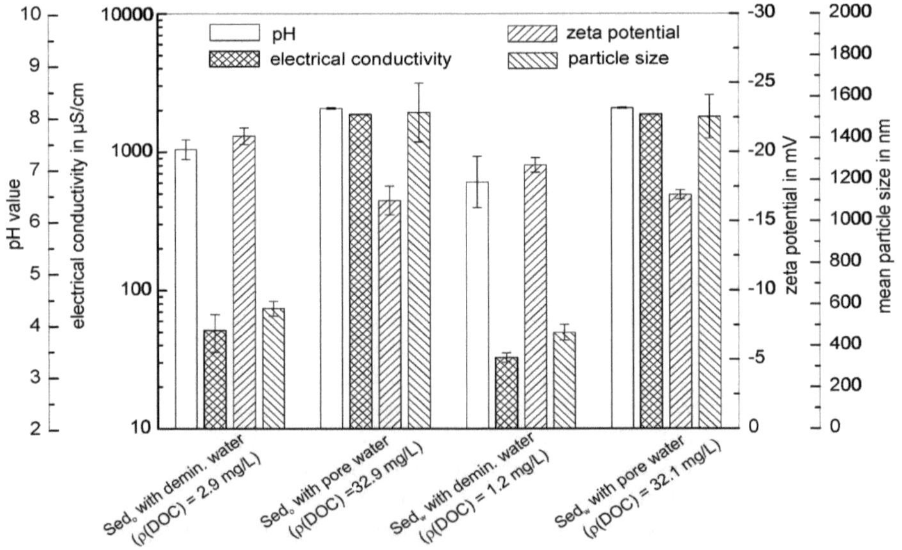

Fig. 6.8. pH value, electrical conductivity, particle size and zeta potential of original and washed river sediment (Sed_o and Sed_w) dispersed in demineralized water and pore water (DOC concentration in pore water: 31.9 mg l^{-1})

Titration experiments. The stability and aggregation behavior of fine sediment particles was investigated in titration experiments. Figure 6.9a shows the zeta potential and particle size of original and washed sediment as a function of the $MgSO_4$ concentration. Titration with $MgSO_4$ solution caused a decrease of the negative zeta potential with increasing salt concentration from –20 mV to –8 mV. The particles remain stable in the concentration range of 0 to 3 mmol l^{-1}. The increase of $MgSO_4$ concentration leads to aggregation of sediment particles. The aggregation for washed sediment particles was stronger than for original sediment particles. During sediment washing with the NaOH solution, a significant part of organic compounds was desorbed from the particle surface. Therefore, it is assumed that the presence of organic compounds on the surface of original sediment particles can cause their stabilization (Kretzschmar et al. 1995; Kretzschmar and Sticher 1997) such that the aggregates are smaller than for the washed sediment.

The influence of the pH value on the stability of original and washed sediment particles was also investigated in the titration experiments (Fig. 6.9b). The zeta potential remained negative (from –42 mV to –7 mV) at the pH range investigated (pH 2 to

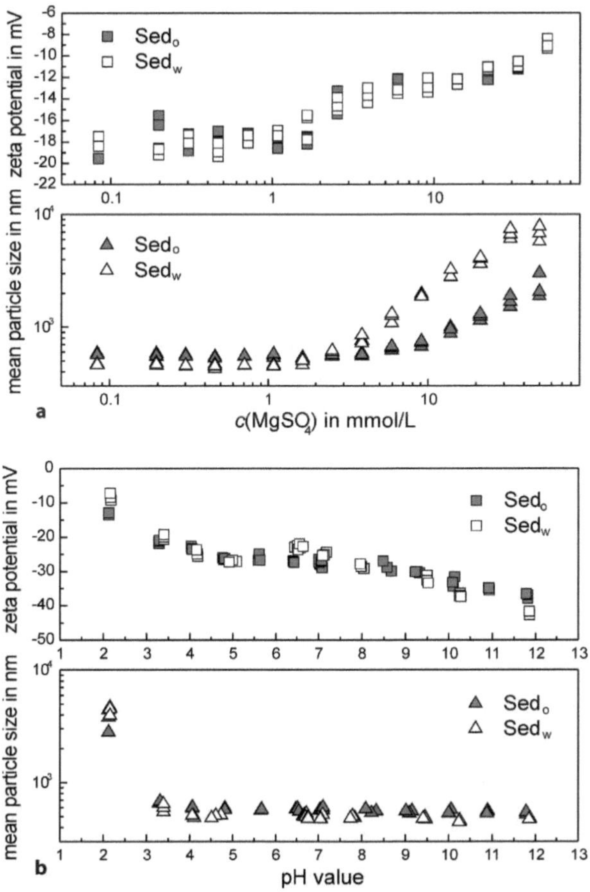

Fig. 6.9.
Influence of $MgSO_4$ concentration (**a**) and pH value (**b**) on the zeta potential and mean particle size of original (Sed_o) and washed sediment (Sed_w)

pH 12) and as expected, it became less negative with decreasing pH value due to the increasing protonation of the particle surface.

The average particle size of the sediment remained stable (about 500 nm) in the pH range of 4 to 12. At lower pH values (pH 4 to pH 2), the particle size increased rapidly and sediment particles agglomerated. At pH 2, agglomerates with an average particle size of about 3 μm to 4 μm were detected. Washing showed no significant influence on the aggregation behavior of the sediment particles.

Stirring tank experiments. Additionally, stirring tank experiments were performed for the characterization of the aggregation behavior of sediment particles. Figure 6.10 shows the median particle size (d_{50}) as a function of the $MgSO_4$ concentration. In the stirring tank experiments, an aggregation of the sediment particles was observed. Increasing $MgSO_4$ concentration led to an increase of the median particle size. At high salt concentrations (100 mmol l^{-1} to 160 mmol l^{-1}), a maximal value of aggregate size was reached (about 40 μm).

Column Leaching Experiments and Size Exclusion Chromatography

In Fig. 6.11, SEC chromatograms of different eluate fractions from the column leaching test are shown. The samples were taken after elution with

1. 750 ml of demineralized water
2. 600 ml of $MgSO_4$ solution (0.05 mol l^{-1})
3. 800 ml of $MgSO_4$ solution (0.1 mol l^{-1}), and
4. 800 ml of NaOH solution (0.1 mol l^{-1})

The concentration of dissolved organic carbon (DOC) in the leachate samples was 77 mg l^{-1} (1), 12 mg l^{-1} (2), 15 mg l^{-1} (3), and 83 mg l^{-1} (4).

Elution with $MgSO_4$ solution led to lower DOC concentrations in the eluates compared to the elution with demineralized water. An increasing Mg^{2+} concentration obviously intensified electrostatic interactions between inorganic sediment and organic matter and

Fig. 6.10.
Aggregation behavior of sediment particles in a stirring tank. Influence of $MgSO_4$ concentration

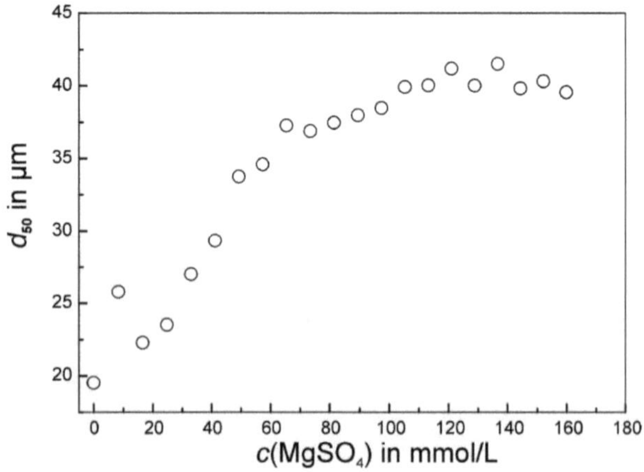

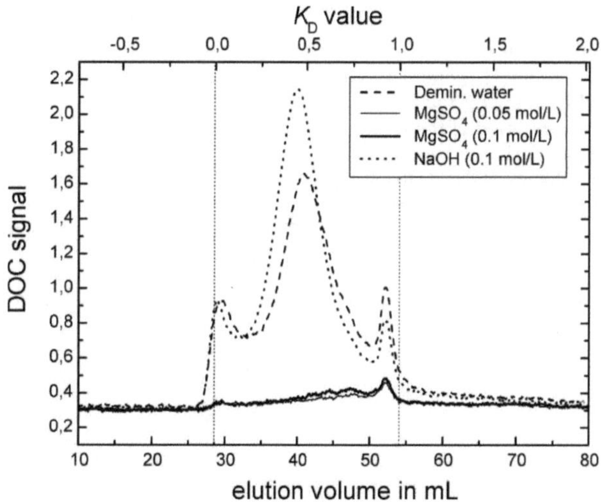

Fig. 6.11. SEC/DOC chromatograms of selected eluate fractions from the column leaching experiment

therefore, organic matter was retained in the column. Furthermore, it was noted that the availability of leachable organic matter decreased with increasing time of elution.

During elution with NaOH solution, DOC mobility increased. This was due to dissolution and release of organic matter from sediment surface.

Sorption of Heavy Metals

In the sorption experiments, the influence of organic matter on the sorption of heavy metals (Cu, Pb and Zn) onto the fine sediment particles (washed) and kaolin particles was investigated. Sediment showed a better sorption capacity for all heavy metals investigated than the kaolin (Fig. 6.12). The river sediment contained a range of different minerals (as shown in Table 6.2) with different properties and surface groups which could account for the high sorption of heavy metal cations.

Figure 6.12 shows the sorption kinetics of zinc onto sediment and kaolin in the presence and absence of organic matter. At pH 5, sorption of Zn was stronger in the absence of organic matter than in its presence. At pH 7, a positive influence of the organic matter on the zinc sorption could be determined especially in the case of kaolin. The zinc cations were probably sorbed onto the solid phase as metal-organic complexes, which are usually more stable at high pH values (Schmitt et al. 2003). The results for copper and lead which are not shown here were similar to those obtained for zinc.

6.2.4 Conclusions

The titration and stirring tank experiments have shown that the aggregation behavior of the sediment particles is mainly influenced by the ionic strength in the aqueous phase. The pH value has little significant influence over the natural riverine range.

In rivers with low salt concentrations, it has to be expected that fine particles are stable in dispersion and thus, they remain mobile. The strong sorption of heavy metals

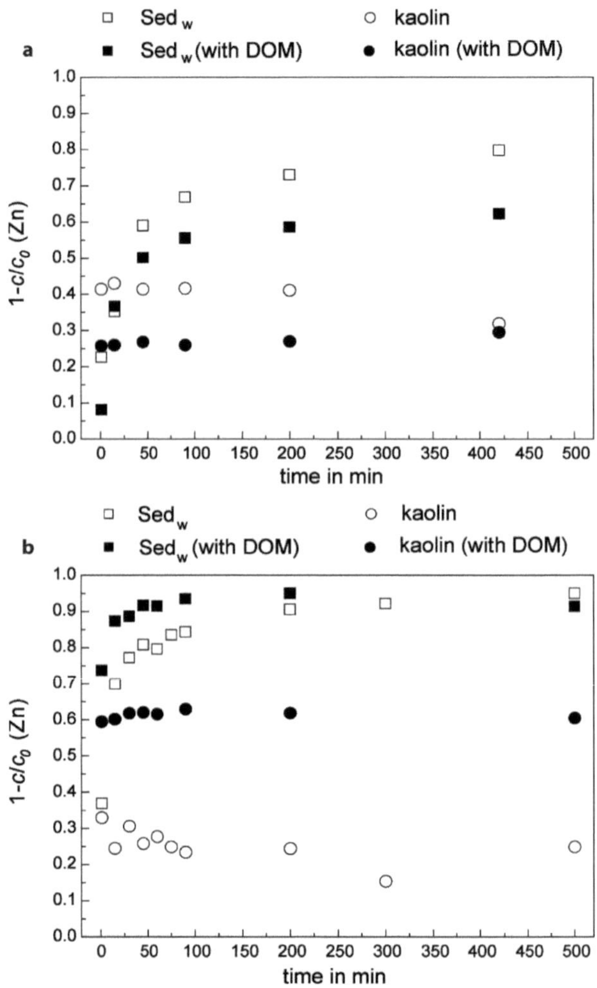

Fig. 6.12.
Sorption kinetics of zinc ($c_0 = 50\ \mu mol\ l^{-1}$) onto washed sediment ($Sed_w$) and kaolin (concentration of the solid phases: $5\ g\ l^{-1}$) in the presence and absence of DOM ($40\ mg\ l^{-1}$) at pH 5 (**a**) and pH 7 (**b**)

on the fine sediment particles can lead to an environmental risk in the context of particle mediated contaminant transport.

An increase of ionic strength (in natural systems: e.g., estuaries with mixing of river and sea water) enables particle aggregation and their sedimentation. As a consequence, contaminants would be accumulated, and stored in the river channel sediments.

Acknowledgment

The authors thank the German Ministry for Research and Education for financial support (02WF0468). The authors also thank Ulrich Reichert, Reinhard Sembritzki, and Matthias Weber for their experimental assistance. Prof. G. H. Jirka and Dipl.-Ing. G. Kühn are gratefully acknowledged for the successful cooperation.

References

Amos CL, Grant J, Daborn DA, Black K (1992) Sea carousel – A benthic annular flume. Esturaine, Coastal and Shelf Science 34:557–577

Buffle J, Leppard GG (1995a) Characterization of aquatic colloids and macromolecules. 1. Structure and behavior of colloidal material. Environ Sci Technol 29:2169–2175

Buffle J, Leppard GG (1995b) Characterization of aquatic colloids and macromolecules. 2. Key role of physical structures on analytical results. Environ Sci Technol 29:2176–2184

Grolimund D, Borkovec M, Barmettler K, Sticher H (1996) Colloid-facilitated transport of strongly sorbing contaminants in natural porous media: a laboratory column study. Environ Sci Technol 30:3118–3123

Huber S, Frimmel FH (1992) A liquid chromatographic system with multi-detection for the direct analysis of hydrophilic organic compounds in natural waters. Fresenius' Z Anal Chem 342:198–200

Kretzschmar R, Borkovec M, Grolimund D, Elimelech M (1999) Mobile subsurface colloids and their role in contaminant transport. Advances in Agronomy 66:121–193

Kretzschmar R, Robarge WP, Amoozegar A (1995) Influence of natural organic matter on colloid transport through saprolite. Water Resources Research 31(3):435–445

Kretzschmar R, Sticher H (1997) Transport of humic-coated iron oxide colloids in a sandy soil: influence of Ca^{2+} and trace metals. Environ Sci Technol 31:3497–3504

Lick W (1982) Entrainment, deposition and transport of fine-grained sediments in lakes. Hydrobiologica 91:31–40

Müller RH (1996) Zetapotential und Partikelladung in der Laborpraxis: Einführung in die Theorie, praktische Meßdurchführung, Dateninterpretation. Wissenschaftliche Verlagsgesellschaft, Stuttgart

Roy SB, Dzombak DA (1997) Chemical factors influencing colloid-facilitated transport of contaminants in porous media. Environ Sci Technol 31:656–664

Schmitt D, Saravia F, Frimmel FH, Schuessler W (2003) NOM-facilitated transport of metal ions in aquifers: importance of complex-dissociation kinetics and colloid formation. Water Research 37:3541–3550

Annekatrin Fritsche · Hilmar Börnick · Eckhard Worch

6.3 Equilibrium and Kinetics of Sorption/Desorption of Hydrophobic Pollutants on/from Sediments

6.3.1 Introduction

The input of anthropogenic substances into the rivers of Germany has been significantly decreased over the past few years. In contrast to the river water, sediments are still heavily polluted by metals and organic contaminants. Therefore, sediments can act as sources for contaminants. Sorption/desorption processes in the system river water/sediment are expected to be dependent on environmental conditions. But at present the knowledge regarding the effects of the hydrochemical and hydrodynamical conditions on remobilization of pollutants out of historical contaminated sediments is still insufficient.

The objectives of this work were to investigate the effect of dissolved natural organic matter (DOM) onto sorption and desorption processes and competition effects among hydrophobic organic contaminants (HOC). Furthermore, sorption and desorption processes of HOCs were investigated under different hydrodynamic conditions.

6.3.2 Experimental Methods

Sediment materials and water. Natural river sediment, taken out of an old anabranch of the river Elbe near Torgau/Germany, was used for the study. As a pretreatment step, the sediment, containing an organic fraction of 0.24% (f_{OC} = 0.0024 g g^{-1} dry weight), was dried at 40 °C, sieved (<2 mm) and homogenized. The experiments were accomplished with natural water possessing different concentrations and consistence of natural DOM (laboratory water: about 1 mg l^{-1}, river water: 6 mg l^{-1}, pond water: 14 mg l^{-1}). The ionic strength of the pond and river water was about 400 µS cm^{-1}. Sodium chloride was added to the laboratory water to set its ionic strength to an equal level as pond and river water. Preliminary analysis of all water and sediment materials used in the study showed no contamination with the model pollutants.

Chemicals. For the investigations three relevant anthropogenic compounds were selected: 1,2,4-trichlorobenzene, pentachlorobenzene and pentachloronitrobenzene (PESTANAL®, purchased from Riedel-de Haën). The criteria of substance selection were their appearance in the list of priority pollutants of the EU (directive 2000/60/EC), their occurrence in rivers, and their biological and chemical persistence within the experimental periods. The structure, aqueous solubility and octanol-water partitioning coefficients (K_{OW}) of the model compounds are shown in Table 6.3.

Experimental methods. All experiments were performed in the dark, in completely mixed batch (CMB) reactors, with equal solid-to-water ratios (10 g dry sediment, 200 ml solution). Within the sorption experiments, the concentration of each investigated model contaminant was in a range of 2 to 500 µg l^{-1}. The CMB reactors were shaken for sorp-

Table 6.3.
Structure and properties of investigated model contaminants (Chemical Fact Sheet 2005)

	Structure	Aqueous solubility S (mg l^{-1}, 20 °C)	$\log K_{OW}$
1,2,4-Trichlorobenzene	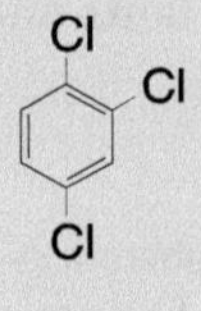	49	3.8 … 4.2
Pentachlorobenzene	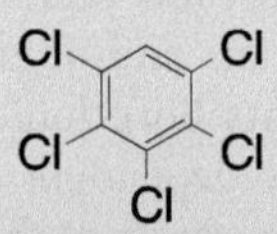	0.5	4.8 … 5.4
Pentachloronitrobenzene	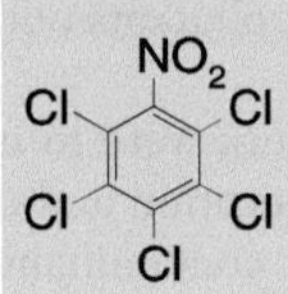	0.8	4.2 … 4.8

tion equilibration. The desorption experiments were done by replacing the supernatant liquid phase with contaminant-free water after the sorption experiments. Afterwards, the CMB reactors were shaken until a state of desorption equilibrium was reached.

Desorption kinetics were studied using equally preloaded sediment. For preloading 200 g dry sediment and 300 ml liquid phase containing 500 μg l^{-1} of each model contaminant were shaken for 170 h reaching the state of sorption equilibrium. Afterwards, the supernatant liquid was decanted off. Desorption experiments were performed with 10 g of the preloaded solid phase and 200 ml contaminant-free water. To determine the increase of liquid-phase concentration during the desorption process, samples were taken after different times. For the investigation of the effect of hydrodynamical conditions on the desorption process, the agitation within the CMB-reactors was varied (0, 100, 150 and 200 rotations per minute).

All experiments were replicated at least twice. The errors including analytical and methodical deviation are about 10% for sorption and up to 23% for desorption equilibrium experiments. The error of kinetic experiments are up to 50%.

Analytical methods. Prior to analysis, the model contaminants were separated from the aqueous phase by liquid-liquid extraction with dichloromethane. Therefor, 50 ml liquid phase, 5 g sodium chloride and 3 ml dichloromethane were shaken for 45 min at 250 rpm. The organic phase was isolated and dried with sodium sulfate. The solid samples were extracted with a mixed solvent consisting of acetone and dichloromethane (1:4 vol:vol). The extracts were concentrated in a rotary evaporator, dried and cleaned in sodium sulfate/Florisil®-cartridges and concentrated to 1 ml in a gentle stream of nitrogen. The analysis of liquid phase and sediment extracts were accomplished in a GC-MS system (Thermo Quadrupol MS Trace DSQ). 1,2,3,4-Tetrachloro-5-nitrobenzene and 1,3,5-tribromobenzene were used as internal standards.

The organic content of solid phases was analyzed by a TOC-Analyser with boat sampler (Rosemount Analytical Dohrmann DC-190) after digestion with hydrochloric acid. The DOM content of the liquid phase was analyzed by a Shimadzu TOC-5000-Analyser and quantified as non-purgeable organic carbon (NPOC). Since volatile organic compounds are not relevant for the investigated water samples, NPOC can be set equal to DOC (dissolved organic carbon).

6.3.3 Results and Discussion

Equilibrium Experiments

To investigate a possible *competition effect among the model contaminants*, isotherms were determined by using single contaminant/DOM and mixed contaminants/DOM systems. The mixed system contained respectively 2 to 500 μg l^{-1} of 1,2,4-trichlorobenzene, pentachlorobenzene and pentachloronitrobenzene in addition to the natural DOM. For this purpose, river Elbe water was used as liquid matrix. The sorption coefficients were acquired from the first linear part of the isotherms and normalized to the organic carbon content of the solid phase (f_{OC} = 0.0024 g g^{-1} dry weight). In Table 6.4, the results of the single contaminant and mixed contaminant sorption isotherms are shown.

Table 6.4. Competition effect – partitioning coefficients of the sorption isotherms

	$\log K_{OC}$ (single contaminant/DOM)	$\log K_{OC}$ (mixed contaminants/DOM)	$\log K_{OC}$ (literature)
1,2,4-Trichlorobenzene	3.5	3.8	3.8 … 4.2
Pentachlorobenzene	4.1	4.7	4.8 … 5.4
Pentachloronitrobenzene	4.3	4.8	4.2 … 4.8

Table 6.5.
Results of the sorption and desorption isotherms in dependence on the DOM concentration in the liquid phase

	DOC (water) (mg l^{-1})	$\log K_{OC,\,Sorption}$ (mixed contaminants/DOM)	$\log K_{OC,\,Desorption}$ (mixed contaminants/DOM)
1,2,4-Trichlorobenzene	1	3.7	4.4
	6	3.8	4.2
	14	3.7	4.3
Pentachlorobenzene	1	4.5	5.0
	6	4.7	4.9
	14	4.7	4.9
Pentachloronitrobenzene	1	4.4	5.2
	6	4.8	5.1
	14	4.5	5.2

No significant competition effects among the investigated model contaminants could be observed for sorption or for desorption processes.

Additional to the competition effect, the *influence of the DOM concentration on the sorption and desorption* was investigated. During the river Elbe flood in 2002, an obvious increase of the dissolved organic carbon (DOC) concentration of the river water was detected (Börnick et al. 2003). According to several authors (e.g., Chiou et al. 1986; Laor et al. 1998), the sorption processes of HOCs is influenced by DOM. Amiri et al. (2005) showed that sorption of nitro-substituted organic compounds was significantly influenced by the DOC content of the liquid phase. In column experiments, Reemtsma et al. (2003) found a linear correlation between the mobility of adsorbable organic halogen compounds and the release of solid organic matter connected with increasing DOC concentrations. The sorption of 1,2,4-trichlorobenzene, pentachlorobenzene and pentachloronitrobenzene was not influenced by the concentration of DOM (Table 6.5). The results of desorption isotherms with different DOM concentration in the liquid phase showed no influence of DOM on the desorption process of the investigated HOCs (Fig. 6.13, Table 6.5).

For all compounds investigated, there is an obvious discrepancy between the partitioning coefficients calculated from the sorption and desorption data. The comparison in Table 6.5 indicates that the partitioning coefficients of desorption isotherms are above the results of the sorption isotherms. The so-called hysteresis effect could be assumed to be the reason for this effect. The differences between sorption and desorption isotherms can be seen in Fig. 6.14. According to literature (Pignatello 2000), hysteresis can be explained by *(a)* an inadequate time allowed for equilibration or *(b)* by

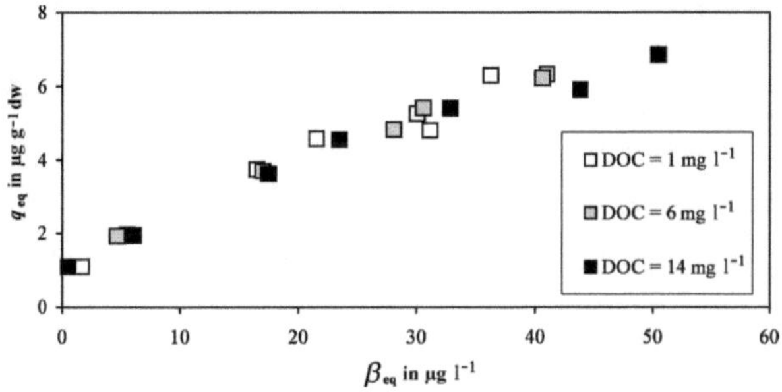

Fig. 6.13.
Desorption isotherms of pentachlorobenzene in water possessing different amounts of DOM (quantified as DOC). q_{eq}: equilibrium loading of sediment; β_{eq}: equilibrium concentration in the aqueous phase

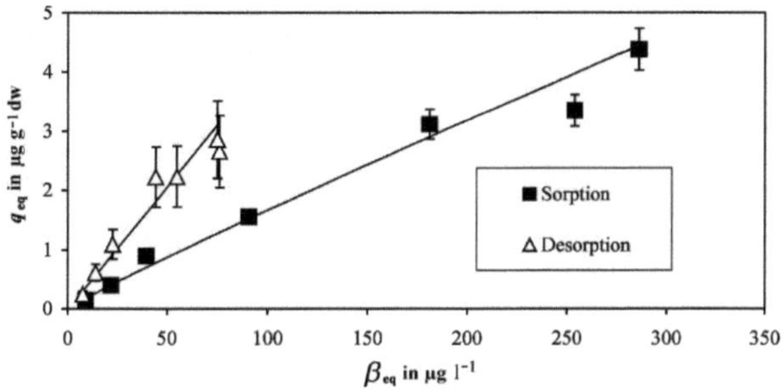

Fig. 6.14.
Comparison of sorption and desorption isotherms of 1,2,4-trichlorobenzene in river Elbe water

the occurrence of irreversibly or resistantly sorbed fractions. Case *(a)* can be excluded because sorption kinetic experiments showed that the systems under consideration reached equilibrium after three days (see kinetic curves in the following section) while the duration of each sorption isotherm experiment was seven days. Furthermore, long-term kinetic tests with duration of four weeks resulted in the same loading of the investigated contaminants on the sediment as the experiments with duration of only three days. Therefore, the observed discrepancy between sorption and desorption isotherms as presented in Fig. 6.14 must be caused by a resistantly sorbed fraction, characterized by very slow desorption kinetics, or irreversibly sorbed fractions (case *(b)*).

Desorption Kinetics

The aims of kinetic experiments were to get information about the influence of DOM and hydrodynamic conditions on the rate and extent of HOC remobilization out of contaminated river sediments. The experiments were conducted with sediments equally preloaded with model compounds. The results were analyzed using an overall kinetic model based on the following equation:

$$\frac{dq}{dt} = -k_{des}(q - q^*)$$

with q: loading of sediment at time t, k_{des}: desorption rate constant, q^*: equilibrium loading of sediment at time t.

The desorption rate constants were estimated by curve fitting using the program DEKIN 1.0. The highest measured standard deviation of the desorption rate constant

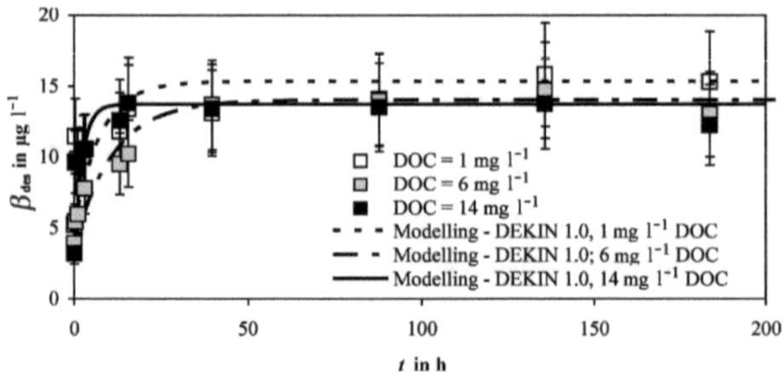

Fig. 6.15. Desorption kinetics of 1,2,4-trichlorobenzene in water possessing different DOM concentrations in a strongly resuspended system

Table 6.6. Desorption rate constants of investigated HOCs for different DOM concentrations in a non-resuspended system

DOC (mg l^{-1})	k_{des} (1,2,4-trichlorobenzene) (s^{-1})	k_{des} (pentachlorobenzene) (s^{-1})	k_{des} (pentachloronitrobenzene) (s^{-1})
1	3.6×10^{-6}	7.6×10^{-8}	2.5×10^{-6}
6	2.6×10^{-6}	8.0×10^{-8}	3.0×10^{-6}
14	3.0×10^{-6}	15.0×10^{-8}	1.3×10^{-6}

within the desorption kinetic experiments were 0.9×10^{-6} s^{-1} for 1,2,4-trichlorobenzene, 5.0×10^{-8} s^{-1} for pentachlorobenzene and 0.9×10^{-6} s^{-1} for pentachloronitrobenzene. Taking in account these standard deviations it can be concluded, that the desorption experiments showed no significant differences in the kinetic curves (Fig. 6.15) and in the desorption rate constants (Table 6.6) determined for different *DOM concentrations*. It can be derived from these experiments that there is no influence of the DOM concentration on the remobilization rate of the investigated HOCs for all investigated hydrodynamic conditions.

To investigate the *influences of the hydrodynamic conditions* on the desorption process the CMB-reactors were exposed to different agitation rates. As can be seen in Fig. 6.16, the desorption rate constants, k_{des}, of pentachlorobenzene and 1,2,4-trichlorobenzene were strongly controlled by the intensity of resuspension within the system. For pentachloronitrobenzene desorption, no obvious impact of the hydrodynamic conditions was noticed. It seems that the influence of hydrodynamics on the desorption rate constant is specific for each substance. This is in accordance with findings of Latimer et al. (1999) who showed for polycyclic aromatic hydrocarbons that remobilization depends on the chemical properties of the adsorbed substance.

In contrast to the desorption rates, the equilibrium concentrations of the model contaminants after desorption were not influenced by the hydrodynamic conditions. On the other hand, there was a strong *impact of the liquid-solid contact time* during preloading on the desorption equilibrium concentration indicating an increase of sorption strength with time. In Fig. 6.17, the kinetic curves of 1,2,4-trichlorobenzene desorption after HOC/sediment contact times of one and four weeks are shown for a moderately resuspended system. Although the mass sorbed onto the sediment was nearly the same in both cases (0.22 µg g^{-1} vs. 0.21 µg g^{-1}), the desorption after a preloading time of four weeks was much lower than the desorption after one week preloading time. Experiments in a non-resuspended system showed that the desorp-

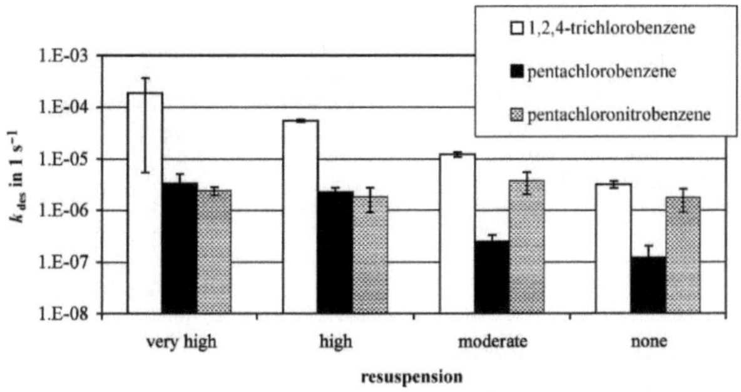

Fig. 6.16. Effect of hydrodynamic conditions on the desorption rate constant, k_{des}

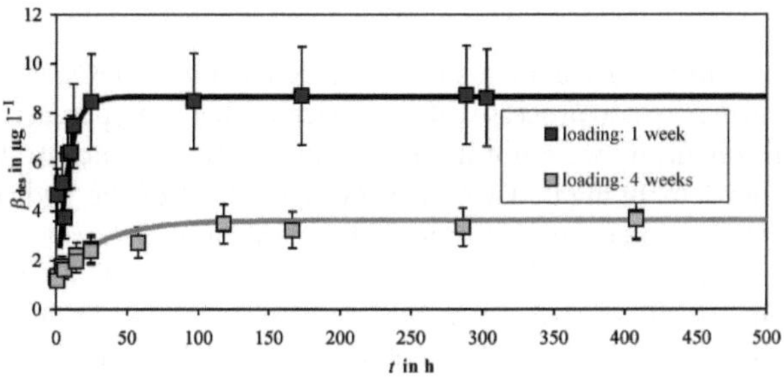

Fig. 6.17. Influence of the HOC/sediment preloading contact time on the desorption process of 1,2,4-trichlorobenzene in a moderately resuspended system

tion of pentachlorobenzene and pentachloronitrobenzene also decreases with preloading time. After a preloading time of 20 days 30% of the sorbed 1,2,4-trichlorobenzene, 14% of pentachlorobenzene and 34% of pentachloronitrobenzene were desorbed. The increasing of preloading time up to 32 days, decreased the reversibly sorbed fractions to 26% for 1,2,4-trichlorobenzene, 4% for pentachlorobenzene and 11% for pentachloronitrobenzene. Since it is known that the investigated HOCs are biodegradable under anaerobic conditions, the experiments were performed under aerobic conditions. Thus, the determined effect could be explained by a change of the ratio of reversibly and irreversibly/resistantly sorbed fractions, which was already described in the discussion of sorption and desorption isotherms (Fig. 6.13). In the literature, this effect is referred to as "aging" (Alexander 2000; Reid et al. 2000; Lee et al. 2004) and is caused by diffusion of molecules into deeper layers of the sorbent particles, whereby the bioavailability of the HOCs and the possibility of remobilization of contaminants out of polluted sediments decreases. On the other hand it is possible that changes in hydrodynamic or in hydrochemical conditions can remobilize HOCs also from historical contaminated sediments (Alexander, 2000).

Taking into account the standard deviations given before, it can be stated that the desorption rate constants were not significantly influenced by the HOC/sediment preloading contact time. It can be assumed that the values of k_{des} given in Table 6.7 represent only the fast desorbing fractions that are not influenced by the HOC/sediment contact time. The run-times of the experiments (up to 400 h) were not long enough to find out if there is an additional slow desorption step, as postulated by Cornelissen et al. (1997). Long-term experiments of about a half year would be necessary to decide if there is a resistant, characterized by a very slow desorption, or an irreversible sorption.

Table 6.7. Desorption rate constants of the model contaminants against sediment/HOC contact time in a non-resuspended system

HOC/sediment contact time *t* (d)	k_{des} (1,2,4-trichlorobenzene) (s^{-1})	k_{des} (pentachlorobenzene) (s^{-1})	k_{des} (pentachloronitrobenzene) (s^{-1})
17	2.7×10^{-6}	8.0×10^{-8}	3.2×10^{-6}
20	3.8×10^{-6}	19.5×10^{-8}	1.0×10^{-6}
32	2.9×10^{-6}	7.2×10^{-8}	1.8×10^{-6}

6.3.4 Conclusions

In contrast to statements in literature for other HOCs, our results showed that sorption and desorption processes of 1,2,4-trichlorobenzene, pentachlorobenzene and pentachloronitrobenzene were not influenced by the DOM concentration in the investigated range from 1 to 14 mg l^{-1} DOC. However, an effect of the hydrodynamic conditions and the HOC/sediment contact time on the desorption process was observed. Higher resuspension rates increased the desorption rate constant of chlorinated aromatic hydrocarbons, whereas the release of pentachloronitrobenzene was not affected by different strength of agitation. Therefore, the remobilization of HOCs out of contaminated sediments seems to be a function of chemical properties. For a better understanding of the influence of chemical properties on the extent of contaminant remobilization further research is needed.

The sorption of the investigated model contaminants was not completely reversible, which might be caused by resistantly and irreversibly sorbed fractions. The resistantly and irreversibly sorbed fractions increased with time ("aging effect"). It can be concluded from our results, that the possibility of remobilization of the investigated contaminants out of polluted sediments decreases with time.

References

Alexander M (2000) Aging, bioavailability and overestimation of risk from environmental pollutants. Environmental Science and Technology 29:2713–2717

Amiri F, Börnick H, Worch E (2005) Sorption of phenols onto sandy aquifer material: the effect of dissolved organic matter (DOM). Water Research 39:933–941

Börnick H, Grischek T, Worch E (2003) Ausgewählte Untersuchungsergebnisse von Wasser- und Schlammproben aus dem Raum Dresden während des Elbe-Hochwassers im August 2002. Proc. Jahrestagung der Wasserchemischen Gesellschaft, Stade, pp 211–217

Chemical Fact Sheet (2005) www.speclab.com/compound/chemabc.htm

Chiou CT, Malcolm RT, Brinton TI, Klie DE (1986) Water solubility enhancement of some organic pollutants and pesticides by dissolved humic and fulvic acids. Environmental Science and Technology 37:5657–5664

Cornelissen G, van Noort PCM, Govers HAJ (1997) Desorption kinetics of chlorobenzenes, polycyclic aromatic hydrocarbons, and polychlorinated biphenyls: Sediment extraction with Tenax(R) and effects of contact time and solute hydrophobicity. Environmental Toxicology and Chemistry 16:1351–1357

Directive 2000/60/EC of the European Parliament establishing a framework for community action in the field of water policy (2000)

Laor Y, Farmer WJ, Aochi Y, Strom P (1998) Phenanthrene binding and sorption to dissolved and to mineral-associated humic acid. Water Research 32:1923–1931

Latimer JS, Davis WR, Keith DJ (1999) Mobilization of PAHs and PCBs from in-place contaminated marine sediments during simulated resuspension events. Estuarine, Coastal and Shelf Science 49:577–595
Lee S, Kommalapati RR, Valsaraj KT, Pardue JH, Constant WD (2004) Bioavailability of reversibly sorbed and desorption-resistant 1,3-dichlorobenzene from a Louisiana superfund site soil. Water, Air and Soil Pollution 158:207–221
Pignatello JJ (2000) The measurement and interpretation of sorption and desorption rates for organic compounds in soil media. Advances in Agronomy 69:1–73
Reemtsma T, Savric I, Jekel M (2003) A potential link between the turnover of soil organic matter and the release of aged organic contaminants. Environmental Toxicology and Chemistry 22:760–766
Reid BJ, Jones KC, Semple KT (2000) Bioavailability of persistent organic pollutants in soils and sediments – a perspective on mechanisms, consequences and assessment. Environmental Pollution 108:103–112

Andreas Kleeberg · Michael Hupfer · Giselher Gust

6.4 Phosphorus Entrainment Due to Resuspension, River Spree, NE Germany

6.4.1 Introduction

Both resuspension and transport of suspended particulate matter (SPM) are driven largely by hydrodynamics at the sediment water interface, initiating and controlling the particle exchange between river sediment and water column (e.g., Black et al. 2002; El Ganaoui et al. 2004). Transport of suspended sediment is accompanied by the transport of phosphorus (P). It is an important mechanism to understand as the transport of sediment-associated P in lowland rivers often constitutes a high percentage (23–61%) of the total annual P load (e.g., Svendsen et al. 1995). Interactions of resuspended SPM and particulate P require knowledge of sediment properties and hydrodynamic conditions, such as the composition, the critical shear stress, the entrainment rate as well as the sinking velocity (El Ganaoui et al. 2004). They are often inferred through the use of flume experiments (Redondo et al. 2001; Witt and Westrich 2003; El Ganaoui et al. 2004) or from direct in situ measurements (Gust and Morris 1989; Black et al. 2002). For rivers, the availability of fine-grained sediments on the stream bed is viewed as a transient depositional feature (e.g., Droppo and Stone 1994; Bungartz and Wanner 2004). This paper reports on an in situ experiment on river bed resuspension in a stretch of lowland river Spree in comparison to a concurrent laboratory experiment. The aim of this study is to *(1)* determine erosion thresholds of sediments and benthic particulate P, *(2)* determine entrainment rates, and *(3)* address problems in applying results from laboratory experiments to riverine conditions.

6.4.2 Material and Methods

Study Site

River Spree, NE Germany, is a medium-sized lowland river with a catchment area of about 10 000 km^2. The river emerges at an elevation of 580 m, and flows for 400 km to Berlin through several shallow lakes affected by river regulation (Köhler et al. 2002; Bungartz and Wanner 2004). The experimental site Kossenblatt (14° E, 52°07' N) is part of a 21.1 km long sixth-order section of the river and represents a slow flowing (0.1–0.3 m s^{-1}) stretch of the Krumme Spree. The slope of the trapezoid profile is 0.01%, and the runoff (Q) usually varies between 12 and 16 m^3 s^{-1} at water depths of 1.5–2.5 m (Bungartz and Wanner 2004).

Methods

In both the in situ and the laboratory experiments, an erosion chamber (Gust 1990), hereafter called microcosm MC, with spatially homogeneous bottom stress τ, has been used (Fig. 6.18). Calibration of shear velocity u^* (= $(\tau/\rho)^{1/2}$, ρ = density) of the MC has been conducted using a variety of approaches (Gust 1990; Gust and Müller 1997; Tengberg et al. 2004).

For the in situ experiment (INS), on 19 May 2005, the MC (i.d. 19.2 cm) was deployed by means of a tripod. A scuba diver adjusted the distance between stirring disk and sediment surface (8 cm). Required river water feed through was provided continuously by a peristaltic pump at $Q = 245 \pm 1$ ml min^{-1}.

For the laboratory experiment (LAB), on 31 May 2005, 20 undisturbed sediment cores taken with a sediment corer (UWITEC©, Mondsee, Austria) were sliced in the field downcore into two horizons (0–3, 3–8 cm), pooled and placed stratified into the MC. The MC was gently filled with ~4 l unfiltered Spree water and kept for 12 days in the dark at room temperature for consolidation and conditioning, respectively.

In the two runs of the experiment INS and LAB u^* ranged from 0.57 to 1.67 cm s^{-1}. It was incrementally increased every 20 minutes (see Fig. 6.19). Data on turbidity were recorded continuously (0.1 min^{-1}) by a commercial Forward Scatter Turbidimeter TF

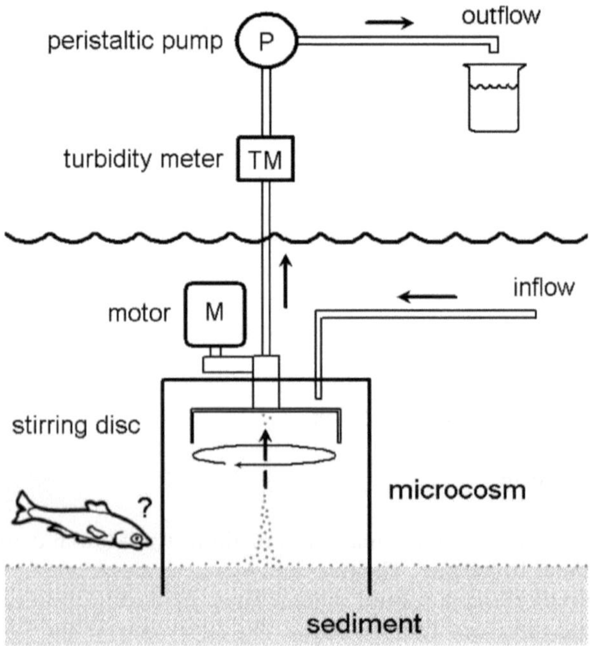

Fig. 6.18. Experimental arrangement for the in situ determination of entrainment rates due to resuspension. The microcosm was driven as an open system. For the laboratory experiments a closed microcosm with a reservoir supplying continuously river water was used instead of the open river inflow

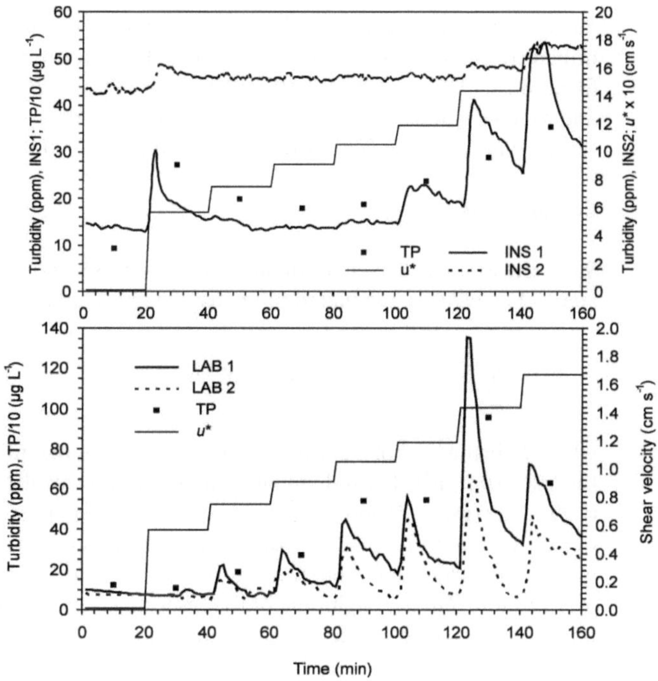

Fig. 6.19. Course of turbidity for the first (INS 1) and second run (INS 2) of the in situ experiment at Kossenblatt, 19 May 2005 (*upper panel*), and for the first (LAB 1) and second run (LAB 2) of the repetition of the in situ experiment under laboratory conditions, 31 May 2005 (*lower panel*) at an incrementally enhanced shear velocity (u^*). The changing concentration of total P (*TP*) is plotted for INS 1 and LAB 1. Note that in INS u^* was multiplied by 10, and *TP* divided by 10, and in LAB *TP* divided by 10

10-512 (Optek-Danulat, Essen 1, Germany) with Labview®. An integrated water sample for chemical analysis was taken for each u^* interval. SPM was determined in triplicate by filtering water samples through pre-weighed cellulose acetate filters (0.45 µm, Sartorius, Göttingen, Germany), drying at 105 °C, and weighing again. Means of turbidity ($n = 10$) provided the calibration curve for SPM (mg l^{-1}) with (ppm) $\times \sim 1.5 =$ (mg l^{-1}). 50 ml suspension aliquots of LAB 1 were taken at selected u^* to determine the sinking velocity spectra of the resuspended particles. By pipetting 0.65 ml subsamples into a settling column of 50 mm i.d. and 600 mm length, sinking velocity w_s of particle ensembles from $u^* = 0$, 0.85 and 1.44 cm s^{-1} were filmed by a video camera (Sony DCR-PC110E PAL) and subsequently processed by MATLAB software (Version 2006a) providing in addition to spectra of w_s and aggregate size other derived variables utilizing Stoke's Law. For the statistical treatment 14 commensurate w_s classes from 50 to 400 m d^{-1} with 25 m d^{-1} each were used.

Soluble reactive P (SRP) was determined photometrically (Murphy and Riley 1962) using a segmented flow analyzer (Skalar San^{plus}). Total P (TP) in water samples was measured as SRP after wet digestion using peroxodisulfate for 2 h at 121 °C and 0.12 MPa. Total iron (TFe) concentration was analyzed with a flame atomic absorption spectrometer (Perkin Elmer 3300, Rodgau-Juegesheim) after aqua regia digestion.

To study the P distribution for particles size classes and sinking behavior a stepwise filtration scheme following Shand et al. (2000) was carried out. Four glass settling tubes (i.d. 4.7 cm, height 30 cm) were filled with river water from four sampling stations near Kossenblatt (Radinkendorf, 27/28 April 2005) and stored vertically in a large container filled with river water (for details see Bungartz and Wanner 2004). Uppermost 435 ml water were withdrawn at 0 h, after 1, 6, and 24 h. TP and TFe were determined from an unfiltered aliquot. TP was also determined in filtrates after a step-wise filtration with 1 µm (PC, Whatman), 0.8 µm (PC, Whatman), 0.45 µm (CA, Schleicher and Schuell), and 0.2 µm (CA, Whatman) membrane filters giving total dissolved P (DTP). Particulate P (PP) is the difference between TP and DTP.

Calculations

Since the MC was driven as an open system in INS and LAB, entrainment rates (E) per step (x) of u^* were calculated in terms of a mass balance as follows:

$$E_x = \sum_{i=1}^{n} \left(\frac{L_{x_{\text{out}}} - L_{x_{\text{in}}}}{A} \right) \quad (\text{mg m}^{-2}\,\text{h}^{-1}) \tag{6.2}$$

where L_x is the load of x at the outflow (out) and inflow (in) (mg h^{-1}) and A is the area of the MC (0.037 m^2). The individual L_x in Eq. 6.2 was calculated by multiplying the concentration of x (mg m^{-3}) and the water discharge of the peristaltic pump (m^3 h^{-1}). The period of time within the amount of sedimentary resuspendable matter has been exhausted for a given u^* was calculated from the time difference from when the concentration of SPM was at maximum and the beginning of the respective stepwise increase in u^*. The theoretical particle transport distance was calculated according to Thomas et al. (2001).

6.4.3 Results

In the first run of the in situ experiment (INS 1) there was a remarkable turbidity peak at the beginning of the first step of u^* (0.57 cm s^{-1}). This yielded an entrainment rate of 1.177 mg m^{-2} h^{-1}, equal to the amount of SPM (9.2% of total SPM) generated in the fifth step of u^* (1.19 cm s^{-1}) with 11.3% of total SPM of this run (Fig. 6.19, upper panel). Located only about 1 m adjacent to the first site (INS 1), the sediments of the second in situ run (INS 2) revealed large differences in all parameters determined, yet with a similar trend in the entrainment curve. The first peak in INS 2 at $u^* = 0.57$ cm s^{-1} is identical, but at a much lower entrainment rate (251 mg m^{-2} h^{-1}). It provided approximately the same proportion of SPM (12.3% of total SPM) as at the fifth step of u^* (1.19 cm s^{-1}) with 10.1% of total SPM. The stock of easily resuspendable matter forming surface sediments was exhausted after 9 min (INS 2) and 11 min (INS 1), respectively. The time series of TP were identical in their major trend in both experimental cases, except that the concentration in LAB 1 was higher by a factor of two than in INS 1. The fraction of particulate P in River Spree inflow water at Kossenblatt amounted to 74.5% TP. In the course of INS this ratio remained relatively constant, i.e. 66.5 ±7.9% TP in INS 1 and 67.2 ±7.9% TP in INS 2.

In contrast to the INS runs, no peak in turbidity arose in the laboratory (LAB) runs at the lowest u^*, indicating the absence of flocculent, easily resuspendable material at the sediment surface (Fig. 6.19, lower panel). The subsequent time series of turbidity under incremental increases of u^* was similar except for the maximum peak of LAB 1 at u^* 1.44 cm s^{-1}. For both experimental cases the peak always coincided with the highest

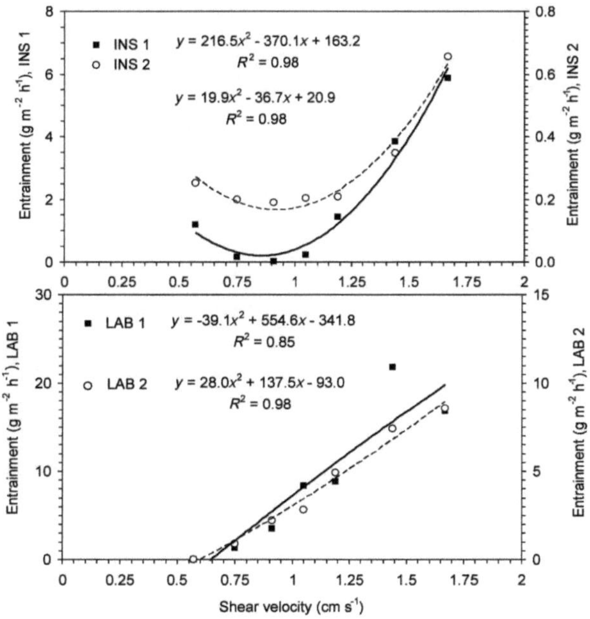

Fig. 6.20.
Entrainment rates (SPM dry weight) vs. shear velocity for the in situ experiment (INS, *upper panel*), and the laboratory experiment (LAB, *lower panel*)

concentration of TP. The fraction of particulate P in inflowing river water was only 21.8% TP as a result of fast particle settling in the reservoir. In comparison, particulate P averaged 79.9 ±14.2% in INS 1 and 85.0 ±9.9% in INS 2.

In the experiment LAB, an analog major increase of entrainment with an increasing u^* is found (Fig. 6.20). The entrainment rates of LAB 1 are about two times higher than those of LAB 2, indicating a reduced heterogeneity compared to the INS data since aliquots from a pooled sediment sample are used. The shape of the best fit, almost a straight line, illustrates the absence of easily resuspendable matter at low u^*.

Phosphorus entrainment rates also increased with an increasing u^*; those of INS 1 and INS 2 differed by a factor of approx. 6, and those of LAB 1 and LAB 2 by a factor of 1.5 (Fig. 6.21).

The mean values of particle parameters (Table 6.8) did not change significantly with increasing u^* (Fig. 6.19, LAB 1). In the w_s frequency distribution the percentage of particle number was constant with increasing u^*. Whereas, in the particle masses the percentage of light particles in the fractions with lower w_s (75–200 m d^{-1}, $n = 6$ w_s classes) decreased from 94.9% at $u^* = 0$ cm s^{-1}, to 86.3% at $u^* = 0.85$ cm s^{-1}, and to 74.3% at $u^* = 1.44$ cm s^{-1}. Inversely, the percentage of heavier particles in the fractions with higher w_s (275–400 m d^{-1}, $n = 6$ w_s classes) increased from 3.3% at $u^* = 0$ cm s^{-1}, to 4.9% at $u^* = 0.85$ cm s^{-1}, and to 17.9% at $u^* = 1.44$ cm s^{-1}.

The eroded particles, represented as SPM from experiments LAB with their relatively high w_s (Table 6.8) cannot move downstream very far prior to subsequent resuspension (Fig. 6.22). For example particles with a w_s of 225 m d^{-1} would be transported over 200 m. Since preferentially particles with a higher w_s were resuspended at $u^* = 1.44$ cm s^{-1} the higher SPM concentration compared to that at $u^* = 0.85$ cm s^{-1} decreases faster towards the level of the river Spree water.

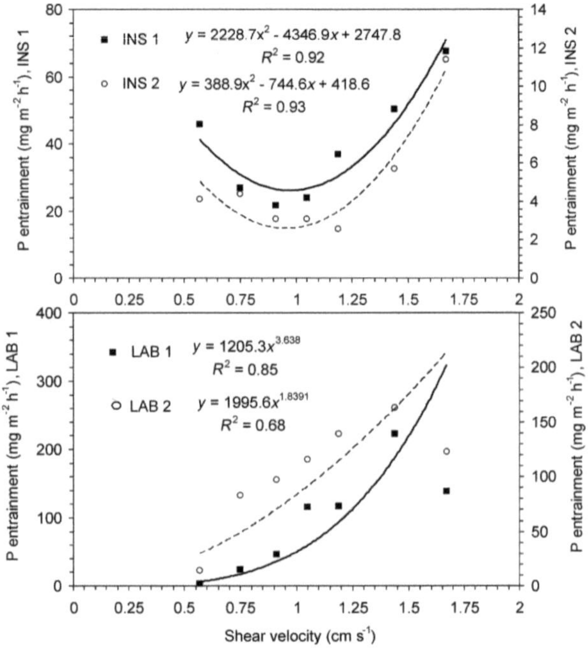

Fig. 6.21. Phosphorus entrainment rates vs. shear velocity for the in situ experiment (INS, *upper panel*), and the laboratory experiment (LAB, *lower panel*)

Table 6.8. Mean values ± standard deviation for characteristics of particles resuspended in the first laboratory run (see Fig. 6.18, LAB 1) at an increasing shear velocity

Particle parameter	Unit	Shear velocity ($cm\ s^{-1}$)		
		0	0.85	1.44
Number		497	11010	1454
Sinking velocity	$m\ d^{-1}$	113.0 ± 37.7	114.5 ± 53.6	112.6 ± 41.6
Area	mm^2	0.023 ± 0.03	0.016 ± 0.01	0.026 ± 0.05
Diameter[a]	μm	134.3 ± 71.7	119.2 ± 42.9	133.4 ± 81.2
Mass[a]	mg	2.8 ± 5.5	1.4 ± 2.1	3.6 ± 13.1
Density[a]	$g\ cm^{-3}$	1.154 ± 0.09	1.151 ± 0.08	1.149 ± 0.09

[a] Calculated from Stoke's law.

As a result of the enhancement of SPM concentration (Fig. 6.22) at an u^* of 0.85 and 1.44 cm s^{-1} the in situ TP concentration of river Spree would theoretically enhance from 122 μg l^{-1} by 6 and 36 μg l^{-1}, respectively.

The stepwise filtration reveals that most P (62.9 ±16.6%, $n = 12$) is associated with larger particles (>1 μm) representing the 'coarse particulate P' (Fig. 6.23) which is settling relatively fast. TP decreased at decreasing rates of P sedimentation from 3.68 mg $m^{-2} h^{-1}$ (after 1 h), to 0.54 mg $m^{-2} h^{-1}$ (after 6 h), and 0.17 mg $m^{-2} h^{-1}$ (after 24 h). The same applies to TFe (Fig. 6.23). The percentage of 'coarse particulate P' (>1 μm) in the sus-

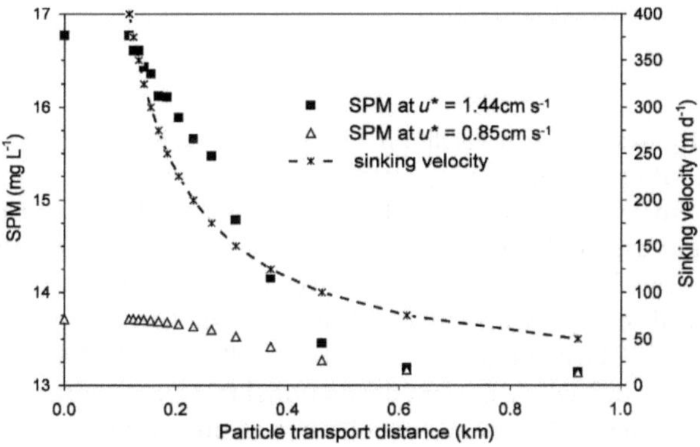

Fig. 6.22. Theoretical particle transport distance for 15 commensurate groups of a sinking velocity spectrum at $u^* = 1.34\ \mathrm{cm\ s^{-1}}$ from LAB 1 (Tab. 6.8) and the decrease of suspended particulate matter (SPM) concentration after a resuspension event in the water column of river Spree (discharge = 16 $\mathrm{m^3\,s^{-1}}$) back towards the initial SPM concentration of 13.14 mg $\mathrm{l^{-1}}$

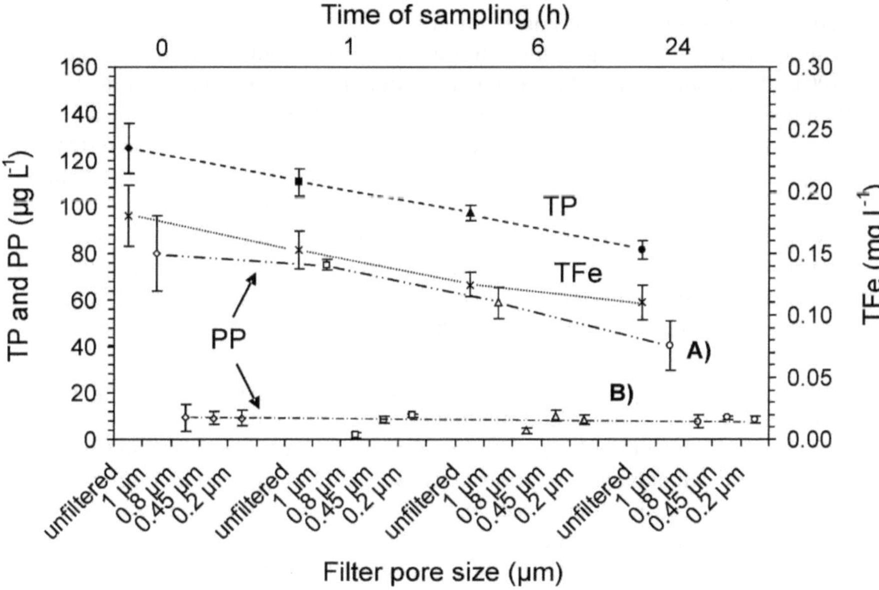

Fig. 6.23. Results from a step-wise filtration of a water sample taken after 0, 1, 6 and 24 h out of settling tubes. Since the fractional P distributions among the four sampling stations of river Spree (Radinkendorf, 27/28 April 2005) did not significantly differ, means and standard deviations of $n = 4$ are shown. *Filled symbols* represent total P (*TP*), *cross symbols* total Fe (*TFe*), and *open symbols* particulate P (*PP*). *PP* fractions are operationally defined as follows: >1 µm = 'coarse particulate P' (case A), <1 … >0.8 µm = 'fine particulate P', <0.8 … >0.45 µm = 'coarse colloidal P', <0.45 … >0.2 µm = 'fine colloidal P' (case B). Phosphorus <0.2 µm = 'dissolved P'

pended flocs decreased within 24 h down to 49.3 ±12.9% (case A). Theoretically, the smaller the particle size in the filtrate the smaller the PP concentration and the slower its decrease. However, there were no changes in the concentration from the 0 h to the 24 h sample of the 'fine particulate P' (16.9–16.1%), the 'coarse colloidal P' (23.7–27.6%) as well as the 'fine colloidal P' (34.5–35.7%) which remained on the same level (case B). The dissolved P (<0.2 µm, not shown) was only 5.9 to 10.2% of TP.

6.4.4 Discussion

Besides the great variability in the time series and rates within both runs of INS and LAB experiments (Fig. 6.19), comparison reveals that entrainment of sediment and associated P is controlled by u^* (Figs. 6.20 and 6.21) as well as sediment heterogeneity and consolidation depending on river bed conditions. Run INS 1 confirms the existence of a fine-grained fluffy layer at the sediment surface with high water contents easily eroded, with a low critical u^* threshold for resuspension (e.g., Droppo and Stone 1994). The comparison of INS 1 and INS 2 and the LAB runs reveals that such a boundary is indeed a transient depositional feature (Doppo and Stone 1994), related to flocculation within the water column and not easy to mimick in laboratory experiments. In a river experiment, Rhône River, France, it was shown that an observed fluff layer was mainly representative of recent deposits of originally suspended particles, probably trapped in the overlying water column during sediment sampling (El Ganaoui et al. 2004). Thus sampling and transporting sediments undisturbed from field to lab and then inserting them as cores into flumes poses a difficult experimental task. Average mass eroded in the LAB runs at same u^* was six times higher than that of the INS runs (Fig. 6.20), indicating a different degree of consolidation of the undisturbed Spree sediment. A new method of preparing flume cores has to be devised.

Table 6.8 indicates a homogenous particle distribution in the sediment layers experiencing resuspension since homogenization by the peristaltic pump was ruled out. However, the increasing dominance of heavier particles at an increasing u^* suggests a selective transport and the tendency of higher w_s as longer the sediment are exposed to higher u^*.

A high P entrainment arises at low u^* (Fig. 6.19, INS 1) from flocculated particles, initiating batch-wise P burdens and maintaining high percentages of particulate P in the flow (Fig. 6.23). In concordance with the increase of SPM concentration (Fig. 6.22) TP concentration can enhance temporarily. However, the resuspended, reprocessed or remoulded particles, such as detritus or feces, show significantly higher sinking velocities (Table 6.8) and masses than those from SPM (Fig. 6.22). These findings explain the large potential distances of particle and P transport known as 'P spiralling' varying between distances of 1 and 1000 m depending on w_s and the deposition stress (Gust and Kozerski 2000) of the sinking-particle distribution (e.g., Sharpley et al. 2003).

6.4.5 Conclusions

Laboratory microcosm experiments using sliced and pooled cores ensure a high reproducibility for particle cycling experiments, whereas in situ experiments describe the resuspension and deposition behaviors closer to reality. The latter require a larger sample size due to small-scale sediment heterogeneity. Low flow-generated shear velocities u^* can lead to batch-wise P burdens when resuspending freshly deposited materials of the fluff layer. Such events would be underestimated in laboratory experiments lacking this transient storage feature of rivers.

Acknowledgments

The authors are grateful to I. Henschke (BTU Cottbus) for technical assistance and scuba diving, to C. Herzog, A. Lüder, H.-J. Exner as well as M. Reiche (all IGB Berlin) for laboratory assistance, and D. Hoffmann and V. Müller (TUHH) for processing the particle images. This manuscript benefited from two anonymous reviewers. This study was financially supported by the Federal Ministry of Education and Research (BMBF, FKZ 02WF0469).

References

Black KS, Tolhurst TJ, Paterson DM, Hagerthey SE (2002) Working with natural cohesive sediments. Journal of Hydraulic Engineering 128:2–8

Bungartz H, Wanner SC (2004) Significance of particle interaction to the modelling of cohesive sediment transport in rivers. Hydrological Processes 18:1685–1702

Droppo I, Stone M (1994) In-channel surficial fine-grained sediment laminae (part I): physical characteristics and formational processes. Hydrological Processes 8:101–111

El Ganaoui O, Schaaff E, Boyer P, Amielh M, Anselmet F, Grenz C (2004) The deposition and erosion of cohesive sediments determined by a multi-class model. Estuarine, Coastal and Shelf Science 60:457–475

Gust G (1990) Method of generating precisely-defined wall shear stresses. US Patent Number: 4,973,1651990

Gust G, Morris MM (1989) Erosion thresholds and entrainment rates of undisturbed in situ sediments. Journal of Coastal Research 5:87–99

Gust G, Müller V (1997) Interfacial hydrodynamics and entrainment functions of currently used erosion devices. In: Burt N, Parker R, Watts J (eds) Cohesive sediments. Wiley, Chichester, UK, pp 149–174

Gust G, Kozerski H-P (1997) In situ sinking particle flux from collection rates of cylindrical traps. Marine Ecology Progress Series 208:93–106

Köhler J, Gelbrecht J, Pusch M (Hrsg.) (2002) Die Spree – Zustand, Probleme, Entwicklungsmöglichkeiten. Limnologie Aktuell, Bd. 10, E. Schweizerbart'sche Verlagsbuchhandlung, Stuttgart, pp 384

Murphy J, Riley JP (1962) A modified single solution method for determination of phosphate in natural waters. Analytica Chimica Acta 27:31–36

Redondo JM, Durrieu de Madron X, Medina P, Sanchez MA, Schaaff E (2001) Comparison of sediment resuspension experiments in sheared and zero-mean turbulent flows. Continental Shelf Research 21:2095–2103

Shand CA, Smith S, Edwards AC, Fraser AR (2000) Distribution of phosphorus in particulate, colloidal and molecular-sized fractions of soil solution. Water Research 34(4):1278–1284

Sharpley A, Krogstad T, McDowell R, Kleinman P (2003) Phosphorus transport in riverine systems. Encyclopedia of Water Science, Marcel Dekker Inc. New York

Svendsen LM, Kronvang B, Kristensen P, Graesbol P (1995) Dynamics of phosphorus-compounds in a lowland river system – importance of retention and nonpoint sources. Hydrological Processes 9(2):119–142

Tengberg A, Stahl H, Gust G, Müller V, Arning U, Andersson H, Hall POJ (2004) Intercalibration of benthic flux chambers I. Accuracy of flux measurements and influence of chamber hydrodynamics. Progress in Oceanography 60:1–28

Thomas SA, Newbold JD, Monaghan MT, Minshall GW, Georgian T, Cushing CE (2001) The influence of particle size on seston deposition in streams. Limnology and Oceanography 46(6):1415–1424

Witt O, Westrich B (2003) Quantification of erosion rates for undisturbed contaminated cohesive sediment cores by image analysis. Hydrobiologia 494:271–276

Ralf Siepmann · Frank von der Kammer · Wolfgang Calmano

6.5 Determination of Heavy Metal Mobility from Resuspended Sediments Using Simulated Natural Experimental Conditions

6.5.1 Introduction

Heavy metal loads in river flows continue to decrease due to better control and pre-treatment of wastewater. However metal release still occurs, e.g., from abandoned mining sites, diffuse input, and contaminated floodplains. Metals discharged into aquatic systems are mostly adsorbed on suspended particles and fine grained riverine sediments which are predominantly deposited in groyne fields and harbor basins. Distribution, mobility and bioavailability of heavy metals in rivers do not simply depend on total concentrations but, critically, on their chemical and physical associations and on transformation processes they undergo. Gradually, contaminant potentials are formed in the sediments, from which under changing chemical conditions heavy metals can be mobilized (Hong et al. 1991; Peiffer 1997). Understanding possible mobilization and transformation effects of metals bound to sediments requires detailed studies of release mechanisms and how they are affected by hydrodynamic and biogeochemical processes (Calmano et al. 2005).

Anoxic sediment deposits are the main source for a secondary release and the spread of contaminants (Förstner et al. 1999). The main reason for reduction processes in limnic and marine sediments is the degradation and mineralization of organic substances through microbial activity. Oxygen consumption in an anaerobic environment results in formation of metal sulfides and/or carbonates. These minerals are stable in the sediments until oxygen becomes available, e.g., by dredging, flood events or tidal streams. These disturbances are followed by several simultaneous processes: desorption, diffusion of contaminants into the water phase from pore waters, mineral dissolution processes, relocation of sediment-bound contaminants, and oxidation processes. Oxidation of sulfides produces sulfuric acid which may lower the pH in micro zones of the sediment, dissolve other mineral phases, e.g., carbonates, and can lead to the further release of heavy metal ions (Calmano et al. 2005). A summary of the main factors is shown in Fig. 6.24.

Criteria for the prognosis of the middle- and long-term behavior of metals in sediments should include the acid production capacity (APC) of the system as well as abilities for neutralizing acidic constituents. In a sediment/water system, the most important reactions producing hydrogen ions are oxidation of inorganic and organic sulfur-, nitrogen- and iron-species (Hong et al. 1994). The buffering or acid neutralizing capacity (ANC) may be attributed to the acid neutralization capacity of the solids and that of the dissolved phase.

Most river sediments are well buffered against acidification, as can be shown by acid titrations. Dissolution of carbonates occurs at neutral to slightly acidic pH, ion exchange capacities of clay minerals maximize at pH 5, dissolution of aluminum hydroxides begins at pH 4, and the dissolution of iron hydroxides at pH 2. Under certain conditions mobilization and transfer of sediment bound heavy metals may also occur at neutral pH, e.g., at changes of the ion concentration (displacement by ion exchange and/or complexation) or by microbial reactions (methylation).

Dissolved metal concentrations are not only controlled by release reactions but also by re-adsorption processes on suspended material. In a multi-chamber system the metal

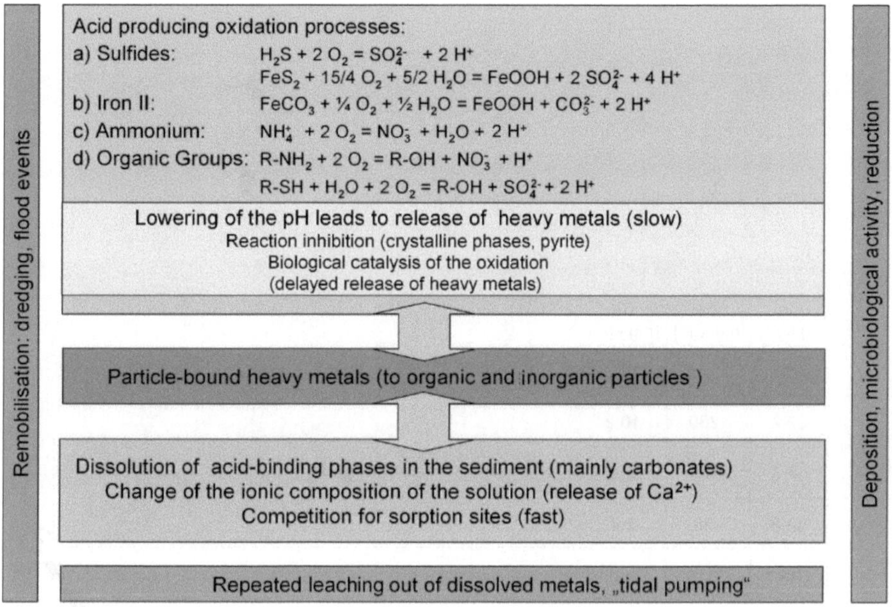

Fig. 6.24. Overview of chemical processes that follow sediment resuspension until the sediment is re-deposited

transfers from anoxic sediments to competitive re-adsorption sites on different sediment model compounds were studied during oxidation experiments (Calmano et al. 1988). These studies showed that e.g., the transfer of copper to quartz sand and clay was negligible, a small fraction was transferred to Mn- and Fe-oxides, and the main fraction could be found adsorbed to algal cell walls representing the organic part of the sediment.

Calculation of the solubility of heavy metals in reduced sulfidic sediments indicates supersaturation for most metals, but they were nevertheless available to thin film gradient (DGT) measurement (Naylor et al. 2004, 2006). This suggests that even under anoxic conditions, there is a dynamic pseudo-steady state where metals are not bound as sulfides in an inert, insoluble phase, but are released into the pore water and re-adsorbed onto suspended matter. In the in situ perturbation experiment from Naylor at al. the pore-water equilibrium was disturbed by inducing a flux from the pore water to the probe, showing the dynamic response of sediments in the transfer of metals.

6.5.2 Experiments

Sediment Composition

Sediment samples were taken from different locations at the river Elbe, upstream of the Hamburg Harbor, and in the channels of the Hamburg Harbor. The sampling sites are shown in Fig. 6.25.

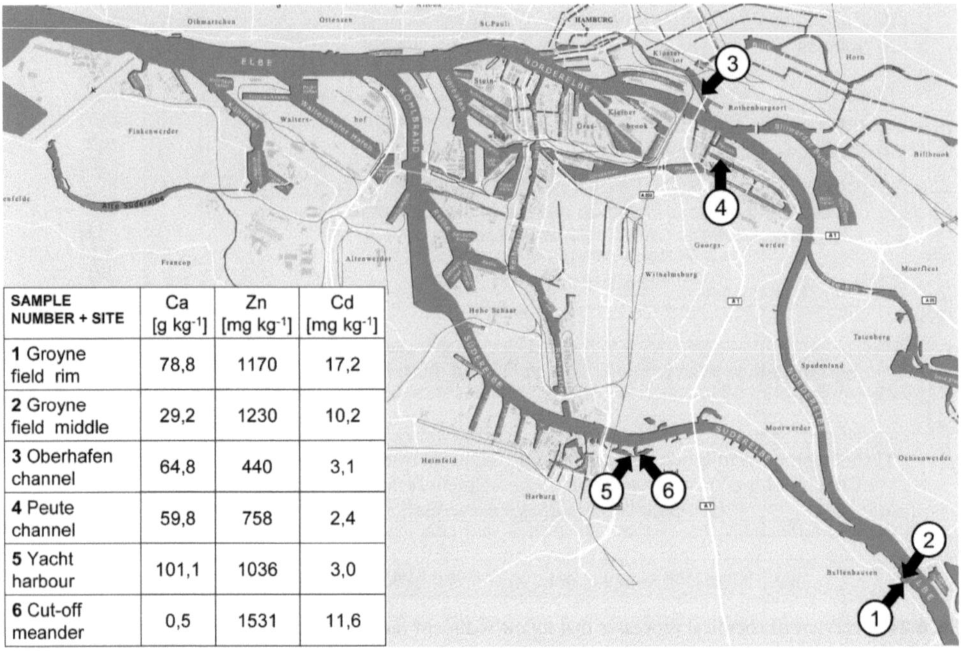

SAMPLE NUMBER + SITE	Ca [g kg-1]	Zn [mg kg-1]	Cd [mg kg-1]
1 Groyne field rim	78,8	1170	17,2
2 Groyne field middle	29,2	1230	10,2
3 Oberhafen channel	64,8	440	3,1
4 Peute channel	59,8	758	2,4
5 Yacht harbour	101,1	1036	3,0
6 Cut-off meander	0,5	1531	11,6

Fig. 6.25. Overview of the different sampling sites at the river Elbe and in the Hamburg Harbor and some selected metal contents of the sites. (*1, 2*) Groyne field at Elbe-km 607.7 that was sampled monthly by several workgrups of the SEDYMO Project. (*3*) "Oberhafenkanal", harbor channel close to the Norderelbe. (*4*) "Peutekanal", harbor channel in Veddel. (*5*) Yacht Harbor at the Süderelbe. (*6*) "Schweenssand", a cut-off meander at the Süderelbe

The sediments showed very different calcium contents ranging from 0.5% (site 6) to 10% (site 5). Most of the sediments are rather rich in calcium. As there are no major sources of calcite in the catchment area, the Elbe sediments were expected to be poor in calcium. The high contents, therefore, can only be explained by biogenic decalcification of the supernatant waters. X-ray diffraction analyses of the upper 6 cm of an Elbe River sediment core demonstrated calcite and quartz as the main constituents (Schwartz 2006).

Metal contents were high in the groyne field (sites 1 and 2), but showed slight variations due to different sampling locations. The measured values in the harbor channels (sites 3 and 4) were approximately half of the contents directly at the river for all metals.

Sediment Depth Profiles

The sediment used for the experiments described below was retrieved from the sampling site "Schweenssand" (6). It is connected to the main stream that supplies frequent tidal flooding of the area with short intermediate dry periods. A sediment core was taken at the sampling site for the later resuspension experiments and cut into cm-layers to obtain a depth profile of the metal contents of the complete sediment (Fig. 6.26). Dry sieving of the sediment slices resulted in 3 grain size fractions: Small slug shells

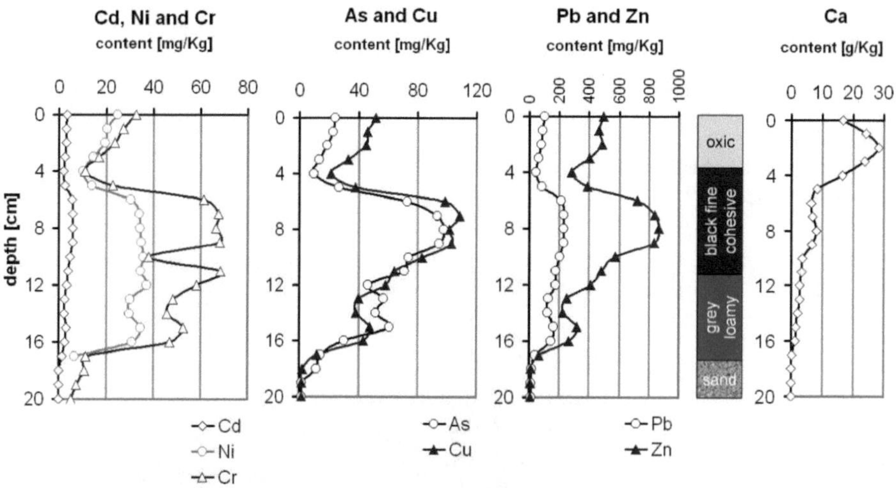

Fig. 6.26. Depth profile of metal contents at the sampling site for resuspension experiments. The metal contents were measured without separating grain size fractions

above 1 mm, quartz sand from 1 mm to 0.063 mm, and the cohesive fine fraction smaller than 0.063 mm. Metal contents were much higher in the fine-grained anoxic zone of the sediment (see 4–12 cm), whereas Ca-concentrations were highest close to the sediment surface.

Experimental Setup for the Resuspension Experiments

To study both short- and long-term processes during sediment resuspension and to simulate erosion by a flooding event we took an undisturbed groyne field sediment core from sampling site (6) and transferred it directly from the field into an erosion apparatus or "water column simulator", that was developed and supplied by the Institute of Marine Engineering at Hamburg University of Technology (see Müller et al. this volume). The sample was taken in December at winter conditions. The water temperature was 6 °C, so the phytobenthic activity was very weak. Freshly collected river water was used as a supernatant. The chloride concentration of the supernatant water was 150 mg l^{-1}. The apparatus, equipped with an adjustable stirring unit was placed in a climatic chamber at a constant temperature of 12 °C in the dark to prevent phytobiotic activity (Fig. 6.27).

After a consolidation phase of 2 weeks, the upper, mainly oxic layer of the sediment was eroded and kept in suspension above the remaining sediment core.

The apparatus was open to allow diffusion of oxygen during the experiment.

After 27 days, the total supernatant suspension was removed and transferred to a separate vessel where it was kept in suspension by continuous stirring and monitoring. Following this, fresh river water was refilled to simulate the water change that normally occurs frequently, and a further part of the remaining anoxic core layer was resuspended to simulate another erosion event without the presence of the upper oxic sediment layer. This part of the experiment lasted for 14 days.

Fig. 6.27.
"Water column simulator" with inserted DGT devices before start of the resuspension, and during resuspension of the anoxic layer. The inner diameter of this device is 29 cm and the used height is 59 cm

With the use of the "water column simulator" as an erosion device for natural sediment, a linking of hydrodynamic and chemical factors was possible. The critical shear force of erosion for both the oxic and anoxic layer was determined. The suspended mass was quantified and analyzed, as well as the solute composition. Monitoring of metal release during erosion provided the amount of metals bound to suspended particles, the amount of metals dissolved in solution, and by the use of thin film gradient technique (DGT) the dissolved amount that might be readily bioavailable.

The samples were taken in the upper part of the "water column simulator", so that only the fine fraction of the resuspended sediments was measured. During the experiment eroded sand grains or those released from the breakdown of the suspended agglomerates resettled quickly and therefore were not sampled.

6.5.3 Results

Controlled Resuspension of the Oxic Sediment Layer

Critical shear velocity for the initial erosion of 3.4 cm of the oxic sediment layer within 12 hours was $u^* = 2.23$ cm s^{-1}. This shear velocity was kept constant during the first phase of the erosion (27 days). The metal content of the resuspended sediment from the oxic layer (Table 6.9) were much higher than the bulk concentrations found in the corresponding layers of the sediment core. This is due to the higher fine grain content of the sediment suspension at the top of the stirred column.

The redox potential of the solution as well as oxygen saturation and pH (6.7) remained constant throughout the experiment. Nevertheless, we observed an increase in sulfate concentration in solution, as well as a decrease of hydrogen carbonate (see

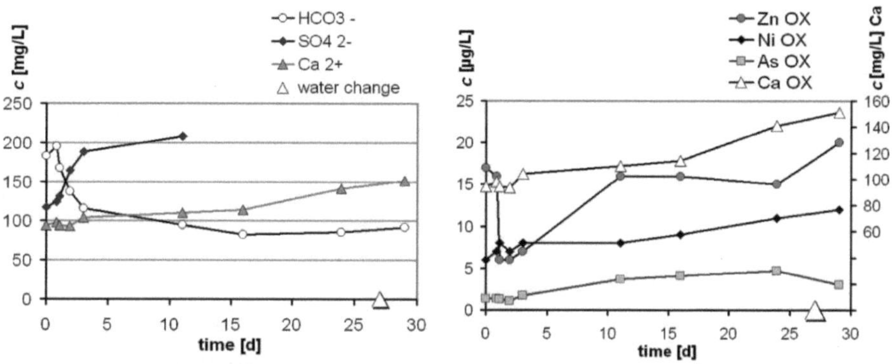

Fig. 6.28. Sulfate, hydrogen carbonate, calcium and metal contents of 0.2 µm-filtrates obtained during the oxic layer resuspension

Fig. 6.28), which indicates the presence of reduced sulfur compounds close below the oxidic sediment water interface. These processes occur quickly in the first three days of erosion. Obviously the initial small amount of acidity produced is neutralized by the water alkalinity, since no release of Ca was observed during the first three days. Further slow oxidation supplies more H^+ indicated by the dissolution of carbonates and the release of Ca, as noted after day 3.

The Ni and As concentrations in the supernatant solution did not increase during the actual oxidation phase (first three days), then increased slowly (see Fig. 6.28). This release is correlated to the carbonate dissolution process. Zn present in solution at the beginning of the experiment was partially adsorbed on the suspended particles and removed from the solution. However, Zn concentration in solution increased again after 5 days. We suppose that Zn is released either from the dissolution of mineral phases during the experiment, or from the breakdown of aggregates during the prolonged stirring process.

Controlled Resuspension of the Anoxic Sediment Layer

Critical shear velocity for the secondary erosion of 4.8 cm of anoxic sediment layer within 12 hours was $u^* = 2.80$ cm s^{-1}, indicating a strongly compacted older sediment layer. Here, the redox value of the solution as well as the oxygen content dropped immediately after the start of the suspension from an initial value of 220 mV to 20 mV, but recovered within a day and slowly approached the initial values within 4 days. Oxidation of metal sulfides is indicated by a strong release of sulfate, whose formed acid consumed part of the hydrogen-carbonate buffer and lead to a simultaneous release of Ca. The pH value did not change. These reactions naturally terminated after day 4 (see Fig. 6.29).

Ca and Ni measured in solution, with the given detection limits of the analytical instruments, showed a steep increase in concentration following the oxidation reaction, whereas Zn was removed from the solution during the first phase of the oxidation reaction, as shown in Fig. 6.29. After 24 hours more Zn is released than was measured in the initial solution concentration. This supports the suggestion of a secondary Zn release by dissolution of mineral phases and breakdown of aggregates during the stirring process. Arsenic was released from the pore water.

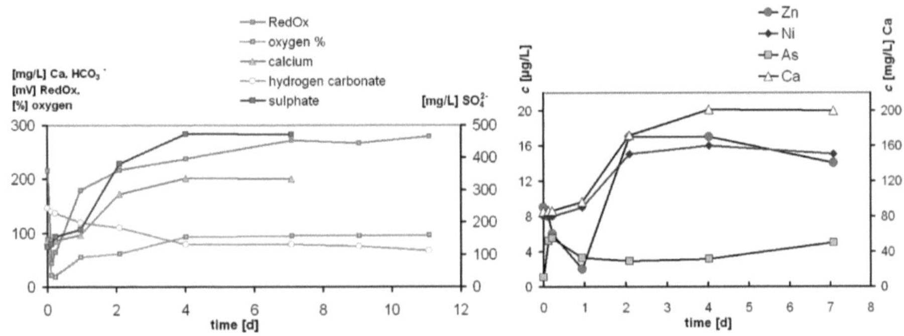

Fig. 6.29. Redox potential, oxygen content and significant ion concentrations during resuspension of the anoxic layer. The *right graph* shows the metal contents of 0.2-µm filtrates

Fig. 6.30.
Metal contents of the particles filtered during oxic and anoxic layer resuspension

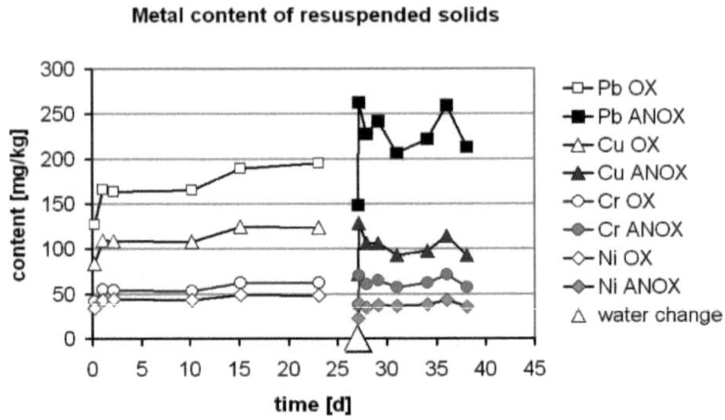

Resuspension of Particle Bound Metals

Resuspension of the *oxic* layer yielded a concentration of 20 g l^{-1} solids at the top of the column measured by filtration. During resuspension of the *anoxic* layer, 30 g l^{-1} solids were suspended. While the core analysis showed lower metal contents in the oxic layer, the metal contents of the suspended sediments were nearly equal in both resuspension experiments, as shown in Fig. 6.30. This can be explained by a grain size fractionation: Only the strongly-contaminated fine fraction of the sediment is retrieved at the top of the suspension chamber.

Table 6.9 summarizes the amount of resuspended solids and their metal contents. Compared to the dissolved concentrations, the particle-bound amount was a factor of 1 000 larger than the amount in solution for Zn, and still a factor of 100 larger for Ni.

DGT Measurements

In a "differential gradient in thin films" (DGT) probe an array of polyacrylamide hydrogel with a layer of Chelex 100 ion exchange resin, a diffusion gel layer of defined thickness and a filter membrane cover is held in a rigid plastic housing that provides a defined exposure area. Dissolved metal compounds and those bound in small complexes can diffuse through the filter and gel layer and are adsorbed completely at the

Table 6.9. Solid concentrations and concentration of metals bound to suspended material

	Solids ($g\,l^{-1}$)	Zn ($mg\,l^{-1}$)	Pb ($mg\,l^{-1}$)	Cu ($mg\,l^{-1}$)	Cr ($mg\,l^{-1}$)	Ni ($mg\,l^{-1}$)	Cd ($mg\,l^{-1}$)
Oxix layer	20	22	3	2	1	0.8	0.16
Anoxic layer	30	40	7	3	2	1.2	0.20

resin layer, where hence the solution concentration is zero. This results in a linear concentration gradient through the diffusive layer. The mass of metal accumulated in the resin is obtained by leaching the Chelex gel with acid and analyzing the leachate. With the given immersion time of the gel, the exposure area, the thickness of the diffusion layer and the diffusion coefficient for each metal in the layer the time-average solution concentration of not or only weakly bound metal species can be calculated (Davison and Zhang 1994).

DGT can be used to determine the "bioavailable" fraction of metals, since it is assumed that only free dissolved metal ions and those released from labile complexes pass through the diffusion layer (Tusseau-Vuillemin et al. 2003). Newer studies show that meta-stable organic complexes can also be determined with DGT, so that the results have to be interpreted with caution concerning predictions about bioavailability (Buzier et al. 2006).

The thin film gradient devices were immersed in the suspension for specific time spans, mostly 4 days. While only Zn, Ni and As concentrations in the suspension filtrates exceeded the detection limits of the analytical instruments, more metals could be detected by the accumulation in DGT devices.

Part of the metals is bound in stable complexes and colloidal particles that pass the filter but are not detected by the DGT devices, so the concentrations measured with the DGT method were only 25 to 50% of those determined in the 0.2-μm filtrates.

The DGT-measured metal species showed different behavior (see Fig. 6.31). Zn follows the solution concentration, whereas the amount of Ni remained constant even though the solution concentration is doubled in the anoxic suspension.

Cu concentrations increased slowly during the experiment, but more Cu was released during resuspension of the oxic layer. An increase could also be measured in solution after day 11.

Cd showed a similar release behavior as Zn, but without the initial re-adsorption from solution. This strengthens the suggestion that the release process of Zn is slow due to mineral dissolution by oxidation or de-agglomeration.

6.5.4 Discussion and Conclusions

Erosion of the oxic and anoxic layer of the studied Elbe River sediment needed shear velocities of $u^* = 2.20\text{–}2.80\ \text{cm s}^{-1}$. These values were rather high compared to other, non-cohesive inland river sediments normally ranging from $u^* = 0.3\text{–}1.8\ \text{cm s}^{-1}$ (Gust, pers. comm.). It can be assumed that these compacted groyne field sediments can only be mobilized by dredging operations.

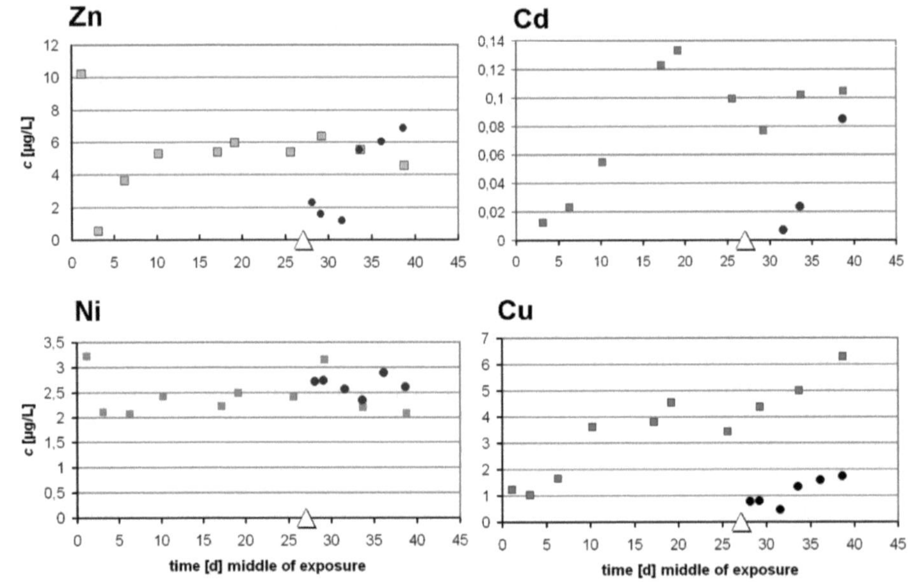

Fig. 6.31. Accumulated metal concentrations during 4 days measured with thin film gradient technique (DGT). The *square light gray dots* represent the values of the oxic layer resuspension, the *round dark dots* show the values of the anoxic layer resuspension

During resuspension in the "water column simulator" particle aggregates are broken down and a separation of grain sizes arises in the column. The gradient of grain sizes favors the short-term resettling of coarser sand particles and long-term suspension of the highly contaminated fine fraction of the sediment ("washing effect").

During the whole experiment the buffer capacity of the sediment was not exceeded and the pH remained constant at 6.7. The release of SO_4^{2-}, and the consumption of HCO_3 indicates acid formation from sulfides during the resuspension of the oxic layer.

Remobilization of heavy metals was affected by different effects. In both resuspension experiments about 70% of the initially dissolved Zn was immediately adsorbed on the resuspended sediment particles. During the first day of resuspension of the oxic layer increase of sulfate concentration indicates the release of acidity, but no rise in the concentration of dissolved metals was found. Later a very slow increase of Zn and Ni concentrations were observed in solution. This may be due to the breakdown of aggregates, following weak oxidation processes and dissolution of some metal compounds. Ca was also released slowly but the carbonate buffer of the sediment was barely influenced by the initial minor acid production.

During resuspension of the anoxic layer Zn and Ni were mobilized parallel to the production of sulfate and the release of Ca until day 4. While in the case of oxic sediment the release of metals is slow, in anoxic sediment it is fast and enhanced by the effects discussed above.

Even at pH-neutral conditions resuspension of the anoxic layer leads to a minor release of Zn and Ni during the actual oxidation. While the total dissolved concentrations, determined in the 0.2-µm filtrates, were higher during resuspension of the *an-*

oxic sediment layer, the amount of free hexaquo ions and labile complexes measured by DGT were in the same range for *both* cases, and for Cu they were only 1/10 of the oxic case.

The results of this study indicate that even at shear velocities that affect only the oxic layer of sediment and may be caused by tidal movements of the water body in river estuaries, small amounts of free Zn, Cd and Cu ions are released into the water body. These release processes are not due to the initial resuspension process, but develop in a time span of more than a week. It can be assumed that this slow release is mediated by the breakdown of aggregates, desorption and dissolution of incorporated metal compounds.

This results can be seen as a link to further delayed release processes (Fengler et al. 1999). In case of resuspension of anoxic sediment layers, the release process of heavy metals is governed by the initial oxidation process. Although no pH change could be measured, more acid is produced, followed by ion exchange processes at the sediment matrix. The oxidation of sulfides to sulfate happens rather fast, the main part is oxidized within two days. The metals release is low, because there is no change in the pH during the process. The complete release of heavy metals from suspended sediments can be summarized in three steps: *(1)* Immediate release from oxidation and matrix exchange processes. *(2)* Mid-term release from the oxidized resuspension, probably caused the physical breakdown of agglomerates. *(3)* Delayed long-term release by acidification if no sufficient buffer capacity is available (Fengler et al. 1999).

Although the first two processes only lead to very low dissolved concentrations of heavy metals, compared to the amount bound on resuspended sediment particles, a repeated leaching of sediments that undergo frequent suspension-deposition-cycles, such as the upper oxic layers of harbor basin sediments, may contribute significantly to the heavy metal load released and spread in the water body.

Acknowledgments

We would like to thank Prof. G. Gust (Hamburg University of Technology) and his coworkers for fast technical support during the experiments, Dr. G. Fengler (University of Greifswald) and A. Matthäi (Hamburg University of Technology) for sampling and additional analysis, and the German Federal Ministry of Education and Research for financial support.

References

Buzier R, Tusseau-Vuillemin M-H, Mouchel J-M (2006) Evaluation of DGT as a metal speciation tool in wastewater. Sci Tot Environ 358:277–285

Calmano W, Ahlf W, Förstner U (1988) Study of metal sorption/desorption processes on competing sediment components with a multichamber device. Environ Geol Water Sci 11:77–84

Calmano W, von der Kammer F, Schwartz R (2005) Characterization of redox conditions in soils and sediments: heavy metals. In: Lens P, Grotenhuis T, Malina G, Tabak H (eds) Soil and Sediment remediation, IWA Publ. London UK, pp 102–120

Davison W, Zhang H (1994) In situ speciation measurements of trace components in natural waters using thin-films gels. Nature 367:546–548

Fengler G, Förstner U, Gust G (1999) Verification experiments on delayed metal release from sediments using a hydrodynamically controlled erosion apparatus (in German). Abstract Annual Meeting German Society of Water Chemistry, Regensburg, pp 240–243

Förstner U, Calmano W, Ahlf W (1999) Sedimente als Schadstoffsenken und -quellen: Gedächtnis, Schutzgut, Zeitbombe, Endlager. In: Frimmel FH (Hrsg.) Wasser und Gewässer – Ein Handbuch, Spektrum Akademischer Verlag Heidelberg, pp 249–279

Hong J, Calmano W, Wallmann K, Petersen W, Schroeder F, Knauth H-D, Förstner U (1991) Change in pH and release of heavy metals in the polluted sediments of Hamburg-Harburg and the downstream Elbe during oxidation. In: Farmer JG (ed) Heavy Metals in the Environment, vol. 2. CEP Consultans, Edinburgh, pp 330–333

Hong J, Förstner U, Calmano W (1994) Effects of redox processes on acid-producing potential and metal-mobility in sediments. In: Hamelink JL, Landrum PF, Bergman HL, Benson WH (eds) Bioavailability – Physical, Chemical and Biological Interactions, Lewis Publishers, Boca Raton, pp 119–141

Naylor C, Davison W, Motelica-Heino M, Van Den Berg GA, Van Der Heijdt LM (2004) Simultaneous release of sulfide with Fe, Mn, Ni and Zn in marine harbour sediment measured using a combined metal/sulfide DGT probe. Sci Tot Environ 328:275–286

Naylor C, Davison W, Motelica-Heino M, Van Den Berg GA, Van Der Heijdt LM (2006) Potential kinetic availability of metals in sulphidic freshwater sediments. Sci. Tot. Environ. 357:208–220

Peiffer S (1997) Umweltgeochemische Bedeutung der Bildung und Oxidation von Pyrit in Gewässersedimenten. Bayreuther Forum Ökologie, Universität Bayreuth, vol. 47

Schwartz R (2006) Geochemical characterization and erosion stability of fine-grained groyne field sediments in the Middle Elbe River. Acta Hydrochim Hydrobiol 34:223–233

Tusseau-Vuillemin M-H, Gilbin R, Taillefert M (2003) A dynamic numerical model to characterize labile metal complexes collected with diffusion gradient thin films devices. Environ Sci Technol 37:1645–1652

Chapter 7

Transport Indicators

Wolfhard Symader

The transport of cohesive sediments is poorly understood and difficult to model. Four papers in this chapter highlight quite different aspects of the transport phenomena, show the need of future research and present some approaches, how a better insight can be gained using natural tracers.

The first paper provides a working theory that is based on more than fifteen years of intensive fieldwork in small mountainous basins, where most of the controlling factors could be studied in detail. The three following papers focus on the aspects of pre-event conditions, floodplains as particle sinks and the situation in the estuarine zone.

Most of the cohesive material is event and supply controlled, and therefore reflects properties of its sources. The most important groups of sources are top soil and river bottom sediments. Deposition occurs when the bottom shear stress is less than the critical shear stress. There are at least three quite different controlling situations for this process: *(i)* The kinematic wave effect shifts material under transport from the peak of the flood wave to the recession limb, while the rising limb picks up new material. Therefore transported cohesive sediments can change their properties completely. *(ii)* A decrease in runoff and an increase in roughness result in immediate deposition of material, as it is the case on floodplains. Roughness is the explanation for the importance of the microrelief. *(iii)* A general decrease of flow velocity, which helps the rivers to build up a delta at the coast or accumulates mud in river harbours, with potentially high loads of contaminants.

Deposited particles can both be disconnected from the system for centuries, and build up a new source for the next event. That is one of the reasons, why the temporal variation of particle characteristics is so high, and why it is so important to understand the pre-event conditions.

Most floodplains represent the long term memory of the particle transport of a river. But situations might change, and floodplains may be eroded again.

The situation on floodplains is more an exception than the normal case. As floodplains are only reached by extreme events and fall dry thereafter, the material can be embedded into the matrix by plants of cultivation and is thoroughly fixed. In estuarine environments the deposited material can be remobilized merely by a change of the flow direction, which can happen twice a day. This material is hardly fixed and belongs to the fluffy layer on top of the bottom sediments. This situation is similar to the particle transport in small rivers during dry weather flow.

There is a common bond of all four papers. The authors used chemical characteristics e.g., trace metals, major ions or organic contaminants to understand the behavior of cohesive sediments.

Wolfhard Symader · Reinhard Bierl · Andreas Kurtenbach · Andreas Krein

7.1 The Relevance of River Bottom Sediments for the Transport of Cohesive Particles and Attached Contaminants

7.1.1 Introduction: Source and Transport Indicators

The understanding of the transport of cohesive sediments in flowing waters is one of the major tasks in fluvial hydrology. The bulk of material is transported during events as suspended particulate matter, but a continuous exchange with the river bottom and its interaction with the transport mechanisms of coarse material make it difficult to distinguish between different processes. From the water chemistry point of view highest concentrations of dissolved solids normally occur under dry weather flow conditions, when the concentrations of suspended particulate matter are lowest and the river bottom is supposed to behave as a sink. What is insignificant for the transport of most of the material can be crucial for understanding the fluxes between water body, suspended particles and river bottom.

Based on a simplified concept of sources, sinks and transport, the first detailed investigations that went beyond quantitative aspects concentrated on the sources of material. Bierl et al. (1996) and Walling (1996) pointed out that suspended particle concentrations are mostly supply controlled. Therefore identifying the type of material leads to the source, and knowing the source gives the opportunity to investigate the mechanisms of activation. By comparing selected properties of sediment samples with potential parent material it was assumed that both type and location of particle sources can be identified. An excellent review on tracing suspended sediment sources is given by Walling (2005).

Fingerprinting Source Materials

Peart (1989) cites three main difficulties in using fingerprinting techniques. These are *(i)* enrichment of suspended matter in fines and organic material relative to its source, *(ii)* transformation of sediment properties within the fluvial system, and *(iii)* storage and subsequent remobilization of material. However, the main challenge is to select appropriate characteristics.

Characteristics suggested for the determination of type and location of suspended particles can be appointed to three groups of variables. The best characteristics are those which play the role of a key fossil. It must be possible to attach them clearly either to a known location or to a defined type of sediment source. They should be time invariant and not influenced by intermediate factors such as grain size distribution. The mineralogical composition, which is used for the determination of the location of sources, belongs to this group. This approach has a long tradition in geology and helps to understand the structure of profiles. Müller and Förstner (1968) analyzed clay minerals in suspended material and could detect paragonite in twelve samples of the rising limb of a summer flood in the river Alpenrhein, which came from a local source. The application of mineralogical fingerprints, however, is restricted to heterogeneous catchments with differences in the subsoil lithology.

A promising set of indicators for different bedrocks can be gained from natural radioactivity, e.g. Thorium, but access to an efficient laboratory is required (Olley and Murray

1994). The color of suspended sediments was used by Grimshaw and Lewin (1980) and Imeson et al. (1984). In spite of some problems in quantifying local properties particle color is a useful parameter for examining variations. Using the colors brown and gray Grimshaw and Lewin (1980) could distinguish two different soil types, whereas Imeson et al. (1984) attached a gray color to material which entered the river together with interflow water. Brown (1985) suggested the use of pollen. This method has advantages as well, because pollen and spores have high decay rates. If pollen are found in suspended sediment, the material probably comes from eroded topsoil and if the vegetation cover shows no mosaic pattern but can be regionalized, it must be possible to locate the sediment source. On the other hand this method takes a lot of effort and experience. A common disadvantage of all these variables is that they can only be used under favorable conditions.

A second group of characteristics refers to chemical properties well aware of the possibility that the material may be considerably transformed during the transport downstream. Particle associated major ions, heavy metals, organic trace substances or biological characteristics as the number of germs (Matson et al. 1978), chlorophyll (Strunk 1993), biomass or the identification of certain types of detritus (Gilvear and Petts 1985) can be used in all kinds of catchments. Gallé et al. (2004) adopted the FTIR-DRIFT technique to document changes in sediment composition. The main difficulty in using chemical characteristics is their high temporal variation and the poor knowledge of the processes that are responsible for this.

Relations between Sources and Hydrological Processes

The third group of variables utilizes the investigation of hydrological processes to identify sediment sources. The relation of dissolved zinc, iron and manganese to the first increase of a flood wave due to the remobilization of fluvial sediments belongs to this category. Grimshaw and Lewin (1980) investigated details of the sediment-discharge relationship, while Imeson et al. (1984) used the position of the sample on the hydrograph. This approach looked very promising, but can only be applied, when kinematic wave effects can be excluded.

Investigating particle properties provides an excellent insight into the dynamics of particle transport. It pointed out the importance of sources and showed how complex the temporal patterns during a flood event can be. However, striking results on a local level could not compensate the lack of finding general structures that are needed for modeling the process. It took additional ten years that hydrologists and sedimentologists started to look at the major sinks, especially the floodplains. The need for changing the angle of research was enhanced by the increasing importance of valuable data on sediment budgets and a growing awareness about the role of floodplains in material fluxes, contaminant storage, ecological response and carbon cycling. Even river basin management cannot be accomplished without a better understanding of floodplains.

Floodplain Sediments

Again scientific research started with a quite simple concept. Floodplains are supposed to be natural sinks with low deposition rates and small, more or less continuous spatial variations. Walling et al. (1992) reported of deposition rates between 0.14 and 0.5 cm per year. Howard (1992) modeled floodplain deposition on the basis of three main assumptions: coarse material is deposited near the riverbanks, deposition decreases with distance from the river and with the altitude of the floodplain. Nanson and Croke (2002) reviewed some of the main challenges in floodplain research.

Floodplains are not always in a quasi equilibrium with their environment, which means that periods of sedimentation are interrupted by floodplain erosion. The question of equilibrium is connected to the problem of extreme events, which can build up or destroy floodplains.

But scale problems are not restricted to temporal aspects. Investigations of sedimentation rates on field plots near the river banks often show constant sedimentation rates dominated by sedimentation during high winter floods. However, sedimentation may start at the tributaries and not at the main channel. In this case the whole basin has to be considered and highest rates can be found in the backwater areas.

High sedimentation rates do not correspond always to high inputs in fines and nutrients. Thoms et al. (2000) reported highest concentrations of organic carbon, total nitrogen and phosphorus in distal areas and not near the rivers. The main source is light organic debris forming a "bath tub ring" (this excellent descriptive term for the phenomena was given by Thoms at the IAHS symposium 2000 in Waterloo) around the borders of a floodplain area. The distribution and structure of vegetation has a marked effect on transport and sedimentation of solid material also. Besides vegetation is a major source of organic material.

Today floodplain research is an interdisciplinary topic that reaches far beyond the aspect of a major sink for material. Contrary to the aspect sources, where research progress has become slow, floodplain research is a highly active field.

Transformation during Transport

There is no dispute about the general question, whether channel sediments play an important role in the transport of particle associated contaminants, the rates of adsorption and desorption or the counteracting processes of accumulation and remobilization within a river. The spatial distribution of sediment associated environmental pollutants is well documented in a large number of case studies. However, comparatively little attention has been paid to the temporal variations of river bottom sediments. Even in a primer on sediment-trace element chemistry (Horowitz 1991) information about the spatial aspects of sampling design only can be found. Horowitz (1991) states that one advantage of sampling bed sediments is its less marked spatial and temporal chemical variability in comparison to suspended sediments. Data of several large rivers prove this point. Knox (1989) claims that sediments only move during major events, separated by periods of high stability. Very often it is suggested that sediment samples for environmental analyses should be taken after a long period of low flow at the end of the summer, because pollutants tend to accumulate in the sediment during the year. Förstner and Wittmann (1979) point out that only two important factors must be taken into account, when stream samples are used for the identification of pollution sources: the conditions of high water discharge and the influence of particle size. Although movements of bed load are both investigated in the field and in the laboratories and described by a number of hydraulic models, it seems to be a common assumption in water quality studies that the situation at the river bottom remains more or less constant or changes only gradually.

In order to replace assumptions by real facts an extended sampling program was started with a pilot study in 1991 and established in 1993 in order to understand the temporal behavior of river bottom sediments in small mountain rivers.

7.1.2 Case Studies in Small Mountain Rivers

Area under Investigation

The basins of the Kartelbornsbach, Olewiger Bach and Ruwer cover an area of about 3, 30 and 300 km^2, respectively (Fig. 7.1). They are heterogeneous in land use, and to a certain degree in geology as well. The Kartelbornsbach is a limestone basin. Bedrock of the Olewiger Bach and the Ruwer is dominated by Devonian shalesand quartzites. Each river is polluted by sewage that comes from effluents of small villages and dispersed clusters of houses. As the basins are representative for many mountainous landscapes in Germany and even in West-Europe, the results can be transferred to other regions.

Results and Discussion

One of the general assumptions concerning the temporal changes of sediment characteristics at river bottom was, that a complete mixing during winter removes most of the fine material and dilutes the concentrations of sediment associated contaminants. During summer and even more during the low flow periods in autumn sediment material is supposed to be deposited and built up, until the first high floods in winter start a new cycle.

The time series of the hydrophobic polycyclic aromatic hydrocarbon (PAH) benzo(ghi)perylene, measured at a station in the upper part of the Olewiger Bach

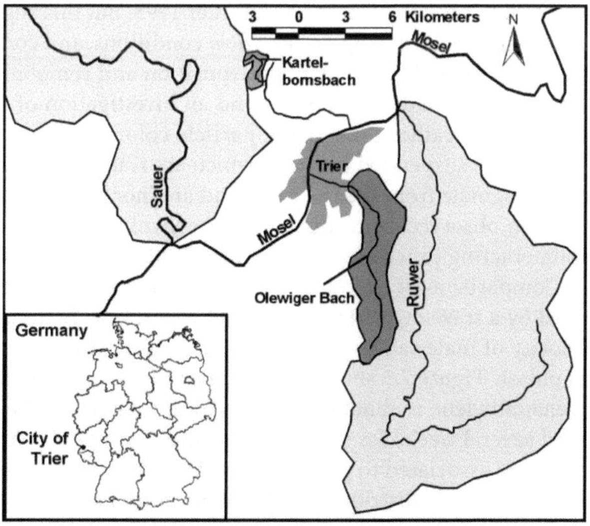

Fig. 7.1.
Area under investigation

Fig. 7.2.
Weekly sediment samples (<63 µm) of the Olewiger Bach

(Fig. 7.2) do not fit to these assumptions at all (Schorer 1988). To begin with the temporal variation is high and makes it difficult to detect an annual cycle. Secondly, in 1994 one can observe a level of higher contents from May to December which is interrupted by several short-term peaks. One year later in 1995 higher concentrations of the PAH are confined to a period from May until September. A continuous deposition and an enrichment during autumn could not be found.

The difference between the two graphs goes back among others to a severe reduction of waste water effluents in summer 1995. But this suggests that polluted sediments can be mobilized during dry low flow conditions, and consequently the plateau in 1994 reflects an equilibrium between deposition and remobilization. This interpretation is supported by field observations and an investigation of suspended particulate matter during dry weather flow. Using particle color, Udelhoven et al. (1997) found out that most of the suspended particles which are transported under dry weather flow conditions originate from the river bed and are inorganic. Symader et al. (1997) could show that the observed concentrations of organic pollutants are the result from the two counteracting processes of depositing and removing polluted particles.

Comparisons of time series at two stations of the Kartelbornsbach, which are separated by a traveling distance of about 200 m gave first information about the traveling velocity of material. Organic contaminants mostly enter the river and are transported as pulses. Figure 7.3 shows that in most of the cases the traveling velocity of the PAH acenaphthylene is about one week. However, most of the major ions and heavy metals need several weeks for the same distance. So it can be assumed that organic contaminants are associated to particles of a fluffy high mobile upper layer, while heavy metals are more evenly distributed in the vertical profile.

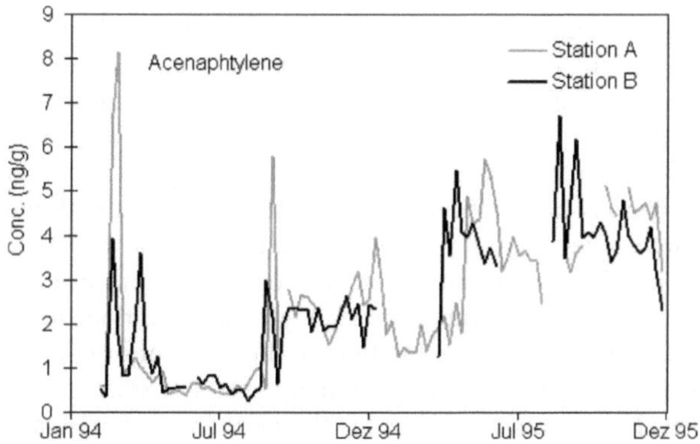

Fig. 7.3.
Acenaphtylene in weekly sediment samples of the Kartelsbornsbach

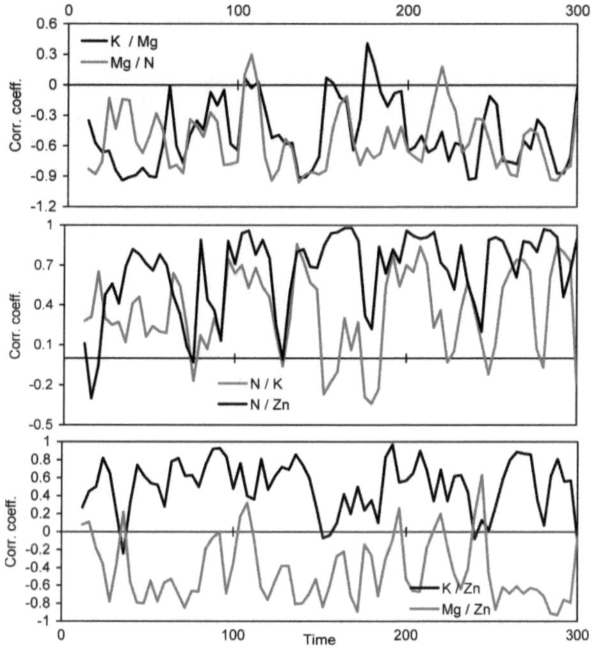

Fig. 7.4.
Moving window correlations between major elements in sediment samples of the Kartelbornsbach

Koll and Dittrich (1998) carried out laboratory experiments in a tilting flume, where they fed fine gravel over an armored layer. They could show that the feeding material replaced parts of the old material, and up to 25% of the feeding material remained embedded within the bed surface. The armored layer was not destroyed during this process. These experiments indicate that replacement is an alternative to mixing processes, and need less transport energy. But there are still two open questions. The high temporal variability of particle associated solids in channel sediments have to be explained either by changing particle sources or by changing transport mechanisms (Symader and Bierl 2000).

Calculating the coherence spectra between different elements reveals several groups of elements with similar behavior (Symader and Bierl 2000). They are Calcium and Magnesium, as was expected in a limestone area, organic carbon, nitrogen and phosphate with its dominant proportion coming from the sewage plant, and iron and potassium and only for the long periods copper. For a detailed analysis of short term variation Pearson correlation coefficients were calculated using a moving window of twelve weeks that moves forward in steps of four weeks. Some of these results are shown in Fig. 7.4.

Calcium and magnesium show an inverse relationship to all other solids. Their medium concentrations are characteristic for bedrock material. The lowest concentrations correspond either to the maxima of iron and potassium or a sediment highly enriched in organic material. Top concentrations are found in autumn and late summer, when water temperature is high and conditions for bioprecipitation are favorable.

Detailed analysis of the temporal pattern of organic carbon and nitrogen indicated that the major source is periphyton and not sewage, as was previously assumed. But periphyton acts as a filter for the very fine particles and that is why there are good correlations with phosphate and zinc. Iron and potassium originate predominantly from top soil but are enriched by periphyton as well.

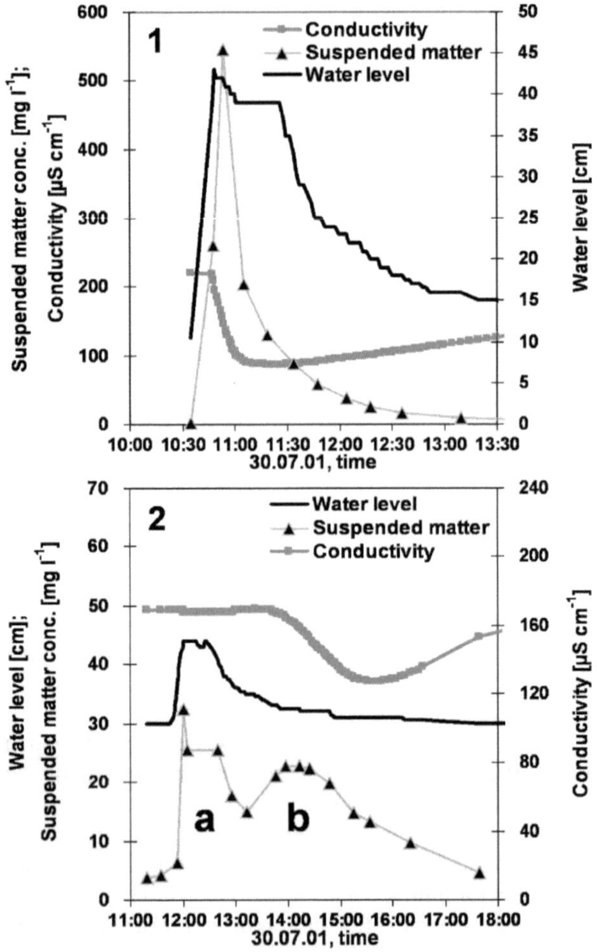

Fig. 7.5.
Artificial flood wave at two sampling stations (*1*, *2*) of the river Ruwer

The identification of sources explained the varying composition of material, but not the temporal variation. A valuable insight provides the investigation of Kurten-bach et al. (2005) about the kinematic wave effect on the transport of particulate matter. In 2001 the water works of Trier released water from their drinking water reservoir into the river Ruwer inducing a flood wave (Fig. 7.5). At the sampling station (1) the flood wave consists of released water, as the sharp decrease in electric conductivity indicates. The impact of the wave destroyed the armored river bed and mobilized some material from the river bottom. At station (2), which was situated 3 km downstream, flood wave and water body have decoupled and the wave consists of old water.

But more interesting is the graph of the particle concentrations. The sharp single peak has vanished and is replaced by two minor peaks with one on the rising and one on the falling limb of the wave (Fig. 7.5 *a*,*b*). Due to the kinematic wave effect the material that is transported lags behind the wave and part of it is deposited, when the transport capacity falls below the critical value to keep the material in suspension. At the same time the front of the flood wave can pick up new material. Although this is a continuous process the heterogeneity of the river bed with areas rich and poor in fine material leads to peaks or pulses of material.

The temporal variation of river bottom sediments under the fluffy layer is dominated by flood events. It is both a matter of sources within the channel and within the basin, the conditions for deposition, e.g. occurrence of macrophytes and periphyton, and it goes back to the kinematic wave effect. Particulate material is not transported downstream over a long range. The material that is collected comes from the direct vicinity of the sampling station. The dominant process of transport is deposition and remobilization at short intervals.

To understand the microstructure of the process a continuous monitoring is required. Therefore an acoustic monitoring device was developed, where submersed microphones were

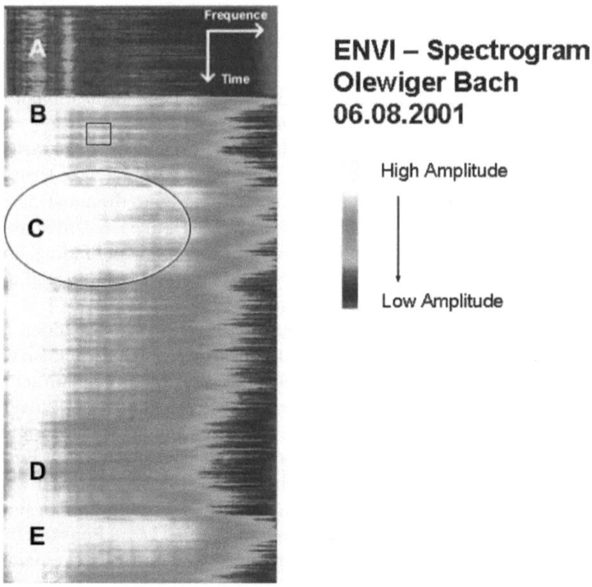

Fig. 7.6.
Temporal variations of the spectrograms of acoustic signals during an artificial flood wave

attached under steel plates that could be fastened to the river bottom. A location was chosen, at which the steel plate was neither buried nor dug out. In Fig. 7.6 the results of an artificial flood wave at the Olewiger Bach are shown. Each line represents a spectrum of the overall noise. In order to handle the data monitoring was condensed to five spectra per second.

Figure 7.6 indicates that the transport of material starts very abruptly with the front of the wave. The macrostructure consists of three main centers of activity (B, C and E). The main burst of material in section C consists of at least three subunits, and each subunit is built up by a change between high and medium activity. This gives the whole figure the resemblance of a layered structure. During natural events this mechanism is overlaid by all the material that is washed in from the basin. Time series of turbidity measurements indicate that this a continuous process with changes due to the exhaustion of sources and the tapping of new sources with changing flow paths during the runoff generation process.

Conclusions

In mountainous regions high temporal variations of particle properties and sediment associated contaminants show, that the channel sediment is a highly dynamic agent of transport, where deposition and remobilization take turns at short time intervals. Consequently, to get a comprehensive insight into the transport of particles and particle associated contaminants, sediments should be investigated together with the water body and suspended matter. There is a vertical gradient at the river bottom with a fluffy layer on top, a well mixed layer below and bedrock material that is only moved during catastrophic events. The transport of particles is mainly controlled by the activation of particle sources and the kinematic wave effect that leads to a discontinuous process, when most of the transport happens during single pulses. These results should be transferable to most of the basins in low mountain regions, but probably not to lowland rivers.

References

Brown AG (1985) The potential use of pollen in the identification of suspended sediment sources. Earth Surface Processes and Landforms 10:27–32

Bierl R, Symader W, Gasparini F, Hampe K, Udelhoven T (1996) Particle associated contaminants in flowing waters – the role of sources. Arch Hydrobiol Spec Issues Adv Limnol 47:229–234

Förstner U, Wittmann GTW (1979) Metal pollution in the aquatic environment. Springer-Verlag, Berlin

Gallé T, Van Lagen B, Kurtenbach A, Bierl R (2004) An FTIR-DRIFT study on river sediment particle structure: Implications for biofilm dynamics and pollutant binding. Environ Sci Technol 38: 4496–4502

Gilvear DJ, Petts GE (1985) Turbidity and suspended solids variations downstream of a regulating reservoir. Earth Surface Processes and Landforms 10:363–373

Grimshaw DL, Lewin J (1980) Source identification for suspended sediments. J Hydrol 47:151–162

Horowitz AJ (1991) A primer on sediment-trace element chemistry. Lewis Publ, Chelsea

Howard AD (1992) Modeling channel migration and floodplain sedimentation in meandering streams. In: Carling PA, Petts GE (eds): Lowland floodplain rivers. Wiley, New York, pp 1–42

Imeson AC, Vis M, Duysings JJ (1984) Surface and subsurface sources of suspended solids in forested drainage basins in the Keuper region of Luxembourg. In: Burt TP, Walling DE (eds) Catchment experiments in fluvial geomorphology. Geo Books, Norwich UK, pp 219–233

Knox J (1989) Long- and short-term episodic storage and removal of sediment in watersheds of southwestern Wisconsin and northwestern Illinois. IAHS Publ 184:157–164

Koll K, Dittrich A (1998) Sediment transport and erosion in mountain streams. IAHS Publ 249:309–316

Kurtenbach A, Krein A, Symader W (2005) The significance of channel flow processes for the coupling of runoff generation with dissolved and particulate transport – an analysis based on artificial flood waves in two meso-scale middle mountain catchments. Hydrologie und Wasserbewirtschaftung 49(4):172–181, (in German)
Matson E, Hornor SG, Buck JD (1978) Pollution indicators and other microorganisms in river sediment. J Water Pollut Control Fed 50:13–20
Müller G, Förstner U (1968) Sedimenttransport im Mündungsgebiet des Alpenrheins. Geol Rundschau 58:229–259
Nanson GN, Croke JC (2002) Emerging issues in floodplain research. IAHS Publ 276:271–278
Olley JM, Murray AS (1994) Origins of variability in the ^{230}Th/^{232}Th ratio in sediments. IAHS Publ 224:65–70
Peart MR (1989) Methodologies currently available for the determination of suspended sediment source: a critical review. Proc 4th Int Symp River Sedimentation, Beijing, pp 150–157
Schorer M (1998) Raumzeitliche Dynamik von anorganischen und organischen Schadstoffen in Sedimenten eines Fliessgewässers. Dissertation Universität Trier
Strunk N (1993) Schwebstoffcharakteristika und Hochwasserdynamik – Eine Untersuchung zur Identifikation und Aktivierung partikulärer Stoffquellen. Dissertation Universität Trier
Symader W, Bierl R (2000) Time series analysis of chemistry in bottom sediments of the Kartelbornsbach, Germany. IAHS Publ 263:175–181
Symader W, Schorer M, Bierl R (1997) Space-time pattern of organic contaminants in river bottom sediments. IAHS Publ 243:37–44
Thoms MC, Foster JM, Gawne B (2000) Flood-plain sedimentation in a dryland river: the river Murray, Australia. IAHS Publ 263:227–236
Udelhoven T, Nagel A, Gasparini F (1997) Sediment and suspended particle interactions during low water flow in a small heterogeneous catchment. Catena 30:135–147
Walling DE (1996) Suspended sediment transport by rivers: a geomorphological and hydrological perspective. Arch Hydrobiol Spec Issues Adv Limnol 47:1–27
Walling DE (2005) Tracing suspended sediment sources in catchments and river systems. Sci Tot Envion 344:159–184
Walling DE, Quine TA, He Q (1992) Investigating contemporary rates of floodplain sedimentation. In: Carling PA, Petts GE (eds) Lowland floodplain rivers. Wiley, New York, pp 165–184

Andreas Kurtenbach · Andreas Krein

7.2 Pre-Event Hydrological Conditions As Determinants for Suspended Sediment and Pollutant Transport during Artificial and Natural Floods

7.2.1 Introduction

Aquatic fine particles control river water quality both chemically and physically, because they carry a major amount of heavy metals and hydrophobic organic contaminants (Evans et al. 1990; Gallé et al. 2004; Yu et al. 2001) and their accumulation can have a negative impact on benthic habitats (Broekhuizen et al. 2001; Greig et al. 2005). As the transport of these fine particles and the associated pollutants is mainly supply controlled (Bierl et al. 1996; Walling 1996), the focus should be on the investigation of flood events and discharge related transport phenomena. The numerous processes which determine the suspended matter and particle bound pollutant transport during natural floods can be classified into those acting in the river basin and those operating in the channel system. The superimposition of both groups during natural floods impedes the interpretation of water quality dynamics at monitoring sites. However, extracting significant processes for fine particle and pollutant mobilization and assessing the relevance of in-channel and basin processes are prerequisites to elaborate sustainable water quality management strategies.

Thus, the objective of our study is to analyze the in-channel related processes on their own and to characterize the ruling factor "pre-event hydrological conditions" for suspended matter and particle bound pollutant transport. Therefore, two strategies have been adopted:

1. First, apart from analyzing numerous natural floods, artificial flood events were generated in order to investigate in-channel transport dynamics. The outstanding advantage of these artificial flood waves is that some of the governing processes can be excluded or steered by the experimental design. This includes, for instance, hydraulic boundary conditions such as the total and peak discharge or the duration of the event. Additionally, the variability of chemographs and hydrographs during the artificial floods can be traced back to the preceding base flow, the activation and exhaustion of a few in-channel sources as well as the composition and amount of the introduced water.
2. In a second step, maximum pollutant concentrations, suspended matter transport and particle characteristics measured during single artificial and natural flood events were related to hydrological and hydro-meteorological pre-event conditions such as previous number of floods and antecedent precipitation.

7.2.2 Area of Investigation

The study area is the heterogeneous Olewiger Bach basin which covers an area of 35 km^2. The basin is situated in the northern Hunsrück Mountains nearby the city of Trier, southwest Germany (Fig. 7.7). Devonian schist in combination with quartz and diabase veins dominates the underlying bedrock of the basin. Pleistocene terraces of the river Mosel overlie the geological structures in the northern part of the basin. Land use is mainly agriculture with permanent pasture in the valleys and arable farming on the broad ridges. North- and east facing slopes are predominantly forested and some south-facing slopes in lower reaches of the basin are cultivated with vineyards. Settlement

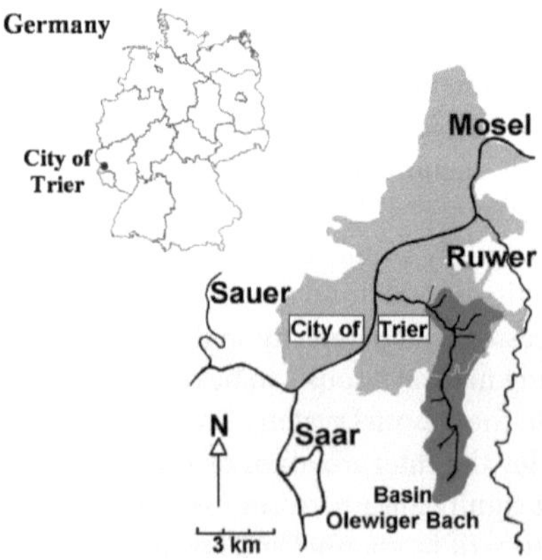

	Olewiger Bach
catchment size [km^2]	35
geology	quartzite and devonian schist
altitude [m]: lowest / highest	153.5 / 508.3
slope [°]: mean / maximum	9.9 / 46.1
mean annual temperature [°C]	8 (higher ridges)[1] 9.1 (lower river valley)[1]
mean annual precipitation [mm]	784 (lower river valley)[1] 791 (higher ridges)[1]
land use [%][2]	
crop land	47.2
green land	14.4
vine yards	4.9
forest	23.6
settlement area	9.9
water	0
Data from Kurtenbach et al. (2005)	[1] Krein (2000), time period 1961-1990 [2] year of land use classification 1990

Fig. 7.7. Map and characteristics of the investigated Olewiger Bach basin

areas cover around 10% of the basin (Fig. 7.7). Runoff from several roads, effluents from small industries and untreated waste water from solitary farms influence the river water quality. The cross-sections of the Olewiger Bach, some of them anthropogenically influenced, are predominately rectangular shaped with a large width/depth ratio and vertical river banks in argillaceous material. Channel slope between the waterworks and station Weingut (Fig. 7.8) amounts to approximately 2.0% and between station Weingut and gauging station Olewig-Kloster to 1.4%.

7.2.3 Material and Methods

Natural and artificial flood events have been investigated over a wide range of hydrological conditions during various years (e.g., Krein 2000; Kurtenbach et al. 2005). Discharge and precipitation were recorded continuously in the basin. In cooperation with the Trier municipal waterworks, the artificial floods were generated by a waterworks in the channel of the Olewiger Bach (Fig. 7.8). The discharged water was pumped through a pipeline from a drinking water reservoir in the neighboring Ruwer basin. It is characterized by low electrical conductivity values (approximately 75 μS cm^{-1}) and low suspended matter concentrations (approximately 2 mg l^{-1}).

During the natural floods water samples were taken midstream by two liter polyethylene bottles and suspended matter was collected by 20 liter containers at the main gauging station (Fig. 7.8). During the artificial floods this sampling was extended to additional stations along the brook axis (Fig. 7.8).

The measurement program encompasses particle associated heavy metals, polycyclic aromatic hydrocarbons (PAH), nutrients as well as suspended matter concentration, organic carbon content, C/N ratio and grain size distribution. During the sampling procedure, temperature-compensated (at 25 °C) electrical conductivity and temperature were measured in situ using a WTW conductivity meter. Suspended matter concentrations were determined gravimetrically by filtering a water sample over a WHATMAN GF/F glass fiber filter. The suspended matter was separated by centrifugation, freeze dried and homogenized in an agate mortar. For heavy metal analysis,

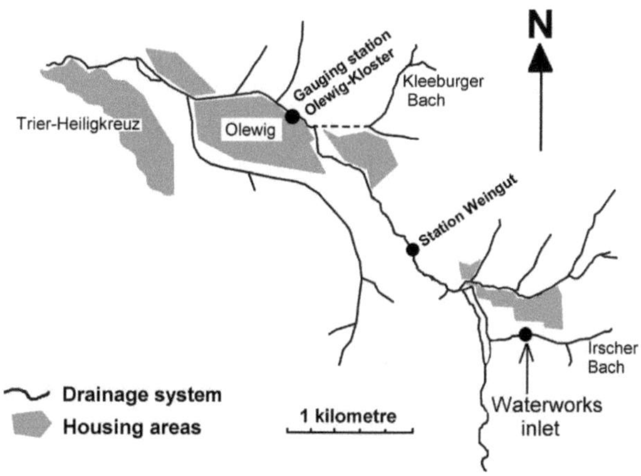

Fig. 7.8. Map of the sampling locations along the brook axis of the Olewiger Bach

bulk suspended matter was digested with concentrated nitric acid for 6 hours under pressure at 170°. Subsequently, selected particle bound heavy metals (lead, copper, zinc, iron, manganese) were measured with atomic absorption spectroscopy (VARIAN-SpectrAA-10 and VARIAN-SpectrAA-640 GTA100). The analysis of the effective particle size area distributions were performed without pre-treatment directly out of the water samples using the stream laser system GALAI CIS-1 (Aharonson et al. 1986). Nitrogen and total carbon were determined by a LECO CHN 1000, the C_{org}-content with the LECO RC-412 element analyzer. The polycyclic aromatic hydrocarbons (PAHs; 16 US-EPA) were solvent extracted and measured with gas chromatography/mass-spectrometry (GC HP5890II – MSD HP5970-B) in SIM Mode. A detailed description of the organic analyses is provided by Krein and Schorer (2000). Reference materials were analyzed for both inorganic and organic substances to meet analytical quality assurance standards.

7.2.4 Results and Discussion

The results of both the artificial and the natural floods reveal the considerable role of pre-event hydrological conditions in quantitative and qualitative suspended sediment transport dynamics in the Olewiger Bach basin.

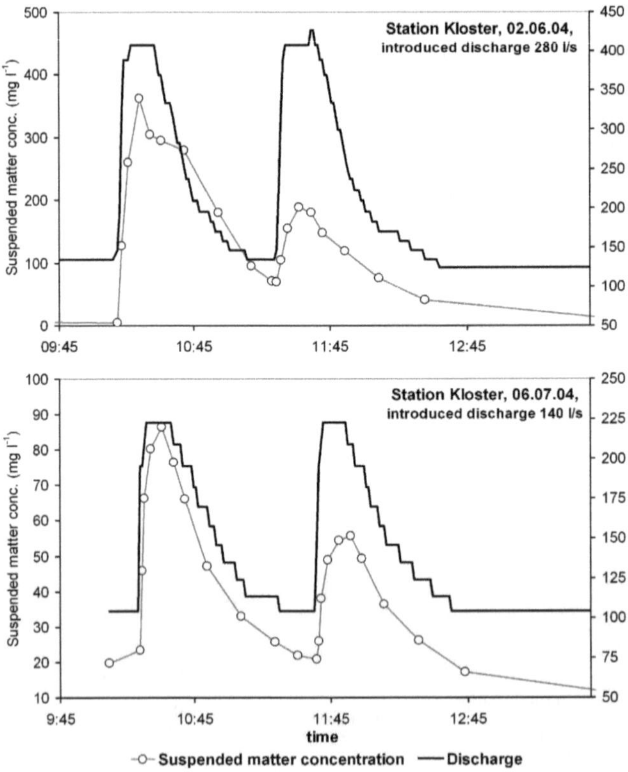

Fig. 7.9. Sediment exhaustion processes during double peaked artificial floods in the Olewiger Bach

Figure 7.9 depicts the dynamics of suspended matter transport during two double peaked artificial flood events at gauging station Olewig-Kloster, 3.2 kilometers downstream of the waterworks inlet. The introduced discharge during the events was 280 l s^{-1} for 30 min on 02.06. 2004 and 140 l s^{-1} for 30 min on 06.07. 2004. During the first discharge peaks the suspended matter concentration increases up to 360 mg l^{-1} and 86 mg l^{-1} respectively due to an in-channel erosion of sediments and bank material. Weakly consolidated sediments and easily erodable particles are exhausted during these first flood waves. Thus, despite the same shear stress during the second discharge peaks, the suspended matter concentrations increase only up to 188 mg l^{-1} and 55 mg l^{-1} respectively.

These results indicate that dependent on the location of available erosion material the quantity of suspended matter transport at a specific monitoring site is strongly influenced by in channel pre-flood conditions. This impedes a simple correlation between suspended matter concentration and discharge considerably.

The increasing discharge during the artificial floods is responsible for the mobilization of coarser particles, coming along with a decreasing percentage of transported particles <2 µm (Fig. 7.10a). This reduction in the relative amount of the finest particle size fraction is an important key factor for the typical decrease of specific heavy metals observed during numerous artificial floods, as shown in Fig. 7.10b by particle bound copper and zinc. However, during both events minimal copper and zinc contents vary marginally and lie more or less within the analytical relative standard deviations (between 5 to 6% for both metals), indicating that in contrast to the quantity of suspended matter, in-channel pre-floods are less important for suspended matter quality dynamics.

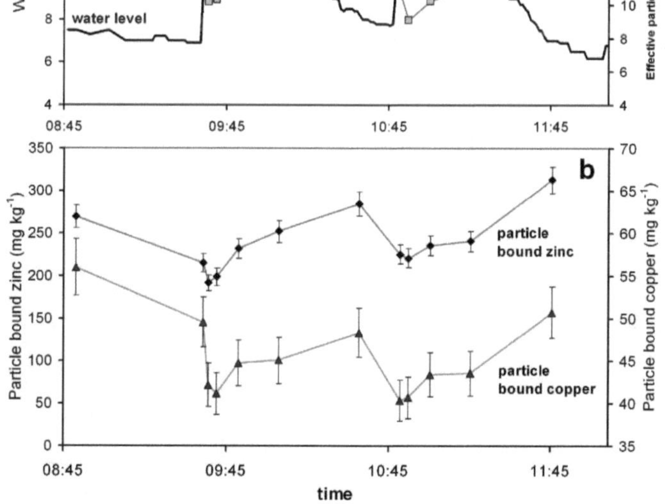

Fig. 7.10. Dynamics of the effective particle size fraction <2 µm (**a**) and the particle bound copper and zinc content of bulk suspended matter (**b**) during the double peaked artificial flood on 02.06. 2004, sampling site Weingut, 1.5 km downstream of the waterworks

Natural flood waves are the result of many interrelated processes (Krein 2000). According to the ecotoxicological risk potential, maximum pollutant contents during such natural floods are of particular importance. Lee et al. (2002) for instance could not identify a correlation between the first flush, during which the concentration of pollutants is substantially higher than during later flood stages, and the antecedent dry weather period. Blake et al. (2003) and Hewitt and Rashed (1992) however identify a higher dependency among concentrations of specific heavy metals during storm events and antecedent hydrological conditions.

The importance of basin wide antecedent conditions for the suspended matter quality in the Olewiger Bach basin is illustrated in Fig. 7.11. Every dot in Fig. 7.11a characterizes the relationship of the maximum particle bound zinc concentration during an artificial or a natural flood event to the ten day pre-rain amount. During natural floods the maximum zinc concentrations decrease with increasing pre-rain amount. In contrast, during the artificial floods the decrease of heavy metals owing to the mobilization of larger particles as shown in Fig. 7.10 always leads to low zinc concentrations, even in times of low pre-rain-amounts (Fig. 7.11a). Thus, it can be assumed that the sources for high zinc concentrations during the natural floods are located outside the river network. Particle-bound zinc originates predominately from streets and roof surfaces (Ellis and Revitt 1982; Hillenbrand et al. 2005). These sources are often hydraulically connected by storm sewer systems to the main tributary and probably be characterized by a rapid source exhaustion. However, the decrease mainly occurs if the source of the examined water quality constituent is spatially constricted and easily erodable. Additionally, the concentrations of the source pool must be significant higher than the typical background contents within the catchment area. A good example for an element of mainly lithogenic origin is iron with small local enrichments or dilution after dry weather flow conditions and geochemical background values following wet hydrological conditions (Fig. 7.11b).

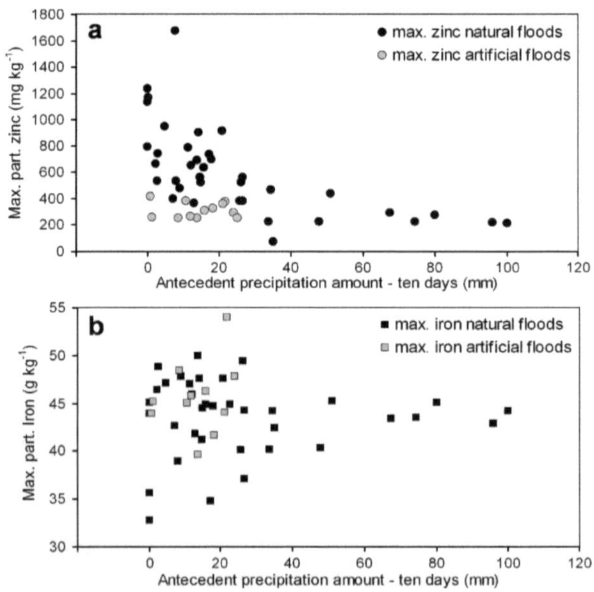

Fig. 7.11. Maximum particle bound zinc (**a**) and iron (**b**) concentrations during single artificial and natural floods versus antecedent precipitation amount for 10 days, gauging station Olewig-Kloster, 3.2 km downstream of the waterworks

Figure 7.12 shows, that the dynamics of the maximum particulate carbon content stands between zinc and iron dynamics with a high variability during artificial floods and a decrease during natural floods following wet antecedent conditions. This behavior can be attributed to possible in-channel sources of organic carbon comprising for instance the seasonal dependent and anthropogenically influenced growth of natural biofilms (Gallé et al. 2004; Schorer and Eisele 1997). During natural floods however, the low density organic matter is rapidly eroded and exhausted with increasing pre-rain.

Apart from inorganic pollutants such as the heavy metal zinc, pre-event hydrological conditions influence the dynamics of organic contaminants too. Every dot in Fig. 7.13 illustrates the relationship of the maximum suspended matter bound Benzo(ghi)perylene concentration during an artificial or natural flood to the previous number of flood events. The maximum Benzo(ghi)perylene concentrations decrease with increasing number of prior floods. The lower concentrations during the artificial floods again indicate the importance of specific basin sources for this pollutant. However, one natural flood differs from the general behavior (marked by the letter a in Fig. 7.13). Prior to this specific natural wave a drizzle with 3 mm pre-rain within 24 hours occurred which did not result in a flood at the gauging site. Such minimal events in a river basin are often only recognized as a suspended matter cloud with very high contents of pollutants. The first flush mobilized the sources in the catchment area and partially removed the toxic Benzo(ghi)perylene. Sealed anthropogenic areas – sources for PAHs (Brown and Peake 2006; Evans et al. 1990; Krein and Schorer 2000) – are cleaned and only small concentrations were identified in the following flood events.

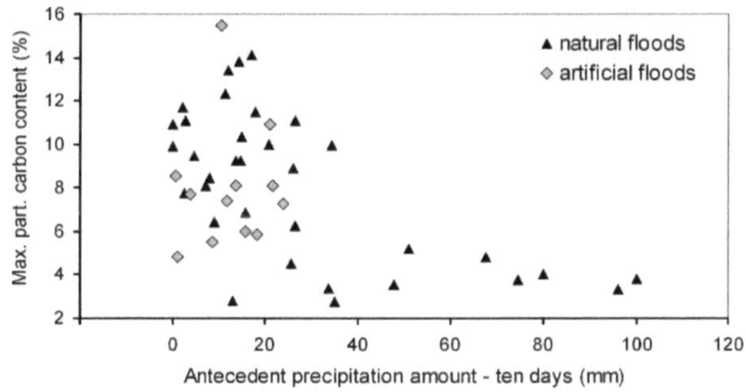

Fig. 7.12. Maximum particulate carbon content of bulk suspended matter during single floods versus antecedent precipitation amount for 10 days, gauging station Olewig-Kloster

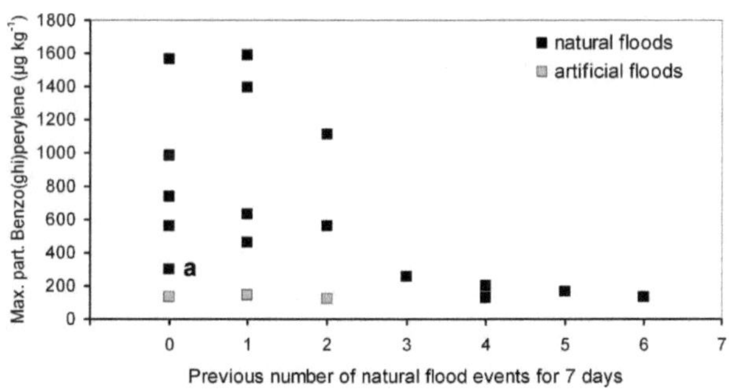

Fig. 7.13. Maximum concentrations of the particle-bound PAH Benzo(ghi)perylene in single flood events versus the previous number of natural flood events for 7 days, gauging station Olewig-Kloster

7.2.5 Conclusions

The detailed analysis of artificial and natural floods was proved to be a powerful approach to elucidate the relevance of both in-channel and basin wide pre-event hydrological conditions for transport dynamics of suspended matter and particle bound quality constituents.

Three central conclusions can be drawn:

a In-channel antecedent hydrological conditions control the amount of available material. River basin pre-event conditions additionally affect suspended matter composition and particle bound contaminants.
b By comparing suspended particles from natural and artificial floods it can be shown that specific heavy metals and PAHs are flushed into the river and are less remobilized from river bottom sediments.
c Changing pre-event conditions impede simple relationships between discharge and suspended matter concentration and composition.

Acknowledgments

The research projects were supported financially by the German Research Foundation. The authors would like to thank the Trier municipal waterworks for generating the artificial floods. Additional thanks are also extended to the employees in the laboratory of the department as well as the numerous student assistants. The authors thank the "Landesamt für Umwelt, Wasserwirtschaft und Gewerbeaufsicht Rheinland-Pfalz" for supplying precipitation data sets for the Olewiger Bach basin.

References

Aharonson EF, Karasikov N, Roitberg M, Shamir J (1986) GALAI CIS-1, a novel approach to aerosol particle size analysis. J Aerosol Sci 17:530–536
Bierl R, Symader W, Gasparini F, Hampe K, Udelhoven T (1996) Particle associated contaminants in flowing waters – the role of sources. Arch Hydrobiol Spec Issues Advanc Limnol 47:229–234
Blake WH, Walsh RPD, Barnsley MJ, Palmer G, Dyrynda P, James JG (2003) Heavy metal concentrations during storm events in a rehabilitated industrialized catchment. Hydrological Processes 17:1923–1939
Broekhuizen N, Parkyn S, Miller D (2001) Fine sediment effects on feeding and growth in the invertebrate grazers *Potamopyrgus antipodarum* (Gastropoda, Hydrobiidae) and *Deleatidium* sp. (Ephemeroptera, Leptophlebiidae). Hydrobiologia 457:125–132
Brown JN, Peake BM (2006) Sources of heavy metals and polycyclic aromatic hydrocarbons in urban stormwater runoff. Science of the Total Environment 359:145– 155
Evans KM, Gill RA, Robotham PWJ (1990) The source, composition, and flux of polycyclic aromatic hydrcarbons in sediments of the river Derwent, Derbeyshire, U.K. Water, Air and Soil Pollution 51:1–12
Ellis JB, Revitt DM (1982) Incidence of heavy metals in street surface sediments: solubility and grain size studies. Water, Air and Soil Pollution 17:87 – 100
Gallé T, Van Lagen B, Kurtenbach A, Bierl R (2004) An FTIR-DRIFT Study on River Sediment Particle Structure: Implications for Biofilm Dynamics and Pollutant Binding. Environ Sci Technol 38(17):4496–4502

Greig SM, Sear DA, Carling PA (2005) The impact of fine sediment accumulation on the survival of incubating salmon progeny: Implications for sediment management. Science of the Total Environment 344:241–258
Hewitt CN, Rashed MB (1992) Removal rates of selected pollutants in the runoff waters from a major rural highway. Water Research 26(3):311–319
Hillenbrand T, Toussaint D, Böhm E, Fuchs S, Scherer U, Rudolphi A, Hoffmann M, Kreißig J, Kotz C (2005) Einträge von Kupfer, Zink und Blei in Gewässer und Böden – Analyse der Emissionspfade und möglicher Emissionsminderungsmaßnahmen. Umweltbundesamt Texte 19/05, Dessau
Krein A, Schorer M (2000) Road runoff pollution by polycyclic aromatic hydrocarbons and its contribution to river sediments. Water Research 34:4110–4115
Krein A (2000) Stofftransportbezogene Varianzen zwischen Hochwasserwellen in kleinen Einzugsgebieten unter Berücksichtigung partikelgebundener toxischer Umweltchemikalien. Ph.D. Thesis, University of Trier, Aachen
Kurtenbach A, Krein A, Symader W (2005) The significance of channel flow processes for the coupling of runoff generation with dissolved and particulate transport – an analysis based on artificial flood waves in two meso-scale middle mountain catchments (in German). Hydrologie und Wasserbewirtschaftung 49(4):172–181
Lee JH, Bang KW, Ketchum LH, Choe JS, Yu MJ (2002) First flush analysis of urban storm runoff. Science of the Total Environment 293:163–175
Schorer M, Eisele M (1997) Accumulation of inorganic and organic pollutants by biofilms in the aquatic environment. Water, Air and Soil Pollution 99:651–659
Walling DE (1996) Suspended sediment transport by rivers: a geomorphological and hydrological perspective. Arch Hydrobiol Spec Issues Advanc Limnol 47:1–27
Yu KC, Tsai LJ, Chen, SH, Ho ST (2001) Correlation analyses on binding behaviour of heavy metals with sediment matrices. Water Research 35(10):2417–2428

Kurt Friese · René Schwartz · Frank Krüger

7.3 Transport and Storage of River Sediment and Associated Trace Metals into Floodplains of the Elbe

7.3.1 Introduction

One of Europe's most polluted rivers in the last half century was the river Elbe with its tributaries, which was severely affected by human activities. The worst inputs came from the former GDR and Czechoslovakia, where environmental precautions were almost non-existent. For many years industrial effluents and sewage were discharged into the river untreated, resulting in a strong reduction of the ecosystem. Owing to the construction of sewage treatment plants and especially the closure of several factories, the water quality of the river Elbe has greatly improved since German reunification in 1990 (e.g., Wilken et al. 1994; Schwartz et al. 1999). Nevertheless, the river sediments still contain pollution from the past, and will continue to be contaminated until the material is remobilized by high floods and transported downstream (Förstner et al. 2004). In contrast to most other European rivers, the river Elbe still has extended floodplains that are valuable biotopes and reduce the consequences of high flooding. The floodplains are regularly flooded, mainly in winter and spring because of heavy rainfalls and/or the snow melt in the low mountain range. River sediments are deposited on the floodplains during flooding, forming the substratum for soil formation. After being covered with plants the sediments are in general resistant to erosion, so that deposited contaminated material is retained in the floodplain much longer than in the river itself. Therefore, the alluvial soils within the floodplain can be regarded as to be the river's long-term 'memory' of sediment pollution.

As part of the river-floodplain system, wetlands have a valuable function in controlling surface and subsurface hydrological processes, such as by controlling flood water and groundwater recharge. During inundation an exchange of water, sediments, chemicals and biota takes place between the main channel and the floodplain. Organic and inorganic toxicants in surface waters can be transferred to floodplains and the groundwater, and can lead to unfavorable changes in aquatic and terrestrial communities. Furthermore, as floodplains are often used in agriculture as pastures and for cultivation, pollutants can be introduced into the food web via the contamination of soil, water or plants (Gröngröft et al. 2005).

Although floodplains show similarities to terrestrial and aquatic ecosystems, there are several differences, particularly the importance of the continuous storage of polluted sediments (Krüger et al. 2003, 2005). Chemicals are hydrologically transported to floodplains via high floods, precipitation and groundwater. Alluvial soils in floodplains can become highly reduced when submerged, but usually have a narrow oxidized surface zone that allows aerobic processes (Gröngröft et al. 2000). The biotic and abiotic transformations that occur within the floodplain environment may change the effect and reaction of several heavy metals and organic compounds (Calmano et al. 2005) some cause toxic conditions, while others induce the loss of chemicals to the atmosphere or the underground. The unique and diverse hydrological conditions in floodplains result not only in changes in the chemical forms of materials but also in the spatial movement of material within the floodplains, such as in water–sediment exchange and plant uptake (Friese et al. 2000; Gröngröft et al. 2005).

Using the floodplains for agriculture entails knowledge of flooding frequency and of the input and distribution of sediments and associated pollutants. Typical elements and organic components for the river Elbe catchment are Hg, Cd, and As (Krüger et al. 2005) as well as PCDD/F, PCB, HCB and with a little bit lower importance HCH and DDX (e.g., Witter et al. 1998, 2003). Miehlich (1983) gave a first description of heavy metal contamination profiles of typical Elbe River floodplain soil. Meißner et al. (1994) and Krüger et al. (1997) investigated topsoil in the middle course of the river Elbe and worked out correlation coefficients between heavy metals and organic carbon. Krüger et al. (2005) investigated the heavy metal pollution of topsoil along the German river stretch of the Elbe between the Federal States of Saxony and Lower Saxony, and showed correlation between heavy metal contents and organic carbon, elevation, river kilometration and other factors.

This section describes studies carried out on the transport and deposition of arsenic (As), cadmium (Cd), nickel (Ni), chromium (Cr), copper (Cu), lead (Pb) and zinc (Zn) by high floods at a floodplain demonstration site of the river Elbe catchment basin (Schönberg) representing a typical alluvial site far away of the most pollution areas of concern.

7.3.2 Study Site

The investigation area with the study site is shown in Fig. 7.14. The research area is located at the Elbe main stream at km 435–485 (German kilometration) and represents a typical downstream region where most of the pollutants are well mixed. The sharp morphological changes of the floodplain Schönberg, which are of both natural and anthropogenic origin (e.g., due to its use for livestock farming and pasture, as

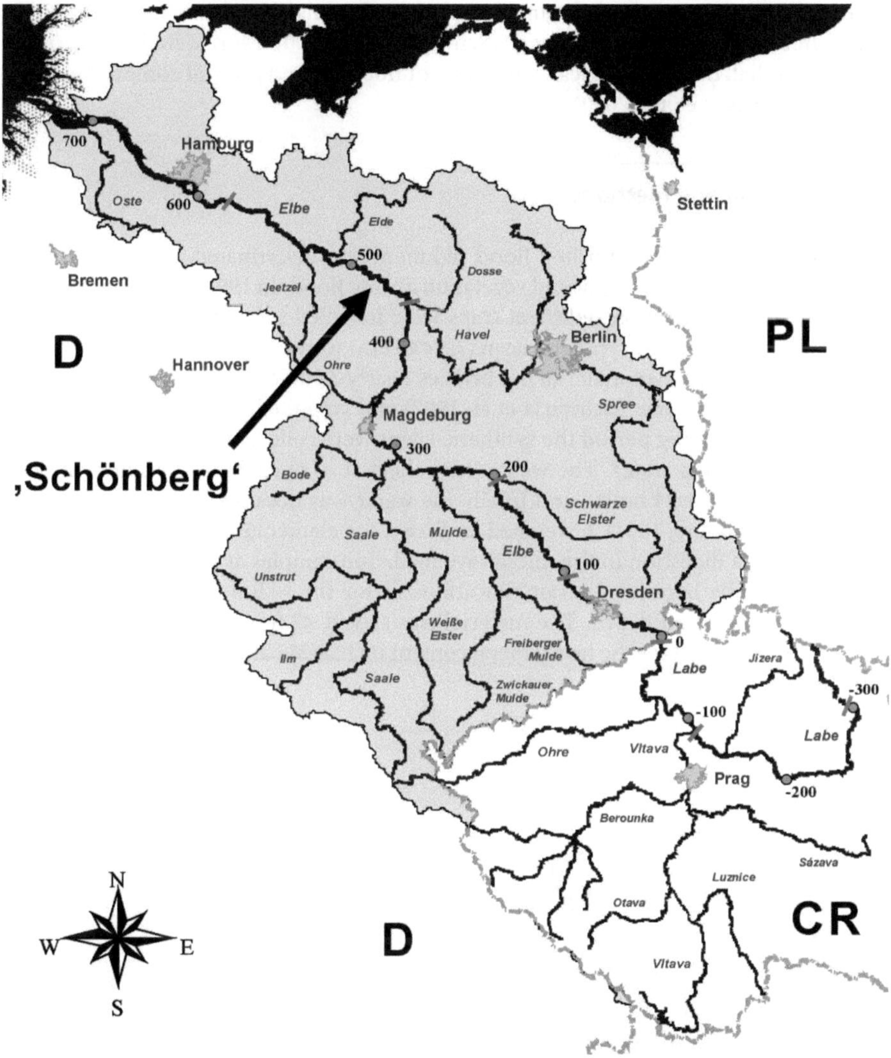

Fig. 7.14. Map of the river Elbe basin with the location of study site Schönberg

well as the presence of dikes), results in special flooding characteristics. This floodplain area shows a distinct micro-relief not only at the surface but also in the subsoil, as ascertained by geodetic measurements (Büttner et al. 2006). Several parts of the area (hollows, old arms) are located at geodetic elevations below the mean water level (MWL). The geodetic relief elevation varies between 1.3 m below and 3.5 m above mean water. Owing to the morphological heterogeneity and different flooding frequencies, soil composition varies sharply, ranging from sandy soils via those with large proportions of silt or clay, to soils with high levels of organic substance. Soil moisture measurements have indicated that high areas only rarely flooded (if at all) might still contribute to contaminant transport into lower soil horizons and the groundwater after heavy rainfall via seepage (Schwartz et al. 2000). More details concerning the micro-relief, the hydrological conditions of flooding and the soil composition can be found in Friese et al. (2000).

7.3.3 Material and Methods

At Schönberg site fresh deposited flood sediment was investigated at 11 locations in the floodplain varying in height and vegetation during floods in 1997 and 1998 (Fig. 7.15, Table 7.1). At each location sediment traps were installed to study the sedimentation process directly. Pieces of synthetic lawn (30 × 40 cm) were exposed in the recent floodplain. The synthetic lawn material has bristles nearly 3 cm long to simulate the roughness of wetlands surfaces (Schwartz et al. 1997) and trap the high flood sediments. At the end of the flooding period the synthetic lawns were collected. The sediments were rinsed out using tap water. The water was collected in open bins, the sediment was allowed to settle for 24 hours, and finally the water was decanted. The samples were oven-dried (24 hours, 105 °C), weighed and used for elemental analysis with ICP-MS after aqua regia digestion in the microwave field. Soil samples at study site Schönberg were collected by hand at the same locations as for the sediments representing the topsoil down to 10 cm depth. The material was sieved <2 mm, oven-dried (24 hours, 105 °C), and analyzed for the heavy metal content by ICP-MS after microwave digestion with aqua regia.

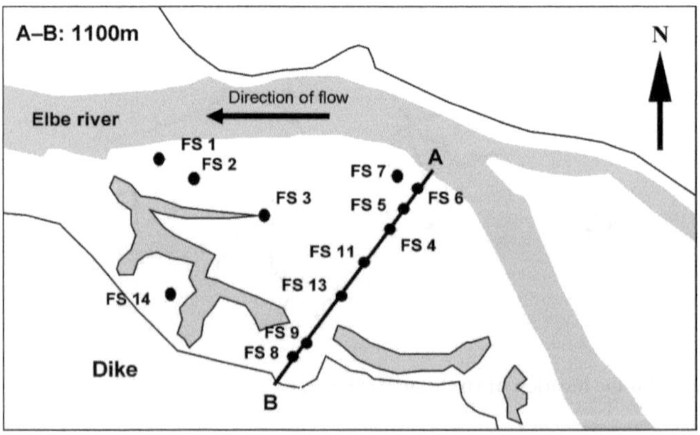

Fig. 7.15. Schematic map of the floodplain area of the study site Schönberg with sample locations

Table 7.1. Sample locations for flood sediment and topsoil at the Schönberg study site

Location	Description	Distance to the Elbe (m)	Elevation about MWL (m)
FS 1	Riverbank	25	0.4
FS 2	Plain, river near	135	0.5
FS 3	Old flood channel	315	–0.3
FS 4	Flood channel	180	0.5
FS 5	Flood channel	173	0.1
FS 6	Flood channel	60	0.3
FS 7	Flood channel openning	45	0.4
FS 8	Hollow, dike near	944	–0.3
FS 9	Plateau, dike near	905	1.1
FS 13	Central erosion channel	518	0.9
FS 14	Plain, dike near	748	0.7

7.3.4 Results and Discussion

The highest mass inputs at the study site Schönberg occurred with the highest discharge (Friese et al. 2000). Differences in mass input at sites of similar elevation are caused by the different roughness of landscape surfaces and by the larger distance between river and the sampling sites. The transport capacities of river water decrease depending on the sinking water discharges.

The input of dry matter varies over a wide range – from 38 000 kg ha^{-1} at the riverbank (FS 1) to 47 kg ha^{-1} (FS 13) in areas with high velocity flows over a large stretch of the river. The average dry matter input (median) was about 2 000 kg ha^{-1} (Table 7.2). According to Friese et al. (2000), the most important factor concerning the dry matter input is the distance between site and river.

Between the spring flood and summer flood in 1997 the amount of mass deposited within the floodplain area decreased dramatically as a result of strong erosion of sediments in the upstream catchment area during the spring flood. This reduced the amount of material available for erosion and re-suspension during the following summer flood. Additionally, the maximum discharge of the summer flood (1 321 m^3 s^{-1}) was lower as for the spring flood (1 850 m^3 s^{-1}) and consequently for some locations high above the MWL no flooding and material deposition occurred (FS 9, FS 13). Winter flood 1998/1999 showed the highest discharge with maximum 2 356 m^3 s^{-1} and associated high deposition of flood sediment (Table 7.2).

In general, the element concentration in fresh deposits from the first high flood period (spring 1997) are much higher than in the following flooding period of the same year (summer 1997). This can be expected because the time span between the erosion of contaminated material from the upstream sources (mainly the upstream groyne fields, see Baborowski et al. 2004, 2005 and this volume, Sect. 3.3) during the

Table 7.2. Arsenic and trace metal mass concentrations in the fresh deposited flood sediment of the study site Schönberg (values for As and trace metals in mg kg^{-1} dw)

Flood event	Location	Mass of material deposited ($g\ m^{-2}$)	As	Cd	Ni	Cr	Cu	Pb	Zn
Spring 1997	FS 1	3810	26.4	4.1	43.3	74.0	77.1	75.7	699
Summer 1997		413	20.7	5.8	52.3	106	116	116	1141
Winter 1998		714	23.8	4.6	37.8	90.1	99.1	98.9	819
Spring 1997	FS 2	230	36.1	7.2	64.9	126	112	137	1202
Summer 1997		21	10.6	3.8	30.4	69.3	74.6	86.0	1193
Winter 1998		904	26.4	7.4	59.5	128	137	132	1025
Spring 1997	FS 3	798	107	6.7	55.8	109	130	124	663
Summer 1997		65	108	7.1	39.1	78.8	104	114	1259
Spring 1997	FS 4	204	55.9	6.0	46.0	54.1	69.1	61.7	793
Summer 1997		18	16.8	6.6	34.0	46.1	69.8	92.1	1358
Winter 1998		242	29.1	8.4	52.2	130	143	139	1064
Spring 1997	FS 5	113	103	10.5	42.6	35.5	86.0	62.5	1733
Summer 1997		13	25.3	13.2	33.0	47.3	182	102	2910
Winter 1998		192	29.9	8.0	52.5	118	141	141	985
Spring 1997	FS 6	973	41.7	7.5	68.1	118	136	128	1161
Summer 1997		76	68.1	18.6	138.8	78.2	95.4	93.1	2178
Winter 1998		967	24.8	7.5	45.4	115	141	124	1083
Spring 1997	FS 7	1607	43.5	7.6	75.6	134	146	136	1139
Summer 1997		29	14.1	4.6	38.9	52.9	60.3	102	1276
Winter 1998		1035	27.5	7.4	61.7	121	140	135	1015
Spring 1997	FS 8	192	39.6	5.9	61.5	111	118	111	1079
Summer 1997		12	21.0	2.6	40.1	47.6	54.9	93.9	518
Winter 1998		205	26.8	6.0	41.6	110	123	131	673
Spring 1997	FS 9	152	43.7	6.4	65.8	121	128	188	1240
Winter 1998		243	18.0	4.2	43.0	82.5	86.9	96.8	586
Spring 1997	FS 13	5	33.1	6.1	46.5	75.0	112	420	4380
Winter 1998		183	9.1	2.7	22.3	55.0	54.0	74.2	351
Spring 1997	FS 14	298	31.0	4.2	50.4	104	86.3	126	651
Summer 1997		48	47.7	4.3	33.8	61.8	88.7	94.6	2534
Winter 1998		397	20.1	5.1	50.6	129	138	155	802

spring flood in 1997 (March) and the occurrence of the summer flood 1997 (July) was too short to accumulate new contaminated material in the groyne fields. But for zinc the opposite behavior is obvious at specific locations in the floodplain (FS 1, FS 4, FS 5, FS 6, and FS 14). Several explanations for this observation exist. First, the origin of the spring flood and of the summer flood 1997 were slightly different although all three flood events considered in this paper were fed mainly from the headwaters of the river Elbe (Anonymous 1997, 1998). The spring flood 1997 was mainly the result of heavy precipitation throughout the whole upper catchment of the river Elbe associated with snow melt whereas the summer flood 1997 was related to the heavy rainfall events in the triangle region of Czechoslovakia, Poland, and Germany which led to the dramatic high flood of the river Odra. Consequently, the composition of pollutants might be somewhat different. It might be also an effect of erosion and transport of zinc-contaminated material from upstream sources which were exposed to the surface during the erosion effects of the spring flood. Higher mass concentrations in the flood sediment of the summer flood 1997 compared to the spring flood 1997 were also observed at location site FS 1 for the elements Cd, Cr, Cu, Mn, Ni, and Pb. Probably, re-suspension of higher contaminated material from the spring flood within the floodplain site and re-deposition in position FS 1 might be an explanation. Unusual high mass concentrations for Pb and Zn were observed in flood sediment from spring flood 1997 at location FS 13. This location, nearly 1 m above MWL is the central erosion channel and only low amount of material was deposited during spring flood 1997 (approx. 5 g m^{-2}). Since the amount of material on the sediment trap was very low the risk of contamination during sampling and preparation of the sample is quite high. On the other hand the extremely high mass concentrations of Pb and Zn in this sample could also be a result of a sorting effect if only very fine and strongly contaminated material was deposited. The winter flood 1998/1999, approximately 15 months after the summer flood 1997 eroded and transported again high contaminated material from upstream sources into the floodplain site Schönberg. Since the discharge of the winter flood was considerably higher than that of the spring flood 1997 it is not clear if only new deposited contaminated material was eroded from the upstream sources (e.g., upstream groyne fields) or if additionally, also deeper contaminated material was eroded from that groyne fields by the higher shear forces.

The high mass concentrations of cadmium, zinc, lead and copper in fresh deposited flood sediments (Table 7.2) showed that the enrichment of heavy metals in recent Elbe River floodplains will continue, despite continued improvement in the purification of sewage and industrial effluents as long as historical contaminated sites still exist as pollutant sources.

The topsoil of the sample locations at the study site Schönberg show with one exception high amounts of organic carbon (5.2–10% dw). Arsenic and the trace metals are highly enriched compared to regional background values estimated by Krüger et al. (1998, 1999) (Table 7.3). Enrichment factors are 2.3 to 6 for As, 19.4 to 64 for Cd, 1 to 3 for Ni, 0.9 to 3 for Cr, 4.1 to 12 for Cu, 5.8 to 14 for Pb, and 3.9 to 15 for Zn excluding site FS 6 which exhibit unusual low values for all elements measured.

In general, the arsenic and metal mass concentrations of the topsoil are in same order as the mass concentrations of the fresh deposited flood sediments at the same location. At location sites 1, 2, 4, and 14 (here with the exception of Zn) the mass con-

Table 7.3. Trace metal and arsenic mass concentrations in the alluvial topsoil (0–10 cm) of the study site Schönberg (fraction < 2 mm) (all values are given in mg kg^{-1} dw)

Location	C_{org} (% dw)	As	Cd	Ni	Cr	Cu	Pb	Zn
FS 1	7.7	42.8	7.4	87.0	149	154	127	1222
FS 2	7.0	66.4	12.9	109	273	298	219	1503
FS 3	6.6	69.5	4.9	68.9	184	205	213	526
FS 4	7.8	111	10.4	78.5	202	261	269	1113
FS 5	6.5	91.8	9.4	74.7	197	238	241	914
FS 6	0.5	4.3	0.4	19.4	11.6	16.2	13.8	62.6
FS 7	5.2	78.0	7.0	69.2	140	164	222	917
FS 8	7.4	57.2	4.9	52.5	141	172	220	453
FS 9	5.2	48.7	4.2	51.2	91.2	99.9	132	452
FS 13	6.2	49.5	3.9	42.9	87.7	97.8	133	401
FS 14	10.0	82.5	9.7	79.4	214	237	309	983
Geogenic[a]		19	0.2	41	94	24	22	103

[a] Regional geogenic background values (fraction < 20 μm) after Krüger et al. (1999).

centrations of the elements are considerably higher in the topsoil whereas at location sites 5, 7, 8, 9, and 13 the elements exhibit similar mass concentrations, some elements are slightly enriched in the topsoil and others in the flood sediments.

There are two remarkable observations which could not be explained yet. At location FS 2 the topsoil showed the highest mass concentration for the trace metals Cd, Ni, Cr, Cu, and Zn although the deposited flood sediments during all three floods exhibit medium ranged mass concentrations for these elements. The highest mass concentrations for As and Pb in the topsoil were detected at location FS 4 where again the flood sediment did not show the highest mass concentrations for these both elements. On the other hand the topsoil of location FS 6 showed the lowest values for all elements presented, far below the regional geogenic background. At this location Cd and Ni exhibit the highest mass concentrations observed in flood sediment from the summer flood 1997.

7.3.5 Conclusions

The results clearly demonstrate that the floodplains of the lowland areas of the river Elbe with their alluvial soils are still an important sink for manifold pollutants. The results are evident that not only the far transport of pollutants is important for the level of contamination but also local conditions like the micro relief (e.g., elevation in relation to MWL; distance to the stream) have a great influence. Anthropogenic pollutant inputs are adsorbed on fresh high flood sediment, which contain a high level of organic carbon and a large proportion of fines. Whereas the subsoil generally have uncritical amounts of heavy metals (Friese et al. 2000), critical levels may be reached in the topsoil.

Acknowledgments

The investigation of flood sediments and soils at the study site Schönberg were partly funded by BMBF project "Entry and deposition of pollutants by high floods into floodplains and soils used for agriculture of the rivers Oka and Elbe" (no. 02 WT 9617/0). We are grateful to Prof. Dr. R. Meißner and Dr. H. Rupp for their substantial cooperation.

References

Anonymous (1997) Deutsches Gewässerkundliches Jahrbuch – Elbegebiet, Teil I. Landesamt für Umweltschutz Sachsen-Anhalt (Hrsg), Halle (Saale)

Anonymous (1998) Deutsches Gewässerkundliches Jahrbuch – Elbegebiet, Teil I. Landesamt für Umweltschutz Sachsen-Anhalt (Hrsg), Halle (Saale)

Baborowski M, Claus E, Friese K, Pelzer J, von der Kammer F, Kasimir P, Heininger P (2005) Comparison of different monitoring programs of the 2002 summer flood in the river Elbe. Acta Hydrochim Hydrobiol 22:404–417

Baborowski M, von Tümpling W, Friese K (2004) Behaviour of suspended particulate matter (SPM) and selected trace metals during the 2002 summer flood in the river Elbe (Germany). Hydrol Earth Syst Sc 8:135–150

Büttner O, Otte-Witte K, Krüger F, Meon G, Rode M (2006) Numerical modelling of floodplain hydraulics and suspended sediment transport and deposition at the event scale in the middle river Elbe, Germany. Acta hydrochim hydrobiol 34: in press

Calmano W, von der Kammer F, Schwartz R (2005) Characterization of redox conditions in soils and sediments: heavy metals. In: Lens P, Grotenhuis T, Malina G, Tabak H (eds) Soil and Sediment Remediation, IWA Publishing, London, pp 102–120

Förstner U, Heise S, Schwartz R, Westrich B, Ahlf W (2004) Historical Contaminated Sediments and Soils at the River Basin Scale – Examples from the Elbe River Catchment Area. Journal of Soils and Sediments 4:247–260

Friese K, Witter B, Brack W, Büttner O, Krüger F, Kunert M, Rupp H, Miehlich G, Gröngröft A, Schwartz R, van der Veen A, Zachmann DR (2000) Distribution and fate of organic and inorganic contaminants in a river floodplain – results of a case study on the river Elbe, Germany. In: Wise DL, Trantolo D, Cichon EJ, Inyang HI, Stottmeister U (eds) Remediation Engineering of Contaminated Soils. Marcel Dekker, New York, Basel, pp 375–428

Gröngröft A, Schwartz R, Miehlich G (2000) Wirkung eines Winterhochwassers auf Grundwasserstand, Luftgehalt und Redoxspannung eines eingedeichten Auenbodens. In: Bundesamt für Naturschutz (ed) Renaturierung von Bächen, Flüssen und Strömen. Angewandte Landschaftsökologie 37:277–282

Gröngröft A, Krüger F, Grunewald K, Meißner R, Miehlich G (2005) Plant and soil contamination with trace metals in the Elbe floodplains: A case study after the flood in August 2002. Acta Hydrochim Hydrobiol 33:466–474

Krüger F, Büttner O, Friese K, Meißner R, Rupp H, Schwartz R (1997) Lokalisation der Schwermetallbelastung durch Simulation des Überflutungsregimes. DBG-Mitteilungen 85:949–952

Krüger F, Prange A, Janzten E, Trejtnar K, Miehlich G (1998) Geogene Hintergrundwerte. Wasserwirt Wassertechnik 7:16–19

Krüger F, Prange A, Jantzen E (1999) Ermittlung geogener Hintergrundwerte an der Mittelelbe und deren Anwendung in der Beurteilung von Unterwassersedimenten. In: Gröngröft A, Schwartz R (Hrsg) Eigenschaften und Funktionen von Auenböden an der Elbe. Hamburger Bodenkundliche Arbeiten 44:39–51

Krüger F, Schwartz R, Stachel B (2003) Quecksilbergehalte in Sedimenten und Aueböden der Elbe und deren Beurteilung unter besonderer Berücksichtigung des Sommerhochwassers 2002. Vom Wasser 101:213–218

Krüger F, Meißner R, Gröngröft A, Grunewald K (2005) Flood induced heavy metal and arsenic contamination of Elbe River floodplain soils. Acta hydrochim hydrobiol 33:455–465

Meißner R, Guhr H, Rupp H, Seeger J, Spott D (1994) Schwermetallbelastung von Böden und Elbsedimenten in ausgewählten Gebieten Ostdeutschlands. Z f Kulturtechnik und Landentwicklung 35:1–9
Miehlich G (1983) Schwermetallanreicherung in Böden und Pflanzen der Pevestorfer Elbaue. Abh naturwiss Verein Hamburg 25:75–89
Schwartz R, Duwe J, Gröngröft A (1997) Einsatz von Kunstrasenmatten als Sedimentfallen zur Bestimmung des partikulären Stoffeintrags in Auen und Marschen. DBG-Mitteilungen 85:353–356
Schwartz R, Nebelsiek A, Gröngröft A (1999) Das Nähr- und Schadstoffdargebot der Elbe im Wasserkörper sowie in den frischen schwebstoffbürtigen Sedimenten am Messort Schnackenburg in den Jahren 1984–1997. Hamburger Bodenkundliche Arbeiten 44:65–83
Schwartz R, Gröngröft A, Miehlich G (2000) Charakterisierung typischer Böden im Überschwemmungsbereich der unteren Mittelelbe und Ergebnisse zu deren Wasserhaushalt. In: Friese K, Witter B, Miehlich G, Rode M (Hrsg) Stoffhaushalt von Auenökosystemen, Springer-Verlag, Berlin, pp 65–78
Wilken RD, Simon M, Guhr H (1994) Die Elbe: zur früheren, heutigen und zukünftigen Belastungssituation. In: Wagner R (Hrsg) Wasserkalender 1995 – Jahrbuch für das gesamte Wasserfach, Erich Schmidt Verlag, Berlin, pp 3–96
Witter B, Francke W, Franke S, Knauth H-D, Miehlich G (1998) Distribution and mobility of organic micropollutants in river Elbe floodplains. Chemosphere 37:63–78
Witter B, Winkler M, Friese K (2003) Depth distribution of chlorinated and polycyclic aromatic hydrocarbons in floodplain soils of the river Elbe. Acta Hydrochim Hydrobiol 31:411–422

Fridbert Ackermann · Birgit Schubert

7.4 Trace Metals as Indicators for the Dynamics of (Suspended) Particulate Matter in the Tidal Reach of the River Elbe

7.4.1 Introduction

Over the last decades, numerous studies have been dealing with the characteristics and dynamics of estuarine sediments and suspended particulate matter (SPM) (e.g., Dyer 1979, 1986; Dyer et al. 2001; Huntley et al. 2001; Jay et al. 1997; Kappenberg et al. 1996; Meade 1969, 1972; Perillo 1995; Postma 1955, 1967; Schubel 1984). Among others, these studies discussed the development of the turbidity zone and of the transport of particulate matter in upstream direction. The details of the mechanisms of transport, mixing, deposition, and erosion of fine particulate matter and the contaminants adsorbed thereto in river estuaries are very complex and not well known yet.

Computer-aided simulations have not been able to describe the dynamics of fine-grained, cohesive particulate matter entirely so far. In qualitative terms, the available transport models can describe the formation of the turbidity zone (e.g., Grabemann et al. 1995; Kappenberg et al. 2001; Rolinski 1999). However, details of the dynamics of fine-grained particulate matter, such as its upstream transport (tidal pumping) or its retention time in the estuary cannot be described.

In surface waters, trace metals and certain organic contaminants are preferentially bound to particles. Predominantly, they are associated with the fine-grained <20 µm fraction of suspended matter or sediments (Ackermann et al. 1983; OSPAR 2002). The amounts dissolved in the aquatic phase are of minor importance. This also applies to the transition zone from freshwater to salt water. Intensification of chemical desorption of trace metals with increasing salinity is not of appreciable relevance compared to particle-associated trace-metal concentrations (Müller et al. 1975; Salomons et al. 1981; Zwolsman 1999).

Over the past decades, concentrations of some trace metals (cadmium (Cd), copper (Cu), mercury (Hg), and zinc (Zn)) in SPM and sediments in the Elbe Estuary have decreased in the longitudinal profile from an elevated fluvial level at Geesthacht (tidal limit) to values that are up to one order of magnitude lower near the river mouth at Cuxhaven, where the sediments are predominantly of marine origin (e.g., Ackermann 1998; ARGE Elbe 1980; Banat et al. 1972; Förstner et al. 1990; Knauth et al. 1993; Prange et al. 1997). At any given site in the estuary, these concentrations result from the mixing of fluvial and marine SPM/sediments. Also results of the GKSS Research Centre (Knauth et al. 1993), the Delft Hydraulics Laboratory, and the Technical University of Hamburg-Harburg (Salomons et al. 1988; Förstner et al. 1990) clearly demonstrate a significant transport of highly contaminated solids from locations upstream of Hamburg into the lower Elbe Estuary at high river-discharge rates.

The composition (e.g., grain-size distribution, organic-carbon content) and the concentrations of suspended matter in estuaries show large short term variations within short horizontal and vertical distances, i.e. within a few minutes and meters. Due to, *inter alia,* this high heterogeneity of the water body, a reliable estimate of loads of particle-bound contaminants carried by rivers via estuaries to the sea cannot be achieved by direct measurements in the estuary. It is even doubtful whether a direct estimate of loads will succeed in near future (Ackermann 1998; Ackermann et al. 1998; Dyer et al. 2001; Jay et al. 1997; Huntley et al. 2001; Knauth et al. 1993).

Moreover, several studies using tracers investigated the dynamics of cohesive fine-grained particulate matter in surface waters (rivers, estuaries, coastal zones). These investigations applied both tracers intentionally added to the water body, such as radio-active and inactive isotopes or elements, fluorescent substances, as well as tracers naturally present in the water body, like diverse clay minerals, rare-earth compounds, natural radio-active substances or fall-out products, and the ratios of carbon and oxygen isotopes (literature reviews: Coakley et al. (1990), Olley et al. (2001)). Investigations in estuaries utilize the significant differences in concentrations of such tracers in mainly fluvially influenced particulate matter, i.e. near the tidal limit, and in predominantly marine-influenced particulate matter in the outer estuary. Using several natural tracers, Salomons et al. (1988) could provide evidence of upstream transport of fine-grained particulate matter of marine origin reaching far into the freshwater region of the river. The freshwater limit is located at Elbe-km 683. Also compounds from waste-water effluents, e.g., trace metals, were used as tracers in estuaries. Respective results of the estuaries of the rivers Elbe, Weser and Ems are presented by Ackermann (1998), and of the Humber Estuary by Dyer et al. (2001). Also pigments in sediments were used as marker compounds for the characterization of suspended matter and for the assessment of trace-metal concentrations (Wiltshire et al. 1996).

7.4.2 Measurements and Methods

The SEDYMO sub-project 18b investigates the dynamics of fine-grained cohesive solids, i.e. grain-size fraction <20 µm of suspended particulate matter (SPM) and recently deposited sediments, of both marine and fluvial origins in the Elbe Estuary. The preferentially particle-bound heavy metals Cd, Cu, Pb, and Zn are used as tracers in this

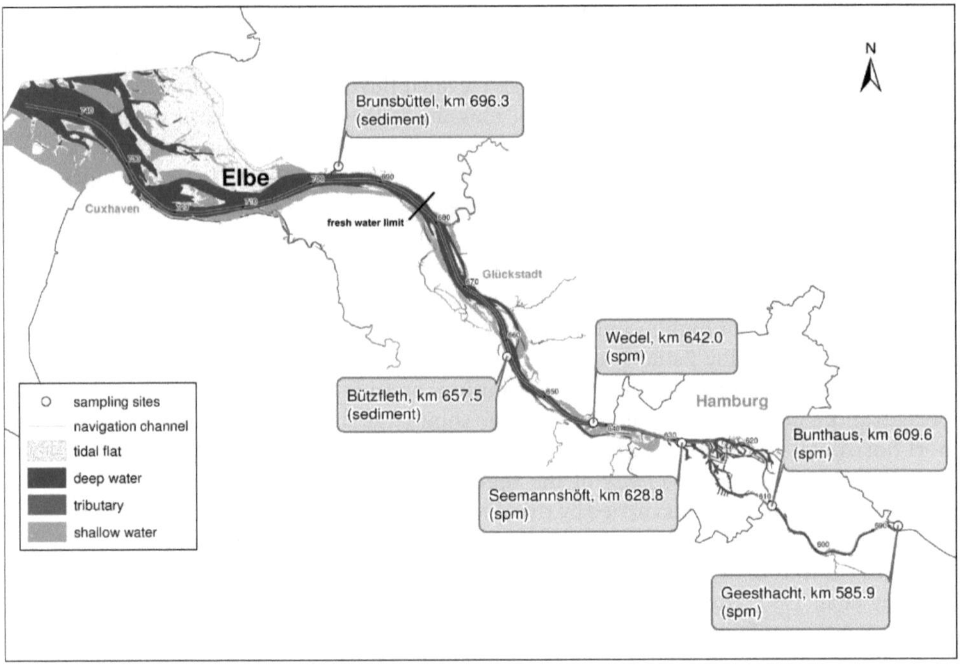

Fig. 7.16. Sampling sites. Samples were collected over one to four weeks

study. Trace-metal concentrations were analyzed after ultra-sonic sieving in the fine-grained fraction <20 µm, where the trace-metal load predominantly accumulates. The transport behavior of coarser fractions (sand) is not subject of this study.

The variations of trace-metal concentrations at the six measuring stations of the Working Committee for the Protection of the Elbe (ARGE Elbe, Arbeitsgemeinschaft für die Reinhaltung der Elbe) and the Federal Institute of Hydrology, Koblenz (BfG) (Fig. 7.16) in the Elbe Estuary are mainly controlled by river discharge, which is measured at Neu Darchau, km 536.4. From these variations, a better understanding of the complex dynamics of SPM may be gained.

7.4.3 Results and Discussion

For all indicator elements, the dependence of concentrations on river discharge is similar at all sampling sites seaward from Hamburg. Results are demonstrated for Cd and Zn (Fig. 7.17, 7.18). At times of high river discharge, high concentrations of trace metals are recorded, i.e. the portion of solids of marine origin is low and vise versa. These patterns were also observed at the stations Bützfleth (Elbe-km 657.5), and Brunsbüttel (Elbe-km 696.3) with the concentration level at Brunsbüttel being distinctly lower.

Knowing the trace-metal contents of fluvial SPM (Geesthacht, Elbe-km 586) and of marine SPM from the North Sea coast (e.g., Koopmann et al. 1993), the trace-metal

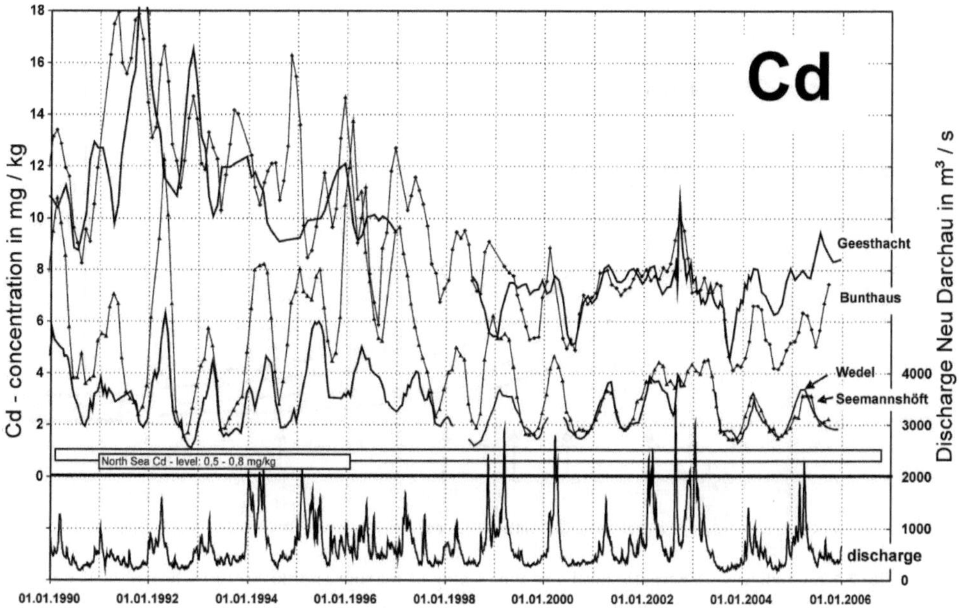

Fig. 7.17. Variations of Cd levels (moving averages over three measurements) in the fraction <20 µm in SPM with river discharge from Neu-Darchau (km 536.4)

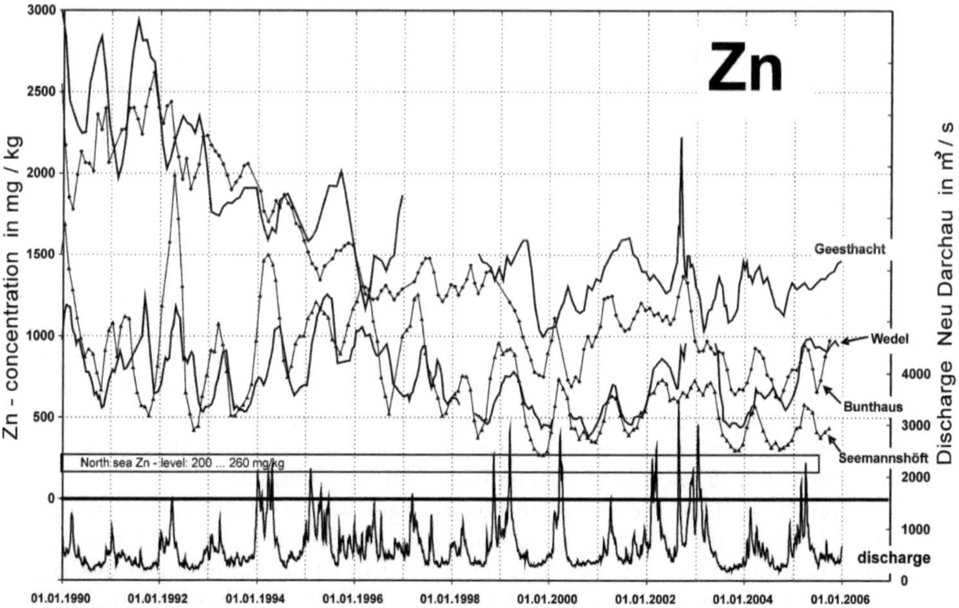

Fig. 7.18. Variations of Zn levels (moving averages over three measurements) in the fraction <20 µm in SPM with river data from Neu-Darchau (km 536.4)

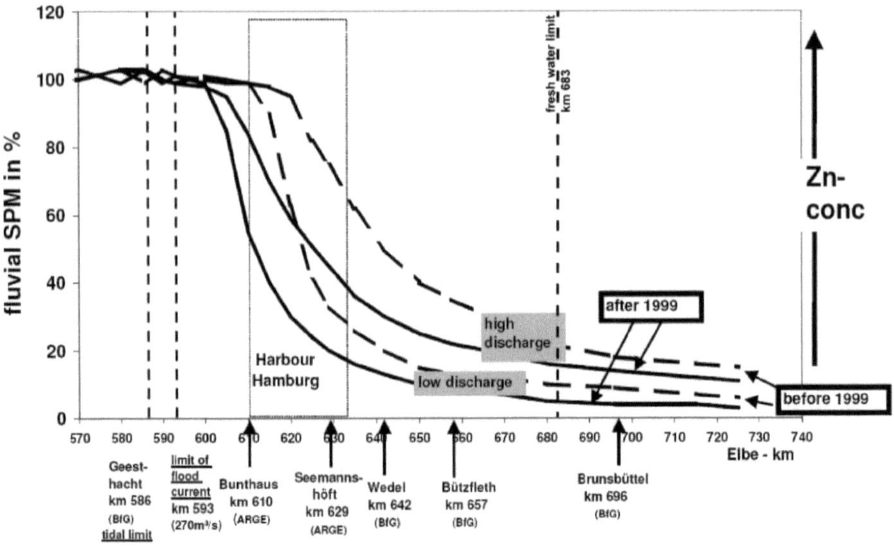

Fig. 7.19. Qualitative schematic model of the mixing of marine and fluvial fine-grained SPM and the adsorbed trace-metal concentrations in the tidal Elbe (example: mixing curves for Zinc)

concentrations at a given sampling site in the estuary at a given time allow to estimate the respective portions of fluvial and marine solids under different river-discharge conditions (cf. mixing curves for Zn in Fig. 7.19). The location of the area of the highest concentration gradient, and thus also of the strongest change in the mixing ratio of fluvial and marine SPM depends strongly on river discharge. This zone has shifted significantly upstream (10–15 km) after the last deepening of the Elbe navigation channel in 1999/2000.

Seemannshöft/Wedel

Until 1999/2000, the variations of the indicator elements Cd, Zn, Cu and Pb at Seemannshöft (Elbe-km 628.8) normally used to be higher than those observed downstream at Wedel (Elbe-km 642). The concentration minima at both stations, however, were mostly at similarly low levels, resulting from comparable portions of SPM of marine origin at both stations (70% to 90%) under low-flow conditions. However, the trace-metal maxima at Seemannshöft were always significantly above those at Wedel due to significantly lower marine portions of suspended solids at Seemannshöft of 20% to 40% compared with 40 to 60% at Wedel under high-flow conditions.

After 2000/2001, the situation has changed fundamentally, so that trace-metal levels at the station Seemannshöft are at the same level (Cd) as or even below (Zn, Cu, Pb) the respective values at Wedel, not only at low river discharges but also at medium to high flows (cf. Fig. 7.17 and 7.18). Obviously, the marine portions of SPM at Seemannshöft now are equal to or even higher than those at Wedel due to an upstream shift of the mixing curve of marine and fluvial solids (cf. Fig. 7.19).

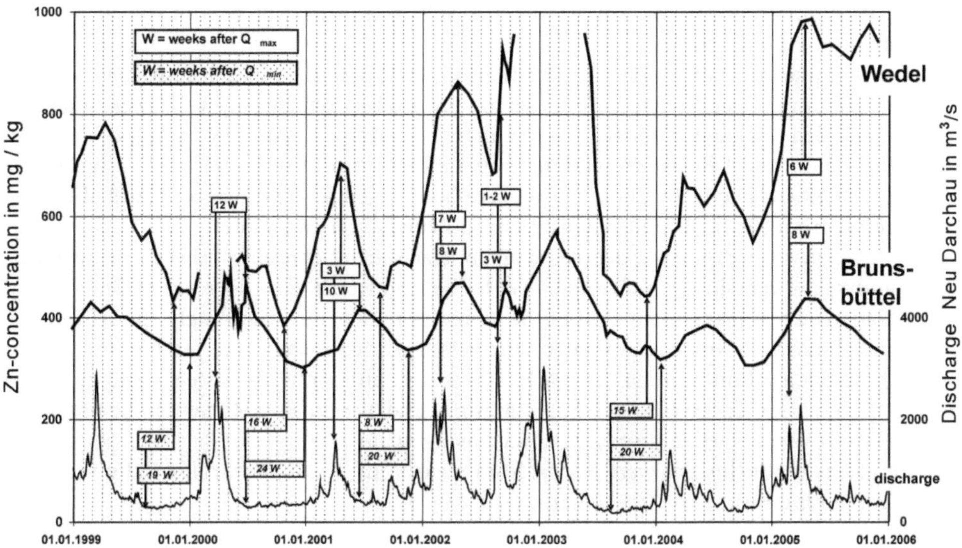

Fig. 7.20. Variations of Zn levels in the fraction <20 μm in SPM at Wedel (km 642.0) and Brunsbüttel (km 696.3) with river discharge from Neu-Darchau (km 536.4)

Bunthaus/Geesthacht

Prior to 1999, at the station Bunthaus (Elbe-km 609.6) upstream of Hamburg, no systematic river-discharge dependent variations in the concentrations had been observed. The mean concentrations of all indicator elements were comparable to those at the station Geesthacht and were thus in the fluvial range. After 1999/2000, the situation changed distinctly. Since 1999, the concentrations of Zn, Cu, and Pb and since autumn 2003 also the concentrations of Cd measured at Bunthaus have been below those of Geesthacht and have varied systematically with river discharge. These observations indicate that marine SPM can now even reach the station Bunthaus upstream of Hamburg. From trace-metal concentrations, a maximum contribution of marine SPM to the fine solids of up to 50% is estimated in 2004/2005. The limit of flood current during low river discharge is located around Elbe-km 593, i.e. 16 km upstream of Bunthaus (cf. Fig. 7.19).

The upstream shift of the mixing curve of marine and fluvial solids may be attributed to river engineering carried out in the Elbe Estuary, e.g., the deepening of the navigation channel in 1999.

Wedel/Brunsbüttel

In addition, the patterns of the Zn concentrations at the stations Wedel and Brunsbüttel (Fig. 7.20), allow to derive the temporal shift of the mixing curve outlined in Fig. 7.19 and therewith of the turbidity plume depending on varying river discharges. Obviously, with rising river discharge, the Zn levels at Wedel begin to increase without delay and those at Brunsbüttel with a short delay, i.e. the turbidity plume and the mixing curve of fluvial/marine solids, respectively, shift in the whole estuary downstream without significant time lag. This shift can result in transport of part of the suspended matter to the North Sea.

The maximum of the Zn concentration, however, is reached significantly later than the maximum of river discharge: at the station Brunsbüttel usually only 8–12 weeks after the maximum discharge, and still 3 weeks after the extreme flood event in August 2002. At the station Wedel, the corresponding temporal shifts are distinctly shorter: 3–7 weeks and 1–2 weeks respectively. These periods are interpreted as the transport times of fluvial solids from the tidal limit downstream to Brunsbüttel and Wedel.

When river discharge decreases again, trace-metal concentrations in Brunsbüttel and Wedel remain at an elevated level for several weeks. The decrease of Zn concentrations, i.e. the upstream shift of the turbidity plume and accordingly, the refilling of the plume with less-contaminated marine suspended matter take place more slowly: the concentration minima are reached at Brunsbüttel only 20–24 weeks and at Wedel 8–16 weeks after the river-discharge minima, and at Brunsbüttel they can be observed even in periods when discharge is increasing again.

Similar conclusions were drawn from measurements of the turbidity at given sites in the Elbe Estuary by Grabemann et al. (1995).

7.4.4 Summary and Outlook

Due to an increasing portion of marine fines in estuarine particulate matter, particle-bound trace-metal concentrations strongly decrease along the longitudinal profile of the Elbe Estuary from Geesthacht (tidal limit, km 586) to Cuxhaven (km 730). In addition, at a given site, concentrations vary strongly up to the factor five with river discharge. These variations are due to a shift of the turbidity plume in the Elbe Estuary depending on river discharge: at high river discharge, the turbidity zone and accordingly the gradient of the trace-metal concentration shift downstream, and vise versa. Therefore, the dynamics of the downstream or upstream shift of the turbidity zone can be derived from trace-metal concentration variations at a given site. In addition, the respective portions of fluvial and marine solids under different river discharges at a given site can be estimated from trace-metal concentrations at this site, provided the fluvial and marine trace-metal concentrations are known.

Computer-aided simulation of solids transport in the estuary were not able to reproduce the observed phenomena satisfactorily, so far. Measurements of trace metals as indicators of SPM dynamics in estuaries provide a valuable support to the development of computer simulations and the verification of their results.

Measurements of trace metals in particulate matter of the Elbe Estuary will be continued in order to keep track of the development of solids transport phenomena. The understanding of the dynamics of fine-grained, cohesive particulate matter in the Elbe Estuary, but also in other estuaries is of increasing interest. For instance, the medium- to long-term destination of fine-grained contaminated sediments in the Elbe Estuary is of importance for the assessment of the ecological impacts of the relocation of dredged material.

Moreover, the regular dredging of large amounts of fine-grained contaminated sediments required for maintaining the navigable depth of estuarine fairways and port basins is very cost-intensive. After the latest deepening of the navigation channel in the

Elbe Estuary in 1999, the amounts of material to be dredged annually in the harbor of Hamburg and in the river reach about 15–20 km downstream of Hamburg increased strongly from 2–3 million $m^3 a^{-1}$ to 7–11 million $m^3 a^{-1}$. Decisions for the sustainable management of sediments and of measures to reduce the amounts of material having to be dredged in the future, can be derived only if the reason for the increased need for dredging is understood. Results of trace-metal analyses indicate that the increased amounts of material to be dredged in Hamburg Harbor since 2000 are not due to the increased upstream transport of marine particulate matter, however, it is supposed that material dredged and relocated downstream of Hamburg drifts back into the harbor within short time. Additional transport of about 5–8 million $m^3 a^{-1}$ of marine sediments to the upper estuary would result in much lower trace-metal concentrations than actually observed.

Acknowledgment

The Waterways and Shipping Agencies Lauenburg, Hamburg, and Brunsbüttel kindly collected the samples over more than 20 years. The data from the stations Bunthaus and Seemannshöft were provided by Mr. Bergemann, *Working Committee for the Protection of the Elbe (ARGE Elbe, Hamburg)* and Mrs. Dr. Sievers of the *Institute of Hygiene and the Environment, Hamburg*. Essential funding was provided by the *Federal Ministry for the Environment, Nature Conservation and Nuclear Safety (BMU)* within the 'Monitoring Program of Water Quality in Transboundary Rivers and in Coastal Waters' and by the *Federal Ministry of Education and Research (BMBF) Coordinated Research Program SEDYMO.*

References

Ackermann F, Bergmann H, Schleichert U (1983) Monitoring of heavy metals in coastal and estuarine sediments – a question of grain-size: <20 µm versus <60 µm. Environ Technol Letts 4:317–328

Ackermann F (1998) Dynamik der Schwermetallbelastung in feinkörnigen Sedimenten und Schweb-stoffen im Tidebereich von Ems, Weser und Elbe. BfG-1188, Bundesanstalt für Gewässerkunde, Koblenz

Ackermann F, Schubert B (1998) Zur Problematik der Bestimmung von Frachten partikulär gebundener Schadstoffe in die Nordsee. Poster 8th Magdeburger Gewässerschutzseminar, 20–23.10.1998, Karlsbad/Czech Republic

ARGE Elbe (Arbeitsgemeinschaft für die Reinhaltung der Elbe) (1980) Schwermetalldaten der Elbe. Bericht über die Ergebnisse der Schwermetalluntersuchungen im Elbabschnitt von Schnackenburg bis zur Nordsee 1979/1980

Banat K, Förstner U, Müller G (1972) Schwermetalle in Sedimenten von Donau, Rhein, Ems, Weser und Elbe im Bereich der Bundesrepublik Deutschland. Naturwiss 59:525–528

Coakley JP, Long BFN (1990) Tracing the Movement of Fine-Grained Sediment in Aquatic Systems. A Literature Review. National Water Research Institute Canada, Centre for Inland Waters, Burlington, Ontario. Scientific Series no. 174

Dyer KR (ed) (1979) Estuarine hydrography and sedimentation. Cambridge University Press. Cambridge

Dyer KR (1986) Coastal and estuarine sediment dynamics. Wiley, Chichester

Dyer KR, Robinson M-C, Huntley DA (2001) Suspended sediment transport in the Humber Estuary. In: Land-ocean interaction – measuring and modelling fluxes from river basins to coastal seas. IWA Publishing, London, pp 169–183

Förstner U, Schoer J, Knauth H-D (1990) Metal pollution in the tidal Elbe River. In: Allen RJ, Campbell PGC, Förstner U, Lum K (eds) Fate and effects of toxic chemicals in large rivers and their estuaries. Sci Total Environ 97/98:347–368

Grabemann I, Kappenberg J, Krause G (1995) Aperiodic variations of the turbidity maxima oft two German coastal plain estuaries. Netherlands J Aquat Ecol 29:217–225

Huntley DA, Leeks GJL, Walling DE (eds) (2001) Land-ocean interaction – Measuring and modelling fluxes from the river basin do the coastal seas. IWA Publishing, London

Jay DA, Geyer WR, Uncles RJ, Vallino J, Largier J, Boynton WR (1997) A review of recent developments in estuarine scalar flux estimation. Estuaries 20:262–280

Kappenberg J, Schymura G, Kühn H, Fanger H-U (1996) Spring-neap variations of suspended sediment concentration and transport in the turbidity maximum of the Elbe Estuary. In: Kausch H, Michaelis W (eds) Suspended particulate matter in rivers and estuaries. Arch Hydrobiol Spec Issues Adv Limnol 47:323–332

Kappenberg J, Grabemann I (2001) Variability of the mixing zones and estuarine turbidity maxima in the Elbe and Weser Estuaries. Estuaries 24:699–706

Koopmann C, Faller J, v Bernem K-H, Prange A, Müller A (1993) Schadstoffkartierung in Sedimenten des deutschen Wattenmeeres, June 1989–June 1992, UBA-R&D-Project 10903377, GKSS 94/E/6

Knauth H-D, Gandraß J, Sturm R (1993) Vorkommen und Verhalten organischer und anorganischer Mikroverunreinigungen in der mittleren und unteren Elbe. German Federal Ministry of Environment, Nature Conservation and Reactor Safety, Research Report 10204363, Erich Schmidt Verlag Berlin

Meade RH (1969) Landward transport of bottom sediments in estuaries of the Atlantic coastal plain. J Sediment Petrol 39:222–234

Meade RH (1972) Transport and deposition of sediments in estuaries. In: Nelson BW (ed) Environmental framework of coastal plain estuaries. Geol Soc Amer, Memoir 133:91–120

Müller G, Förstner U (1975) Heavy metals in sediments of the Rhine and Elbe Estuaries: mobilization or mixing effect. Environ Geol 1:33–39

Olley JM, Caitcheon GG, Hancock G, Wallbrink PJ (2001) Tracing and Dating Techniques for Sediment and Associated Substances – A Consultancy Report for the Sydney Catchment Authority

OSPAR (2002) JAMP Guidelines for monitoring contaminants in sediments. Technical Annex 5 – Normalisation of contaminant concentrations; Ref no. 2002–16

Perillo GME (ed) (1995) Geomorphology and sedimentology of estuaries. Developments in Sedimentology 53, Elsevier, Amsterdam

Postma H (1955) Die Entstehung der Trübungszonen im Unterlauf der Flüsse, speziell im Hinblick auf die Verhältnisse in der Unterelbe. Dtsche Hydrograph Zeitschr 8:138–144

Postma H (1967) Sediment transport and sedimentation in the marine environment. In: Lauff GH (ed) Estuaries AAAS Publication 83, Washington DC, pp 158 – 179

Prange A (1997) Erfassung und Beurteilung der Belastung der Elbe mit Schadstoffen. Teilprojekt 2: Schwermetalle – Schwermetallspezies. Zusammenfassende Aus- und Bewertung der Längsprofiluntersuchungen in der Elbe. BMBF-Forschungsvorhaben: 02-WT 9355/4, GKSS-Forschungszentrum Geesthacht GmbH, Geesthacht

Rolinski S (1999) On the dynamics of suspended matter transport in the tidal river Elbe: Description and results of a Lagrangian model. J Geophys Res 104 C11, 26.043–26.057

Salomons W, Eysink WD (1981) Pathways of mud and particulate trace metals from rivers to the southern North Sea. In: Nio SD, Schüttenhelm RTE, van Weering TCE (eds) Holocene marine sedimentation in the North Sea basin. Blackwell, Oxford, pp 429–450

Salomons W, Schwedhelm E, Schoer J, Knauth JH (1988) Natural tracers to determine the origin of sediments and suspended matter from the Elbe Estuary. Wat Sci Tech 20:89–102

Schubel JR (1984) Estuarine circulation and sedimentation: an overview. In: Hag BU, Milliman JD (eds) Oceanography of Arabian Sea and Coastal Pakistan. Van Nostrand Reinhold Co.Inc., New York, pp 114–136

Wiltshire KH, Geissler C-D, Schroeder F, Knauth H-D (1996) Pigments in suspended matter from the Elbe Estuary and the German Bight: their use as marker compounds for the characterisation of suspended matter and in the interpretation of heavy metal loadings. In: Kausch H, Michaelis W (eds) Suspended particulate matter in rivers and estuaries. Arch Hydrobiol Spec Issues Adv Limnol 47:53–63

Zwolsman G (1999) Geochemistry of trace metals in the Scheldt Estuary. Geologica Ultraiectina, Mededelingen vande Faculteit Aardwetenschappen Universiteit Utrecht no. 171

Chapter 8

Fine Sediment Particles

Ole Larsen

The papers focus on the investigation of the behavior of trace metals in natural sediments and their association with fine sediment particles. The contributions demonstrate how the fine particles and potential contaminants move in the environments (rivers through estuaries to the marine environment) and how the sediment erosion can be studied using contaminants and REEs. The ability to predict the movements of fine particles and their associated trace metals is essential for sustainable sediment management.

One contribution describes the transport in sediments of the Tisza River in Hungary. In the riverine sediments the contaminant transport of mining spill was investigated over 5 years and the results illustrate the dynamic nature of the transport processes. In particular, the mixing of contaminated sediments with non-contaminated sediments due to the erosion and transport of sediments in the river makes it difficult to accurately trace the contamination. The investigated catchment was influenced by severe floods and the sediment transport occurring during the floods have most likely caused the mobility of the contaminants in the sediments. Estuarine sediments are subject to an intense sediment transport due to the tidal interaction. Here, fine sediment particles can be transported in both directions and the particle transport is equivalently difficult to trace. Different methods for investigation of the particle transport were compared in their efficiency to trace the particle transport in the estuarine and marine environment. Hereby a method was developed based on REE labeled sediments to investigate sediment transport. The REE are particularly efficient due to the deviation from the natural background in REE composition. In the marine environment an assessment of the particle transport can be made using various methods. The main difficulty with marine sediments is their low accessibility and often models of sediment transport are needed. The erodability of the surface layers is significantly different from the deeper cohesive sediments. The experiments presented in one of the contributions showed that the transport processes of the upper sediments can be assessed using ^{234}Th as tracer. Furthermore, the processes involved with the entrainment and transport of contaminants within marine sediments were investigated in mesocosm experiments. The remobilization of copper in marine sediments was shown to take place within the uppermost layers of the sediments. The derived fluxes of contaminants over the sediment-water interface are highly related to the sediment management.

Ole Larsen · William Davison · Kyriakos Vamvakopoulos · Flemming Møhlenberg

8.1 Transport and Reactions of Contaminants in Sediments

8.1.1 Introduction

Many of the sediments in our coastal environments are contaminated with various metals. The highest concentrations of contaminants are found in harbours, where antifouling paints and industrial activities are the main sources. In the past years there has been a strong focus on TBT that is known to be highly toxic and to affect the hormonal balance of many animals. Almost all substitutes for TBT are based on Cu-complexes. Copper is known to form strong complexes with natural organic matter and the total Cu-concentration in sediments is found to correlate with the concentration of organic matter (see Fig. 8.1).

Recent investigations have shown that Cu concentrations in the marine environment can be very high in estuaries with heavy boat traffic (Comber et al. 2002) and in many areas exceed the sediment quality guideline values (e.g., CCME 1999; Long et al. 1998), and a few studies have demonstrated that Cu may be one of the most important constraints on benthic fauna. Chapman (1990) suggested an integrated approach (Sediment Quality Triad) to evaluate effects of contaminated sediments consisting of three complementary components including chemical analysis of contaminants, sediment toxicity, and assessment of resident biota such as changes in benthic community structure. An example of the third SQT component is shown in Fig. 8.2. Using ca. 80 synoptic samples, linkages between forcing factors (sediment contaminants and water quality) and effects (benthic invertebrate species abundance) were analyzed using Partial Least Squares regressions (Møhlenberg et al. 2007). Initially, more than 60 potential predictors were included in PLS regressions, and significant predictors were identified by cross-validation or boot-strapping. Overall, the "natural" conditions such as salinity, station depth and sediment organic carbon were the most influential predictors of species richness and diversity. Depending on the location considered or the grouping of data, either nutrient concentrations or sediment contaminants, especially copper and cadmium, were the second most important factor affecting benthic communities, accounting for up to 25% of the variation in species richness (Fig. 8.2). Using field experiments Lenihan et al. (2003) demonstrated that copper mainly caused reductions among crustaceans and echinoderms, while organic enrichment of sediments promoted annelids, but had variable effects on crustaceans and echinoderms. Hence, the effects of copper enrichments in sediment seem to be manifested through elimination of "sensitive" species leading to reductions in species richness, as shown in Fig. 8.2.

Predictive geochemical models, both numerical and conceptual, of elemental transformations and particularly remobilization underpin the assessment of risk posed by heavy metals in many environmental situations. The increased use of metals (e.g., copper) may critically affect the risk to intensively utilized coastal and estuarine environments. This environmental problem raises a question at a more fundamental level: what are the controls on heavy metals in marine sediments? The environmental chemistry of heavy metals has been studied during the past decades, with considerable research interest concentrated on Cd, Pb, and Cu (e.g., Serbst et al. 2003; Burnton et al. 2005; Chapman et al. 1998). For routine monitoring most sediment samples are col-

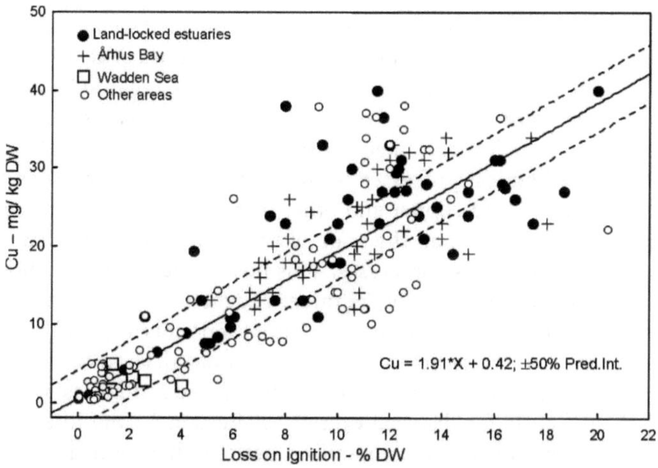

Fig. 8.1.
The bulk copper concentration displayed against the loss on ignition in a variety of estuarine sediments. Loss on ignition corresponds mainly to loss from organic matter although it also includes weight loss from carbonates and hydrated minerals (the data originates from Møhlenberg et al. 2007)

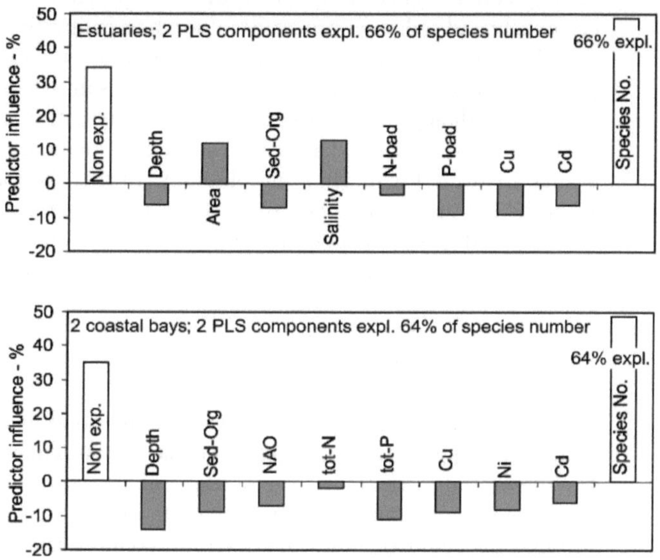

Fig. 8.2.
PLS analysis of 60 data sets from various European estuaries and 38 data sets from 2 coastal bays. Sixty-four and 66% of the variation of total species number could be explained using 2 PLS components. The heavy metals Cu and Cd (estuaries) and Cu, Ni and Cd accounted for 18 and 23% of the variation in species richness. In these coastal waters heavy metals are as important as nutrients for the faunal biodiversity (the data originates from Møhlenberg et al. 2007)

lected directly from dredgers and analyzed as mixed bulk samples. Typically, bulk parameters like content of Corg and total concentrations of heavy metals are measured and the results are taken as representative for a large sediment volume. However, recent studies (Glud et al. 1996; Shuttleworth et al. 1999; Fenchel and Glud 1998) are showing that the geochemical structure of sediments is much more complicated than previously thought. Existing models are based on a 1-dimensional view of sediments with zones of microbial/chemical activity systematically layered. Biological activity is largely seen as a physical process that pertubates this state. This picture has largely developed from measurements based on horizontally slicing sediments and performing measurements on porewaters and the solid phase of the resulting volumetrically averaged sample. Measurements at one thousandth of this volumetric scale (DGT/DET/

optodes/electrodes) have revealed a detailed solute structure, which suggests that sediment processes occur in microniches (Fones et al. 1998). This new perception of sediments suggests that sediments cannot be regarded as horizontally uniform, and consequently the use of one-dimensional models is called into question. Moreover, as microniches are likely to be short lived (days/weeks), the dynamic nature of the geochemical processes must be considered. At present, the transport, reactions and dynamics of copper and other heavy metals in marine sediments are known at scales much larger than the scale of the bacteria and associated microniches (1–50 μm) involved in these processes. This poses a new question: how does analytical scale affect the interpretation of sediment geochemistry in terms of processes and reactions?

The dredged sediment in Europe amounts to more than 200 000 000 $m^3 yr^{-1}$. Although most of the sediment is not contaminated, large amounts contain high concentrations of heavy metals. The most contaminated sediments are treated as chemical waste and the costs for disposal exceed 100 Euro m^{-3}. Due to the high costs of sediment treatment and strong competition between harbours a unified European legislation concerning contaminated sediments is desirable. Such legislation should be based on the ecotoxicity and sediment processes rather than politically defined maximum concentrations in sediments. In response to the lack of understanding of fundamental processes a project under the FP5 of the European Union was initiated to investigate the importance of the individual processes and their spatial distribution within sediments. The project partners were: TU-Delft, Lancaster University, University of Copenhagen and the Max Planck Institute for marine microbiology. In this section we report selected results from the EU project TREAD (EVK3-CT-2002-00081), with the focus on copper.

The investigation includes the study of heavily contaminated harbor sediment whose locality may not be published.

8.1.2 Experimental Approach

Based on the conceptual model of processes in sediments occurring at microniches an analytical program was launched to enable the assessment of local fluxes and concentrations in sediments at high temporal and spatial resolution. Using high resolution techniques we attempted to investigate the fluxes of solutes (within sediments and through the sediment-water interface) in contaminated marine sediments upon disposal in the environment. Two sediments were selected: a fine-grained (silt) harbor sediment with low benthic activity and a permeable sand with high benthic activity. The harbor sediment was heavily contaminated with heavy metals while the sand from Sylt (Germany) represents pristine marine sand. The investigation included, besides a careful field site description, the sampling of several large undisturbed sediment blocks (0.18 m^2, 10 cm depth) that were brought to the laboratory facility in Bremen. The sediments were installed in four mesocosms to mimic disposal of dredged material in the environment: *(1)* dumping at a site with high current velocities (silt was added to the sandy sediment, mimicking resuspension during a dredging event or disposal in a high-current environment – mesocosm A), *(2)* dumping the sediment after reworking the sediment at a low energy environment (the silt was homogenized and slightly oxidized before settling in mesocosm D) and their respective control systems (undisturbed sand –

mesocosm B and undisturbed harbor sediments – mesocosm C). The environmental controls of the 4 mesocosms were varied simultaneously throughout the two years the experiment lasted. Light, temperature and water flow were kept constant while, at different times, fresh supplies of benthic fauna and organic material were added and the salinity adjusted.

Novel techniques were used to detect and quantify key processes at the scale of their occurrence in sediments, including different forms of microsensors (electrodes and optodes), gel-sampling techniques and microscopy. More traditional measuring approaches (bulk flux measurements and extracted pore water profiles), integrating the activity of larger sediment volumes, provided comparisons with the high-resolution measurements.

8.1.3 Heavy Metals at the Field Site

Heavy metal content is routinely measured in harbor sediments by the authorities and is discussed here with kind permission. Literature data on the heavy metal concentration in North Sea sediments and data from the contaminated site are available for the fine fraction (<20 μm) of the sediment. Measurements of Hg, Cd, Pb, Cu and Zn, as well as organic tin from North Sea and Weser and Elbe sediments (BLMP 2002) and the contaminated and investigated site are shown in Table 8.1. The heavy metal concentrations in the harbor sediments are similar to the maximum concentrations reported in the literature while the concentrations of copper are elevated 10–20 fold compared to the North Sea and river sediments. The highest literature value was found in a harbor area in northern Bremen, but the concentrations here were still more than five times lower than the lowest measurements from our study sediments. The butyl-tin concen-

Table 8.1. Sediment data of 10 different stations throughout the North Sea, Weser and Elbe (BLMP 2002) and the sampling location measured by the harbor authorities. Data for the inorganic metal concentrations were provided as minimum, average and maximum data for the 10 locations and are averaged accordingly. The value in brackets indicates the next smaller value. For the harbor site, data from 4 different sites around the sampling location were provided, the measurements from the closest site are presented here averaged over 5 years of monitoring. Concentrations of the inorganic elements are given in mg kg^{-1} dry weight; the concentrations of the butyl-tin species are given in μg kg^{-1} dry weight. The measurements were made on the fine fraction of the sediment <20 μm

	Average	Average maximum	Maximum	Harbor site
Hg	0.38	0.52	1.5	0.9 ± 0.49
Cd	0.66	1.4	4.4	1.4 ± 0.29
Pb	77	111	234	128 ± 10
Cu	25	40	81	558 ± 115
Zn	227	375	903	742 ± 114
Monobutyltin	0.39	–	3.5	0.76 ± 0.28
Dibutyltin	0.45	–	5.5	4 ± 1.5
Tributyltin	0.34	–	1.4	78.8 ± 44.5
Tetrabutyltin	1.24	–	39.4 (6.0)	0.7 ± 0.094

trations measured in the Elbe between 1991 and 1997 also show distinct differences to those at our field site. While the concentrations of the different compounds are very similar in the Elbe, the harbor site shows large differences in the concentrations of the different compounds. The concentrations of butyl-tin vary at the different stations within the same harbor basin and the same year by a factor of up to 40. However, for all measurements the concentration of tributyl-tin was always highest, followed by di-, mono- and finally tetrabutyl-tin. As copper and tin are used in anti fouling paints for ships, it is not surprising to find high concentrations in the harbor basins, especially those close to a shipyard like our site. Similarly the site close to Bremen Harbor and one Elbe site, possibly close to Hamburg Harbor, showed the highest concentrations for Cu. These concentrations are still lower than at our site, but the samples originate from outside the harbor.

A comparison of the metal concentrations measured over a period of more than 20 years (6 years for the butyl-Sn) show that the concentrations decrease (BLMP 2002). This trend is less pronounced for Cd and Pb. Such a trend could not be seen in the measurements from the harbor basin, but the available data only cover 5 years. Any trend would be disguised in the high scatter of the small data set.

Heavy metal data are not available in the literature for the Sylt location or similar sites. This site is, however, very likely typical of the low end of the concentration spectrum.

Comparison of the data obtained within our project with literature data is a little difficult, as the literature data were measured in the fine fraction of the sediment. Our measurements (Fig. 8.2) were carried out in the total sediment, comprising mainly fine sand for the Sylt site, and ~70% fine material for the contaminated site.

8.1.4 Results

Bulk Concentrations of Copper in Sediments

The copper concentrations in all sediments were measured, after extraction with 6M HCl and many fold dilution, by ICP-MS. The concentrations were measured in five different sediment cores from the two control mesocosms (undisturbed sand from Sylt (B) and harbor sediment (C)). The average concentration (line) and the results from each individual sample are displayed in Fig. 8.3.

In both sediments the concentrations are significantly higher in the uppermost sediment layers despite bioturbation mixing the sediments. A further pronounced feature is that the horizontal variability is as large as the vertical variability in both sediments. The total copper concentration in the harbor sediment is about 80 times higher than the concentrations found in the sand.

Bulk Concentrations of Copper in Porewater

The concentrations of dissolved heavy metals in the porewaters of all mesocosms were also measured. The results after about one year of the experiment are showed in Fig. 8.4 (corresponding to the same time as the sediment profiles presented in Fig. 8.3). The pore water from the harbor sediments (C and D) have higher concentrations of dissolved Cu than the pore waters from the sand (A and B) (3 to 9 times, depending on the

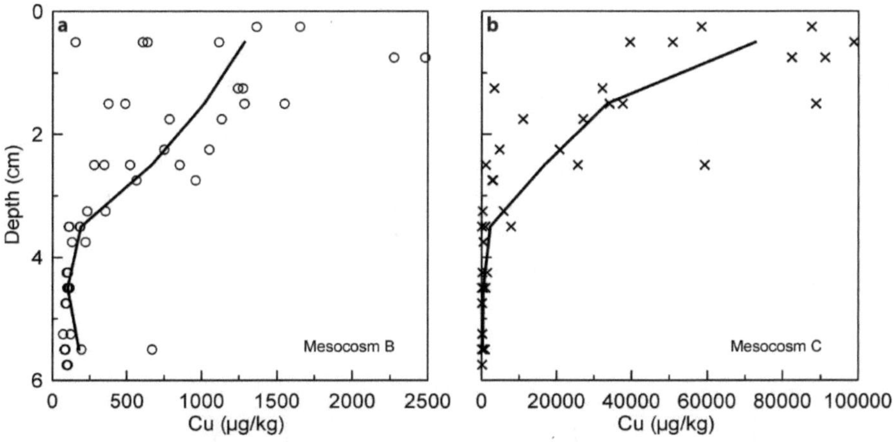

Fig. 8.3. Copper concentration measured in sediments from mesocosm B and C. Each core was sliced in 1 cm slices. The concentration of copper was measured in two samples from each depth and the mean is represented by the symbols in the plot. Five cores was analysed and the mean of all five cores is shown as a line. Mesocosm B (*circles*) is the undisturbed sand from Sylt; C (*crosses*) is the undisturbed harbor sediment. The cores were sampled a few cm apart

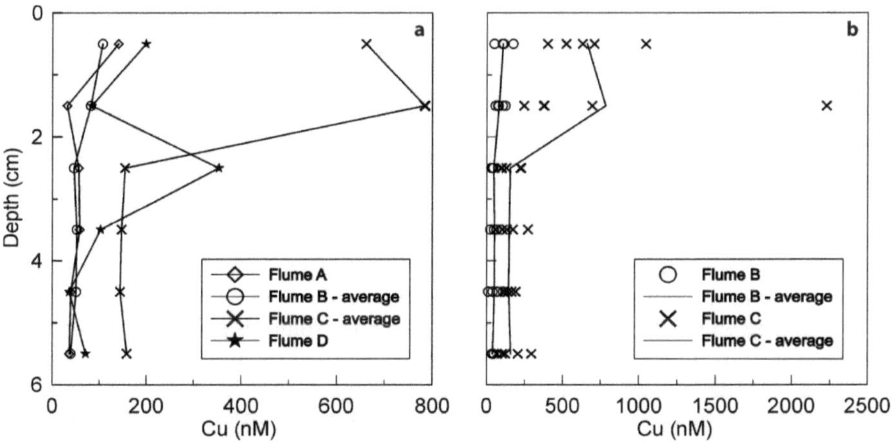

Fig. 8.4. a Copper concentration measured in porewater extracted by centrifugation of the sediment under anoxic conditions through a 0.2-µm filter. Each result is the average of duplicate determinations. Additionally, the results from mesocosm B and C are averaged values of results from 5 different cores. A represents the results from sand with a surface layer of harbor sludge; B is the undisturbed sand from Sylt; C is the undisturbed harbor sediment; D is the homogenized harbor sediment. **b** Results from 5 individual cores of mesocosm B (*circles*) and C (*crosses*) and their average values. The cores were sampled a few cm apart

sampling depth). In all sediments the Cu concentration in the porewater decreases with depth. The variations between the sampled cores are in the same range for the two ecosystems, indicating that the concentrations do not to any large extend depend on the transport path (advective vs. diffusive). The variation between 5 different cores sampled from mesocosms B and C is shown in Fig. 8.4b.

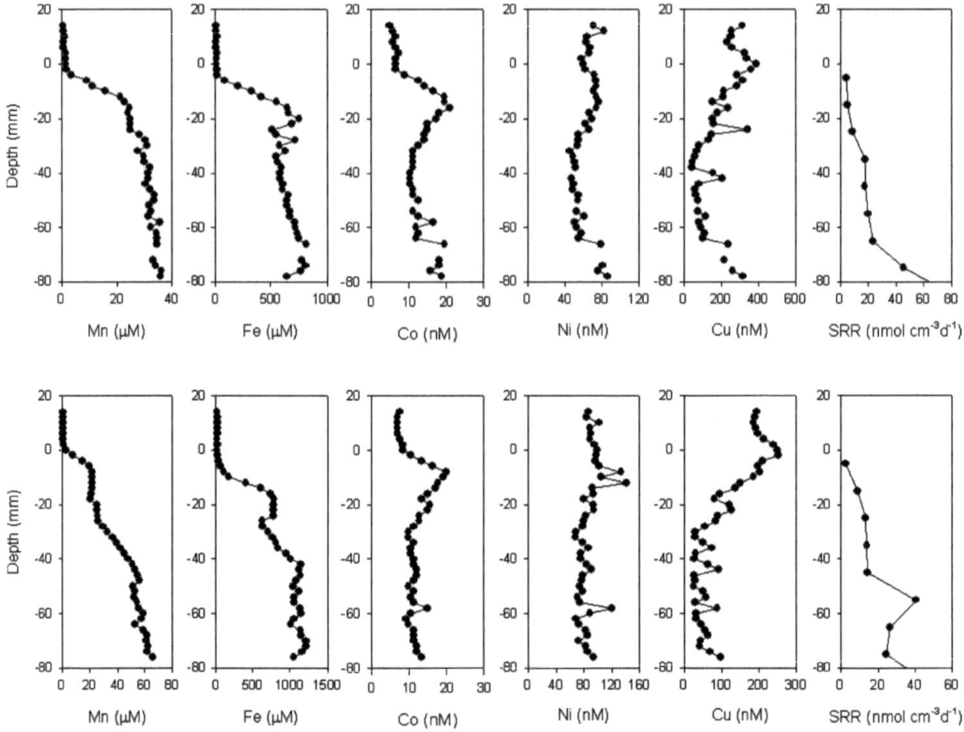

Fig. 8.5. Concentration of metals measured with DET from the harbor sediments. The results presented in the *upper panel* originates were from mesocosm C (control) while the *lower panel* are results from the homogenized harbor sludge (the results were published in Tankere-Muller et al. 2007)

The lack of systematic variation between the individual measurements suggests that any structure in the solute composition is not resolved at a cm scale.

The solute composition was investigated at high resolution using DET (diffusive equilibria in thin-films). Two cm wide strips of gel (0.8–1.2 mm thick), either continuous or segmented and mounted in a plastic support, were inserted into the sediment. Solutes in the porewater of the sediment equilibrate with the water within the gel within the typically 56 hours deployment time. The solute sampled in this way corresponds to a volume of only about 20 µl. Results from DET deployments in the harbor sediments are presented in Fig. 8.5. The location of the sediment water-interface is represented by zero.

The two high resolution porewater profiles are much smoother than the profiles displayed in Fig. 8.4. The high resolution measurements show that the Cu concentration peaks exactly at the sediment water-interface within the 2 mm resolution of the measurement. The high concentrations at the sediment water interface suggest that there is a local source at the surface, which is probably release from the reactive organic material that is rapidly oxidized and from associated oxidants, such as manganese oxides. With this rapid remobilization, Cu diffuses upwards and downwards from the sediment surface layer.

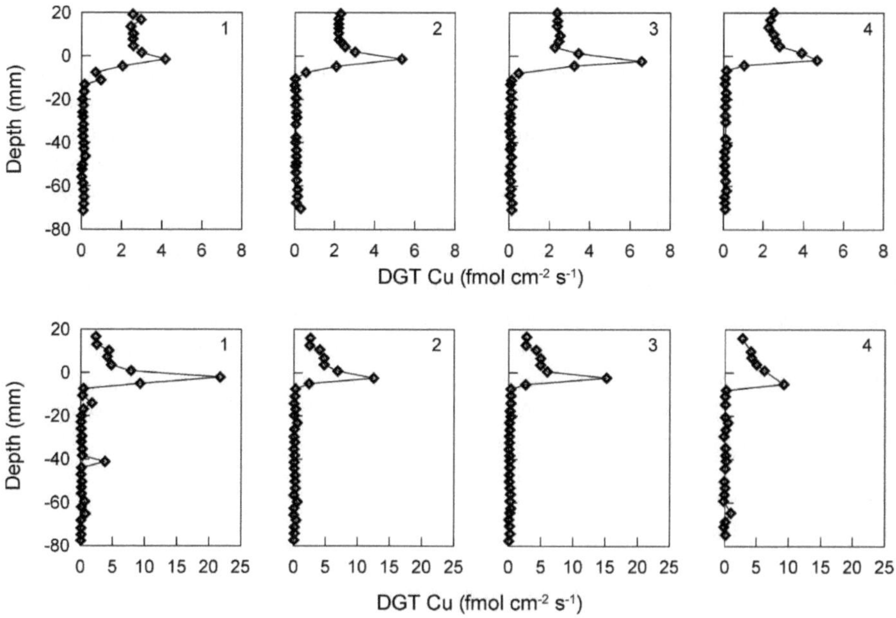

Fig. 8.6. Vertical profiles of DGT-measured fluxes for the harbor sludge: Control (*upper panel*) and homogenized sediment (*lower panel*). Numbers 1 to 4 refer to adjacent 3 mm wide vertical columns within a single DGT probes. The location of the sediment-water interface is represented by zero. Please note the difference in scale. The results were published in Tankere-Muller et al. 2007

Fluxes of Copper in the Sediment

With DGT (Diffusive Gradients in Thin-film) a method is available to quantify the remobilization in sediments. The essential part of DGT probes is a cation exchanger protected behind a thin gel layer. When the probe contacts the sediment, solutes diffuse into the diffusive gel and the cations are trapped in the ion exchanger. After approximately 24 h deployment the probes were retrieved and cut into 3 mm squares that were treated with acid. Cations were measured in the eluate. The measured accumulation of Cu on the resin allows calculation of the average flux from the sediment to the resin over the deployment period. DGT measurements (Fig. 8.6) can be interpreted as an average concentration at the surface of the device during the deployment time. This is usually lower than the bulk concentration because *(a)* supply from solid phase to solution is kinetically limited and *(b)* only part of the Cu organically complexed is measured by DGT when the diffusion coefficient for the free metal ion is used in the calculation. In this work the ratio of the DGT-derived concentration of Cu to the concentration in porewater measured by DGT was between 0.3 and 1.2, indicating that Cu was not dominated by large organic complexes and that there was an appreciable supply from solid phase to solution. It is therefore a good approximation to attribute the DGT-measured flux to the resupply flux of Cu to the porewaters. If anything this is likely to underestimate the magnitude of the highly localized flux, as the DGT maxima were defined by single data points. Measurement at higher vertical resolution than the 3 mm used for DGT would probably reveal higher maxima.

Forms of Copper in the Sediment

Small blocks of undisturbed sediment were sampled in front of the DET and DGT devices at each sampling campaign. Before embedding the sediments with a methacrylate resin, the sediments were stored anoxically in formaldehyde to fix fauna and bacteria. After hardening of the resin the sediments were polished and prepared for microscopy. Semi quantitative elemental analysis was performed using an electron microscope for elemental mapping and measurement of points. The elemental mapping showed that copper is heterogeneously distributed within the sediments. Analyzing the Cu-hotspots revealed that copper is predominantly associated with Fe-oxides and sulfides. The Fe-oxides contained traces of all contaminants and the copper concentrations ranged from 0.2 to 3% Cu (on an atomic basis). The sulfides containing Cu cover a range from a few percentage of Cu substitution in pyrite through pure chalcopyrite to almost pure Cu-sulfide. An example of sediment particles and the obtained analytical results are shown in Fig. 8.7.

8.1.5 Discussion

Collectively these measurements show that contaminants are intensively cycled in marine sediments according to the prevailing biogeochemical conditions. After copper is introduced to the sediments (presumably as organic particles) the copper moves gradually into various inorganic forms. In the upper millimeter of the sediments, where oxic conditions prevail, the copper becomes associated with iron(III) and manganese oxides. When the oxides in the course of diagenesis are used as electron acceptors the

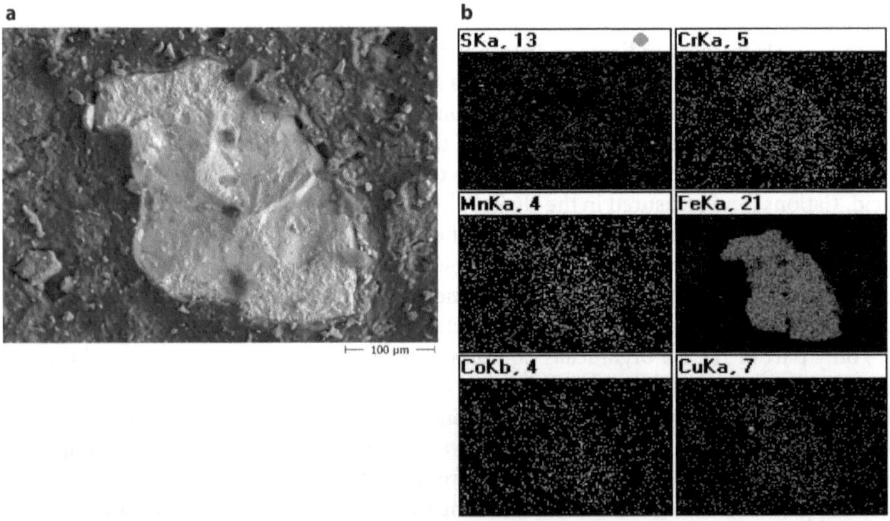

Fig. 8.7. a Back-scattered image of Cu-containing particle in thin section of the harbor sediment. **b** Energy dispersive analysis (mapping) illustrating the distribution of selected elements in the viewed section. Operating conditions: 15 kV accelerating voltage and a total counting time of 18 min

copper is released and entrained in the sulfides. Due to bioturbation and other disturbances the sulfides are brought into the oxic sediments again where they are oxidized and a transitory release of copper can be recorded. As metals are repeatedly cycled through these processes larger aggregates or nodules of iron and manganese oxides form in the upper sediment and sulfides in the deeper sections. With increasing time, copper is found in increasingly pure copper sulfides. This description of processes is in accordance with the findings of other workers (e.g., Parkman et al. 1999).

The high resolution measurements of the solute composition show that there are various factors that influence the small scale structure of Cu in the porewaters of sediments. There is good evidence that the productive silt sediment is laterally homogeneous, which supports the use of a vertical one dimensional model to describe the sequence of electron acceptors involved in the oxidation of org C. However, fluxes to and from the sediment can only be estimated accurately if measurements are made at mm or finer intervals because it appears that Cu is released to solution very close to the interface. The DET measurements show a sharp Cu maximum at the interface, but with DGT it is even sharper. Such an observation is consistent with rapid supply from solid phase to solution at this location. If it is sufficiently rapid to fulfill the DGT demand, while immediately above and below the site of supply the DGT demand can not be sustained, the DGT measurement will indeed have a sharper peak than the corresponding concentration profile. This highly localized remobilization at the interface would be missed by conventional measurements at typically cm resolution, which would underestimate the release flux and consequently the risk posed by the sediment.

For sandy sediments (not all results shown) the surface maximum is not so pronounced, but there is a much more heterogeneous distribution of Cu in the porewaters, both laterally and vertically. Here Cu release will depend on both diffusional fluxes at the interface and the localized chemistry in the convective pathway of the surface sediment. The analysis of the solid phase, which showed that chalcopyrite can be present in highly localized clusters, indicates that there may be some areas ($<cm^2$) of the sediment surface that have much higher Cu release rates than others. For these sediments it is inappropriate to use conventional one dimensional models of diagenesis. To develop a quantitative appreciation of the processes occurring and the resulting fluxes requires the use of a full three dimensional model of the redox reactions associated with the decomposition of organic matter. Such a model, which allows for vertical structures and localized microniches, has recently been developed (Sochaczewski et al. in preparation). While such models can be used to simulate particular events and experimental details, their ability to predict average fluxes is limited by the intrinsic stochastic nature of the system.

There can be no doubt that a full appreciation of the dynamics of metals in sediments requires measurements to be made at fine scales without lateral averaging. Quantitative interpretation of these data requires full 3 dimensional modeling of the reaction and transport dynamics of the system. Such measurements and associated models will be able to identify cases where risks are lower than previously thought (e.g., reduced fluxes of Co and Ni), but they will also show that in some cases risks may be higher (e.g., Cu and Cd). Data sets at the required resolution are still very limited, but as more data become available in the next decade our understanding of metal dynamics in sediments is likely to increase greatly. This new found knowledge is likely to greatly benefit the risk assessment and management of contaminated sediments.

One source of the highly localized flux at the sediment-water interface is likely to be the oxidation of sulfides brought upward in the sediments due to bioturbation or other disturbance. High biological activity or other sediment disturbance is therefore likely to enhance the copper flux through the sediment-water interface, as seen in the investigated sediments.

The fluxes through the sediment-water interface as well as the concentration of copper in the porewater are only to a limited extend related to the bulk concentration of copper. This suggests that the copper compounds formed in the marine environment are rapidly turned over in the upper millimeter of the sediments. The process rate appears in fact to be high enough to ensure partial equilibrium in these sediments. Accordingly, the environmental risk is related to a combination of factors associated with sediment management (disturbance) and the size of the copper pools in the sediments.

Acknowledgments

The authors are grateful for having been invited by Ulrich Förstner and Bernhard Westrich to attend the SEDYMO symposium held in 2006 at Hamburg-Harburg University of Technology to present selected results from the TREAD study. This section represents an overview of work which has involved financial support from various agencies particularly the EU (EVK3-CT-2002-00081), which is gratefully acknowledged.

References

BLMP (2002) Meeresumwelt 1997–1998. Bundesamt für Seeschifffahrt und Hydrographie (BSH), Hamburg und Rostock

Burnton ED, Phillips IR, Hawker DW (2005) Geochemical partitioning of copper, lead, and zinc in benthic, estuarine sediment profiles. J Environ Qual 34:263–273

Canadian Council of Ministers of the Environment (CCME) (1999) Canadian Environmental Quality Guidelines, Winnipeg

Chapman PM (1990) The Sediment Quality Triad approach to determining pollution-induced degradation. Sci Tot Environ 97/98:815–825

Chapman PM, Wang F, Janssen C, Persoone G, Allen HE (1998) Ecotoxicology of metals in aquatic sediments: Binding and release, bioavailability, risk assessment, and remediation. Can J Fish Aquat Sci 55:2221–2243

Comber SDW, Gardner MJ, Boxall ABA (2002) Survey of four marine antifoulant constituents (copper, zinc, diuron and Irgarol 1051) in two UK estuaries. J Environ Monit 4:417–425

Fenchel T, Glud RN (1998) Chemolithotrophic veil architectures enhance fluxes at the marine benthic interface. Nature 394:367–369

Fones GR, Davison W, Grime GW (1998) Development of constrained DET for measurements of dissolved iron in surface sediments at sub-mm resolution. Sci Tot Environ 221:127–137

Glud RN, Ramsing NB, Gundersen JK, Klimant I (1996) Planar optrodes, a new tool for fine scale measurements of two dimensional O_2 distribution in benthic communities. Mar Ecol Prog Ser 140:217–26

Lenihan HS, Peterson CH, Kim SL, Conlan KE, Fairey R, McDonald C, Grabowski JH, Oliver JS (2003) Variation in marine benthic community composition allows discrimination of multiple stressors. Mar Ecol Prog Ser 261:63–73

Long ER, Field LJ, MacDonald DD (1998) Predicting toxicity in marine sediments with numerical sediment quality guidelines. Environ Toxicol Chem 17:714–727

Møhlenberg F, Josefson AB, Vale C (2007) Marine benthic macrofauna – Linkages between chemical and biological quality of surface waters. Deliverable 16 of REBECCA project (SSPI-CT-2003-502158)

Parkman RH, Charnock JM, Bryan ND, Livens FR, Vaughan DJ (1999) Reactions of copper and cadmium ions in aqueous solution with goethite, lepidocrocite, mackinawite, and pyrite. Amer Min 84:407–419
Serbst JR, Burgess RM, Kuhn A, Edwards PA, Cantwell M G, Pelletier MC, Berry W J (2003) Precision of dialysis (peeper) sampling of cadmium in marine sediment interstitial water. Arch Environ Contam Toxicol 45:297–305
Shuttleworth SM, Davison W, Hamilton-Taylor J (1999) Two dimensional and fine structure in the concentrations of iron and manganese in sediment pore-waters. Envi Sci Technol 33:4169–4175
Sochaczewski L, Stockdale A, Zhang H, Tych W, Davison W (2007) A three dimensional model of transport and reaction dynamics in sediments: 3D-TREAD, in prep
Tankere-Muller S, Zhang H, Davison W, Finke N, Larsen O, Stahl H, Glud RN (2007) Fine scale remobilisation of Fe, Mn, Co, Ni, Cu and Cd in contaminated marine sediment. Mar Chem, in press

Michael Kersten · Björn Bohling

8.2 Comparison of Cohesive Sediment Erosion Rates Determined from ^{234}Th Radionuclide Tracer Profiles and Erosion Experiments in the Mecklenburg Bight, Baltic Sea

8.2.1 Introduction

The science of sediment dynamics in aquatic systems has greatly advanced in the last two decades, and quantitative estimates of flood and storm impacts are possible if the appropriate data and site-specific models are available. Unlike sandy sediments, understanding the mechanisms involved in the resuspension processes of cohesive sediments in aquatic systems remains an open case in water-related engineering disciplines. The treatment of cohesive sediment is therefore addressed almost entirely from an empirical standpoint, and field and laboratory experiments are needed to determine site-specific sediment properties (e.g., Lau and Droppo 2000; Whitehouse et al. 2000). Cohesive sediments are a sink for a range of contaminants. Therefore their resuspension and consequent impacts on water quality is a major concern in many rivers, lakes, harbours, estuaries, and coastal areas. Many flume and field experiments have demonstrated that the relationship between bottom current velocities and erosion is nonlinear. Therefore, storm events will produce most of the cohesive sediment erosion for a given year. For coastal areas, few field data exist pertaining to sediment mobilization and transport in storm conditions owing to difficulties in obtaining measurements and to vagaries of the weather at open sea. Nonetheless, a resuspension potential can be assessed by comparing a maximum shear stress modeled for a storm event, and the experimentally determined critical shear stress, using an appropriate erosion rate parameterization (Sanford and Maa 2001; Lick et al. 2006). A modeling approach, however, requires field validation, and here we present an unconventional method which uses sediment profiles of the naturally occurring radiogenic isotope ^{234}Th to verify the model data. The major advantage of this approach is that it provides field data on the net resuspension potential, i.e., resuspension of sediments in the presence of subsequent deposition. Moreover, this information can be derived retrospectively not only for the sediment surface but also for individual layers down to the entire scour depth of the sediment re-established after a storm event.

8.2.2 Sampling Site and Experiments

The present-day morphology of the western Baltic seafloor was formed by repeated advances and retreats of ice during the Pleistocene and Holocene Transgressions. The advancing ice masses dug basins and troughs in the continental shoreline. Mecklenburg Bight is such a trough with a maximum depth of 28 m, i.e. well below the fair weather wave base. Up to 7 m of mud is accumulated below the 20 ±2 m isobath, which includes over half of the Bight (1 400 out of 2 500 km^2, Fig. 8.8), the remainder being fine- to coarse-grained sand. Net sediment accumulation rates of 30–65 mg cm^{-2} yr^{-1} (dry weight) were reported for mud sampling sites in Mecklenburg Bight based on ^{210}Pb radionuclide dating of (albeit severely disturbed) sediment cores (Gellermann et al. 1990). During the course of a year, the SPM composition is controlled by enhanced biological production in spring and summer, and wave-driven resuspension in the autumn and winter periods. Tidal currents in the semi-enclosed Baltic Sea are

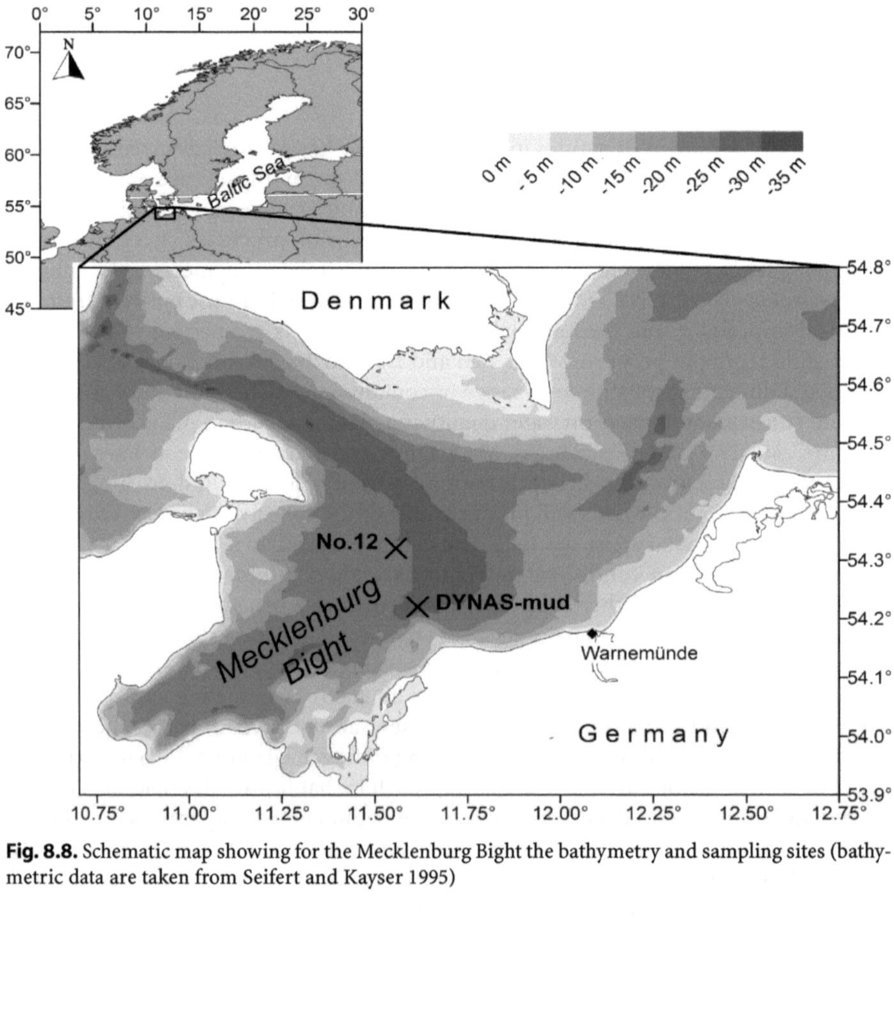

Fig. 8.8. Schematic map showing for the Mecklenburg Bight the bathymetry and sampling sites (bathymetric data are taken from Seifert and Kayser 1995)

negligible, and the hydrographic structure is mostly related to atmospheric forcing. There is a large seasonal and a minor annual wind variation in the area. During summer, stratification results from heating of the upper layer, and leads to anoxia in bottom waters. Winter cooling and winter storms remove much of the upper layer stratification. Since the Baltic Sea is fetch limited (i.e., distance that wind blows over the open water body), the dominant wind direction is important for the maximum wave heights (NE for the Mecklenburg Bight). The brackish salinity conditions also change both vertically and seasonally due to admixture of less saline waters from the east and more saline water from the northwest. Winter storms can introduce sufficient quantities of saltwater from the North Sea into the Bight, adding to the pronounced summer stratification through the buoyancy input. A comparison of the near-bottom hydrodynamic conditions by bottom tripod and current-meter mooring with experimentally derived critical shear stress velocities suggests that particle transport is controlled by storm events in winter, whereas under calm summer season conditions shear stress velocities do not exceed the critical values (Ziervogel and Bohling 2003; Bohling 2005). Bottom current scour leads in the former case to intense resuspension of the surface mud layer to a yet unknown depth, and to elevated near-bottom SPM concentrations in the range of 1–10 mg l^{-1} (Kersten et al. 1998). Light scattering measurements have shown that in winter season an entrained SPM-rich nepheloid layer may extend up to 2–5 m above the bottom, and be transported by the bottom currents counter-clockwise through the Bight to the northeast (Prandke 1986). These transport patterns have recently been modeled using an Eulerian approach (Kuhrts et al. 2004; Bobertz et al. 2005).

Two adjacent sampling sites in the eastern part of the Mecklenburg Bight were studied (Fig. 8.8). Sampling site no. 12 at 54°18.90' N and 11°33.00' E in the outer part of the Bight is the same as the International Baltic Monitoring Program reference sampling site BMP-M2, and was sampled for the radionuclide tracer profile measurements, while the adjacent sampling site "DYNAS-mud" (54°13.26' N, 11°36.96' E) was sampled for the erosion chamber experiments. Both sites are covered by mud (>90% medium silt sized) at a water depth of 25 m, with a clay mineral content of 3.1 ±0.6%, and a mean grain size of about 20 µm. One may argue whether this type of mud is to be classified as cohesive sediment, since Dyer (1986) has set up a threshold where cohesion begins to be significant when sediment contains more than about 5–10% of clay by weight. However, Mitchener and Torfs (1996) define cohesion by the content of the fines fraction (<63 µm) and experimentally derived a threshold of 3–15% fines by weight for the occurance of cohesion. Whitehouse et al. (2000) give a similar figure of 10% fines by weight for the onset of cohesion. According to these definitions the studied mud can be classified as cohesive sediment.

Sediment core sampling, sample pre-treatment, and analysis of both the radiotracer depth profiles (Kersten et al. 2005) and the erosion chamber experiments (Gust and Müller 1997; Bohling 2003; Ziervogel and Bohling 2003; Ziervogel and Forster 2006) have been discussed in detail previously. In our interdisciplinary approach, results of both these studies originally performed independently will be compared. Full details of the methodologies of both studies can be found in Kersten et al. (2005) and Bohling (2003), respectively.

8.2.3 Net Erosion Rate from Radionuclide Tracer Profiles

^{234}Th is produced in seawater from the decay of ^{238}U, and is particularly useful in assessments of suspended particulate matter (SPM) transport in coastal environments due to its relatively rapid decay (Kersten et al. 1998). The parent nuclide ^{238}U is dissolved in seawater under oxidizing conditions as the anionic uranyl carbonate complex, and therefore behaves conservatively with salinity. The daughter nuclide ^{234}Th, on the other hand, is present in the form of the particle-reactive hydrolysis product $Th(OH)_n^{(4-n)+}$. Measured distribution coefficients of the latter are on the order of $K_d \geq 10^6\,l\,kg^{-1}$, while ^{238}U has only a maximum $K_d = 10^2\,l\,kg^{-1}$. ^{234}Th is therefore scavenged rapidly by SPM accumulating in excess of the background fraction supported by the particulate ^{238}U load. The disequilibrium which is induced between ^{234}Th and its parent ^{238}U has for many years been recognized as a radiotracer of particle residence times in marine environments. In coastal systems with no significant river input as in the present case study, scavenging particles originate mainly from biological processes and sediment resuspension. In a previous paper, SPM residence times were determined for the Mecklenburg Bight on the basis of the short-term variability in the partitioning of ^{234}Th in the water column (Kersten et al. 1998). Once scavenged by SPM and deposited to the bottom, particulate ^{234}Th decays too rapidly to accumulate in sediment (half-

Table 8.2. Natural radiotracer activities for a sediment core taken after a storm event in November 8, 1994, at sampling site no. 12

Metric depth (cm)	Water content (wt.-%)	Mass depth M (g cm^{-2})	^{238}U activity (dpm g^{-1})	^{234}Th activity (dpm g^{-1})	Exchange loads J (mg cm^{-2})
0–1	79.8	0.505	2.50	8.16	3.6 ±0.2
1–2	77.1	1.08	2.82	7.73	4.0 ±0.2
2–3	74.0	1.73	2.81	8.25	4.6 ±0.3
3–4	71.9	2.43	2.49	7.51	5.0 ±0.4
4–5	70.7	3.16	2.85	7.73	4.5 ±0.5
5–6	67.6	3.97	3.08	5.65	3.7 ±0.6
6–7	64.3	4.86	3.13	4.95	2.8 ±0.3
7–8	63.6	5.78	3.06	4.88	2.0 ±0.5
8–9	63.4	6.69	2.62	5.37	1.4 ±1.7
9–10	60.1	7.69	2.58	3.34	1.0 ±0.5
10–11	58.5	8.73	2.29	3.90	
11–12	56.7	9.81	2.67	3.04	
12–13	57.2	10.88	2.28	2.75	
13–14	58.3	11.92	2.30	3.17	
14–15	58.2	12.97	2.44	3.00	

life 24.1 days). Any enrichment in surficial sediment layers, if at all detectable, is therefore purely controlled by SPM admixture events.

Specific ^{238}U and ^{234}Th activities on a dry weight basis for the sediment cores collected in November 1994 are listed in Table 8.2. The water content decreases gradually from over 80 wt.-% for the sediment surface layer (0–1 cm), to a discontinuity and thereafter constant 58 ±2 wt.-% at about 10 cm depth (Table 8.2 and Fig. 8.9). Figure 8.9 shows the radiotracer profiles in the non-linear cumulative mass depth scale, M (g cm^{-2}), instead of the conventional linear depth scale (cm). This scale conversion facilitates depositional mass flux estimation in the case of significant post-depositional compaction. Conversion between the linear depth scale z and the mass depth scale m for both sampling sites is given in Table 8.2 as defined by

$$m = \int_0^z \rho_i \mathrm{d}z \tag{8.1}$$

where ρ_i (g cm^{-3}) is the dry bulk density in the ith core section of linear depth interval Δz_i. A normal secular equilibrium with the parent ^{238}U sediment activity is found for the Sept. 1995 cores, but an anomalous disequilibrium was found for the Nov. 1994 cores at the same site, with an up to 3-fold excess ^{234}Th activity in the topmost sediment layers (Table 8.2, Fig. 8.9). Secular equilibrium values were reached only below 10 cm depth. The difference between both sampling campaigns was a storm event (Beaufort Force 8, i.e. wind velocities above 15 m s^{-1}) which occurred just a few days before the Nov. 1994 campaign. A likely scenario is that a relatively quiescent period marked by slow sediment deposition, and (if at all) weak bioturbation seasonally restricted by the summer bottom water anoxia, was followed by an instantaneous wave-driven perturbation event that somehow admixed ^{234}Th-laden SPM into the sediment. Though the exact mechanism is not clear, the radiotracer profile clearly evidences a perturbation depth of up to 10 cm in this cohesive sediment during the storm event. This is also

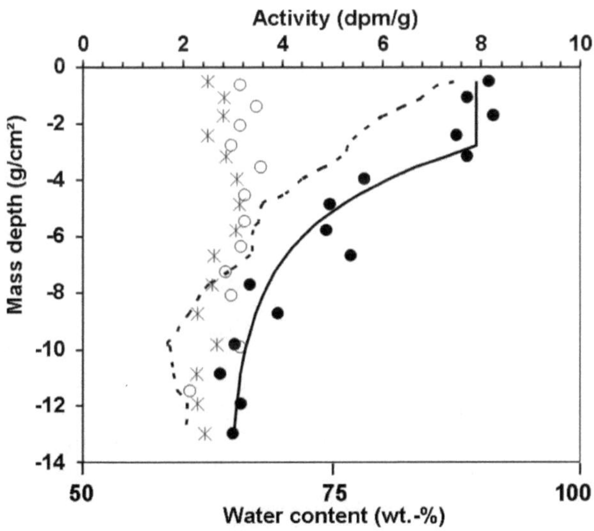

Fig. 8.9. Specific ^{234}Th activity mass depths profiles for sediment cores sampled in November 1994 upon a storm event (●), and in September 1995 (○), and the supporting ^{238}U activity for the latter (∗). The *solid curve* represents a fit by a one-dimensional sediment mixing model (Kersten et al. 2005). The *dotted line* is water content for the Nov. 1994 case

reflected in a two-layer pattern of the sediment porosity, with the upper partially consolidated layer having a lower resistance to erosion than the fully consolidated layer below the porosity discontinuity (Fig. 8.9).

The model for the fit curve for the ^{234}Th tracer distribution depicted in Fig. 8.9 has been previously discussed in detail (Kersten et al. 2005). The exponential decrease is due to ^{234}Th admixture, rather than decay upon steady-state burial. Our conceptual model for the admixture mechanism is based on a time sequence of *(i)* fluidization of the partially consolidated sediment layer, with low ^{234}Th equilibrium inventory, *(ii)* entrainment of the thus produced fluid mud, *(iii)* mixture of ^{234}Th far beyond the equilibrium level supported by the ^{238}U inventory (Kersten et al. 1998), and *(iv)* re-deposition/consolidation of the thus ^{234}Th-laden mud layer upon cessation of the storm event. The radiotracer profile is then mainly a result of a non-instantaneous particle settling process. Coarser flocs are deposited first, but have scavenged a lower amount of particulate and dissolved ^{234}Th due to shorter residence time in the water column. The longer the mud portion kept entrained during the storm event, the more particulate and dissolved ^{234}Th from the water column was scavenged, and the later it was deposited thus forming the characteristic ^{234}Th profile exponentially increasing from bottom to top of the now partly consolidated sediment. The constant but most elevated ^{234}Th activity in the upper-most portion of the core profile may have resulted from some post-depositional perturbation on the days before the measurement campaign. Based on this conceptual model, sediment-water interaction models are appropriate for an assessment of the net sediment mass effectively eroded, which appears (if at all accountable) much less than the total amount of sediment scoured during this complex sediment bed dynamics. Such models can be used if data are available on the ^{234}Th inventory of SPM particulates in the bottom water phase (Rutgers van der Loeff and Boudreau 1997). In the case of an instantaneous admixture of $^{234}Th_{ex}$ in excess of the supported levels down to an effective mixing depth, the amount of material exchanged during the wave event can be simply assessed. For this, the ratio of the $^{234}Th_{ex}$ inventory has to be related to the excess $^{234}Th_{SPM}$ load accumulated in the admixed suspended matter. Activities for the latter of 290 ±10 dpm g^{-1} were measured during the same November 1994 campaign just 3 m above the sea bottom (Kersten et al. 1998). These activities, albeit nearly two orders of magnitude higher, are still in the mean range of what is to be expected in similar dynamic coastal environments (Radakovitch et al. 2003). For a rough balance between input and output (net erosion = net deposition), one may estimate the sediment exchange flux J (net mass of resuspended sediment per unit surface area for the respective sediment depth in mg cm^{-2}), and integrate this from top to down the maximum linear mixing depth:

$$J = \sum_i \frac{^{234}\mathrm{Th}_{ex}}{^{234}\mathrm{Th}_{SPM}} m_i = \sum_i \frac{^{234}\mathrm{Th}_{ex}}{^{234}\mathrm{Th}_{SPM}} \rho_i \Delta z_i \tag{8.2}$$

The estimated exchange loads for each of the measured sediment layer in the Nov. 1994 sediment core are listed in Table 8.2 (with variance representing the scatter in the actual ^{234}Th data). The thus estimated integral sediment exchange flux of 33 ±5 mg cm^{-2} represents a non-steady-state amount of wave-induced short-term (dry-weight) mass movement into/out of the sediment deposit, and can be translated into a particulate contaminant flux as discussed previously by Kersten et al. (2005). This net erosion flux is actually less than the dry bulk density of a 1-cm wet sediment slice. It therefore corresponds to only a few mm of wet sediment depth, in the order of what has been found for *the* yearly net linear sediment accumulation rate in the Mecklenburg Bight (Gellermann et al. 1990), and also in the order of what has been found recently for transport bottoms in a freshwater environment using the particle-reactive tracer ^{7}Be instead of ^{234}Th (Fitzgerald et al. 2001). The activity of the former tracer does not depend on salinity because of atmospheric processes as the source term, and its use is therefore more appropriate at freshwater sites (Feng et al. 1999). For derivation of a net erosion rate, the integral sediment exchange flux has to be divided by the time during which the flux has been effective, but this is difficult to estimate. Nowadays it is believed that the occurrence of "turbulent bursts" (Nelson et al. 1995; Amos et al. 1997) are responsible for the entrainment of sediment under such conditions. This would yield in a net erosion rate in the order of 10^{-3} g m^{-2} s^{-1} for such a scenario, but may change by one order of magnitude more or less depending on the assumed time constraints. The gross erosion rate according to our conceptual model is an order of magnitude higher since it comprises scour of a maximum of 10 cm of partially consolidated sediment depth.

8.2.4 Discussion and Conclusions

Most natural cohesive beds have a fluffy, unstable surface layer. The cores investigated during the erosion chamber experiments (Bohling 2003; Ziervogel and Bohling 2003; Ziervogel and Forster 2006) were covered by such a well-developed albeit thin fluffy surface layer. It exceeded a thickness of a few millimeters in January 2002, and was up to 1 cm thick in September 2000. This loosely bound material does not show any cohesive effects, unlike the underlying silt-sized partially consolidated sediment, and hence there is a shift in the threshold for erosion towards lower values. A mean critical shear stress velocity u_c^* of 0.62 cm s^{-1} was measured for this material, with low variation except for the maximum value of 1.00 cm s^{-1} in July 2001 (Ziervogel and Bohling 2003), and a somewhat lower value of 0.4 cm s^{-1} in September 2000. Literature data of critical shear stress velocities for fluff across a range of environments compiled by Ziervogel and Bohling (2003) are of the same order. The mean u_c^* value of 0.6 cm s^{-1} will therefore be used for subsequent modeling. Erosion of the underlying cohesive sediment can not be observed with the experimental setup, nor with another setup used by Bohling (2005).

Both critical shear stress (τ_c) and critical shear stress velocity (u_c^*) are common threads when evaluating sediment stability due to erosion. Shear stress (τ, in N m^{-2} = Pa = 10 g cm^{-1} s^{-2}) can be calculated using the simple quadratic stress law:

$$\tau = \rho_w u^{*2} \tag{8.3}$$

where ρ_w is the water density (approximately 1 000 kg m^{-3}), u^* is the shear stress velocity (m s^{-1}), and τ is the shear stress. The value of the bed shear stress, at which the threshold of erosion is reached, is defined as critical shear stress τ_c. The critical shear stress depends on the particular sediment being tested and is generally derived empirically using Eq. 8.3. A critical shear stress velocity of 0.6 cm s^{-1} translates by this equation to a critical shear stress of 0.04 Pa (Bohling 2003). Erosion rates depend on the actual difference between the maximum bottom shear stress during a storm and the experimentally derived critical shear stress. In a recent review paper by Lick et al. (2006) different equations for sediment erosion rates reported in literature are examined. They concluded that erosion rates for all sediments sizes can best be approximated with a single, uniformly valid equation:

$$E = M\left(\frac{\tau_b - \tau_{cn}}{\tau_c - \tau_{cn}}\right)^n \tag{8.4}$$

where E is the erosion rate (g cm^{-2} s^{-1}), M is a site-specific erosion constant (s cm^{-1}), τ_b is the bed shear stress occurring during an erosion event which is site-specific and is generally a modeled quantity, and τ_{cn} is the critical shear stress for non-cohesive particles given by:

$$\tau_{cn} = 0.414 \cdot 10^3 d \tag{8.5}$$

where d is the mean particle diameter (m). For coarse-grained, non-cohesive sediments, Eq. 8.4 reduces to the simple formula:

$$E = M(\tau_b - \tau_c)^n \tag{8.6}$$

This equation has been used, e.g., by Puls and Sündermann (1990) for sediments typically encountered in the open North Sea. However, it is not valid for fine-grained, cohesive sediments. For the latter, i.e. when d tends towards 0 and thus $\tau_{c,max} \ll \tau_{cn}$, Eq. 8.4 reduces to the formula:

$$E = M\left(\frac{\tau_{max}}{\tau_c}\right)^n \tag{8.7}$$

The value for the exponent n is a matter of debate. Thorn and Parsons (1980), Puls and Sündermann (1990), and Sanford and Maa (2001) used a value for the exponent $n = 1$. Lick et al. (2006) claim from their work with flumes performed over more than two decades that n in Eq. 8.6 is typically about 2 or smaller, and for Eq. 8.7 is about 2 or greater. They have also shown, that Eq. 8.6 may be used for fine-grained sediment with a quite low n (1.3 in their experiments) but at the cost of a quite poor approximation of the experimental data. The erosion constant M used for Eq. 8.6 depends on the value of n (unit compensation). A benefit of the bracket term in Eq. 8.7 is that it becomes dimensionless which render unit search for M quite simple (same unit as E, e.g., $1 \cdot 10^{-4}$ g cm^{-2} s^{-1} is used by Lick et al. 2006). However, in order to remain compatible with parameters used in the well established general ocean circulation model for the Baltic Sea (Kuhrts et al. 2004), we decided to use Eq. 8.6 with a lowest possible $n = 1$ which gives finally an empirical floc erosion rate of:

$$E = M^* \rho_w \left(u_{max}^{*2} - u_c^{*2}\right) \tag{8.8}$$

Fig. 8.10.
Erosion rates for the DYNAS mud station in the Mecklenburg Bight derived from experiments with an erosion chamber (Bohling 2003; Ziervogel and Forster 2006) in comparison with the erosion rate derived from radionuclide tracer profiles and the calculated value from the conceptual model of this study. The *dashed line* indicates the erosion threshold according to the definition of Ziervogel and Bohling (2003), where $u^* = u^*_c$

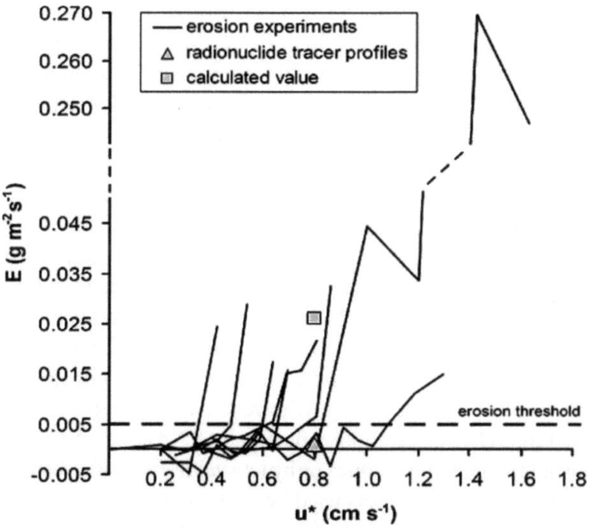

in agreement with the well known fact that the relationship between velocity and erosion is nonlinear. The erosion constant M^* used for Eq. 8.8 is in the order of $1 \cdot 10^{-5}$ s cm^{-1} for the $n = 1$ choosen (Kuhrts et al. 2004). At the studied position in the deepest part of the Mecklenburg Bight a value for the maximum shear stress velocity of 0.8 cm s^{-1} for a NE wind velocity of 15 m s^{-1} were derived from a general ocean circulation model (Kuhrts et al. 2004), which gives a value for the skin friction (bracket) term of 0.26 cm^2 s^{-2}. This in turn gives an erosion potential of 0.026 g m^{-2} s^{-1} during such a storm event which is comparable with the experimentally derived erosion rates at $u^* = 0.8$ m s^{-1} (Fig. 8.10) but an order of magnitude higher than the net erosion rate actually derived from the on-site radiotracer measurements.

One should bear in mind that the thus derived erosion potential is valid only for the fluffy layer, for which experimental data are available. The transport of cohesive sediment is typically characterized by at least two threshold conditions, one for newly deposited sediment, and the other for those that have undergone (partial) consolidation (Amos et al. 1997). Differences in erosion behavior between the topmost fluffy layer and the underlying sediment would best be parameterized using a two-layer model such as, e.g., introduced by Amos et al. (1992):

$$E = \varepsilon_f e^{\alpha[\tau_b - \tau_c(z)]^\beta} \tag{8.9}$$

where ε_f is an empirical floc erosion rate such as determined above, α and β are empirical constants, and z is the ultimate depth of erosion. The latter might reach several cm in depth as evidenced by the radionuclide tracer results discussed above. Depth-dependent shear stress velocities have not been measured but appropriate techniques are available (McNeil et al. 1997; Witt and Westrich 2003). Critical shear stress velocities for consolidated sediment layers underlying the fluffy layer are usually higher (e.g., up a factor of 5) than that measured for the fluffy layer (e.g., Bohling 2005). However, using the latter values would decrease the skin friction term to almost a no-erosion scenario (Kuhrts et al. 2004). On the other hand, our conceptual model of bed dynamics during a storm event includes a fluidization step of the partly consolidated sediment forming an extended fluid mud layer. This would make a model based on Eq. 8.9 obsolete. Moreover, it is traditionally assumed that eroded cohesive sediments are all entrained and take part in the suspension transport, whereas our radionuclide tracer data suggest that bulk erosion of the bed might well be overestimated if the mass of the fluidized mud is only affected by a rather limited bed load transport before being redeposited.

The radiotracer profiles evidence a significant redeposition upon a deep sediment scour event. The occurrence of redeposition is supported also by negative values for erosion rates in the erosion experiments (Fig. 8.10). This would imply that pure resuspension is up to an order of magnitude higher than net resuspension (difference between pure resuspension and pure redeposition) for this mud site, but it may well be characterized by a long-term equilibrium between erosion and deposition (pure resuspension = pure redeposition). Such a situation, however, would have a severe impact on contaminant transport models. For example, a particular SPM concentration (and its contaminant load) can be obtained by high values or by low values of both erosion and deposition, as long as they balance. The parameters used by the two models, however, are quite different, with a direct impact on the prediction of sediment dynamics and choice of remedial action. Large net erosion rates indicate that buried contaminants may be uncovered, be resuspended, and hence will contaminate surface waters which then necessitate dredging or capping as choice of remediation. On the other hand, small net erosion at high shear stresses and sediment scour indicates that contaminants are probably less redistributed over long distances, but permanent sediment reworking would keep them at sediment surface in reach of the epibethon. Natural attenuation may then become effective at much longer time scales than expected as discussed by Kersten et al. (2005) and Leipe et al. (2005).

Acknowledgments

The first author is grateful for having been invited by the convenors Ulrich Förstner and Bernhard Westrich to attend the SEDYMO symposium held in 2006 at Hamburg-Harburg University of Technology. The idea for this interdisciplinary paper was elaborated during preparation of a talk given at this symposium. The erosion chamber experiments were conducted at the Baltic Sea Research Institute Warnemünde (IOW) during the DYNAS project, funded by the German Federal Ministry for Education and Research.

References

Amos CL, Daborn GR, Christian HA, Atkinson A, Robertson A (1992) In-situ erosion measurements on fine-grained sediments from the Bay of Fundy. Mar Geol 108:175–196

Amos CL, Feeney T, Sutherland TF, Luternauer JL (1997) The stability of fine-grained sediments from the Fraser River delta. Estuar Coastal Shelf Sci 45:507–524

Bobertz B, Kuhrts C, Harff J, Fennel W, Seifert T, Bohling B (2005) Sediment properties in the Western Baltic Sea for the use in sediment transport modelling. Journal of Coastal Research 21(3):588–597

Bohling B (2003) Untersuchungen zur Mobilität natürlicher und anthropogener Sedimente in der Mecklenburger Bucht, Ph.D.-thesis, Mathematisch-Naturwissenschaftliche Fakultät, Universität Greifswald

Bohling B (2005) Estimating the risk for erosion of surface sediments in the Mecklenburg Bight (south-western Baltic Sea). Baltica 18(1):3–12

Dyer KR (1986) Coastal and estuarine sediment dynamics. Wiley, Chichester

Feng H, Cochran JK, Hirschberg DJ (1999) ^{234}Th and ^{7}Be as tracers for the transport and dynamics of suspended particles in a partially mixed estuary. Geochim Cosmochim Acta 63:2487–2505

Fitzgerald SA, Klump JV, Swarzenski PW, MacKenzie RA, Richards KD (2001) Beryllium-7 as a tracer of short-term sediment deposition and resuspension in the Fox River, Wisconsin. Environ Sci Technol 35:300–306

Gellermann R, Weiss D, Brügmann L (1990) Datierung von Ostseesedimenten mit ^{210}Pb. Isotopenpraxis 26:375–380

Gust G, Müller V (1997) Interfacial hydrodynamics and entrainment function of currently used erosion devices. In: Burt N, Parker R, Watts J (eds) Cohesive sediments. Wiley, Chichester, pp 149–174

Kersten M, Thomsen S, Priebsch W, Garbe-Schönberg CD (1998) Scavenging and particle residence times determined from ^{234}Th/^{238}U disequilibria in the coastal waters of Mecklenburg Bay. Appl Geochem 13:339–347

Kersten M, Leipe T, Tauber F (2005) Storm disturbance of sediment contaminants at a hot-spot in the Baltic Sea assessed by ^{234}Th radionuclide profiles. Environ Sci Technol 39:984–990

Kuhrts C, Fennel W, Seifert T (2004) Model studies of transport of sedimentary material in the western Baltic. J Mar Systems 52:167–190

Lau Y L, Droppo I G (2000) Influence of antecedent conditions on critical shear stress of bed sediments. Water-Research 34(2):663–667

Leipe T, Kersten M, Heise S, Pohl Ch, Witt G, Liehr G, Zettler M, Tauber F (2005) Ecotoxicity assessment of natural attenuation effects at a historical dumping site in the western Baltic Sea. Mar Pollut Bull 50:446–459

Lick W, Lick J, Jin L, Gailani J (2006) Approximate equations for sediment erosion rates. In: Maa J, Sanford L, Schoellhamer D (eds) Estuarine and coastal fine sediment dynamics. Elsevier (Proc. Marine Sci. 8), New York, pp 106–124

McNeil J, Taylor C, Lick W (1997) Measurements of erosion of undisturbed bottom sediments with depth. ASCE J Hydr Engrg 122:316–324

Mehta AJ, Hayter EJ, Parker WR, Krone RB, Teeter AM (1989) Cohesive sediment transport. J Hydr Engrg 115:504–519

Nelson J, Shreve RL, McLean SR, Drake TG (1995) Role of near-bed turbulence structure in bed load transport and bed form mechanics. Water Resour Res 31:2071–2086

Mitchener H, Torfs H (1996) Erosion of mud/sand mixtures. Coastal Engineering 29(1-2):1–25

Puls W, Sündermann J (1990) Simulation of suspended sediment dispersion in the North Sea. In: Cheng RT (ed) Residual currents and long-term transport. Coastal and Estuarine Studies 38, Springer-Verlag, New York, pp 356–372

Radakovitch O, Frignani M, Giuliani S, Montanari R (2003) Temporal variations of dissolved and particulate ^{234}Th at a coastal station of the northern Adriatic Sea. Estuar Coast Shelf Sci 58:813–824

Rutgers van der Loeff MM, Boudreau BP (1997) The effect of resuspension on chemical exchanges at the sediment-water interface in the deep sea – A modelling and natural radiotracer approach. J Mar Syst 11:305–342

Sanford LP, Maa JPY (2001) A unified erosion formulation for fine sediments. Mar Geol 179:9–23

Seifert T, Kayser B (1995) A high resolution spherical grid topography of the Baltic Sea. Meereswissenschaftliche Berichte – Marine science reports Nr. 9, Institut für Ostseeforschung Warnemünde, pp 72–88

Thorn MFC, Parsons JG (1980) Erosion of cohesive sediments in estuaries: An engineering guide. Proceedings of the International Symposium on Dredging Technology, Bordeaux. International Association of Dredging Companies, Bedford, pp 349–358

Whitehouse R, Soulsby R, Roberts W, Mitchener H (2000) Dynamics of estuarine muds: A manual for practical applications. London, Thomas Telford Publications

Witt O, Westrich B (2003) Quantification of erosion rates for undisturbed contaminated cohesive sediment cores by image analysis. Hydrobiologia 494:271–276

Ziervogel K, Bohling B (2003) Sedimentological parameters and erosion behaviour of submarine coastal sediments in the south-western Baltic Sea. Geo-Mar Lett 23(1):43–52

Ziervogel K, Forster S (2006) Do benthic diatoms influence erosion thresholds of coastal subtidal sediments? J Sea Res 55:43–53

Kate L. Spencer · Tom Benson · Mike Dearnaley · Andy J. Manning
Keiko Suzuki · Jon A. Taylor

8.3 The Use of Geochemically Labeled Tracers for Measuring Transport Pathways of Fine, Cohesive Sediment in Estuarine Environments

8.3.1 Introduction

Understanding the dispersion patterns of the fine sediment fraction (<63 µm) and its associated contaminants is fundamental to the sustainable management of coastal and marine resources. Better predictions of fine sediment transport are of great importance to those involved in seabed mining activities, as well as to port operators and dredging companies who maintain navigation routes and enable continued access to ports, harbours and industrial installations. Additionally, recent trends towards the beneficial re-use of dredged material in coastal protection and habitat creation schemes and the adoption of new dredging techniques, such as water injection dredging, means that understanding fine sediment transport is more pertinent than ever. More accurate measurements of fine sediment transport pathways and fluxes will also aid our understanding of fundamental morphological processes such as coastal response to sea level rise, re-working of sediments within the inter-tidal zone and the interactions between ecology and sediment stability. While a large number of these activities impact directly on the management of the coastal zone and its resources, problems of fine sediment management are by no means restricted to the coastal environment. Offshore drilling operations have in the past released considerable quantities of fine sedimentary material into the marine environment, much of which remains locked into the cuttings piles that have built up beneath the platforms. One of the major issues that has arisen from a review of de-commissioning options for these installations has been the potential liberation of these materials back into the marine environment and their potential impacts on the biota and the surrounding environment.

Prediction of sediment dispersion in the coastal zone relies upon accurate and reliable field techniques for the measurement of sediment transport pathways. Historically, this role has been fulfilled by the use of sediment tracing techniques using radioactive tracers (Courtois and Monaco 1969; Heathershaw and Carr 1978), fluorescent sands (Voulgaris et al. 1998) and sediment trend analysis (McLaren and Bowles 1985). However, radioactive tracer studies are considered unacceptable on environmental grounds and due to the cohesive characteristics of clays the latter two techniques are unsuitable and inaccurate for measuring transport pathways of fine-grained sediments. Therefore, there is an urgent need to develop a field methodology suitable for studying the dispersion and transport of the <63 µm fraction and in particular the clay component. Here, we present an overview of the results of a recent study to examine the potential of using Rare Earth Element labeled sediments as tracers for fine sediment in estuarine and marine environments.

8.3.2 Background

Sediment tracing techniques have been widely used since the 1950s and have huge potential to a range of end-users in the marine environment. Fluorescent sediment tracing has been used to determine sediment transport rates (e.g., McComb and Black 2005; Balouin et al. 2005), to validate sediment transport and re-suspension models (e.g., Cromey et al. 2002) and to assess the morphodynamic response of coastal sediments to changing hydrodynamic conditions (e.g., Michel and Howa 1999). However, due to the cohesive nature of clays and their interaction with their transporting medium these techniques are only suitable for the sand fraction.

Therefore, there has been considerable effort within the scientific community to develop a tracer for the fine sediment fraction (<63 μm) including the use of synthetic tracers and the labeling of clays with a variety of chemical and biological markers. The commercial sector has developed synthetic fluorescent tracers with the same size, density and surface charge characteristics as the flocculated clay and silt fraction. For example, fluorescent paint pigments have been used to track the dispersion of contaminated sediments from combined sewer outfalls and may have some useful applications when studying the behavior of fine sediment deposition in natural waters (Adams et al. 1998). However, these pigments are organic resins, spherical in shape and have high hydrophobicity meaning that their flocculation characteristics and interaction with other natural particles in suspension varies considerably from their natural clay and silt counterparts (Newman et al. 1990). While this generation of tracers has been subject to evaluation on the basis of a number of field studies, these deficiencies have been recognized and hence their use has not been widely accepted by the user community.

Clays, natural sediments and soils have been labeled with DNA (Mahler et al. 1998a), iridium (Yin et al. 1993), short-lived radioactive isotopes (Courtois and Monaco 1969; Heathershaw and Carr 1978; Quine et al. 1999; Kachanoski and Carter 1999) and Rare Earth Elements (REEs) (Krezoski 1985; Mahler et al. 1998b; Matisoff et al. 2001; Zhang et al. 2003). These tracers have then been used to measure particle transport pathways in a number of environments including lakes, groundwater, shallow seas, karst systems and upland catchments. The REEs have been studied in particular as they have a high binding capability with clay minerals, low mobility and low natural background concentrations in sediments.

Even though there is clearly a strong demand for a suitable fine sediment tracer neither synthetic fluorescent particles nor labeled clays have gained widespread usage in the marine environment for a number of reasons. The tracer signal should be easily detectable using routinely available analytical techniques. However, labels such as iridium or DNA are costly to analyze and analytical facilities are not necessarily widely available. Once the labeled sediment is released to the marine environment there is potential for exchange with competing cations in saline solution and hence loss of the tracer signal. Both Mahler et al. (1998b) and Spencer et al. (2007) have found that the retention of REEs in saline conditions is good but the widespread demonstration of these tracer sediments has been largely restricted to freshwaters. Tracer sediment should have the same physical properties as the sediment it is intended to mimic in terms of flocculation and settling characteristics, and erodability. Synthetic fluorescent pigments, while sharing many physical characteristics with silts and clays, display significantly different flocculation behavior and information regarding the physical behavior of labeled natural sediments is not currently available in the literature. Finally, it is critical that that any material discharged to open waters should be environmentally benign and as a result the regulatory authorities are now reticent to issue discharge licenses to release even short-lived radioactive isotopes to the aquatic environment.

8.3.3 Rare Earth Element Labeled Sediment Tracers

A range of REE labeled clay minerals have been developed and their potential as sediment tracers for the <63 μm fraction in estuarine and marine environments has been assessed. The technological constraints highlighted above that have restricted the use of labeled sediments as tracers have been addressed and are reviewed below. Kaolinite, montmorillonite, hydrobiotite and phlogopite have been labeled with La, Sm and Ho. Full details of sediment composition and preparation are given in Suzuki and Spencer (in review) and Suzuki et al. (in prep.).

Detection of the Tracer Signal

Following tracer release in the marine environment, sediment will disperse according to prevailing hydrodynamic conditions, typically resulting in between 10^6 and 10^9 fold dilution of the tracer signal. Therefore, a major technical concern has been the ability to detect the tracer signal once it has entered the coastal environment and to distinguish it from background concentrations. In order to enhance the tracer signal the sorption of REEs to the clay minerals has been optimized by varying pH, solid to solution ratio, ionic strength of solution and equilibration time (Suzuki and Spencer, in review). In addition, phlogopite and hydrobiotite were also pre-treated with sodium tetraphenyl boron ($NaBPh_4$) to enhance cation exchange capacity (CEC) (Suzuki et al. in prep.). Following optimized sorption, the REE content of the labeled clay minerals is between 4.27 and 5.17% dry weight (Table 8.3) and following the dilution of these concentrations the tracer should still be detectable in coastal sediments using ICP-OES and ICP-MS techniques. These analytical techniques are now widely available in most academic and consultancy laboratories. The treatment of expansible phases such as hydrobiotite and phlogopite with $NaBPh_4$ has enhanced their CEC by between 288 and 316% resulting in a significant increase in the amount of REE sorbed to the clay minerals and hence elevating detection levels of the tracer once dispersed in the marine environment.

Retention of the Tracer Signal

Once the labeled sediment tracer is released to the marine environment there is potential for desorption of the REE label through exchange with competing cations and/or complexation with chelating organic compounds present in coastal waters. Significant desorption of the REE label in situ would result in reduced detection of the tracer signal in sediments. Additionally, continuous desorption of the REE label after tracer deployment will render the modeling of tracer sediment dispersion very problematic.

In order to examine the chemical stability of the sediment tracer desorption batch experiments were carried out over varying time, pH, salinity (0.5–35 psu) and organic acid concentrations (Suzuki et al. in prep). The batch desorption experiments were carried out in a 1% solid suspension. The mixture was agitated on an automated rotor table (100 rpm) and 2 ml of the suspension was sampled at regular intervals over 14 days. Concentration of the REE in the supernatant was determined by ICP-OES (Vista-Pro, Varian, UK) after centrifugation (3 000 rpm, 15 minutes).

Table 8.3. Concentration of REEs sorbed onto clay minerals in % dry weight ('original' after optimisation and 'treated' after treatment with $NaBPh_4$)

	La		Sm		Ho	
	Original	Treated	Original	Treated	Original	Treated
Hydrobiotite	2.57 ±0.05	4.58 ±0.31	3.17 ±0.01	5.17 ±0.24	3.64 ±0.07	4.92 ±0.06
Phlogopite	1.14 ±0.10	3.78 ±0.17	1.30 ±0.07	4.27 ±0.12	1.06 ±0.15	4.24 ±0.28
Montmorillonite	3.85 ±0.26		4.37 ±0.24		4.87 ±0.31	
Kaolinite	0.16 ±0.05		0.19 ±0.03		0.34 ±0.20	

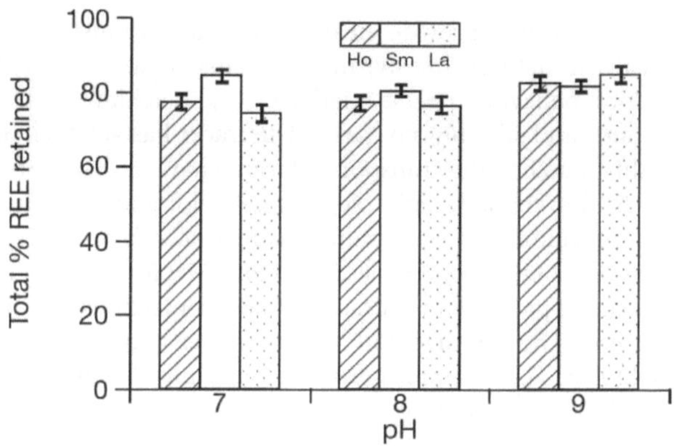

Fig. 8.11.
Rare earth element retention on montmorillonite as a function of pH after 2 weeks exposure to de-ionized water

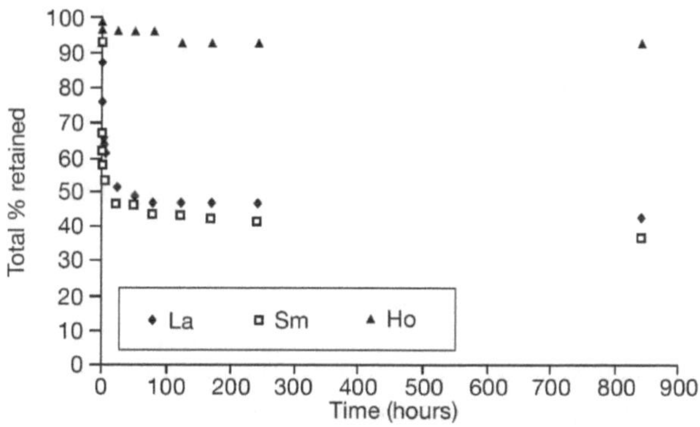

Fig. 8.12.
Retention of REEs on montmorillonite in seawater (36 ppt) as a function of time

Between pH values 7 and 9 in de-ionized water (control), varying concentrations of REE were desorbed from the clay mineral substrate (Fig. 8.11). However, stability was greatest for montmorillonite with between 75–85% of La, Sm and Ho being retained over a 2-week period. Under saline conditions, the La and Sm labeled sediment tracer released over 40% of the sorbed La and Sm within the first 10 hours. However, less than 10% of Ho was desorbed from montmorillonite over a 10-day period and this can be attributed to its small ionic radii and hence high binding capability to clay minerals (Fig. 8.12). Where effective neutralization can be achieved, thermal treatment of clay minerals results in ionic migration to vacant octahedral sites (Miller et al. 1982). Hence, in order to enhance the stability of sorbed Ho the labeled clays were dried at 300 °C during the initial tracer preparation.

Preliminary experiments have also indicated that <10% of Ho was desorbed from montmorillonite over a range of humic acid concentrations, although considerably higher concentrations of La and Sm were desorbed from phlogopite and hydrobiotite. Background concentrations of the REEs and in particular Sm and Ho are low in estuarine sediments (e.g., Weltje et al. 2002). Therefore sediment tracers labeled with these two REEs will be easily distinguishable from background signals.

Physical Behavior of the Tracer Sediment

For any tracer to provide meaningful results it must exhibit the same physical characteristics as natural cohesive mud in terms of particle size, density, erodability and flocculation behavior. This is relatively easy to achieve with non-cohesive materials such as sands and silts, but no particulate tracer has yet been developed which has demonstrably matched the transport relevant characteristics of cohesive sediment.

The floc characteristics, settling behavior and mass flux distributions of Ho labeled montmorillonite were examined using tracer to natural estuarine mud ratios (T:M) of T100, T75:M25, T50:M50, T25:M75 and M100. This was carried out under a range of suspended particulate matter (SPM) concentrations at constant salinity (Manning et al. in prep). The floc size and settling velocity measurements were obtained using the LabSFLOC (Laboratory Spectral Flocculation Characteristics) instrument (Manning et al. 2006). The settling flocs were viewed by a high-resolution video camera and floc images were recorded by an S-VHS video recorder. All individual flocs comprising a population were observed for spherical-equivalent floc size (D) and settling velocity (Ws). Where particle Reynold's numbers were low, the simultaneous direct measurement of D and Ws enabled the calculation of effective density (ρ_e) using Stoke's Law. Floc dry mass, floc porosity and the mass settling flux (MSF) were then calculated for each floc population (Fennessy et al. 1997; Manning 2004). Mass settling flux is defined as the product of the settling velocity and SPM concentration. The experimental work focused on the macrofloc fraction (>160 μm) which represents the majority of flocculated matter in an estuary (Mehta and Lott 1987).

For the entire experimental SPM concentration range, the difference in mean macrofloc diameter for T100 and M100 samples was <16%, while macrofloc porosity consistently ranged from 90–93%. The macrofloc effective densities were <120 kg m^{-3}. This all suggests that both the tracer and natural mud have a similar floc structural composition. The settling velocities, Ws_{macro} for T100 and M100 were 2.9 mm s^{-1} and 2.5 mm s^{-1}, representing a 14% difference (Fig. 8.13) while various T:M ratios produced –4% to +11% deviations from the macrofloc settling velocity of natural mud, suggesting that the tracer readily flocculates with natural cohesive sediments.

The SPM distribution across the complete floc population is also important for accurate MSF simulations. For example, the experimental work revealed that at only a 5% deviation, the mass apportioning across the floc spectrum (for a 500 mg l^{-1} base SPM) was very similar for both 100% tracer (64% macrofloc mass) and pure mud (69% macrofloc mass). This is very similar to values reported by Manning (2004) in a number of European

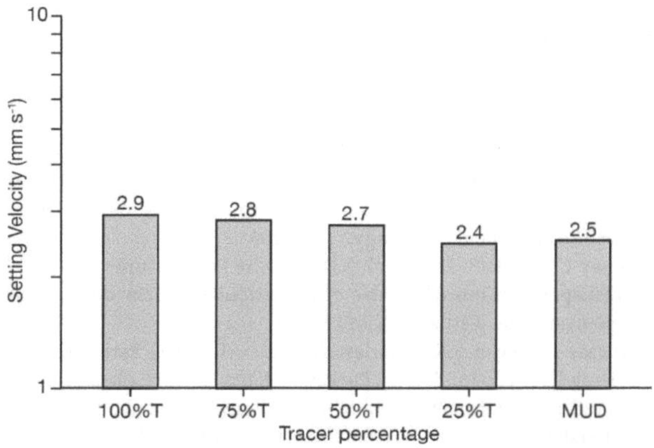

Fig. 8.13.
Mean settling velocity for the floc size fraction >160 µm for tracer to natural estuarine mud ratios (T:M); T100, T75:M25, T50:M50, T25:M75 and M100

estuaries during neap tides. When both mass and Ws factors were combined, the data indicated just an 11% MSF difference between the T100 and M100 suspensions at 500 mg l^{-1}.

Therefore, these preliminary laboratory tests demonstrate that the tracer has very similar dynamics to real mud flocs, while also readily interacting and flocculating with natural mud. Further gravimetric testing of the tracer sediment is currently being performed using a large diameter settling column.

8.3.4 Conclusions

Both commercial and academic end users would benefit from the development of a field-based methodology to measure sediment transport pathways of the <63 µm fraction. To date the development of such a tracer has been problematic. The use of fluorescent sands and other fluorescent particles do not possess the same settling and flocculation characteristics as natural cohesive mud. The use of chemically labeled sediments has been hindered due to problems of tracer detection and potential loss of the tracer signal following release in marine environments. Rare Earth Element labeled sediments have demonstrated a good potential for development as particle tracers. The REE content of the tracer sediment can be optimized by varying solute concentration, pH and ionic strength, while desorption in the saline environment can be minimized by pre-treating the clay minerals. The labeled sediments have similar physical characteristics to natural cohesive sediment. Therefore, they will provide a far more accurate measurement of field sediment transport pathways for the <63 µm fraction than other fine sediment tracers currently available.

In order to gain end-user confidence in this technique there is a need to develop robust sampling protocols and demonstrate the use of the sediment tracer in a number of highly controlled and well-documented field trials. This work is currently in progress.

Acknowledgments

The authors would like to thank the Natural Environment Research Council (NER/D/S/2003/00706), HR Wallingford Ltd. and Harwich Haven Authority for funding this work.

References

Adams EE, Stolzenbach KD, Lee JJ, Caroli J, Funk D (1998) Deposition of contaminated sediments in Boston Harbor studied using fluorescent dye and particle tracers. Est Coast Shelf Sci 46:371–382

Balouin Y, Howa H, Pedreros R, Michel D (2005) Longshore sediment movements from tracers and models, Praia de Faro, South Portugal. J Coast Res 21:146–156

Courtois G, Monaco A (1969) Radioactive Methods for the Quantitative Determination of Coastal Drift Rate. Marine Geology 7:183–206

Cromey CJ, Nickell TD, Black KD, Provost PG, Griffiths CR (2002) Validation of a fish farm waste resuspension model by use of a particulate tracer discharged from a point source in a coastal environment. Estuaries 25:916–929

Fennessy MJ, Dyer KR, Huntley DA, Bale AJ (1997) Estimation of settling flux spectra in estuaries using INSSEV. In: Burt N, Parker R, Watts J (eds) Cohesive Sediments – Proc. of INTERCOH Conf. (Wallingford, England), Chichester: John Wiley & Son, pp 87–104

Heathershaw AD, Carr AP (1978) Measurement of sediment transport rates using radioactive tracers. In: Coastal Sediments '77 (ASCE Symposium, Charleston, South Carolina 1977) New York: American Society of Civil Engineers, pp 399–416

Kachanoski RG, Carter MR (1999) Landscape position and soil redistribution under three soil types and land use practices in Prince Edward Island. Soil Till Res 51:211–217

Krezoski JR (1985) Particle reworking in Lake Michigan sediments: In-situ tracer measurements using a rare earth element. 28th Conference of Great Lakes Research, International Association of Great Lakes Research, Milwaukee, WI

Mahler BJ, Bennett PC, Hillis DM, Winkler M (1998a) DNA-labeled clay: A sensitive new method for tracing particle transport. Geology 26:831–834

Mahler BJ, Bennett PC, Zimmerman M (1998b) Lanthanide-labelled clay: A new method for tracing sediment transport in Karst. Ground Water 36:835–843

Manning AJ (2004) Observations of the properties of flocculated cohesive sediment in three western European estuaries. J Coastal Res SI 41:70–81

Manning AJ, Benson T, Spencer KL, Suzuki K, Taylor JA, Dearnaley M (in prep) Preliminary laboratory tests of the flocculation properties exhibited by a holmium-labelled montmorillonite cohesive sediment tracer

Manning AJ, Friend PL, Prowse N, Amos CL (in press) Preliminary Findings from a Study of Medway Estuary (UK) Natural Mud Floc Properties Using a Laboratory Mini-flume and the LabSFLOC system. Cont. Shelf Res BIOFLOW SI

Matisoff G, Ketterer ME, Wilaon CG, Layman R, Whiting PJ (2001) Transport of Rare Earth Element tagged soil particles in response to thunderstorm runoff. Env Sci Tech 35:3356–3362

McLaren P, Bowles D (1985) The effects of sediment transport on grain-size distributions. Journal of Sedimentary Petrology 55:457–470

McComb P, Black K (2005) Detailed observations of littoral transport using artificial sediment tracer, in a high-energy, rocky reef and iron sand environment. J of Coast Res 21:358–373

Mehta AJ, Lott JW (1987) Sorting of fine sediment during deposition. Proc. Speciality Conf. Advances in Understanding Coastal Sediment Processes. Am Soc Civ Eng, New York, pp 348–362

Michel D, Howa HL (1999) Short-term morphodynamic response of a ridge and runnel system on a mesotidal sandy beach. J of Coast Res 15:428–437

Miller SE, Heath GR, Gonzalez RD (1982) Effects of Temperature on the Sorption of Lanthanides by Montmorillonite. Clays and Clay Minerals 30:111–122

Newmann KA, Morel FMM, Stolzenbach KD (1990) Settling and coagulation characteristics of fluorescent particles determined by flow-cytometry and fluorometry. Env Sci Tech 24:506–513

Quine TA, Govers G, Poesen J, Walling D, van Wesemael B (1999) Fine-earth translocation by tillage in stony soils in the Guadalentin, south-east Spain: an investigation using caesium-134. Soil Till Res 51:279–301

Spencer KL, James SL, Taylor JA, Kearton-Gee T (2007) Sorption of La^{3+} onto clay minerals: A potential tracer for fine sediment transport in the coastal marine environment? Special Publication of the BGS vol. SP274

Suzuki K, Spencer KL (in review) Optimisation of lanthanum (La^{3+}) sorption for development of fine sediment tracers: a preliminary study. Submitted to Chemical Geology
Suzuki K, Spencer KL, Hillier S (in prep) Potassium leaching of clay minerals and its effect on samarium and lanthanum sorption
Voulgaris G, Simmonds D, Michel D, Howa H, Collins MB, Huntley DA (1998) Measuring and modelling sediment transport on a macrotidal ridge and runnel beach: An intercomparison. Journal Of Coastal Research 14:315–330
Yin Y, Chang N, Zhong W, Sun S, Zhang Y, Cui H, Chen S, Feng Y, Sun L (1993) A study of neutron activation tracer sediment technique. Science in China, Series A, 36:243–248
Weltje L, Heidenreich H, Zhu W, Wolterbeek H, Korhammer S, de Goeij JJM, Markert B (2002) Lanthanide concentrations in freshwater plants and molluscs, related to those in surface water, pore water and sediment. A case study in The Netherlands. Science of the Total Environment 286:191–214
Zhang XC, Nearing MA, Polyakov VO, Friedrich JM (2003) Using rare-earth oxide tracers for studying soil erosion dynamics. Soil Science Society of America 67:279–288

Andrea van der Veen · Martina Baborowski · Claudia Kraft · Jörg Kraft
Margarete Mages · Mihály Óvári · Wolf von Tümpling

8.4 Dynamics of Heavy Metals and Arsenic – Hungarian Tisza River Sediments after Mining Spills in the Catchment Area

8.4.1 Introduction

Sediments are an essential compartment of rivers and represent an important basis for aquatic life. In contrast to lakes, they are subject to greater dynamics. Land use, climate and hydrology as well as urban and industrial settlements have a strong influence on the sediment quality and on the sediment's function as a habitat. Anthropogenic activities in river catchment areas, e.g., mining and the related processing of ores, lead to the enrichment of heavy metals and As in river sediments, which has a lasting degrading effect on the sediment quality. The most important fractions of sediments are the fine grain sizes (clay and silt). Their physical and chemical properties, e.g., large specific surface area and high ion exchange capacity, enables them to scavenge contaminants efficiently. Another important factor of the fine grain sizes is their remobilization and their (remote) transport as suspended particulate matter (SPM).

As an example of a river catchment area influenced by strong mining activities, the dynamics of heavy metals and As along the rivers Tisza and Szamos in Hungary after accidental mining spills in Romania are presented and assessed.

In early 2000, two major accidental mining spills caused an ecological disaster at the rivers Tisza, Szamos and some of their tributaries in Romania and Hungary. In both cases, the accidents were triggered by heavy rainfalls and strong snow melts. The first accident occurred near Baia Mare, a gold processing plant in northwestern Romania, which released cyanide and heavy metal containing tailings water (approx. 100 000 m^3; approx. 1 000 t cyanide, approx. 1 000 t heavy metals) into a tributary of the Szamos (WWF 2002). The immediate effect was the death of more than 1 240 t of fish along the Hungarian section of Tisza and Szamos.

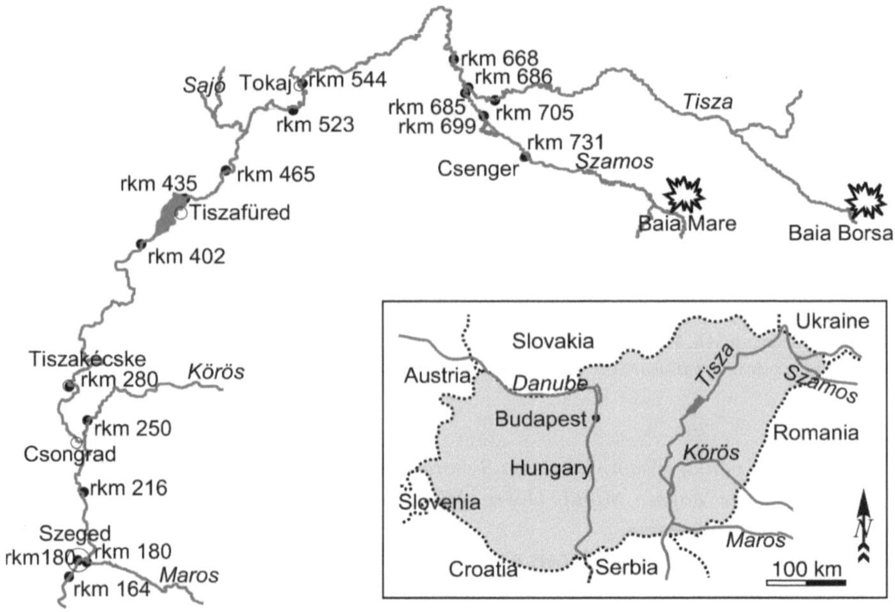

Fig. 8.14. Location of the sampling sites at the rivers Tisza and Szamos (large map; sampling sites labeled according to their distance towards the confluence of Tisza and Danube: river kilometer = rkm = 0) and the spill sites in the Tisza Catchment (*1:* Baia Mare; *2:* Baia Borsa; small map)

The second mining spill happened near Baia Borsa (NW Romania), where a tailings dam broke and released heavy metal containing water (approx. 100 000 m^3) and solid waste (approx. 20 000–40 000 t) into the Vaser River (Romania), which flows into the Upper Tisza (Fig. 8.14). Several studies were performed to investigate the temporal course, the extent and the toxic effects of the accidents on the Tisza River Basin (Black and William 2001; Kraft et al. 2006; Osán et al. 2002; Soldán et al. 2001).

Besides these major accidental inputs, the Tisza catchment is characterized by the strong mineralization of the Maramures County in NW Romania and a long historic record of mining since pre-Roman times with the related environmental impact. Mined elements include non-ferrous metals like Au, Cu, Pb and zinc. Heavy metals in the sediments of the Hungarian section of Tisza and Szamos are strongly enriched compared to the local geogenic background concentrations (Kraft et al. 2003). Water and sediment show higher concentrations especially of Zn in the Szamos than in the Tisza (Woelfl et al. 2004). Metal input into the rivers continues, as Óvári et al. (2004) detected elevated Zn concentrations in the water phase of the Szamos but not of the upper Tisza.

This study investigates the dynamics of As and heavy metals in sediments since the accidents in 2000 along the Tisza and the Szamos in Hungary. The top 5 cm of river sediments (0–5 cm) are used to characterize the surficial trace element contents shortly after the accidents (June 2000), one year later (Feb. 2001) and finally five years after the spills (May 2005). Multivariate statistics, i.e. cluster analysis, are employed to extract information on relations of the sampling sites along the rivers.

8.4.2 Materials and Methods

Sampling

The effects of the mining spills in northwestern Romania on the Hungarian rivers were investigated by sampling the rivers Szamos and Tisza from the Hungarian-Romanian border in the northeast to the Hungarian-Serbian border in the south. The sampled locations are marked in Fig. 8.14. The sites are labeled according to the river (T = Tisza, S = Szamos) and their distance (rkm = river kilometer) towards the confluence of Tisza and Danube (rkm = 0).

In 2000, two locations at the Szamos and 11 locations at the Tisza were sampled. One year later, two additional sites were included in the investigation of the Szamos to attain a better picture of sedimentation processes along the Hungarian river section. The sampling campaign of 2005 included three sites at the Szamos and seven sites at the Tisza.

Surface sediments were sampled along the river bank by carefully scooping a sediment section with a thickness of up to 5 cm with a shovel. The wet samples were kept frozen (–40 °C) in polyethylene bags until further preparation.

Sample Preparation

The frozen samples were freeze-dried at –5 °C for 48 hours (Christ, Alpha 1, Germany). Larger particles, i.e. plant parts and stones, were manually removed with tweezers. Before chemical analysis, the samples were homogenized in an agate mortar and sieved through a 1 mm plastic sieve (Fritsch, Germany).

The chemical composition was determined at the grain size fraction <20 µm, which was separated by wet sieving with ultrasonic treatment and in 2001 by a combination of sieving with ultrasonic treatment and sedimentation described in detail by Bachmann et al. (2001). The applied separation techniques caused differences only in the mass fraction, but no significant changes of the element concentration (Kraft 2002).

Chemical Analysis

The fine fraction of the sediments (<20 µm) was selected for the analysis based on its larger adsorption capacity than coarser grain sizes (Ackermann et al. 1983), thus it represents a major sink for trace elements. The chemical composition was analyzed after the digestion with aqua regia (AR). This extraction releases the environmentally effective element contents in contrast to total digestions.

250 mg sample material were filled in a PTFE microwave vessel with aqua regia (6 ml 37% HCl, 2 ml 65% HNO_3; Suprapure grade, Merck, Germany). The digestion took place in a pressure- and temperature-controlled microwave (Mars 5, CEM, Germany) at the following settings: 1 200 W, T_{max} 180 °C, 32 min. Major elements were determined by inductively coupled plasma optical emission spectrometry (ICP-OES; Optima 3000, Perkin-Elmer, Germany). Trace elements were measured by inductively coupled plasma mass spectrometry (ICP-MS; Agilent 7500c, Agilent Technologies, Germany).

The samples of the campaign in 2001 were digested in open PTFE vessels (200 mg sample, 15 ml 37% HCl, 5 ml 65% HNO_3). The reaction was allowed to proceed for 1 h at ambient temperature before the vessels were heated on a hotplate (= 80 °C) until near dryness. The digests were taken up with 20 ml 1M HNO_3. Major elements and some of the trace elements were determined by ICP-OES (Typ 3520, Bausch and Lomb, ARL; Maxim Fisons Instruments) and ICP-MS (Micromass Platform; As, Cd, Pb).

Statistical Analysis

Statistical data analysis was performed with SPSS 13.0 for Windows (Release 13.0.1). The software was used to compute descriptive statistics and the cluster analysis. The agglomerative hierarchical cluster analysis according to Ward, which uses the squared Euclidean distance, was performed with *Z*-score standardized element concentration data.

8.4.3 Results and Discussion

The sediments can be described as loamy sand to silt that contain in most cases less than 20 wt.-% of the <20 μm fraction (Kraft 2002), while suspended particulate matter of Tisza and Szamos is composed of more than 90 wt.-% of particles <20 μm (Baborowski et al. 2006). Table 8.4 contains the average value and the standard deviation of the trace element concentrations in the fine fraction of the sediments of Tisza and Szamos. The trace element contents (As, Cd, Cu, Pb, Zn) follow a common pattern along the rivers (Fig. 8.15). The Szamos is characterized by high to extremely high concentrations, while the Tisza generally contains lower contents.

Table 8.4. Trace element concentrations ($\mu g\,g^{-1}$) in surface sediments (<20 μm) of the rivers Tisza and Szamos

Element		River Tisza: Year (*n*)			River Szamos: Year (*n*)		
		2000 (9)	2001 (7)	2005 (7)	2000 (2)	2001 (4)	2005 (3)
As	Average	38	40	25	90	138	54
	Std dev	11	10	7	11	17	22
Cd	Average	3	5	3	11	22	6
	Std dev	2	2	1	2	1	2
Cu	Average	152	144	87	387	499	173
	Std dev	64	43	20	49	63	131
Ni	Average	63	64	61	67	72	60
	Std dev	7	6	7	6	4	5
Pb	Average	92	107	83	188	383	141
	Std dev	21	34	31	21	32	25
Zn	Average	573	651	618	1990	2471	1369
	Std dev	125	295	302	28	369	310

The upper course of the Tisza (rkm 705) holds relatively low trace element concentrations in 2000 and 2001. Below the confluence of the two rivers (approx. rkm 683), the concentrations strongly increase in the fine fraction of the Tisza sediments due to the input of contaminated material by the Szamos. In downstream direction, the element concentrations decrease gradually until approximately rkm 435, where the Tisza Lake interrupts the river course. Beyond this point, the concentrations seem to stagnate. The Tisza Lake functions as a trap for SPM in the Tisza as the particle number in general decreased significantly after passing the lake (Baborowski et al. 2002).

In general, it can be stated that the concentrations in the sediments were the highest one year after the mining spills in 2001. Six years after the spills, the contamination level has declined significantly and is also lower than 2000 (2001 > 2000 > 2005). This pattern probably results from the delayed transport of highly contaminated particles from the sources of the spills downstream. In 2000, the year of the spills, the particles had not reached Hungary to the full extent, yet. The following year exhibits extremely high sedimentary element concentrations, which reflect the arrival of the contaminated particles from the spills especially pronounced at the Szamos. Black and Williams (2001) suggest that flood conditions (strong spring flood) could have resulted in the scouring of the upper sediment layers, which lowered the heavy metal concentrations in the surficial sediments to pre-spill levels. The lowest concentrations of the observation period were detected in 2005. They probably result from a cleansing process of the riverbed due to the frequent flooding in the Tisza catchment. Based on mathematical assessment of floods, significant floods occur every five to six years and great floods every ten to twelve years at Danube and Tisza (WHO 2002). During flood situations, sediment is resuspended and transported further downstream and/or deposited into adjacent areas, e.g., floodplains (Förstner et al. 2004). The mixing with "cleaner" material from unaffected tributaries or regions causes a concentration decline (Lewin and Macklin 1987). The improved management of tailing sites (Zinke 2005) must have reduced the input of contaminated material from further upstream to the sediments.

In contrast to the sediments, the concentrations of particulate bound As and Zn increased since the spill, while the concentrations of dissolved Zn display maxima in 2001 but minima in 2005 (Baborowski et al. 2006). The composition of the suspended

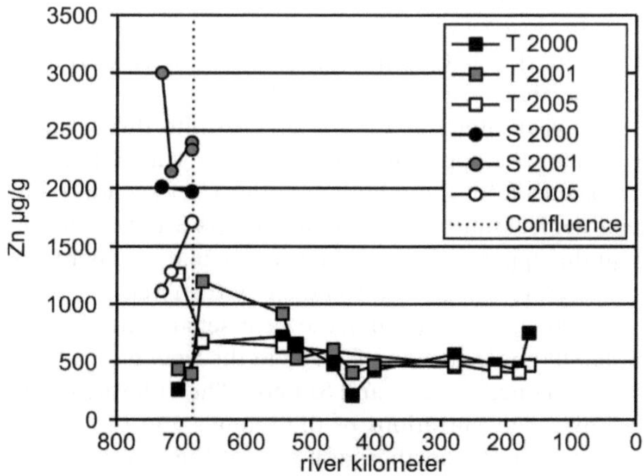

Fig. 8.15. Aqua regia soluble Zn content ($\mu g\ g^{-1}$) in the fine fraction (<20 µm) of the sediments from the rivers Tisza and Szamos in 2000, 2001 and 2005

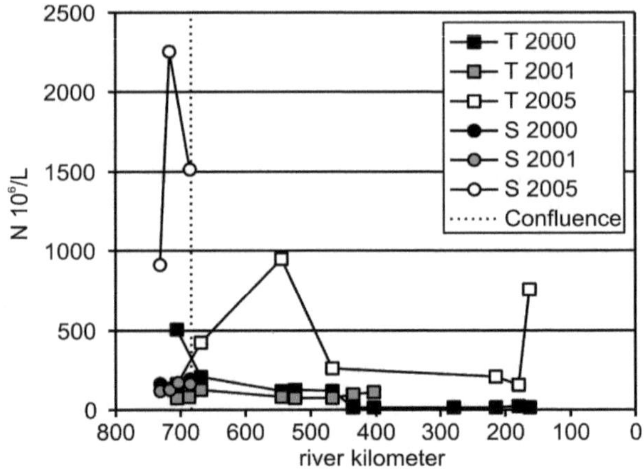

Fig. 8.16. Particle content (<20 µm) in the rivers Tisza and Szamos for the sampling campaigns 2000, 2001 and 2005

particulate matter changed also during the observation period depending on the hydrological conditions (Baborowski et al. 2006). The fine particle content ($N < 20$ µm) of suspended particulate matter (SPM) was the highest in 2005 (Fig. 8.16).

To investigate regional relations of the sampling sites, a cluster analysis was performed for each year based on the As and heavy metal concentrations (Cd, Cu, Pb, Zn; *Z*-score standardized values). These elements were selected based upon the high influence of the mining spills on their concentration. Other elements, e.g., Ni, are not suited for the cluster analysis due to their stable concentration levels along the river course (Table 8.4; Ni: 2005: Tisza: 61 ±7 µg g^{-1}).

The resulting dendrograms (Fig. 8.17) for the individual years share the clean separation of the Tisza from the Szamos. This is in agreement with data on dissolved and particulate bound elements in Tisza and Szamos from Baborowski et al. (2002). Due to the dynamics in the "contamination" level of the sediments, the internal clustering of the Tisza varies between the sampling campaigns.

In 2000 – the year of the spills – the Tisza samples form three clusters (A–C; Fig. 8.17, 2000). The sites rkm 705 and rkm 435 form cluster A due to their low trace element content. Cluster B is composed of sites from the middle and lower section (rkm 465–180). In cluster C, locations from the upper section as well as from the lower section (rkm 280, rkm 164). The latter two are characterized by higher concentrations compared to their neighboring sampling sites.

One year later, the Tisza is divided into three clusters (Fig. 8.17, 2001). Cluster C represents the upper middle of the river course (rkm 668–544), while cluster B comprises sites from the middle to lower river section (rkm 532–280). In contrast to the geographical subsumption of the clusters B and C, cluster A is a combination of samples from the upper and the middle course that are characterized by lower concentrations compared to the general trend along the river.

In 2005 (Fig. 8.17, 2005), the clear separation of the Szamos and the Tisza is interrupted by the inclusion of T-705 in the Szamos cluster. This Tisza location lies above the confluence of Tisza and Szamos. The differing clustering of 2005 is caused by the declining concentrations of the Szamos and probably a similar composition of the sediments in the northeastern part of the investigation area.

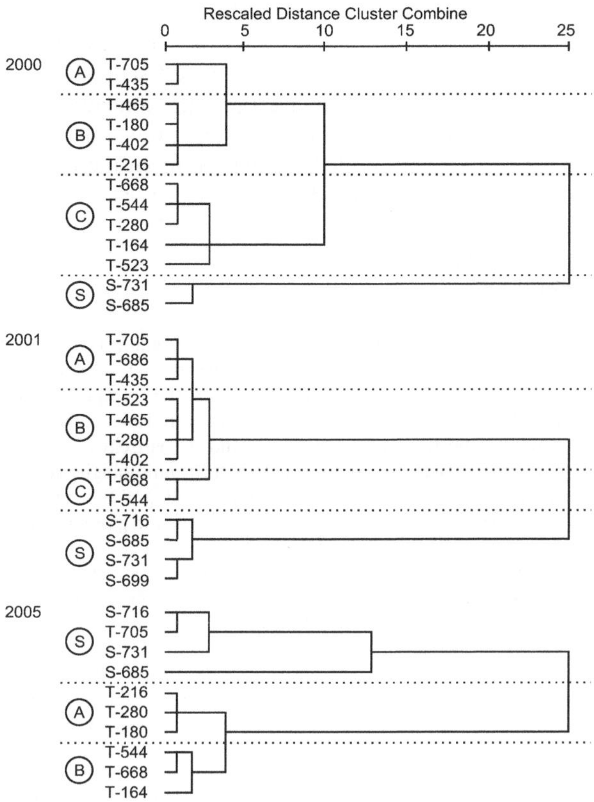

Fig. 8.17. Dendrograms of the cluster analysis for the sampling campaigns in 2000, 2001 and 2005 (concentrations of As, Cd, Cu, Pb, Zn; Z-score standardized values)

The clustering of the Tisza shows the formation of two groups (Fig. 8.17, 2005). Cluster B holds sites from the upper middle section and the lowest site of the Tisza (rkm 164). The other cluster (A) is made up from locations of the lower river course (rkm 280–180).

8.4.4 Conclusions

The contamination level of the sediments in Tisza and Szamos decreased significantly since the heavy metal input from the mining spills in early 2000. The decrease of As and heavy metals is especially pronounced at the extremely contaminated locations of the Szamos.

The concentrations of mining-related elements (As, Cd, Cu, Pb, Zn) display similar dynamics in the sediments: high concentrations in 2000, extremely high values in 2001, lower concentrations in 2005. The temporal delay of the maximum contamination in sediments in distant areas from the source of contamination can be expected in other rivers with a similar hydraulic regime as well. The concentration increase from 2000 to 2001 in the Szamos and to a lesser degree in the Tisza is most likely caused by the transport of sediment with a relatively higher element concentration from further upstream (= closer to the source). Besides erosion processes, the mixing of contaminated with non-contaminated sediment could be responsible for part of the concentration decline since the spill. The non- or lower contaminated material stems from non-mineralized regions. The effects of extreme mining spills can therefore be limited on a larger timescale due to the dilution of contaminated with "cleaner" sediment from unaffected areas.

Further research should investigate the contents of As and heavy metals in adjacent areas of the rivers Tisza and Szamos since the redistribution of contaminated sediment during subsequent floods is connected to an increase in the contaminant concentration in floodplains and sediments downriver.

References

Ackermann F, Bergmann H, Schleichert U (1983) Monitoring of heavy metals in coastal and estuarine sediments – a question of grain-size: <20 µm versus <60 µm. Environmental Technology Letters 4:317–328

Baborowski M, Kraft J, Mages M, Karrasch B, von Tümpling W, Ovari M, Zaray G, Einax JW (2002) Untersuchungen zum Eintrag von gelösten und partikulären Stoffen aus der Szamos in die Tisza (Ungarn). In: Tagungsbericht 2001, Band II, Deutsche Gesellschaft für Limnologie (DGL) Tagungsberichte der DGL, Deutsche Gesellschaft für Limnologie, Tutzing, pp 879–884

Baborowski M, Kraft J, van der Veen A, von Tümpling W, Einax JW (2006) Transport von Spurenmetallen aus der Szamos in die Theiß (Ungarn). In: Jahrestagung der Wasserchemischen Gesellschaft 2006 in Celle, 22.–24.05.2006:159–163

Bachmann T, Friese K, Zachmann DW (2001) Redox and pH conditions in the water column and in the sediments of an acidic mining lake. J Geochem Exploration 73:75–86

Black MC, William PL (2001) Preliminary assessment of metal toxicty in the middle Tisza River (Hungary) flood plain. JSS 4:203–206

Förstner U, Heise S, Schwartz R, Westrich BJ, Ahlf W (2004) Historical contaminated sediments and soils at the river basin scale. Examples from the Elbe River catchment area. JSS 4:247–260

Kraft C (2002) Auswirkungen von Schwermetallemissionen nach Unfällen im rumänischen Bergbau auf das Sediment der Flüsse Szamos und Theiß. (M.S. Thesis), Inst. f. Geowissenschaften, Braunschweig

Kraft C, von Tümpling W, Zachmann DW (2003) Auswirkungen von Schwermetallemissionen nach Unfällen im rumänischen Bergbau auf das Sediment der Flüsse Szamos und Theiß (Ungarn). Zbl Geol Paläont, Teil I:153–169

Kraft C, von Tümpling jr. W, Zachmann DW (2006) The effects of mining in Northern Romania on the heavy metal distribution in sediments of the rivers Szamos and Tisza (Hungary). Acta Hydrochim Hydrobiol 34:257–264

Lewin J, Macklin MG (1987) Metal mining and floodplain sedimentation in Britain. In: Gardiner V (ed) International Geomorphology 1986: Proceedings of the First International Conference on Geomorphology, John Wiley & Sons, Chichester, pp 1009–1027

Osán J, Kurunczi S, Török S, Van Grieken R (2002) X-Ray analysis of riverbank sediment of the Tisza (Hungary): Identification of particles from a mine pollution event.– Spectrochimica Acta Part B: Atomic Spectroscopy 57:413–422

Ovari M, Mages M, Woelfl S, von Tuempling W, Kröpfl K, Záray G (2004) Total reflection X-ray fluorescence spectrometric determination of element inlets from mining activities at the upper Tisza catchment area, Hungary. Spectrochimica Acta Part B 59:1173–1181

Soldán P, Pavonic M, Boucek J, Kokes J (2001) Baia Mare accident – Brief ecotoxicological report of Czech experts. Ecotoxicology and Environmental Safety 49:255–261

Woelfl S, Mages M, Ovari M, Geller W (2004) Determination of heavy metals in macrozoobenthos (chironomid larvae) from the river Tisza by total reflection X-ray fluorescence spectrometry. In: Cser M A, Lászlo IS, Étienne J-C, Maymard Y, Centeno JA, Khassanova L, Collery P (eds) Metal Ions in Biology and Medicine, vol. 8. J. Libbey Eurotext, Montrouge, pp 330–333

WHO Regional Office for Europe (2002) Floods – Climate change and adaptation strategies for human health. WHO-meeting in London, 30 June–2 July 2002, EUR/02/5036813. Copenhagen, Denmark

WWF (2002) The ecological effects of mining spills in the Tisza River system in 2000. WWF, Vienna

Zinke A (2005) Mining risk spot is safe again. Danube Watch 1

Chapter 9

Microbial Effects

Hans-Curt Flemming

Chapter 9 deals with a very important and often neglected component involved in pollutant mobility: the microorganisms. They are ubiquitous. They colonize sediment surfaces in contact with water more or less densely and are present in the water phase, mainly as flocs. Of course, they represent a dynamic phase in the system which has to be considered. However, their influence is very difficult to asses as microorganisms do not represent a chemically homogeneous phase but are spatially heterogeneous and vary in time in terms of population composition, density, activity and presence of extracellular polymeric substances (EPS). This results in a very complex network of interactions as microorganisms respond to the conditions of their environment, and, in addition, can possibly transform pollutants – be it by degradation of organic substances or by changing the species of metals, e.g., by alkylation. Little wonder that such aspects have not been exhaustively addressed scientifically and it is a particular feature of this book to specifically deal with the aspects provided by microorganisms.

The first section deals generally with the most frequent microbial phenomenon which is biofilms. It is important to understand the basic laws of biofilm development and dynamics in order to understand their impact on pollutant mobility, including sediment stability. The nature of EPS will influence biofilm sorption properties as well as biofilm cohesion and "gluing" of sediment particles. The interaction with ions is of particular interest here as bridging effects due to ionic interactions with charged groups in EPS play an important role. Aspects of sorption are addressed as well as the various stabilizing and destabilizing effects on sediments. Sorption properties of biofilms are equally complex as there are different sorption sites such as cells, cell walls, and EPS with different sorption capacities.

A very important step for understanding the effect of microorganisms is the comparison of non-sterile to sterile systems. This is addressed in laboratory systems in Sect. 9.2 in detail, using mucoid and non-mucoid strains of *Pseudomonas aeruginosa* in order to obtain quantitative data. Of course, much care has to be applied when extrapolating these results to natural systems, however, surprisingly consistent parallels could be drawn from slow sand filters samples from a drinking water plant. Interestingly, biofilms can contribute to higher erosion rates when growing directly at the sediment-water interface ("fluffy layer"), disrupting sediment particles from the surface and carrying them into the water phase, contributing to sediment-related pollutant mobility while in greater depth, it could be clearly shown that biofilms increase sediment cohesion.

In Sect. 9.3, again sterile vs. non-sterile experiments were carried out, indicating the cohesive effect of biofilms to sediments and increasing the settlement of suspended particles. In these experiments, sterilization was performed by gamma-irradiation of environmental samples, leaving dead biomass in place which, however, cannot further multiply. When particulate metals are transported in the water phase, bacteria obviously contribute to their accelerated sedimentation which has been confirmed in field investigations: Bacterial cell content, suspended matter content and heavy metal concentrations followed similar patterns during settling after re-suspension.

In general, this chapter contributes an interesting aspect which is far from being fully understood and can represent a problem for modeling as the dynamics of microbial influence are so complex. However, "black-box" approaches neglecting biotic influences cannot the solution of the problem as the sometimes insufficient predictive power of modeling demonstrates.

Hans-Curt Flemming · Martin Strathmann · Carlos Felipe Leon Morales

9.1 Biofilms and Their Role in Sediment Dynamics and Pollutant Mobility

9.1.1 Introduction

Natural sediments are not sterile but inhabited by a large range of microorganisms (Riding and Awramik, 2000) and higher forms of life. As a consequence, these organisms participate in many chemical processes in sediments, in the interaction between sediments and the water phase and in sediment dynamics. In fluvial environments, the interface between the major water body and the sediment, is a very active zone both in physicochemical and biological terms. Especially in highly permeable sediments, the dynamic flux of energy, nutrients, metabolites and particles (including microorganisms) is interdependent with local hydrodynamics (Huettel et al. 2003). Due to their slime matrix, active microbial communities at the water-sediment interface, develop into macroscopic scale structures which modify sediment topography and frictional resistance. These surface alterations have repercussions in fluid flow, shear forces and other physical parameters, especially at the benthic boundary layer. Microbial colonization is not limited to the sediment-liquid interface; equally important on their effect on river sediment hydrodynamics, is their ability to develop at significant sediment depths. At this level, permeability and hydraulic conductivity changes caused by microbial colonization, can have a profound effect on sediment cohesion and sorption/desorption processes (Leon-Morales et al. this vol.).

A key feature for understanding the influence of microorganisms on sediments is their common form of life which is biofilms. Microorganisms in nature do not occur as pure cultures and single organisms but in mixed communities, organized in aggregates. Such aggregates are termed somewhat inaccurate as "biofilms" and include films, colonizing solid surfaces in water but also on liquid-liquid and liquid-gaseous interfaces as well (e.g., at the water-air-interface or water-oil-interface). The term also embraces microbial mats, sludges and flocs (which can be considered as floating biofilms). The justification for this is that all these phenomena have one aspect in common which is

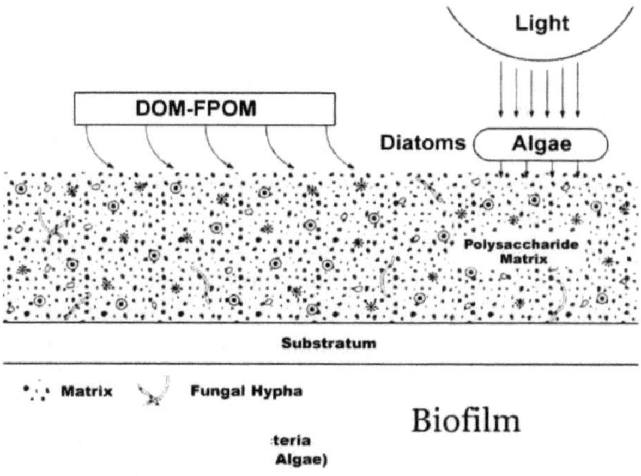

Fig. 9.1. Sediment biofilm (after National Park Service, U.S. Department of the Interior)

that the cells in the aggregates are kept together by a matrix of extracellular polymeric substances (EPS) which provide some basic advantages for the cells, such as a long retention time in a stabile position to each other, allowing for the formation of synergistic microconsortia. Due to their physiological activity, gradients develop in pH-value, redox potential as well as in the concentration of electron acceptors and donators.

Biofilms can form on virtually all surfaces exposed to non-sterile environments, provided enough water and nutrients. They can be found in soil and aquatic environments, on plant surfaces, on tissues of animals as well as in technical systems such as filters, pipes and reactors, and they play a pivotal role in medical context, in particular, in infections (Costerton et al. 1995). Environments in which biofilms have been investigated include not only solid-liquid but also air-liquid (Spiers et al. 2003), liquid-liquid (Macedo et al. 2005) interfaces, among others. Biofilms have to be considered in attempts to understanding transport, immobilization and remobilization processes of sediment particles as well as of pollutants as they form a dynamic interphase between solid surfaces and the water phase. In modeling approaches, this aspect is usually completely neglected as it is difficult to include and to predict, in particular, if the basic rules of biofilm development and properties are not taken into account. In order to account for such influences, experiments on sediment and pollutant mobility, therefore, have to be carried out (see Leon-Morales et al. this volume; Neto et al. this volume). In sediments with light access, diatoms usually contribute the main component of biofilm populations (Stal and de Brouwer, 2003). A typical sediment biofilm is depicted in Fig. 9.1.

9.1.2 Extracellular Polymeric Substances (EPS)

One key aspect for understanding the role of biofilms is the "extracellular polymeric substances" (EPS). These are highly hydrated biopolymers of microbial origin, embedding the biofilm organisms; for a detailed overview see Wingender et al. (1999). EPS are defined as "extracellular polymeric substances of biological origin that participate in the formation of microbial aggregates". EPS represent the construction material of biofilms and the immediate environment for biofilm inhabitants. In general, the proportion of EPS in biofilms can vary between 50 and 90% of the total organic matter. Detailed reviews on EPS can be found in (Wingender et al. 1999) and in (Flemming and Wingender 2002). EPS are mainly responsible for the structural and functional integrity of biofilms and are considered as the key components that determine the physico-chemical and biological properties of biofilms (Allison 2003; Flemming and Wingender 2003). EPS create a microenvironment for sessile cells which is conditioned by the nature of the EPS matrix.

Regardless of their origin, EPS are located at or outside the cell surface. This extracellular localization and the composition of EPS may be the result of different processes: active secretion, shedding of cell surface material, cell lysis, and adsorption from the environment. Microbial EPS are biosynthetic polymers (biopolymers) which consist mainly of polysaccharides and proteins, but can also contain substantial amounts of DNA, lipids, glycolipids and humics. Most bacteria are able to produce EPS, whether they grow in suspension or in biofilms. Cell surface polymers and EPS are of major importance for the development and structural integrity of flocs and biofilms. They mediate interactions between the microorganisms and maintain the three-dimensional arrangement.

Much information has been collected about the chemical nature and physico-chemical properties of extracellular polysaccharides, since they are abundant in many bacterial EPS. Specific polysaccharides (e.g., xanthan) are only produced by individual strains, whereas non-specific polysaccharides (e.g., levan, dextran or alginate) are found in a variety of bacterial strains or species. Non-carbohydrate moieties like acetyl, pyruvyl and succinyl substituents can greatly alter the physical properties of extracellular polysaccharides and the way in which the polymers interact with one another, with other polysaccharides or proteins, and with inorganic ions (Sutherland 1994). The network of microbial polysaccharides displays a relatively high water-binding capacity and is mainly responsible for acquisition and retention of water to form a highly hydrated environment within flocs and biofilms (Chamberlain 1997).

Extracellular polysaccharides are believed to have the main structural function within biofilms by forming and stabilizing the biofilm matrix.

It must be pointed out that polysaccharides are not necessarily the main EPS component. However, not much is known about synergistic gelling of polysaccharides, proteins and humic substances. In many cases of environmental biofilm samples, proteins prevail, and humic substances are also integrated in the EPS matrix, being considered by some authors as belonging to the EPS (Wingender et al. 1999). Although mostly a minor component, lipids can make up a significant proportion of the EPS in some cases. This has been shown in the case of strongly acidophilic organisms, colonizing and leaching pyrite (Gehrke et al. 1998). The role of proteins, however, is mostly considered in terms of their enzymatic activity. Only a few authors speculate that extracellular proteins may also have structural functions. For example, the bridging of extracellular polysaccharides by lectin-like proteins is discussed (Dignac et al. 1998). Furthermore, the role of lectins (proteins with specific binding-sites for carbohydrates) in adhesion of bacterial cells and biofilm formation has been investigated (Tielker et al. 2005). A part of the extracellular proteins has been identified as enzymes. An overview about extracellular enzymes can be found in Wingender and Jaeger (2002). Enzyme activities in biofilms include among others aminopeptidases, glycosidases, esterases, lipases, phosphatases and oxidoreductases (Frølund et al. 1995). Most of these enzymes are an integrated part of the EPS matrix (Frølund et al. 1995). It is believed that their main function is the extracellular degradation of macromolecules into low molecular weight compounds which can then be transported into the cells and are available for microbial metabolism. The degradation and utilization of particulate matter is performed by colonization of the material and the secretion of extracellular enzymes. The EPS matrix prevents the enzymes and the degradation products from loss and keeps them in close proximity to the biofilm cells. Moreover, specific interactions between extracellular enzymes and other EPS components have been observed resulting in the protection and localization of the enzyme (Wingender et al. 1999). It is suggested, that the structure of the EPS matrix might not be purely random but is involved in the regulation and activity of extracellular enzymes. Thus, the cell maintains a certain level of control over enzymes which otherwise are out of reach.

In order to degrade hydrophobic compounds, microorganisms excrete surface-active polymers. An overview of the various types of EPS, their properties and their significance can be found in Neu (1996). The production of those biosurfactants can be induced by hydrophobic carbon sources (Ramsay et al. 1987), indicating the potential of microorganisms to secrete certain EPS when required. It is well known that hydrophobic surfaces can be colonized easily as demonstrated in nature by biofilm formation on leaves during biological degradation.

Adhesion and Cohesion

A major ecological advantage of the biofilm mode of life is that consortia of various organisms can establish and maintain their position over a long period of time, compared to the planktonic form of life. This applies not only to biofilms but also to flocs and allows for the development of synergistic relationships. A classical example is nitrification which takes place in biofilms and allows the spatial closeness of ammonia oxidizers to nitrite oxidizers. The EPS molecules which keep the organisms together and, if they form a biofilm, are responsible for adhesion to a given surface, provide this advantage. There is literally no surface material which cannot be colonized sooner or later, but there are strong differences in the colonization kinetics. In some cases, attachment was found to stimulate the synthesis of EPS (Vandevivere and Kirchman 1993). Both adhesion and cohesion are based on weak physico-chemical interactions and not on covalent bonds.

Three major kinds of forces can be distinguished: electrostatical interactions, hydrogen bonds and London dispersion forces (Mayer et al. 1999; Flemming et al. 2000). This is symbolized in Fig. 9.2 (Flemming et al. 2000).

The individual binding force of any type of these interactions is relatively small compared to a covalent C-C bond. However, the total binding energies of weak interactions between EPS molecules multiply with the large number of binding sites available in the macromolecules and add up to bond values exceeding those of covalent C-C bonds. The matrix network is formed by fluctuating adhesion points and the resulting matrix can behave as a gel as long as a certain shear stress is not exceeded. In this phase, the adhesion points flip back to their original arrangement. Above that point ("yield point"), new adhesion points assemble and the matrix behaves as a highly viscous fluid (Körstgens et al. 2001).

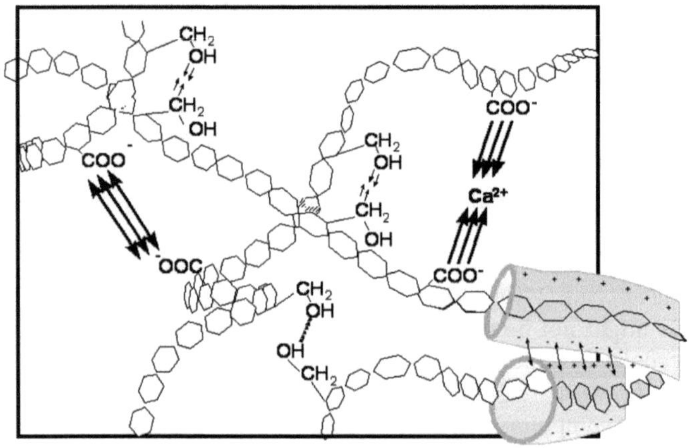

Fig. 9.2.
Weak physico-chemical interactions between polysaccharide strains in the EPS matrix (after Flemming et al. 2000)

Biofilm Architecture and Mass Transport

The architecture of the EPS matrix influences the processes within biofilms profoundly. Costerton et al. (1994) have shown that pores and channels occur in which convective transport is possible to a certain extent. Hoffman and Decho (1999) postulated areas of different density of the matrix, which have been observed experimentally. These features result in an extremely heterogeneous structure. This structure is dynamic; Schmitt et al. (1995) demonstrated in a *P. putida* biofilm which was charged with toluene that a rising concentration of toluene caused the formation of more polysaccharide, and furthermore, these compounds contained more carboxyl groups. In sediment biofilms with access of light, algae are major contributors to EPS production (see Fig. 9.1).

An important question is whether the EPS matrix acts as a diffusion barrier. The major component of that matrix is water. It could be shown by NMR measurements that the self-diffusion coefficient of water within the biofilm is practically the same as in free water, and only a very small fraction, less than 0.1%, displays a significantly lower diffusion coefficient (Vogt et al. 2000). There is evidence that non-charged molecules up to a molecular mass of around 10 000 Dalton experience practically no diffusion limitation. However, if they are consumed, as is the case with oxygen, gradients arise because oxygen consumption by aerobic organisms can occur faster than oxygen can follow the diffusion gradient. This is how anaerobic zones in biofilm arise and why anaerobic organisms can find suitable habitats directly below respiring aerobic colonies. Charged molecules can interact with charged groups of the EPS, which may slow down their mobility to a certain extent. This makes perfect sense from an ecological point of view because the mobility of nutrients, exoenzymes and other products is not restricted within the matrix, which is of great importance for cells located in the center of clusters.

Mass transport is influenced not only by the internal architecture of biofilms but also by their interface to the water phase. Some biofilms have a highly filamentous appearance while others are smooth. It is obvious that a large number of filaments will increase the surface at which interactions with components of the water phase are possible.

It could be demonstrated that in this biofilm matrix which is dominated by polysaccharides with carboxyl groups, calcium acts as an important bridging ion which increases the stability of the network significantly. This is also the case for copper and iron but not for magnesium. In such cases, surfactants will not contribute to the dissolution of biofilms. However, if other biopolymers dominate, it is possible that surfactants have a more significant effect. Hydrogen bonds are also part of the overall binding force. They can be influenced by so-called chaotropic agents which have a high affinity for water, thus interfering with the water shell around the biopolymers. In some cases, this type of bond dominates the binding forces. The extent to which each bond contributes to the cumulative binding force depends strongly of the nature of the EPS molecules. As different strains can produce different EPS, the variety is considerable, suggesting that not all biofilms can be dissolved by means of only one cleaning formulation. This coincides well with observations from practice. EPS are not totally insoluble in water. A certain amount of EPS is continuously lost to the water phase. In wastewater, this contributes to important measures of process parameters such as chemical oxygen demand (COD).

9.1.3 Biofilm Role on Sediment Stability

In sediments, sand grains are at least partially colonized by biofilms (Decho 1994). However, the extent of colonization can range from minute specks (Fig. 9.3) to complete coverage and clogging of the sediment particles (Fig. 9.4). It is long known that biofilms influence the entrainment of sand (Dade et al. 1990), mainly due to the cohesive forces between biofilm-covered sand grains.

Massive colonization of sediments can result in clogging and has been reported by Battin and Sengschmitt (1999). Here, algae were the main causes for copious EPS production which resulted in clogging. In addition, detrital material can accumulate in sediment biofilms, contributing to further clogging (Neu and Lawrence 1997).

When light has access to sediments, algae develop and contribute to biofilm mass. De Brouwer et al. (2005) have investigated in particular the role of the benthic diatoms *Nitzschia* cf. *brevissima* and *Cylindrotheca closterium*. They determined critical shear stress in presence and absence of these organisms and found a significant correlation of extracellular carbohydrate to critical shear stress for *N.* cf. *brevissima* but not for

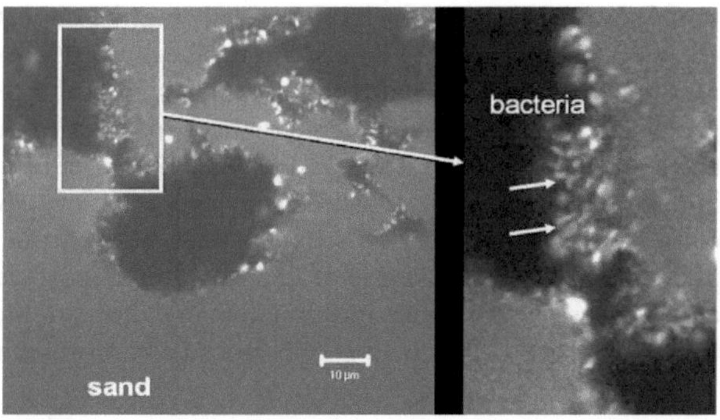

Fig. 9.3. Patchy biofilm on the surface of sand grains. The cells were stained with the DNA specific fluorochrom SYTO 9 (molecular probes). Epifluorescence microscopic image; *bar* = 10 µm (after Strathmann et al. in press)

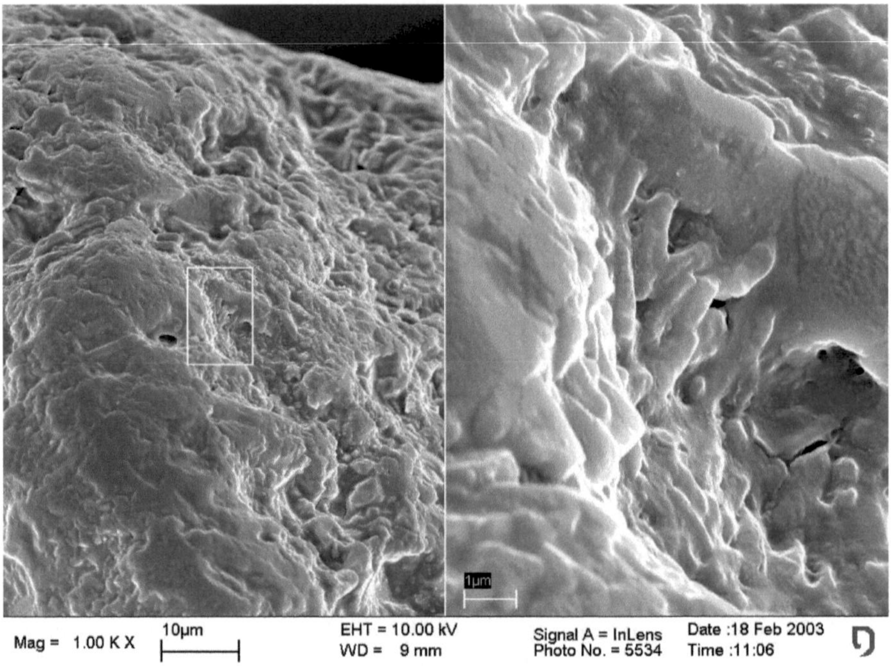

Fig. 9.4. SEM of a continuos biofilm on sand grains, formed by the mucoid strain *Pseudomonas aeruginosa* SG 81; the cells can be seen embedded in a thick alginate matrix (Leis unpublished)

C. closterium, although they could not give an explanation but only the coinciding observation that both organisms formed biofilms of different morphology as visualized by confocal laser scanning microscopy. However, the finding illustrates the fact that the carbohydrate content alone cannot be taken as an indicator for sediment stability; therefore, diatoms cannot be fully accountable for biostabilization in natural sediments as suggested by some authors (Madsen et al. 1993; Yallop et al. 2000). Neto et al. (this vol.) showed that settling of sand was slower with sterilized sediments, indicating the biotic component in settling. De Brouwer et al. (2005) conclude that "it is clear that stability of the sediments is poorly explained by simple indicators such as chlorophyll a and the extracellular carbohydrates. This again indicates that compositional characteristics of the biofilm matrix may be important to explain the stabilizing effect exerted by diatom biofilms". It has been generally considered that the action of EPS in stabilization of sediments involves chemical interactions between functional groups in the EPS and sediment particles in order to physically bind sediment particles together (Paterson 1997; Yallop et al. 1994). De Brouwer et al. (2005) found clear indications that Ca^{2+} ions favored the adsorption of EPS to sediment particles when compared to Na^{+} ions. This suggests that cation divalent bridging is an important process mediating adsorption of EPS to sediment particles (Decho 1994). In Table 9.1, an overview on stabilizing and destabilizing effects of biological components is given (after Black et al. 2002, with results from Leon Morales, this vol.).

Table 9.1. Aspects of stabilizing and destabilizing biological effects on natural cohesive sediment (after Black et al. 2000, with results from Leon-Morales, this vol.)

Stabilizing	Destabilizing
EPS secretion: EPS by bacteria and microphytobenthos enhances cohesion, promotes flocculation and, hence, deposition	Blistering: Trapping of oxygen bubbles in biofilms increases buoyancy of the biofilm to such an extent that it pulls away from sediment
Sediment compaction: Burrowing macrofauna increase sediment density and hence stability	Grazing: Organisms feeding on biofilms cause physical disturbance and reduce cohesion of sediment
Increased drainage: Burrow and channel formation promotes dewatering and higher density of sediment material	Burrow cleaning: Some benthic fauna clean tubes they inhabit in the sediment, giving rise to a localized benthic flux
Flow effects: Plants and animal tubes from dense fields that induce skimming flow in the overlying water, protecting the sediment bed from erosion	Boundary layer effects: Burrows, tubes and tracking of the sediment surface increase bed roughness and hence near bed turbulence so enhancing erosion
Sediment compaction: In deeper sediment layers, EPS lead to stronger cohesion	Fluffy layer: Loose parts of biofilms at the sediment surface are easily sheared away, taking bound superficial sediment material with them

9.1.4 Role of Biofilms As Sink and Source of Pollutants

Dissolved and particulate matter in water is continuously interacting at the interfaces with other phases in terms of sorption and desorption. If a biofilm is present, it will participate in such processes either by accumulation or by biochemical transformation of substances – basically, the ability to sequester matter from the water phase is the key mechanism of nutrient acquisition for biofilm organisms. But even if substances trapped in the EPS network do not interact with biofilms at all, they will have to pass through the biofilm when interacting with the underlying surfaces. In the case of large molecules, they can experience significant diffusion resistance. The term "sorption" refers to *adsorption*, *absorption* and *desorption. Adsorption* implies the retention of a solute on the surface of the particles of a material. *Absorption* in contrast involves the retention of a solute within the interstitial molecular pores of such particles (Skoog 1996). Biofilms are involved in all of them. A closer look at the dynamic role of biofilms in terms of sorption sites reveals a complex system (Fig. 9.5).

When a dissolved or particulate substance, transported by the water phase, meets a biofilm, it will not encounter a uniform structure but a highly heterogeneous hydrogel with very different sorption sites. These include:

- *Extracellular polymer substances* (EPS), mainly consisting of polysaccharides and proteins:
  - Charged groups, e.g.,: $-COO-$, $-SH-$, $-SO_4^{2-}$, $-H_2PO_4-$, $-NH_4^+$, $-NRH_2^+$
  - Apolar groups, e.g.,: aromatics, aliphatics such as found in proteins; also: hydrophobic regions in polysaccharides

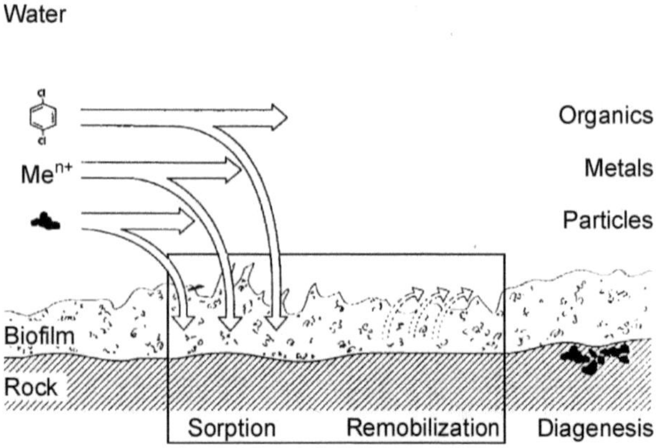

Fig. 9.5.
Role of biofilms in sorption and desorption processes (modified after Flemming and Leis 2002)

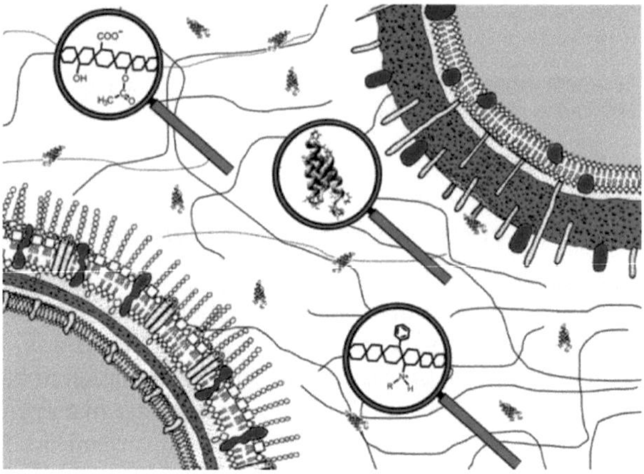

Fig. 9.6.
Various sorption sites in the EPS matrix (from top to bottom): *(i)* charged groups of polysaccharides, *(ii)* proteins with polar and apolar sites, *(iii)* apolar and anionic substituents on polysaccharides, additional sorption sites are the cell walls, the membranes and the cytoplasm (from Strathmann et al. in press)

- Cell walls
  - Outer membrane of Gram-negative cells (lipids)
  - Murein or teichoic acid layer of Gram-negative resp. Gram-positive Bacteria
  - Cytoplasmatic membrane (lipids)
- Cytoplasm

The EPS represent the major component of biofilm organic carbon. In Fig. 9.6, the sorption sites within the hydrogel matrix are schematically depicted.

It is obvious that each of these sites has different sorption mechanisms and capacities. Furthermore, the system is dynamic (Sutherland, 2001). Sorption characteristics in bacteria as living organisms can change depending on a great number of factors (Langley and Beveridge 1999). For example, the extent of sorption by some heavy metals will depend on nutritional factors. Nickel uptake in *Pseudomonas aeruginosa* can be increased or lowered depending on the carbon source supplied during growth (Sar 1998). Another example: a biofilm of *Pseudomonas putida* which was exposed to toluene responded in an increase of charged groups in the EPS, providing more ionic binding sites (Schmitt et al. 1995).

Many authors do not differentiate between various sorption sites in biofilms when investigating metal sorption (e.g., Mages et al. this vol.). EPS seem to be a highly plausible binding site from a mechanistic point of view, considering the charged groups of the EPS and their ionic binding capacity. However, in spite of the large body of references confirming this, Späth et al. (1998) separated EPS from cells after charging activated sludge with Cd^{2+}. Most of the metal was bound to the cell surfaces and not to the EPS as could have been expected considering the charged nature of many EPS components. The same was true for Ni^{2+} and Zn^{2+}. These findings demonstrated that bacterial cell walls can act as templates for metal deposition.

9.1.5 Microbial Mineralization and Sediment Formation

Numerous bacteria in aquatic sediments encounter and bind a wide variety of metals in their environment. As the sediments accumulate, they and their bacterial components become subject to geological forces that eventually result in rock formation. During this time, chemical and physical changes occur within the sediments as diagenetic processes. The cell walls of bacteria, present in these sediments, make suitable biological templates for the concentration of metals and the nucleation of crystals and can often greatly influence the initial mineralization process (Krumbein et al. 1994). Carbonate formation in bacteria is another important sorption process mediated by bacterial surfaces. It happens due to the production of alkaline microenvironments near the cell surfaces as a consequence of physiological activities of the cell (for example, bicarbonate use as carbon source, yielding OH^- groups) and also due to the ability of S-layers to bind available Ca^{2+} which together allows gypsum ($CaSO_4 \cdot 2\,H_2O$) and/or calcite ($CaCO_3$) to precipitate. Silicate formation by bacteria has been a very common phenomenon since the beginning of life. It is very likely that the direct participation of bacteria in the formation of these compounds rather than just a silicate after-formation binding with bacterial surfaces. The phenomenon of formation of *microfossils* is thought to be related to silicate interaction with bacteria (Schultze-Lam et al. 1996). Biofilm cells in drinking water pipes seem to be suitable nucleation sites for mineral deposition. Figure 9.7 is a scanning electron micrograph showing microorganisms as templates for the deposition of iron oxides.

Fig. 9.7. Iron oxides precipitated on bacteria in corrosion products of a drinking water pipe (modified from Flemming and Leis 2002)

The external environment and internal metabolism of living bacteria often exert a profound influence on the chemistry of bound metals. These influences include changes in oxidation state, formation of organometallic compounds and formation of precipitates due to detoxification or energy-yielding mechanisms of the bacterial protoplast. Alternatively, the metals may be affected indirectly by the production of metabolic end-products such as SO_4^{2-} and S^{2-} or an alteration in the local pH and/or E_h. The results of microbial activity may ultimately lead to metal immobilization, remobilization and/or the formation of metal aggregates.

9.1.6 Desorption Processes

Bacterial surfaces and biofilms are not inert chemical structures. They represent a dynamic system in which the various components are synthesized, assembled, modified and finally broken down by autolysins and sloughed off into the environment. Thus, they may contribute to the remobilization of the sorbed substances.

By nature, the immobilization of metal ions in biomass cannot be irreversible. The biological binding sites sooner or later will be degraded. Fate and transport of the metal is directly related to the fate and transport of the bacterial cell. When the cell dies the metal is released. In some instances, this process will lead to mineral formation and is responsible for the deposition of large ores. However, in other cases the sorbed metal ions will return in their more soluble form and be remobilized. Experiments with cell walls of *Bacillus subtilis* and *E. coli* envelopes adsorbed to kaolinite and smectite clays and with the corresponding organic material-clay aggregates showed the complexicity of remobilization processes. Bound to these substances were Ag(I), Cu(II) and Cr(III). The sorbed metals were then leached with HNO_3, $Ca(NO_3)_2$, EDTA, fulvic acids and lysozyme at several concentrations. The findings on remobilization of the sorbed metals, in general showed the order $Cr^{2+} < Ag^+ < Cu^{2+}$. In the wall, clay and composite systems, Cr^{3+} was very stable; at pH 3, 500 micromolar EDTA, 120 ppm fulvic acid and 160 ppm Ca^{2+} released less than 32% (wet weight) of the sorbed chromium. Ag (45–87%) and Cu (up to 100%) were readily removed by these agents. The organic chelators were in general less effective at mobilizing certain metals than elevated Ca^{2+} or low (acidic) pH values. Lysozyme digestion of *Bacillus* walls remobilized Cu^{2+} from walls and Cu-wall-kaolinite composites. Ag^+ and Cr^{3+} smectite inhibited enzyme activity to some extent, and the metals remained insoluble.

Interesting is the comparison of the stability of metal complexes of low molecular weight complexing agents with the stability of metal-EPS-complexes. The EDTA complexes have stability constants of up to 10^{18}, the complex of Cu^{2+} with 1-hydroxyethane-1, 1-diphosphonic acid (HEDP) is about 10^{19}. If these complexing agents occur in the water phase, it should be expected that eventually sorbed metal ions are rapidly complexed. However, the remobilization in this case is unexpectedly low.

As desorption does usually not occur very fast, biofilms have a "memory effect" for pollutants with which they were in contact. This fact has been utilized for localizing industrial and municipal discharges in sewers, using sorption data from sewer biofilms. Using systematic upflow analysis it is possible to reveal the points at which pollutants

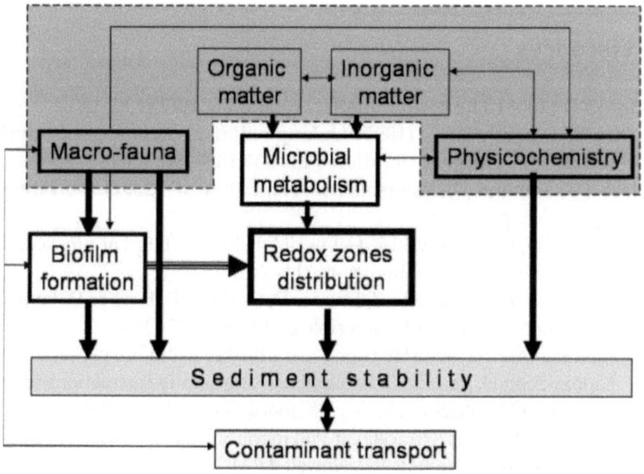

Fig. 9.8.
Complete overview of biogeochemical factors to be taken into account when making predictions on sediment transport in natural aquatic environments. Shaded area: covered by conventional studies

were discarded into the sewer system. This method is applied in order to identify point sources for waste water pollution.

9.1.7 Conclusions

From the information presented above, the complex influence of biofilms on sediment stability and pollutant mobility is obvious (Fig. 9.8). The problem for all attempts to model biofilm influence arises from the vast spatial and temporal heterogeneity of biofilms. In all non-sterile systems, biofilms are present. However, their influence on sediment stability and sorption/desorption of pollutants depends upon many factors such as:

- Nutrient availability (for heterotrophic organisms: biodegradable substances, for phototrophic organisms: access of light and limiting factors such as phosphate), and, as a consequence, biomass concentration
- Composition of the population (predators can change composition dramatically)
- Amount and nature of EPS
- Hydrodynamic conditions (higher shear stress will select for more cohesive EPS and specialized organisms)
- Sediment compartment considered (hydrodynamic influence can change with sediment depth)

It is clear that this is not an exhaustive list of biological factors involved but it is also clear that they cannot be neglected. Unless they are understood and acknowledged for, predictive modeling will be always crude and not really effective. In this respect, Sects. 9.2 and 9.3, explore both the influences of a biogenic component on the stability and settling characteristics of sediments as well as in terms of retention-remobilization processes. Section 9.4, in contrast, deals with a specific case of biofilm influence on the distribution of bound and dissolved metal species in environments directly linked with fluvial systems.

References

Allison D (2003) The biofilm matrix. Biofouling 19(2):139–150

Battin TJ, Sengschmitt D (1999) Linking sediment biofilms, hydrodynamics, and river bed clogging: evidence from a large river. Microb Ecol 37:185–196

Black KS, Tolhurst DJ, Paterson DM, Hagerthey SE (2002) Working with natural cohesive sediments. J Hydraul Eng, pp 2–8

Costerton JW, Lewandowski Z, Caldwell DE, Korber DR, Lappin-Scott HM (1995) Microbial Biofilms. Annual Review of Microbiology 49:711–745

Costerton JW, Lewandowski Z, Debeer D, Caldwell D, Korber D, James G (1994) Biofilms, the Customized Microniche. Journal of Bacteriology 176(8):2137–2142

Chamberlain AHL (1997) Matrix polymers: the key to biofilm processes. In: Wimpenny J, Handley PS, Gilbert P, Lappin-Scott H, Jones M (eds) Biofilms: Community Interactions and Control., UK, BioLine, pp 41–46

de Brouwer JFC, Wolfstein K, Ruddy GK, Jones TER, Stal LJ (2005) Biogenic Stabilization of Intertidal Sediments: The Importance of Extracellular Polymeric Substances Produced by Benthic Diatoms. DOI 49:501–521

Dade WB, Davis JD, Nichols PD, Nowell ARM, Thistle D, Trexler MB, White DC (1990) Effects of bacterial exopolymer adhesion on the entrainment of sand. Geomicrobiol 8:1–16

Decho AW (1994) Molecular-scale events influencing the macroscale cohesiveness of exopolymers. In: Krumbein WE, Paterson DM, Stal LJ (eds) Biostabilization of sediments, BIS-Verlag, Oldenburg, pp 135–148

Dignac MFU, Rybacki D, Bruchet A, Snidaro D, Scribe P (1998) Chemical description of extracellular polymers: implication on activated sludge floc structure. Wat Sci Tech 38:45–53

Flemming HC, Leis A (2002) Sorption properties of biofilms. In: Bitton (ed) Encyclopedia of environmental microbiology, Vol. 5. John Wiley & Sons, Inc., New York, pp 2958–2967

Flemming HC, Wingender J (2002) Extracellular polymeric substances (EPS): Structural, ecological and technical aspects. In: Bitton G (ed) Encyclopedia of Environmental Microbiology, vol. 4. John Wiley & Sons, Inc, New York, pp 1223–1231

Flemming HC, Wingender J (2003) The crucial role of extracellullar polymeric substances in biofilms. In: Wuertz S, Bishop P, Wilderer P (eds) Biofilms in wastewater treatment. An interdisciplinary approach., IWA Publishing, London, pp 401

Flemming H-C, Wingender J, Mayer C, Körstgens V, Borchard W (2000a) Cohesiveness in biofilm matrix polymers. In: Lappin-Scott H, Gilbert P, Wilson M, Allison D (eds) Community structure and co-operation in biofilms, SGM symposium 59, Cambridge University Press, pp 87–105

Flemming H-C, Wingender J, Griebe T, Mayer C (2000b) Physico-chemical properties of biofilms. In: Evans LV (ed) Biofilms: recent advances in their study and control, Harwood academic publishers

Frølund BG, Nielsen PH (1995) Enzymatic activity in the activated-sludge floc matrix. Appl Microbiol Biotechnol 43:755–761

Gehrke T, Telegdi J, Thierry D, Sand W (1998) Importance of extracellular polymeric substances from Thiobacillus ferrooxidans for bioleaching. Appl Environ Microbiol 64(7):2743–2747

Hoffman M, Decho AW (1999) Extracellular enzymes within microbial biofilms and the role of the extracellular polymeric matrix. In: Wingender J, Neu TR, Flemming HC (eds) Microbial extracellular polymeric substances. Springer-Verlag, pp 217–230

Huettel M, Røy H, Precht E, Ehrenhauss S (2003) Hydrodynamical impact on biogeochemical processes in aquatic sediments. Hydrobiologia 494:231–236

Körstgens V, Flemming H-C, Wingender J, Borchard W (2001) Uniaxial compression measurement device for the investigation of the mechanical stability of biofilms. J Microbiol Meth 46:9–16

Krumbein WE, Paterson DM, Stal LJ (1994) General discussion. In: Krumbein WE, Paterson DM, Stal LJ (eds) Biostabilization of sediments. BIS Oldenburg, pp 433–435

Langley S, Beveridge TJ (1999) Metal binding by Pseudomonas aeruginosa PAO1 is influenced by growth of the cells as a biofilm. Canadian Journal of Microbiology 45(7):616–622

Leon Morales CF, Leis AP, Strathmann M, Flemming HC (2004) Interactions between laponite and microbial biofilms in porous media: implications for colloid transport and biofilm stability. Water Research 38(16):3614–3626

Macedo AJ, Kuhlicke U, Neu T, Timmis KN, Abraham W-R (2005) Three stages of a biofilm community developing at the liquid-liquid interface between polychlorinated biphenyls and water. Applied and Environmental Microbiology 71(11):7301–7309

Madsen KN, Nilsson P, Sundbäck K (1993) The influence of benthic microalgae on the stability of a subtidal sediment. J Exp Mar Biol Ecol 170:159–177

Mayer C, Moritz R, Kirschner C, Borchard W, Maibaum R, Wingender J, Flemming HC (1999) The role of intermolecular interactions: studies on model systems for bacterial biofilms. Int J Biol Macromol 26(1):3–16

Neu T (1996) Significance of bacterial surface active compounds in interaction of bacteria with surfaces. Microbiological Reviews 60:151–166

Neu TR, Lawrence JD (1997) Development and structure of microbial biofilms in river water studied by confocal laser scanning microscopy. FEMS Microbiol Ecol 24:11–25

Paterson DM (1997) Biological mediation of sediment erodibility. In: Burt N, Pareker R, Watts J (eds) Cohesive sediments. Wiley, New York, pp 215–229

Ramsay BR, de Tremblay M, Chavarie C (1988) A method for the quantification of bacterial protein in the presence of Jarosite. Geomicrobiol J 3:171–177

Riding R, Amrawik SM (2000) Microbial sediments. Springer-Verlag, New York, Heidelberg, 331 pp

Sar P, Kazy SK, Asthana RK, Singh SP (1998) Nickel uptake by Pseudomonas aeruginosa: role of modifying factors. Current Microbiology 37:306–311

Schmitt J, Nivens D, White DC, Flemming H-C (1995) Changes of biofilm properties in response to sorbed substances – an FTIR-ATR study. Water Science and Technology 32(8):149–155

Schultze-Lam S, Fortin D, Davis BS, Beveridge TJ (1996) Mineralization of bacterial surfaces. Chem Geol 132:171–181

Skoog DA, West DM, Holler FJ (1996) Fundamentals of analytical chemistry, Saunders college publishing

Späth R, Flemming HC, Wuertz S (1998) Sorption properties of biofilms. Water Science and Technology 37(4–5):207–210

Spiers AJ, Bohannon J, Gehrig SM, Rainey PB (2003) Biofilm formation at the air-liquid interface by the Pseudomonas fluorescens SBW25 wrinkly spreader requires an acetylated form of cellulose. Mol Microbiol 50(1):15–27

Stal LJ, de Brouwer JFC (2003) Biofilm formation by benthic diatoms and their influence on the stabilization of interdital mudflats. Ber Forsch.-Zentr., Terramare 12:109–111

Strathmann M, Leon Morales CF, Flemming H-C (2006) Influence of biofilms on colloid mobility in the subsurface. In: Frimmel HF., Flemming H-C, Förstner U (eds) Colloid mobility, Springer-Verlag, Heidelberg, in press

Sutherland IW (1994) Structure-function relationships in microbial exopolysaccharides. Biotech Adv 12:393–448

Sutherland IW (2001) The biofilm matrix – an immobilized but dynamic microbial environment. Trends in Microbiology 9:222–227

Tielker D, Hacker S, Loris R, Strathmann M, Wingender J, Wilhelm S, Rosenau F, Jaeger K-E (2005) *Pseudomonas aeruginosa* lectin LecB is located in the outer membrane and is involved in biofilm formation. Microbiology 151:1313–1323

Vandevivere P, Kirchman DL (1993) Attachment stimulates exopolysaccharide synthesis by a bacterium. Appl Environ Microbiol 59(10):3280–3286

Vogt M, Flemming HC, Veeman WS (2000) Diffusion in *Pseudomonas aeruginosa* biofilms: a pulsed field gradient NMR study. J Biotechnol 77(1):137–146

Wingender J, Jaeger K-E (2002) Extracellular enzymes in biofilms. In: Bitton G (ed) Encyclopedia of Environmental Microbiology, vol. 3. John Wiley & Sons, Inc, New York, pp 1207–1223

Wingender J, Jäger K-E, Flemming H-C (1999) Interactions between extracellular enzymes and polysaccharides. In: Wingender J, Neu T, Flemming H-C (eds) Microbial extracellular polymer substances, Springer-Verlag, Heidelberg, Berlin, pp 231–251

Wingender J, Neu TR, Flemming H-C (1999) What are bacterial extracellular polymeric substances? In: Wingender J, Neu TR, Flemming H-C (eds) Microbial Extracellular Polymeric Substances. Springer-Verlag, Berlin, pp 1–19

Yallop ML, de Winder B, Paterson DM, Stal LJ (1994) Comparative structure, primary production and biogenic stabilization of cohesive and non-cohesive marine sediments inhabited by microphyto-benthos. Estuar Coast Shelf Sci 39:565–582

Yallop ML, Paterson DM, Wellsbury P (2000) Interrelationships between rates of microbial production, exopolymer production, microbial biomass, and sediment stability of intertidal sediments. Microb Ecol 39(2):116–127

Carlos Felipe Leon Morales · Martin Strathmann · Hans-Curt Flemming

9.2 Role of Biofilms on Sediment Transport – Investigations with Artificial Sediment Columns

9.2.1 Introduction

Sediments are a sink for contaminants which can be accumulated for long periods of time. The so-called sediment transport cycle in rivers starts with erosion caused by the action of the flowing water body. Normally, this is a constant but slow process, drastic events however, can result in the sudden remobilization and distribution of accumulated contaminants. Subsequent deposition of aggregates in the river bed closes the cycle. The subject can be approached from a wide variety of disciplines (Förstner 2004) and, traditionally physical and engineering sciences have helped understanding many key physical parameters on the study of sediment transport such as settling and erosion rates, boundary layer shear stresses, etc. In natural sediments, biological influence must be considered too, which applies for almost every aspect of the sediment transport cycle. Sediment ecology is complex with many trophic levels involved. Biofilms are a common way of microbial aggregated life in sediments. Biofilm microorganisms are embedded in a matrix of extracellular polymeric substances (EPS) (Flemming and Wingender 2002). Hydrodynamics determine greatly mass transfer processes in river sediment biofilms as well as their structure. Biofilms are not confined to the sediment-liquid interface but are also distributed along considerable sediment depths (Battin and Sengschmitt 1999). Benthic biofilms, composed mainly of microalgae (but also bacteria) have been recognized as important factors on sediment stability (Sutherland et al. 1998). In terms of retention-remobilization processes, biofilms have considerable and complex sorption properties and capacities. Biofilms contribute to the net capacity of the sediment to retain, transform, and liberate migrating substances. Anionic groups in EPS such as carboxyl, phosphoryl, and sulfate groups can represent important cationic exchange sites. It is not entirelly clear however, how changes on fluid dominant ionic species which commonly occur in subsurface and other environments, contribute to the influence exerted by biofilms. This is especially true in highly permeable sediments were advection and dispersion can represent an important mixing and transport mechanism. The aim of this work is to investigate the influence of biofilms on sediment stability and retention-remobilization processes in model sediment columns, at two commonly found fluid ionic conditions. Biofilm influence is expressed in terms of the relative contribution of two major EPS components: extracellular carbohydrates and extracellular proteins. The collected data is expected to contribute to the understanding of complex interaction processes which determine biofilm influence in sediment transport.

9.2.2 Materials and Methods

Organisms and Influent Solutions

Pseudomonas aeruginosa SG81 and its alginate deficient mutant, *P. aeruginosa* SG81R1 were used as model biofilm forming microorganisms. *P. aeruginosa* SG81 is a mucoid, highly EPS producing bacterium (Grobe et al. 1995). The organism was kept on Pseudomonas Isolation Agar, PIA (Difco^{TM}). Liquid cultures were maintained in Triptic Soy Broth, TSB (Merck). For biofilm growth inside sediment columns (1, 2 or 3 weeks), a defined salts medium, APM50 was used. This medium consisted of 50 mM sodium gluconate, 1 mM KNO_3, 0.1 mM $MgSO_4 \cdot 7\,H_2O$, 0.05% yeast extract and 0.2 M NaCl.

Rhodamine 6G was used as a model organic pollutant. It is used in a wide variety of applications going from ground water tracer experiments to fingerprint detection technology in forensics. The compound was detected by fluorescence (λ_{ex} 480 nm λ_{em} 541 nm) (SFM 25, Kontron Instruments fluorometer). $CaCl_2$ or NaCl based solutions were used at 70 mM concentration as influent solutions either when APM50 media or the model contaminant were not the influents. For bacterial transport experiments various NaCl concentrations were used, ranging from 0.6 mM up to 1 M. Desionized water was the lowest ionic strength influent solution used.

Sediment Columns and Main Experimental Setting

To determine biofilm influence on sediment transport processes, experiments were performed in columns representing both the sediment-liquid interface ("flume columns") and the sediment porous matrix ("PM columns") (Fig. 9.9), in absence (sterile) and presence of biofilms (non-sterile). Organic-free (550 °C, 2 h) sand F_{36} with an average diameter of 0.1 mm was used as model sediment. Flume columns were horizontally positioned, glass columns, half-filled with the model sediment. PM columns in contrast, were completely filled and consisted on vertically positioned glass columns as well as microscopy flow cells which were used for direct visualization and quantification of transport processes inside the sediment porous matrix. The columns were con-

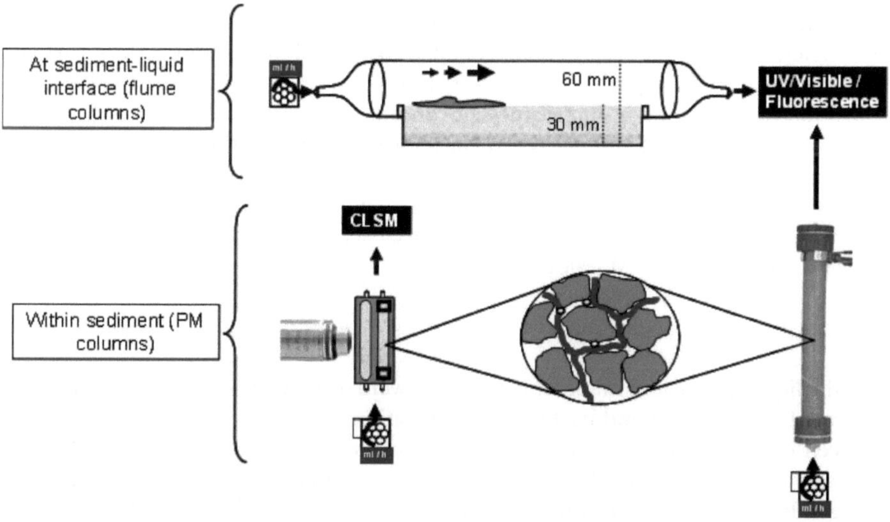

Fig 9.9. Main experimental setting. *Upper section:* "Flume columns" representing the sediment-liquid interface. *Lower section:* "PM columns" both sediment-filled microscopy flow cells (*left*) and completely filled sediment columns (*right*) representing the sediment porous matrix

nected using PVC or tygon tubing (Novodirect, Kehl, Germany) to peristaltic pumps (Ismatec SA, Switzerland). Columns and accompanying tubing were sterilized by autoclaving (121 °C, 20 min) or by dry heat (250 °C, 4 h). Columns for biofilm experiments were inoculated with *P. aeruginosa* suspensions (approx. 10^9 cells ml^{-1}) and then fed with APM50 as specified before. After the completion of the experiments (except for the microscopy flow cells), the remaining sand was removed from the column for further analysis. All described experiments were done at least in triplicate and mean values are presented. Error bars represent the standard deviation around these mean values. Pearson correlation coefficients, r, were calculated with the open source spreadsheet Gnumeric and considered only significant for $p < 0.05$ (t-test distributions).

Stability and Retention-Remobilization Experiments

Stability experiments were done in "flume columns" only. Shear stress was applied to the sediment bed by water flow (flume-like stress) (704S, Watson Marlow peristaltic pump). The starting point of sediment particle movement was assessed at increasing flow velocities using a video camera. The velocities ranged from 1 cm s^{-1} up to 21 cm s^{-1} set at equivalent intervals. *P. aeruginosa* SG81 was used for these experiments. APM50 as influent medium was replaced the day before by either $CaCl_2$ or NaCl solutions as background solutions. The ionic strength of these solutions was kept constant all the time.

For retention-remobilization experiments the flow rate was maintained constant and mobility was stimulated by decreasing the ionic strength of the background solutions. Flume columns were exposed to the same amount of contaminant influent at a concentration of 6×10^{-6} M. This was followed by pollutant-free influent until effluent pollutant concentration was stable and low. *P. aeruginosa* SG81 as well as its mutant were used in these experiments. For experiments with "PM columns", colloid-associated pollutant and bacterial mobility through the sediment were assessed as collision efficiencies. Collision efficiency, α, is defined as the probability of a migrating particle to attach, upon collision with sediment particles. It was obtained as the ratio between deposition rate constants at increasing salt concentrations (up to 70 mM NaCl for colloid-associated pollutant and up to 1 M NaCl for bacteria) and those at high salt concentrations (= 1M NaCl) where deposition is independent of the salt concentration. Due to the reason that deposition is always higher at high salt concentrations, α values closer to 1 indicate high deposition while those closer to 0 indicate low deposition, for further details see (Leon Morales et al. 2004).

Analyses

Both the liquid phase and the sediment matrix were used to obtain information about the model contaminant and bacterial concentrations. Cell enumeration was done by total cell counts (TCC) using thoma cell chambers. EPS material was quantified as total extracellular carbohydrates and total extracellular proteins as described in (Wingender et al. 2001). Biofilms were separated from the sand matrices by mechanical shear stress (stomacher® 400 circulator) in the presence of a cation exchange resin, CER (DOWEX 50 × 8, Fluka) in a modified version of the procedure described by (Frølund et al. 1996).

A UV-VIS spectrophotometer (Cary 50, Varian Inc) was used to monitor bacterial and contaminant concentrations in columns effluents. Confocal laser scanning microscopy, CLSM (LSM510, Zeiss) was used on the microscopy flow cells to observe at the pore scale, the fluorescence of labeled bacteria and colloidal tracers as described in (Leon Morales et al. 2004).

9.2.3 Results

Sediment-Liquid Interface – "Flume Columns"

Sediment stability. The influence of biofilm growth and the ionic nature of the background solution on sediment stability were investigated. Higher shear stress (created by step-wise increasing influent flow rates) was necessary for the initiation of sediment particle movement in the presence of 5 day-old biofilms as compared with sterile, organic-free columns (Fig. 9.10). After 15 days of biofilm development, a fluffy layer developed on top of the sediment which was sheared off at low flow rates (1–5 cm s^{-1}). The underlying sediment, however, displayed an elevated stability. This stabilizing effect could be correlated with increasing protein and carbohydrate concentrations.

An increase on the critical shear stress was observed when the background solution was $CaCl_2 + NaHCO_3$ instead of $NaCl + NaHCO_3$ even though all other parameters including time of biofilm growth were kept constant. There was a higher correlation of carbohydrates to the critical shear stress ($r = 0.89$) as compared with protein ($r = 0.60$). The correlation of carbohydrates to the critical shear stress is further increased and found to be significant ($p = 0.02$), when the background solution was $CaCl_2 + NaHCO_3$.

The ratio carbohydrate/protein, found in the extruded sediment, increased with time of biofilm growth: from 1.58 after 5 days of growth to 2.63 after 15 days of growth and after Na^+ based background solutions and to 2.27 for Ca^{2+} based background solutions.

Retention-remobilization. Model pollutant retention-remobilization was investigated in the presence of Ca^{2+} based background solutions, for columns inoculated with *P. aeruginosa* SG81 and its alginate deficient mutant. After model contaminant column saturation (Fig. 9.11) and the subsequent stimulated remobilization, fractions were

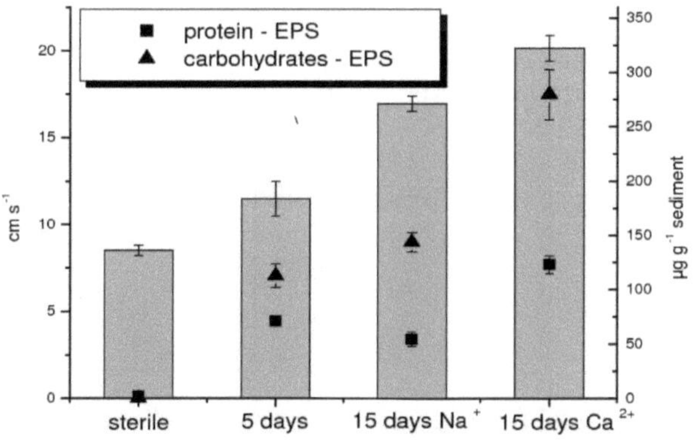

Fig 9.10. Biofilms induced sediment stability. The *bars* indicate the linear velocity (*left y-axis*) at which sediment particle movement started. *Squares* and *triangles* represent micrograms of extracellular proteins and extracellular carbohydrates respectively per gram of extruded wet sediment

collected from sediment columns effluents at fixed time intervals and flow rates. Relative fluorescence of each fraction was plotted against time of elution. Integration of these breakthrough curves was used to quantify the amount of remobilized pollutant.

Sterile columns resulted in the higher remobilization rates in triplicate columns as compared to the columns inoculated with the mutant and the alginate producing bacterium (Fig. 9.12).

Figure 9.12 shows that pollutant remobilization in the presence of both *P. aeruginosa* SG81 and SG81R1 biofilms is significantly decreased. Taking the integrated remobilization of the sterile columns as a starting point, model pollutant remobilization from the alginate-deficient mutant inoculated columns was 52% while it was 23% in presence of a biofilm with the alginate-producing wild strain. The correlation of EPS to pollutant retention is high for both proteins and carbohydrates ($r = 0.90$ and $r = 0.97$ respectively), however, the ratio carbohydrate/protein is higher (2.83) in the mucoid columns as compared to the non-mucoid ones (0.70).

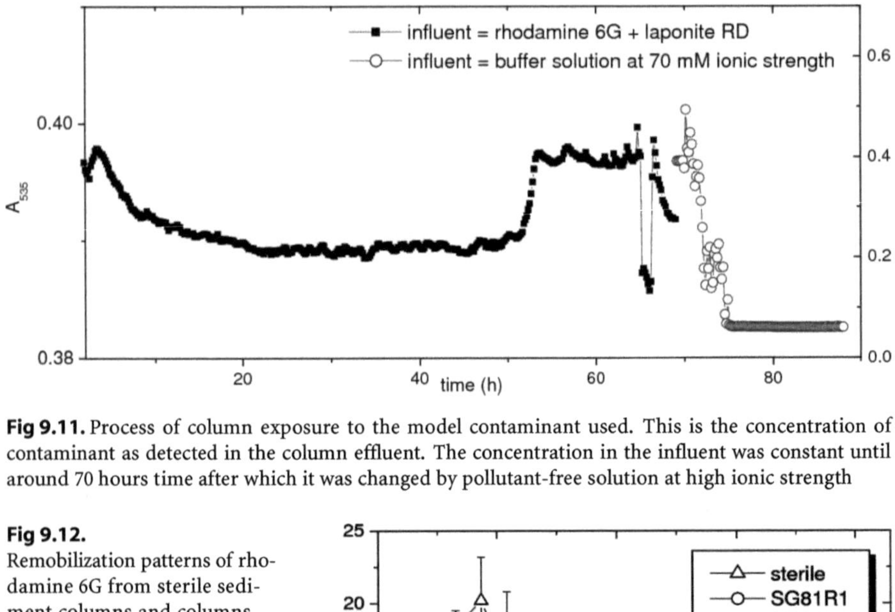

Fig 9.11. Process of column exposure to the model contaminant used. This is the concentration of contaminant as detected in the column effluent. The concentration in the influent was constant until around 70 hours time after which it was changed by pollutant-free solution at high ionic strength

Fig 9.12. Remobilization patterns of rhodamine 6G from sterile sediment columns and columns grown with *P. aeruginosa* SG81 and *P. aeruginosa* SG81R1 biofilms

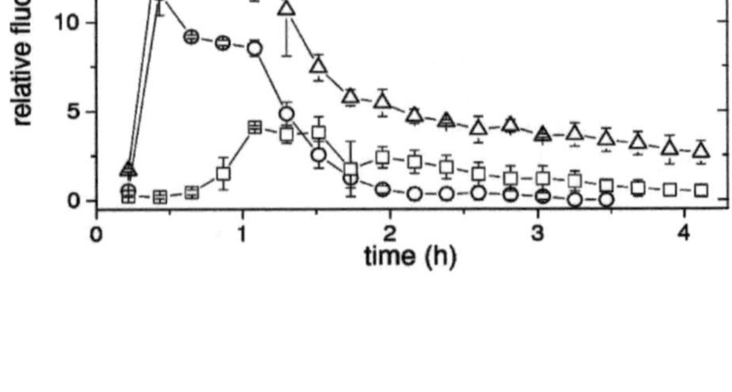

Processes within the Sediment – "PM Columns"

Bacterial transport and biofilm formation in porous media. Biofilms developed also between sediment particles even at considerable sediment depths. These experiments therefore investigate the role of biofilms in pollutant transport and hydrodynamics at this level. For these experiments, only *P. aeruginosa* SG81 was used. Data from the completely saturated columns and from the sand-packed microscopy flowcells showed that the saturated hydraulic conductivity decreased over time when columns were inoculated and fed with nutrients. After an instability period on the measured hydraulic conductivity, it decreased steadily until a more or less constant lower plateau was reached for the rest of the observation period. This plateau was observed after approximately 7 days of constant nutrient influent. In non-inoculated columns a reduction in saturated hydraulic conductivity was observed after the first hours of column packing. The measured hydraulic conductivity remained more or less stable at a higher value than obtained with the biofilm growing columns (Fig. 9.13).

The attachment or collision efficiencies of biofilm bacteria give information on the first steps of biofilm formation in a porous medium. Relatively high collision efficiencies were found for the model microorganism at a wide range of salt concentrations as compared with clay colloids (Fig. 9.14).

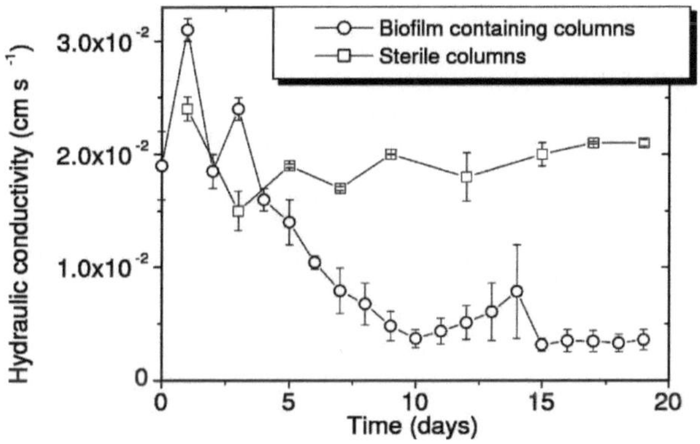

Fig 9.13.
Changes in porous matrix hydraulic conductivity induced by biofilm growth. Hydraulic conductivity was measured using a constant head permeameter

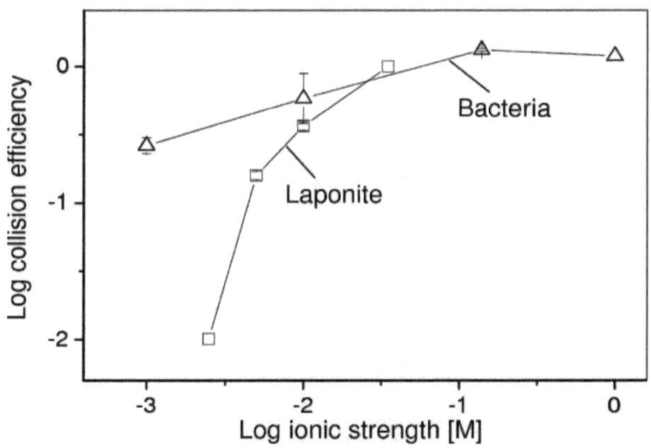

Fig 9.14.
Collision efficiencies obtained from image analysis of bacterial and colloid transport experiments done with sediment packed microscopy flowcells and CLSM

Bacterial attachment occurred even at low ionic strengths (1 mM in Fig. 9.14 and 9.15), however, drastically decreasing ionic strength in the presence of Na^+ ions, results in remobilization of retained microorganisms (Fig. 9.16b). In the presence of clay-like colloids, in contrast, sudden increments on ionic strength resulted in clogging of the porous matrix and remobilization of attached bacteria (Fig. 9.16c). Additionally, biofilm growth was confirmed in the surface of sand grains by CLSM (Fig. 9.16a).

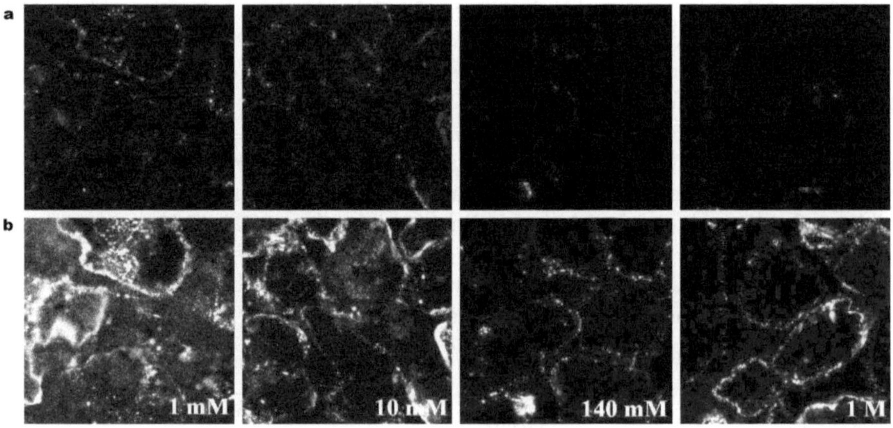

Fig 9.15. Pore scale visualization of bacterial attachment to sediment grains. Bacteria were labeled with SYTO 9 and injected as a pulse. The pictures represent before (**a**) and after (**b**) bacterial elution at the same distance from the flow-cell inlet and at the same range of ionic strengths as depicted in Fig. 9.14

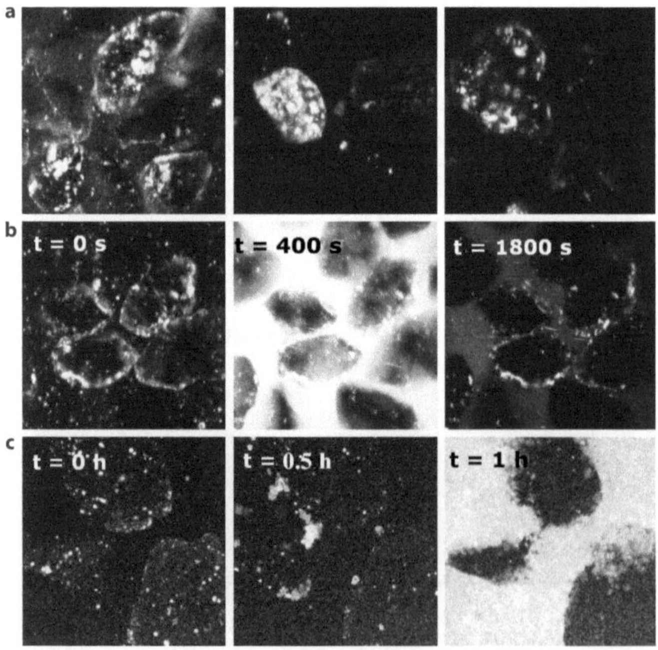

Fig 9.16. Visualization of porous medium hydraulic conditions. **a** Biofilm formation on top of sand grains evidenced by detection of fluorescently labeled (SYTO 9) bacterial cells using CLSM. Uncovered sand grains remain dark. **b** Remobilization of bacterial cells (lighter zones) after drastically decreasing ionic strength in the influent solution. **c** Clogging of the porous matrix with colloidal aggregates at high ionic strength conditions. The aggregates are stained with rhodamine 6G which is also detected by fluorescence (lighter zones)

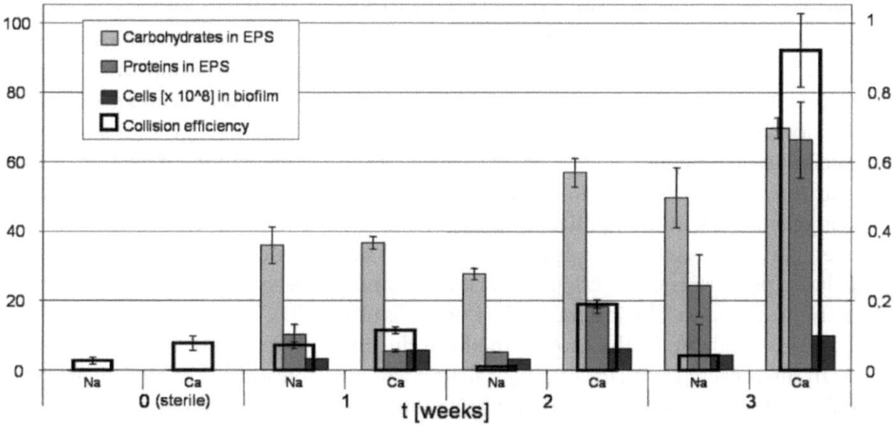

Fig 9.17. Collision efficiencies (*right y-axis*) of colloid-associated rhodamine 6G and corresponding EPS and cell counts (*left y-axis*) remaining in saturated sediment columns (porous matrix) after colloid mobility experiments. *x-axis* represents weeks of biofilm growth

EPS, Electrochemical Conditions and Pollutant Retention

After characterization of biofilm formation and bacterial attachment and transport through the target sediment, biofilm influence on the mobility of the model pollutant was investigated using background solutions dominated by two different ionic species. The amount of EPS especially proteins and cell numbers increased with time in inoculated columns (Fig. 9.17). The impact of biofilm growth on colloid and colloid-bound model pollutant transport depended, however, greatly on the type of cation dominating the background solution previous to pollutant injection. This is especially evident after 3 weeks of biofilm growth (Fig. 9.17).

Pollutant retention increased with time and EPS content in columns with Ca^{2+} dominated background solutions. There was some correlation of carbohydrates to pollutant retention ($r = 0.47$) in these systems, however, the correlation of proteins to retention was significantly higher ($p = 0.02$). In the case of columns with Na^+ dominated background solutions, pollutant retention did not increase with time. The combined EPS production in the porous matrix environment was lower (as much as 1 order of magnitude) as compared to the liquid sediment interface. In contrast to what happened at the sediment-liquid interface, inside the porous medium, carbohydrate/protein ratio decreased with time of biofilm growth (from around 6 in the first week to around 1 in the third week) when Ca^{2+} dominated the background solution. It remained constant, however, when Na^+ was the dominant ion in the background solution.

9.2.4 Discussion

Sediment Stability

At the sediment water interface, biofilm growth had a clear influence on sediment stability, i.e. on the stress necessary to start sediment particle movement. Plateau phase biofilm growth at this interface (15 day biofilms) had a stabilizing effect only after a layer of low density EPS was eliminated at low flow rates. Visual observations showed that this "fluffy" layer carried with it associated sediment material but this was not quantified. It was found that sediment stability is positively correlated with an increased production of EPS. Especially in the case of Ca^{2+} dominated background solutions, extracellular carbohydrates were highly correlated to critical shear stress (Fig. 9.10). The carbohydrate/protein ratio increased with time at the sediment-water interface. This suggests a calcium stabilizing and cementing effect, within carbohydrates and between carbohydrates and sediment particles. The influence of calcium ions in the viscoelastic properties of *P. aeruginosa* SG81 biofilms has been demonstrated already in other studies using rheological methods (Körstgens et al. 2001). In the mentioned studies, calcium was available in the growth medium during all the time of biofilm development. In the present study, in contrast, calcium was available only after biofilm growth. This shows that calcium can be incorporated and can exert an influence in relatively short periods of time. In natural sediments, de Brouwer et al. (2002) found a strong interaction between extractable carbohydrate and sediment particles. Part of this EPS was irreversibly bound to sediments (not re-extracted by 0.1M EDTA). Furthermore, the amount of irreversibly bound EPS increased 50% in the presence of Ca^{2+} ions.

EPS production in PM columns was generally lower than in flume sediment columns suggesting isolated and patchy biofilms and demonstrating the importance of the habitat on biofilm development. In these systems, carbohydrate/protein ratio decreased with time and at constant ionic strength conditions, biofilms affected the hydraulic conductivity of the porous medium. The effect on sediment cohesiveness was not assessed but the retention capacity particularly of clay-like minerals suggests a positive effect on sediment stability. EPS production within the sediment and mobile EPS fractions permeating sediment voids can increase the cohesiveness of sediment grains which can be bonded in bigger aggregates. Under these conditions, rates of sediment mobilization will depend not only on the size of the sediment grains but also on EPS content. At changing electrochemical conditions (e.g., decreasing ionic strength), especially in the presence of monovalent cations, biofilm stability as well as rates of microbial attachment can be drastically affected. This will have obvious consequences on sediment stability. Furthermore, in natural environments, biofilm distribution in the porous medium is not homogeneous and can change depending on depth, ecology and on the biogeochemistry of the site (Yallop et al. 2000). The type of biofilm and the relatively fast changes that can occur in biofilm stability and EPS production rates at changing electrochemical conditions, are determinant for sediment transport in natural environments. This dynamic behavior and the resulting distributions in erosion rates cannot be easily included in sediment transport prediction models.

Retention Remobilization Processes

At the sediment-liquid interface, the absence of alginate (*P. aeruginosa* SG81 mutant) resulted in higher pollutant remobilization as compared with the alginate producing wild strain. In both cases, however, there was less remobilization than in the sterile columns. Although, it is very likely that the excess of carbohydrate was responsible for lower remobilization rates in the wild type inoculated columns, it is not possible to rule out the role of proteins in retention. This is evidenced in the fact that pollutant retention was also highly correlated with protein content. Previous studies (data not shown) have demonstrated that after predominance of Ca^{2+} ions, remobilization of retained particles and biofilm components occurs very slowly. Only after several pore volumes of deionized water, remobilization events start to take place. It is plausible then to assume, that after prolonged periods of a very low ionic strength influent, accumulated pollutant, not remobilized in the time frame of our experiments (Fig. 9.12), could start to be remobilized. This can also be truth in the case of monovalent ion exchange which is realistic in nature e.g., sea water infiltration near coasts.

In PM columns, the correlation of proteins to retention was significantly higher than carbohydrates and Ca^{2+} played an important role both in biofilm stability and pollutant retention. Furthermore, the ratio carbohydrate/protein decreased with time which suggests an important role of proteins in pollutant retention within the sediment matrix. In the case of Na^{+} background solutions, the lack of retention could be attributed to biofilm instability, detachment and co-elution effects.

In summary, carbohydrates are probably more important than proteins in sediment stability at the sediment-liquid interface. In terms of retention-remobilization processes at this level there was no evidence of a predominant carbohydrate role. Within the sediment, in contrast, proteins played a predominant role when conditions were appropriate for pollutant retention (i.e. dominance of Ca^{2+} in the background solutions).

9.2.5 Conclusions

The experimental data from sterile/non-sterile sediment systems show that:

- *P. aeruginosa* biofilms increase the stability of model sediments. The extent of this effect increases with time and with the presence of calcium ions.
- The remobilization of model contaminants is inhibited in the presence of biofilms. This inhibition is higher in the alginate producing mucoid *P. aeruginosa* SG81 strain as compared to its alginate-deficient mutant.

In the porous matrix, transport can be highly influenced by biofilm formation and fluid ionic composition, both in terms of changes in hydraulic conditions (permeability, hydraulic conductivity, dispersivity, etc.) and retention-remobilization of migrating pollutants.

Biofilm influence on sediment transport processes is not limited to the major sediment-water phase interface. Processes occurring in the depth of the sediment must also be taken into account.

Acknowledgments

This work is part of the cooperative research project SEDYMO – "Sediment Dynamics and pollutant Mobility in river basins" financed by the German Federal Ministry of Education and Research (BMBF). We thank all our project partners especially Professors U. Förstner and B. Westrich and their groups, for constructive input.

References

Battin TJ, Sengschmitt D (1999) Linking sediment biofilms, hydrodynamics, and river bed clogging: evidence from a large river. Microb Ecol 37:185–196

de Brouwer JFC, Ruddy GK, Jones TER, Stal LJ (2002) Sorption of EPS to sediment particles and the effect on the rheology of sediment slurries. Biogeochemistry 61(1):57

Flemming H-C, Wingender J (2002) Extracellular polymeric substances (EPS): Structural, ecological and technical aspects. In: Bitton G (ed) Encyclopedia of Environmental Microbiology, vol. 4. John Wiley & Sons, Inc., New York, pp 1223–1231

Förstner U (2004) Sediment dynamics and pollutant mobility in rivers: an interdisciplinary approach. Lakes and Reservoirs: Research and Management 9:25–40

Frølund B, Palmgren R, Keiding K, Nielsen PH (1996) Extraction of extracellular polymers from activated sludge using a cation exchange resin. Water Res 30(8):1749–1758

Grobe S, Wingender J, Trüper HG (1995) Characterization of mucoid *Pseudomonas aeruginosa* strains isolated from technical water systems. J Appl Bacteriol 79:94–102

Körstgens V, Flemming H-C, Wingender J, Borchard W (2001) Influence of calcium ions on the mechanical properties of a model biofilm of mucoid *Pseudomonas aeruginosa*. Wat Sci Tech 43(6):49–57

Leon Morales CF, Leis AP, Strathmann M, Flemming H-C (2004) Interactions between laponite and microbial biofilms in porous media: implications for colloid transport and biofilm stability. Water Res 38(16):3614–26

Sutherland T, Grant J, Amos C (1998) The effect of carbohydrate production by the diatom *Nitzschia curvilineata* on the erodibility of sediment. Limnology Oceanography 43:65–72

Wingender J, Strathmann M, Rode A, Leis A, Flemming H-C (2001) Isolation and Biochemical Characterization of Extracellular Polymeric Substances from *Pseudomonas aeruginosa*. Methods Enzymol 336(25):302–314

Yallop ML, Paterson DM, Wellsbury P (2000) Interrelationships between rates of microbial production, exopolymer production, microbial biomass, and sediment stability in biofilms of intertidal sediments. Microb Ecol 39(2):116–127

Manuelle Neto · Jean-Philippe Bedell · Rémy Gourdon · Guy Collilieux Antonio Bispo

9.3 Role of Bacteria in Heavy Metal Transport during the Dredging in the Rhône River

9.3.1 Introduction

Dredging activities have a wide range of applications, permitting to: limit flooding, restore water flow, restore navigation and improve the quality of water when contaminated sediments are extracted. The number of dredging is increasing due to human activities. Dredging activities have potential economic and environmental impacts, such as ecological or health risk since the pollution detected in these sediments comes from various origins.

Between 1990 and 2000, the quantity of sediments dredged in France amounted to 2.8 million $m^3\,yr^{-1}$ (Hardy 2002). Dredging activities on the Rhône River represent 46% of the total volume of sediments dredged in France.

Over the 1997–2002 period, the average volume of sediments dredged from the Rhône River was around 885 000 $m^3\,yr^{-1}$. The volume of silt dredged in that period was estimated to 3.9 million m^3, i.e., an average of 665 000 $m^3\,yr^{-1}$ (ca. 70 to 75% of the total volume dredged annually in the Rhône River). Nevertheless, this volume represents only about 6% of the total suspended matter transported by the Rhône during a year between Génissiat (department 01, France) and Vallabrègues (department 30, France), which are, respectively, the upstream and downstream reference points.

Dredging on the Rhône River is usually done with a suction dredge followed by a discharge into the water flow. A mixture of water (70 to 90%) and sediment (10 to 30%) is pumped from the sediment and re-suspended in the water flow. The concentration of suspended matter downstream a suction dredge typically ranges between 100 and 300 $g\,l^{-1}$ depending on the material dredged and the type of machine used.

In situ measurements were realized during a re-suspension dredging operation (on the Vaugris site, Isère, 38, France). It was observed that: *(i)* the plume generated was detectable over 1 000 m downstream, and *(ii)* a high bacterial population was present over 500 m downstream.

Bacteria, bio-colloids, are important heavy metal carriers (see the review of Sen and Khilar 2006) whose study appears important in works related to heavy metals transport.

The questions tackled in this document are:

- Does the re-suspended matter transport pollutants or are pollutants rather dissolved?
- What is the role of bacteria in the transport of polluants?

The present study focuses on the processes of re-suspension and settling of dredged sediments. Laboratory experiments were conducted with the sediment dredged in the St. Vallier site. This sediment was firstly suspended in a known volume of water, and then allowed to settle. The concentration of suspended particles in the water column was time-monitored during the settling process. Samples were extracted at different times and to separated in distinct grain size compartments, in order to measure bacterial cell density and heavy metals concentration in each of them. Results obtained with sterilized and non-sterilized sediments were compared to assess the role of bacterial populations.

9.3.2 Materials and Methods

Protocol Development and Experimental Protocol

The experimental approach followed the static protocol described below.

The sediment/water ratio used (in %) is 30/70. This ratio is the one to be used in situ, as indicated by the National Rhône Company. A dry mass of 240 g of sediment was mixed with distilled water (filtered at 0.22 μm on nitrate and acetate filters) in a final volume of 1 l using a rotary shaker, during 24 h. Preliminary experiments were done in triplicates to evaluate the reproducibility of settling and the method of sampling. Sampling from water column during settling was done every 2 min in these experiments and suspended matter concentrations analyzed. These results showed that the suspended matter concentration was the same throughout the whole water column (excepted the water surface). We decided to sample at half the height of the water column.

On the basis of these preliminary experiments, the following sampling times were considered for the different settling phases in the subsequent experiments:

- Initial mix ($t = 0$).
- Rapid settling ($t = 4$ min).
- Progressive settling monitored at times 10, 15 and 30 minutes.
- Approach of stationary phase at times 60 and 120 minutes.

The suspensions were analyzed for suspended matter, bacterial concentration, and heavy metals concentrations.

The mass of sediment and volume of water used in the experiments were defined so that the suspended matter concentration at $t = 0$ was similar to that obtained at the re-suspension point in the in situ experiment (ca. 74 300 mg l^{-1}). The rapid settling phase (obtained at $t = 4$ min) was considered to reproduce in situ suspended matter concentrations obtained 3 meters downstream from the re-suspension point (ca. 2 640 mg l^{-1}).

Characteristics of Selected Sediment

The sediment used was dredged from the St. Vallier station (department 26, France) on the Rhône River.

Some experiments were run with sterilized sediment to assess the role of microbial biomass.

Two different types of sterilizations were tested: one physical sterilization (by gamma radiation) and one chemical sterilization (by formaldehyde). These two sterilizations are commonly used on soil and sediment studies (McNamara et al. 2003). The sterilization by gamma rays is usually considered as the least destructive mode of sterilization (McNamara et al. 2003; Trevors 1996). The sterilization by formaldehyde is used to confirm or infirm some of the results obtained with gamma rays sterilization.

These different sterilization stages were carried out as follows:

- *Gamma rays:* gamma rays sterilization was done by the "Ionisos" company (Dagneux, department 01, France) at 40 KGy, followed by a 3-week rest at 4 °C to destroy residual enzymatic activities.
- *Formaldehyde:* the sterilization by formaldehyde was realized by mixing (with several vigorous shakings) 0.5 ml of formaldehyde solution at 30% per dry g of sediment, followed by a 3-week rest at 4 °C.

Analytical Methods

All experiments were triplicated (unless otherwise stated). The glassware used was washed with HNO_3 at 5% overnight, and rinsed several times with distilled water.

Suspended matter concentrations were measured according to the French standard NF EN 872, by filtration at 1.2 μm on microfiber filter (Whatman).

Filtrations of aqueous suspensions were done with nitrate and acetate filters of 8 µm pore-size as a pre-filtration step to avoid clogging in subsequent filtrations, followed by 1.2 and 0.45 µm pore-size filtrations for particle size fractionation (Millipore filters SCWP, RAWP and HAWP, respectively). Before filtration, an aliquot of suspension was sampled and kept for analysis.

Heavy metal concentrations

- In the suspensions, concentrations were determined by mineralizing 3 to 20 ml of suspension (depending on suspended matter concentration) with 2 ml of HNO_3 at 65% (Merck Suprapur) and 6 ml of HCl at 30% (Merck ultrapur) following the French standard NF EN ISO 15587-1. The suspensions were filtered before being analyzed by atomic absorption (Hitachi 28200) (detection limit for Cd: 0.1 $\mu g\,l^{-1}$ and for Zn: 0.1 $mg\,l^{-1}$).
- In the filtered solutions, Zn concentrations were measured by ICP/AES (Perking Elmer) following the French standard NF EN ISO 11885 (detection limit: 0.05 $mg\,l^{-1}$ for Zn).

Bacterial cell density was measured using *LIVE/DEAD*® Bac*Light*™ protocol: a sample of 100 to 150 µl was incubated with 150 µl of propidium iodide and 200 µl of SYTO 9 during 15 minutes in a total volume of 5 ml completed with water filtered at 0.22 µm. After incubation, the solution was filtered on Millipore isopore filter GTBP (0.22 µm). The filter was placed on a slide between two drops of low fluorescence immersion oil (Zeiss Immersol 518N) and covered with a clear glass cover slip. The observation was then realized on a Zeiss microscope with filter set 09 (Zeiss no.: 488009-0000). Twenty microscopic fields were numbered on each filter. Volumes filtered were performed to 10 at 100 cells per fields. The counting operation was realized on 3 different filters without distinction between viable and not viable cells.

9.3.3 Results

Characteristics of Selected Sediment

The physico-chemical characteristics of the sediment are considered as being representative of the Rhône River's sediments (Table 9.2).

Relatively high concentrations of Zn, Cd, Cu and Fe were observed as shown in Table 9.2.

The bacterial concentration was also found to be relatively high (Table 9.2). Such population exhibited dehydrogenase and denitrification activities. The first one indicates a good respiratory potential and a possible role in the carbon cycle (Engelen et al. 1998). The second one plays a part in the carbon and nitrogen cycles (van Rijn et al. 2006). These results suggest that the pollutants present in the sediment do not significantly inhibit bacterial population, dehydrogenase and denitrifying activities.

Table 9.2. Physico-chemical and microbiological characteristics of selected sediment

Parameters	Protocol used	Values			
pH (water)	NF X31-103	7.28 ±0.16			
pH (KCl)	NF X 31-104	7.16 ±0.11			
Water content (%)	Afnor X31-102	77.46 ±0.01			
TOC (%)	NF ISO 10694	1.52 ±0.1			
Total hydrocarbons (mg/kg DW)[a]		71			
Heavy metal concentrations ($mg\ kg^{-1}$)[a]	ICP/AES after mineralization by micro waves NF EN ISO 11885	Zn 154	Cd 1	Fe 28382	Cu 46
Bacterial enumeration (cells/g DW)	Bac light (Soil extraction according to Ranjard et al. 1997)	7.43×10^9			
Dehydrogenase activity (µg of formed formazan/24 h/g DW)	Brohon et al. 1999	7550 ±245			
Denitrifying activity (ppm of N_2O formed/h/g DW)	Poly 2000	6.85×10^2			

[a] Measures realized by a laboratory certified COFRAC (Comité Français d'Accréditation).

Time Evolution of Suspended Matter Concentration

The evolution of suspended matter concentration in the water column (at half-height of water column) during settling is shown in Fig. 9.18 for the different experiments carried out with non-sterilized (namely "biotic") sediment and with sediments sterilized chemically (using formaldehyde) or by gamma irradiation. In all experiments, initial concentrations of suspended matter were roughly the same (ca. 1.6×10^5 mg l^{-1}).

pH values in the water column for biotic and gamma rays experiments were not differentiated (around 6.9 $\pm$0.02) and relatively steady during all the settling.

Settling of Biotic (Untreated) Sediment

Figure 9.18 shows that the settling of biotic sediment is faster than that of sterilized sediments. More than 98.8% of suspended matter was found to settle over the 4 first minutes (suspended matter concentrations were 1.59×10^5 mg l^{-1} and 1.91×10^3 mg l^{-1} at times 0 and 4 min, respectively).

Settling of Sterilized Sediments

Two types of sterilization were tested, using gamma rays or formaldehyde (physical and chemical treatment). The forms of the settling curves obtained with the 2 sterilized sediments are similar to that of the "biotic" sediment, but the speed of settling over the first 30 minutes was found to be significantly different (Fig. 9.18). Sediment sterilized by gamma rays exhibited the slowest settling, and sediment sterilized with formal-

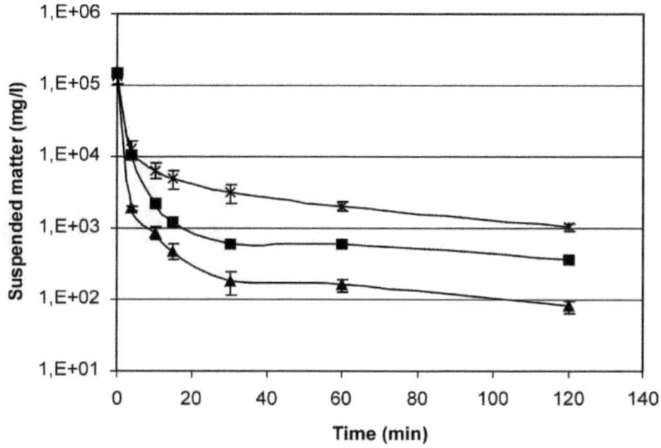

Fig. 9.18. Suspended matter concentration in water column with different treatments. The biotic experiment is represented by a *triangle*, the formaldehyde experiment by a *square*, and the gamma rays experiment by *star*

dehyde showed an intermediate behavior between the "biotic" and the "gamma rays" sediments (Fig. 9.18).

Although the differences observed between the "biotic" and "gamma rays" or "formaldehyde" experiments cannot be exclusively attributed to the presence of microorganisms, results suggest that microorganisms play a role in settling and its velocity. The grain size particles and stability aggregates of biotic sediment and sediment sterilized by gamma radiation showed few differences as underlined before in the review of McNamara (2003) (data not shown). The gamma rays treatment doesn't break aggregates of the sediment. The microbial population via through microbial flocs is known to be involved in the formation of particle aggregates in sediments (Stemmer et al. 1998; van Elsas et al. 1997). The microbial exopolysaccharides bind clay mineral and humic components into microaggreagates (van Elsas et al. 1997). The first hypothesis is that cells death could weaken aggregates. Thus these fragile aggregates could be broken after a long and vigorous shaking as in our experimental protocol. It is assumed that the partial destruction of aggregates results in the formation of smaller particles, which settle at reduced velocity.

Gamma radiation creates free hydrogen and hydroxyl radicals that react as reducing and oxidizing agents and cleave C–C bonds (Trevors 1996). The second hypothesis is that the organic matter of the sediment is transformed, inducing a modification in the settling.

The difference between "biotic" and "gamma rays" sediment may be attributed to these two distinct effects: aggregation and organic matter. This hypothesis is confirmed by the results obtained with the formaldehyde experiment.

Bacteria Distribution

Bacterial counts were done both in the water column sampled at different settling times, and in the different particle size fractions obtained by successive filtrations of the water column at 1.2 μm and 0.45 μm).

Cell counts done in the experiments with sterilized sediments showed that no viable microbial cells were found (data not shown), thereby confirming the efficiency of sterilization treatments.

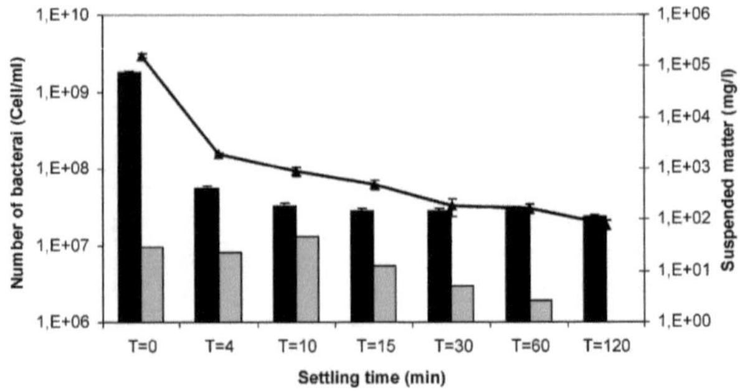

Fig. 9.19.
Bacterial population in the water columns (*black bars*) and in the particle size fraction $F < 1.2$ µm (*gray bars*), and the suspended matter concentration in water column (*line*) during settling

Results obtained with "biotic" sediment are shown in Fig. 9.19 where the histograms and the line represent the bacterial cells concentrations and the suspended matter concentration, respectively.

Bacterial concentration was very high at $t = 0$ (initial mix), then decreased rapidly during settling (1.8×10^9 at $t = 0$ to 2.34×10^7 cells ml^{-1} at $t = 120$, see Fig. 9.19), following the same pattern as suspended matter. Heavy metals concentrations also decreased following a similar pattern (see below and Fig. 9.20).

The number of bacteria in the solutions filtered at 1.2 µm varied only slightly during the settling process (from 1.87×10^6 to 1.30×10^7 cells ml^{-1}). This population (in $F < 1.2$ µm) was found to represent only 0.5% and 6% of the total population at $t = 0$ and at $t = 60$ min, respectively (Fig. 9.19). The cells counted in this fraction correspond to free cells, or cells associated to microparticles or colloïds (particle size less than 1.2 µm).

Cell counts carried out in the solutions filtered at 0.45 µm showed the presence of bacteria at a concentration ranging between 1×10^5 to 1×10^6 cells ml^{-1} (data not shown).

Heavy Metals Distribution

Zn, Fe, Cd and Cu concentrations were monitored in the water column and in the solutions filtered at 1.2 µm and 0.45 µm (*i.e.* "Total fraction", "$F < 1.2$ µm" and "$F < 0.45$ µm"). Table 9.3 shows the results obtained in "biotic" and "gamma rays" experiments. Figure 9.20 illustrates the typical data obtained for Zn, which exhibits a behavior representative of that of the other monitored metallic elements. Measurements, of heavy metal distribution, for "formaldehyde experiment" are underway. These complementary results will permit us to confirm or infirm the tendency obtained with the biotic and gamma rays experiments.

Table 9.3 and Fig. 9.20 show a fast decrease in heavy metals concentration both with biotic and sterilized sediments, with a time course very similar to the drop in suspended matter concentration (Fig. 9.18) and bacterial population (Fig. 9.19). Zinc concentration in the water column was divided by 75 during the first 4 minutes in the biotic experiment, and by 12 in the same time for the abiotic experiment (Table 9.3).

Figure 9.20 shows that although zinc concentration was initially slightly higher in the biotic experiments, the concentrations dropped drastically over the 4 first minutes to become 4 times smaller than in the "abiotic" experiments. After the 4 first

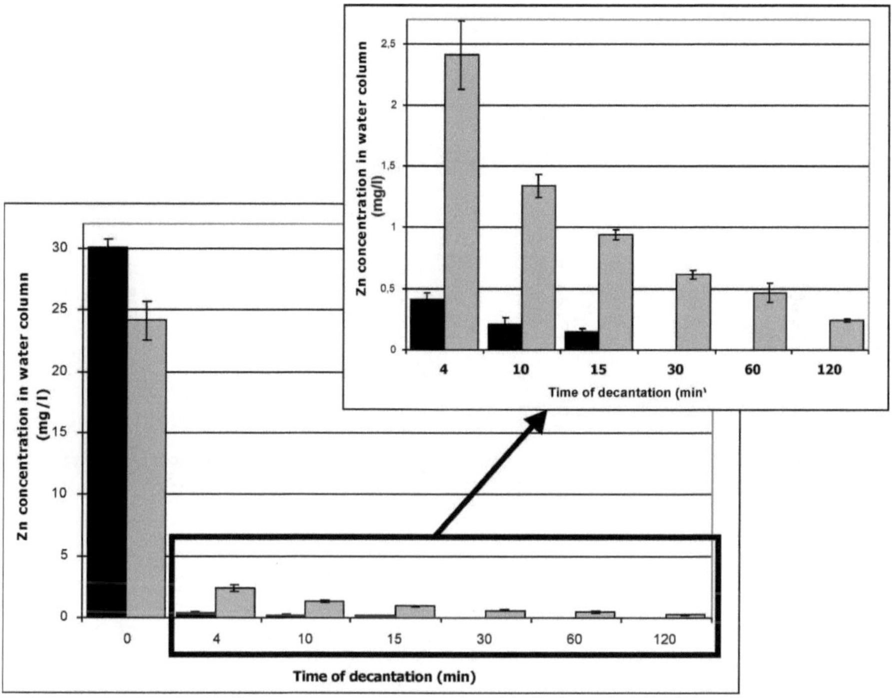

Fig. 9.20. Zinc concentration in the water column during settling in "biotic" (*black bars*) and "abiotic" (*gray bars*) experiments

minutes, the focus in Fig. 9.20 shows that the concentration still continued to decrease in both experiments (below detection limit in biotic experiments), but at a much smaller rate.

The same observations can be done for Cd, Cu and Fe. Consequently, Table 9.3 shows that during the first 4 minutes of the settling process, Cd concentration was divided by 50 and 11 in the biotic and abiotic experiments, respectively. For Cu (respectively Fe), the reduction factors of the concentration in the first 4 minutes are 74 (respectively 62) in the biotic experiments, versus 10 (respectively 7) in abiotic experiments.

Heavy metal concentration in the water column during settling after re-suspension of biotic sediment were therefore much smaller than with sterilized sediment.

Correlation analyses (linear correlation) showed that Zn, Cu, Cd and Fe concentrations were significantly correlated with suspended matter concentration in water column ($0.96 < r < 1$ and $p < 0.01$). This result is consistent with those of Carpentier et al. (2002) and Pettine et al. (1994).

Indeed, experiments done with biotic sediment showed that both suspended matter concentration and heavy metals concentrations were significantly smaller than with sterilized sediments.

Oxidation of anoxic sediment (influx of dissolved oxygen) results in positive change in the redox potential and in a decrease in sediment pH (mainly due to the oxidation of sulfide). This decrease in pH doesn't happen in buffered sediment. In our experiment,

Table 9.3. Zn, Fe, Cu, Cd concentrations in water column (mg l^{-1}) during settling (time is given in minutes). Details on Zn concentration for $F < 1.2$ μm and $F < 0.45$ μm (other metals show similar trends). *ND:* Not detected

			$T=0$	$T=4$	$T=10$	$T=15$	$T=30$	$T=60$	$T=120$
Zn	Biotic	*F* total	30.056 $\pm 7 \times 10^{-1}$	0.408 $\pm 5.77 \times 10^{-2}$	0.208 $\pm 5.2 \times 10^{-2}$	0.150 $\pm 2.50 \times 10^{-2}$	ND	ND	ND
		$F < 1.2$	0.11	ND	ND	ND	ND	ND	
		$F < 0.45$	0.12	ND	ND	ND	ND	ND	
	Abiotic	*F* total	24.139 $\pm 1.56 \times 10^{0}$	2.408 $\pm 2.81 \times 10^{-1}$	1.342 $\pm 9.46 \times 10^{-2}$	0.942 $\pm 3.82 \times 10^{-2}$	0.617 $\pm 3.82 \times 10^{-2}$	0.467 $\pm 7.64 \times 10^{-2}$	0.242 $\pm 1.44 \times 10^{-2}$
		$F < 1.2$	0.15	0.11	ND	ND	ND	ND	ND
		$F < 0.45$	0.16	0.13	ND	ND	ND	ND	ND
Fe	Biotic		4013.9 $\pm 8.86 \times 10^{1}$	63.83 $\pm 1.08 \times 10^{1}$	38.79 $\pm 1.43 \times 10^{1}$	20.25 $\pm 5.41 \times 10^{0}$	11.83 $\pm 2.67 \times 10^{0}$	7.17 $\pm 7.64 \times 10^{-1}$	4.17 $\pm 7.64 \times 10^{-1}$
	Abiotic		3100 $\pm 2.14 \times 10^{2}$	456.46 $\pm 5.21 \times 10^{1}$	267.33 $\pm 2.70 \times 10^{1}$	190.42 $\pm 1.23 \times 10^{1}$	128.42 $\pm 1.22 \times 10^{1}$	85.33 $\pm 1.57 \times 10^{1}$	36.33 $\pm 5.55 \times 10^{0}$
Cd	Biotic		0.100 $\pm 7.74 \times 10^{-3}$	0.002 $\pm 2.78 \times 10^{-4}$	0.001 $\pm 1.50 \times 10^{-4}$	0.001 $\pm 1.61 \times 10^{-4}$	ND	ND	ND
	Abiotic		0.091 $\pm 2.20 \times 10^{-3}$	0.008 $\pm 7.1 \times 10^{-4}$	0.006 $\pm 3.54 \times 10^{-3}$	0.003 $\pm 1.3 \times 10^{-4}$	0.002 $\pm 1.28 \times 10^{-4}$	0.002 $\pm 3.62 \times 10^{-4}$	0.001 $\pm 1.32 \times 10^{-4}$
Cu	Biotic		7.443 $\pm 3.35 \times 10^{-1}$	0.101 $\pm 1.91 \times 10^{-2}$	0.043 $\pm 1.18 \times 10^{-2}$	0.031 $\pm 5.80 \times 10^{-3}$	0.026 $\pm 5.31 \times 10^{-3}$	0.018 $\pm 3.28 \times 10^{-3}$	0.013 $\pm 2.44 \times 10^{-3}$
	Abiotic		6.534 $\pm 3.19 \times 10^{-1}$	0.658 $\pm 5.50 \times 10^{-2}$	0.433 $\pm 8.80 \times 10^{-2}$	0.266 $\pm 3.81 \times 10^{-2}$	0.163 $\pm 2.28 \times 10^{-2}$	0.113 $\pm 3.05 \times 10^{-2}$	0.056 $\pm 6 \times 10^{-3}$

the pH of sediment was 7.2 vs. 6.9 in the water column. We can consider that the low pH variation during the oxidation phase was probably due to the dissolution of mineral carbonate (166 mg kg^{-1} of calcareous in sediment) as described by Caille et al. (2003). Metals co-precipitated with or adsorbed to FeS and MnS are rapidly oxidized. The released Fe and Mn are rapidly re-precipitated and deposited as insoluble oxides/ hydroxides to which newly released metals can become adsorbed at varying rates and extents (Calmano et al. 1993; Stephens et al. 2001; Caetano et al. 2003). Caille et al. (2003) showed that during an extended aeration and after a rapid release of metals (Zn,Cu), a decrease in the metal solubility is observed probably due to their co-precipitation with carbonates or oxides. The low concentration in Zn, Cu and Cd in $F < 1.2$ µm and $F < 0.45$ µm may be attributed to the same phenomena as described above.

9.3.4 Conclusions

A laboratory protocol was developed to simulate dredging by re-suspension and investigate the role of microbial cells in settling and heavy metals distribution. Bacterial cell concentration, suspended matter concentration, and heavy metals concentrations followed similar patterns during settling after re-suspension. A very fast settling phase was observed over the first 4 minutes after re-suspension, followed by a slower phase, approaching stability within ca. 30 minutes. Moreover, we show a significant bacterial population in the studied grain size compartments ($F < 1.2$ µm and $F < 0.45$ µm), and a low Zn concentration.

Settling was found to occur at a smaller rate with sterilized sediments. Zn, Cu, Cd and Fe concentrations in the water column showed a faster decrease during the settling of biotic sediment as compared to the sediment sterilized by gamma rays.

The concentrations of Zn, Cu, Cd and Fe were correlated to the suspended matter concentration in water column in both treatments (biotic and sterilization by gamma rays).

These first results describe the influence of bacteria on settling of suspended matter. The texture and structure of sediment was not modified by the gamma rays treatment (data not shown). However, from the difference in settling we can conclude that the absence of bacteria, which keep the soil structure "cement", induced a fragility of the aggregates after a long and vigorous agitation (the fragility was not observed with a stability test which is sweeter than our 24 h of vigorous agitation). The absence of bacteria does not permit a constant concentration in exopolysaccharides and thus a keep in stability. Therefore the role of bacteria is to maintain aggregation by the production of polysaccharides. This aggregation induces a rapid settling of particles with heavy metals attached.

In order to better elucidate the mechanisms by which microbial cells influence settling of suspended matter and heavy metals distribution, further studies are underway to investigate particle size distribution of "biotic" vs. "abiotic" experiment and evaluate bacterial diversity by DGGE or SSCP in order to characterize the evolution of the bacterial population during settling. These experiments will permit us to better know the microorganisms part in the particles settling by physico-chemical or biological processes, and so the metals settling during the sediment dredging in the Rhône River.

References

Bedell JP, Neto M, Pressiat F (2005) Opérations de dragage, pollution potentielle et enjeux environnementaux. Les sédiments du Rhône Grands enjeux, premières réponses. Journée ZABR, Valence, France. 10 juin 2005. pp113–121

Brohon B, Delolme C, Gourdon R (1999) Qualification of soils through microbial activities measurements: influence of the storage period on INT-reductase, phosphatase and respiration. Chemosphere 38:1973–1984

Caetano M, Madureira MJ, Vale C (2003) Metal remobilisation during resuspension of anoxic contaminated sediment: short-term laboratory study. Water Air Soil Poll 143:23–40

Caille N, Tiffreau C, Leyval C, Morel JL (2003) Solubility of metals in an anoxic sediment during prolonged aeration. Sci Total Environ 301:239–250

Calmano W, Hong J, Forstner U (1993) Binding and mobilization of heavy metals in contaminated sediments affected by pH and redox potential. Water Sci Technol 28:223–235

Carpentier S, Moilleron R, Beltran C, Herve D, Thevenot D (2002) Quality of dredged material in the river Seine basin (France). II. Micropollutants. Sci Total Environ 299:57–72

Engelen B, Meinken K, von Wintzingerode F, Heuer H, Malkomes H-P, Backhaus H (1998) Monitoring impact of a pesticide treatment on bacterial soil communities by metabolic and genetic fingerprinting in addition to conventional testing procedures. Appl Environ Microbiol 64:2814–2821

Hardy D (2002) Historique National des opérations de curage et perspectives, Ministère de l'écologie et du développement durable, pp17

McNamara NP, Black HIJ, Beresford NA, Parekh NR (2003) Effects of acute gamma irradiation on chemical, physical and biological properties of soils. Appl Soil Ecol 24:117–132

Pettine M, Camusso M, Martinotti W, Marchetti R, Passino R, Queirazza G (1994) Soluble and particulate metals in the Po River: Factors affecting concentrations and partitioning. Sci Total Environ 145:243–265

Poly F (2000) Réponses des communautés bactériennes telluriques à des perturbations chimiques complexes: Activités potentielles et empreintes génétiques, Université Claude Bernard Lyon 1, Lyon, pp159

Ranjard L, Richaume A, Jocteur-Monrozier L, Nazaret S (1997) Response of soil bacteria to Hg(II) in relation to soil characteristics and cell location. FEMS Microbiol Ecol 24(4):321–331

Sen TK, Khilar KC (2006) Review on subsurface colloids and colloid-associated contaminant transport in saturated porous media. Advances in Colloid and Interface Science 119:71–96

Stemmer M, Gerzabek MH, Kandeler E (1998) Organic matter and enzyme activity in particle-size fractions of soils obtained after low-energy sonication. Soil Biol Biochem 30:9–17

Stephens SR, Alloway BJ, Parker A, Carter JE and Hodson ME (2001) Changes in the leachability of metals from dredged canal sediments during drying and oxidation. Environ Pollut 114:407–413

Trevors JT (1996) Sterilization and inhibition of microbial activity in soil. J Microbiol Methods 26:53–59

van Elsas JD, Trevors JT, Wellington EMH (1997) Modern soil microbiology. Marcel Dekker, INC, pp 683

van Rijn J, Tal Y, Schreier HJ (2006) Denitrification in recirculating systems: Theory and applications. Aquacult Eng 34:364–376

Chapter 10

Sediment Toxicity Data

Wolfgang Ahlf

In concepts for the assessment of contaminated sediments ecotoxicological test systems are a present line of evidence among others. Sediment toxicity could be pragmatically defined as any toxic effects observed in laboratory toxicity tests. These methods indicate bioavailability and the adverse effect of environmental samples by exposing them directly to test organisms. Due to the heterogeneity of the matrix (sediment) and the genetic variability of test organisms, toxicity tests are often undervalued in relation to their statistical power. In most cases standardized tests are performed with the same requirements for quality control as chemical analyses have. Consequently, toxicity testing can only be done with a limited number of species, tested under controlled laboratory conditions, thus limiting their ecological relevance.

Section 10.1 deals with the quality assurance of ecotoxicological testing. The variation and standard deviation of each bioassay has to be known in order to categorize toxicity reliable. Another crucial point is the quantitative comparison of toxic effects. Sensitivity of toxicity tests is not only species and contaminant-specific. It also varies depending on the endpoint measured. Information on the responsiveness of a test system is hence of high importance for the interpretation and the degree of confidence that would be involved in resulting management decisions.

The results of a test battery can be reduced to toxicity classes. The classification system worked well to detect changes in spatial and temporal sediment quality. A longitudinal survey along the Elbe confirmed the higher toxicity of upstream sediments partly due to historic sediment contamination from the river basin (Sect. 10.2).

A river basin is a dynamic system, where interactions between different compartments (river-sediment-soil-groundwater) greatly influence the ecological quality. The limitations of the single methods for quality evaluation are overcome by using chemical, ecotoxicological and in situ ecological data as different lines of evidence, but combining them for further interpretation. In such a "Weight of evidence"-concept, the more lines of evidence support the conclusion, the stronger the weight of evidence. Section 10.3 features the urgent need to cross disciplinary boundaries in order to derive a realistic assessment regarding the erosion risk of old deposited sediment layers as well as the bioavailability and hazard potential of their associated contaminants at different aquatic sites. Especially the combination of hydrodynamic and ecotoxicological methods will give *(i)* comprehensive insights into the effects of flood events on biota and ecosystems and *(ii)* allow evaluation of sediment and thus water quality with regard to the global climate change and the expectations of more severe floods in the near future.

Wolfgang Ahlf · Susanne Heise

10.1 Quality Assurance of Ecotoxicological Sediment Analysis

10.1.1 Introduction

The overall goal of a well-designed and well-implemented sampling and analysis program is to measure accurately what is really the status of the area studied. Environmental decisions are made on the assumption that analytical results are, within known limits of accuracy and precision, representative of site conditions. Many sources of error exist that could affect the analytical results. These sources of error may include sample collection methods, sample handling, preservation, and transport; personnel training; analytical methods; data reporting; and record keeping. Therefore, a quality assurance program has to be designed for each sediment quality evaluation to minimize these sources of error and to control all phases of the monitoring process.

A summary of errors which can occur during the assessment process are indicated in Table 10.1. Experts agree that 10 to 20 per cent of resources, including manpower, should be directed towards ensuring the quality of analytical determinations for common water quality variables (Anon. 1987). When trace pollutants (e.g., pesticides and trace elements) are measured, the resources required for quality control may reach 50 per cent (Meybeck et al. 1992).

Major problem areas have been identified and discussed by the European thematic framework "Metropolis" (Metrology in Support of Precautionary Sciences and Sustainable Development Policies; Anon. 2004):

- lack of harmonization of the procedures applied by laboratories (starting with the sampling procedure, but also including the approach adopted for the calculation of the uncertainty); this lack of harmonization makes the data obtained from different sources difficult to compare;
- lack of representativeness: data that do not reflect the reality that we want to represent are simply not fit for purpose;
- a too high level of uncertainty associated with the data collected makes the process of decision-making critical (on the other hand, in some cases the uncertainty is not expressed at all!);
- lack of metadata: information about the data (what, how and when measurements were made, who owns the data, etc.) and the way they are reported/used is an essential requirement to allow the use of the data for other purposes (e.g., compilation of databases);
- lack of traceability: The concept of traceability implies that measurement data are linked to stated references through an unbroken chain of comparison, all with stated uncertainties (Quevauviller 2004).

Criticism on the certainty of sediment data led to the lack of environmental quality standards (EQS) in the Water Framework Directive (EC 2000) as well as in the Daughter Directive on priority substances from 17 July 2006. Crane and Babut state that "For sediments, consultation on the setting of EQS raised many concerns about *(i)* the fac-

Table 10.1. Some possible sources of errors in the water quality assessment process with special reference to chemical methods (Meybeck et al. 1992)

Step	Operation	Possible source of error	Appropriate actions
Monitoring design	Site selection	▪ Station not representative (e.g. poor mixing in rivers)	Preliminary surveys
	Frequency determination	▪ Sample not representative (e.g., variations between samples)	
Field operations	Sampling	▪ Sample contamination (micropollutant monitoring)	Decontamination of sampling equipment, containers
	Filtration	▪ Contamination or loss	Running field blanks
	Field measurement	▪ Uncalibrated operations (pH, conductivity, temperature)	Field calibrations Replicate sampling
		▪ Inadequate understanding of hydrological regime	Hydrological survey
Sample shipments to laboratory	Sample conservation and identification	▪ Error in chemical conservation	Field spiking
		▪ Error in biological conservation	Appropriate field pre-treatment
		▪ Break of container	Field operator training
Laboratory	Preconcentration	▪ Contamination or loss	Decontamination of laboratory equipment and facilities
	Analysis	▪ Contamination	Quality control of laboratory air, equipment and water
		▪ Lack of sensitivity	Quality assurance tests
		▪ Lack of calibration	(Analysis of control sample or standards)
		▪ Error in data reports	Check internal consistency of data (e.g., with adjacent sample, ionic balance, etc.)

tors affecting, or at least influencing, sediment toxicity and *(ii)* compliance checking. The former group of concerns stems from sediment heterogeneity, confounding factors such as ammonia, uncertainties related to assessment or testing approaches, and ultimately to the lack of unambiguous relationships between toxic effect endpoints and (individual) chemical concentrations in sediment." (Crane and Babut 2006).

While the traceability concept for quality control of chemical sediment analysis has been discussed to some extent – also in the frame of the discussion on EQS in the Water Framework Directive (Quevauviller 2004; Foerstner et al. 2004), the issue of quality control of (eco)toxicological data of sediments (see also Fig. 2.9, Sect. 2.2) appears to be a diffuse and confusing question for decision makers, especially because variability of biologial systems seems to result in less reliable data compared to assumably exact chemical data. This assumption does not take into account *(a)* the ecological relevance of data, and *(b)* the often overestimated precision of chemical analysis which especially for low concentrations of organic contaminants may be fairly poor (Heise et al. 2005).

The following sections will revise shortly the discussion on sediment chemical data quality, and describe more detailed requirements on ecotoxicological data quality, followed by suggestions for integrating uncertainty of data into environmental decision.

10.1.2 Sediment Chemical Data Quality

From a practical view, the traceability concept for quality control of chemical sediment analysis comprises three categories of investigations (Fig. 10.1), as described by Förstner (2004), Heise et al. (2005, 2004) (see also Sect. 2.2):

- Memory effect, mainly in dated sediment cores from lakes, reservoirs and marine basins, as historical records reflecting variations of pollution intensities in a catchment area. In order to confirm the presence and extent of sediment contamination, the traceability concept addresses "Sampling and sample preparation" (Mudroch and Aszue 1995), "Analysis" (Quevauviller 2002), "Grain size normalization" (Förstner 1989), and assessment of uncertainties, resulting from variations of typical matrix constituents.
- Basic characterization, i.e., sediment as ecological, social and economic value, as an essential part of the aquatic ecosystem by forming a variety of habitats and environments. A system approach is needed comprising biotests and effect-integrating measurements due to the inefficiency of chemical analysis in the assessment of complex contamination. In this scope, issues of bioavailability (Dickson et al. 1994),

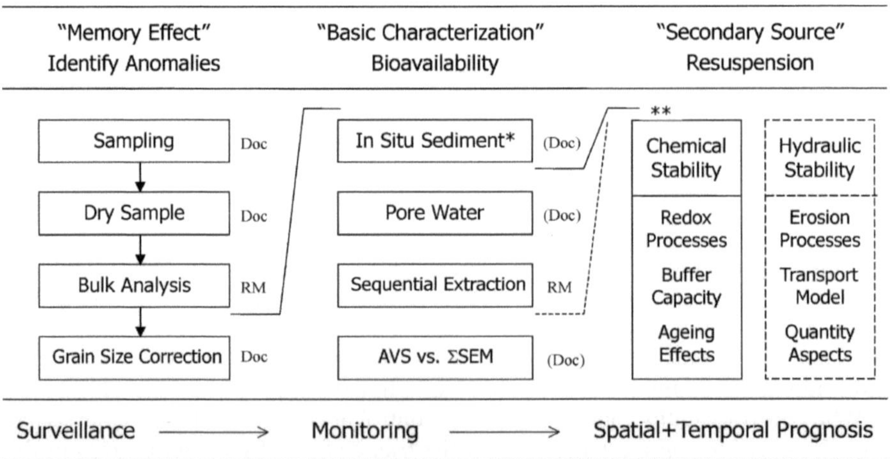

Fig. 10.1. Schematic overview on traceability aspects of chemical sediment analysis (for details see Förstner 2004). *RM:* Reference material; *Doc:* documented procedure; *AVS/ΣSEM:* acid volatile sulfide/sum of simultaneously extractable metals (DiToro et al. 1990); **Wet Sample:* sub-sampling for tests under oxygen-free atomsphere (pore water, sequential extraction, etc.)

porewater chemistry (Carr and Nipper 2001; Tessier and Campbell 1987) and the question how findings within the chemical proportion relates to results of the biological studies need to be addressed.

- Secondary source, mobilization of contaminated particles and release of contaminants after natural or artificial resuspension of sediments. On a river-basin scale, i.e., when applied in a conceptual river basin model (CBM), chemical and ecological information need a strong basis of sediment quantity data. In a dynamic system, this assessment should include not just those materials that are currently sediments, but also materials such as soils, mine tailings, etc. that can reasonably be expected to become part of the sediment cycle during the lifetime of a management approach (Apitz and White 2003). The assessment of such spatial and temporal processes again requires specific attention to sampling and reference material, as well as a good estimation of the uncertainties in the assessment of erosion risk and chemical mobilization studies (hydrodynamic data quality as described by Westrich in Heise et al. 2004).

10.1.3 Sediment Ecotoxicological Data Quality

If several substances in a sample are toxic, the total toxicity can be determined only by a toxicity test. For environmental testing, bioassays provide an integrated picture of the overall toxicity of a pore water, sediment elutriate, or sediment from a contaminated site. Various aquatic organisms, such as vertebrates, invertebrates, protozoa, algae and bacteria are used to test environmental samples. The idea behind these toxicity tests is that the test organisms will react in a predictable way to various types of environmental contaminants.

In principle, the influencing factors affecting the test results and the uncertainty are the same as for chemical analyses:

- human factors (operator carrying out the studies must be competent in the field of work under study and have practical experience related to the work to be able to make appropriate decisions from the observations made as the study progresses)
- environmental factors (control of oxygen conditions, geochemical composition)
- instrumental and technical factors (equipment within specification, working correctly, properly calibrated, procedures established for operational control and calibration, traceability of measurement to the criteria for test validity).

In addition, Quality Assurance/Quality Control (QA/QC) requirements for the biological tests are highly specific, because a calibration using reference material does not cover all aspects of biological variability (Simpson et al. 2005).

Quality Objectives

Accuracy criteria are not applicable to toxicity testing endpoints, because there are no standard organism responses against which to compare test results. In place of an absolute measurement of accuracy for toxicity tests, reference toxicant tests are performed to determine whether organism response is within prescribed acceptability criteria.

QA/QC requirements for sediment toxicity tests generally deal with ensuring that test conditions remain within control limits during the tests and do not contribute to observed effects and thereby confound interpretations regarding the toxicity of the sediments. For sediment toxicity tests, there are control limits for temperature, dissolved oxygen, salinity or pH. Monitoring of sulfides and ammonia in the test chambers may be appropriate for sediments where either of these chemicals is suspected as being a problem, and may be useful for interpreting test results. The sediment toxicity test protocols also require the testing of control samples as negative controls, positive controls, and reference sediments (see next paragraph). The criteria for determining test validity are an essential component of all standardized bioassays and are specific for each toxicity test species. The test results of control samples have to be compared to performance standards, which are used to validate acceptability limits, like mortality in a control sediment. All generated data should be presented in a report or in a standard operating procedure (SOP).

The SOPs should cover all aspects of the assay from the time the sample is collected and reaches the laboratory until the results of the bioassay are reported. A description of experiments concerning the validation conducted to determine variability, limit of quantification, and the quality controls should be documented for data audit and inspection; the traceability is a requirement for good analytical practice. Any deviations from SOPs should be documented with justifications for deviations.

Quality Control Procedures

Laboratory quality control procedures for sediment bioassays are listed in Ecology (2003). Here we will give a brief overview how control procedures ensure the quality of ecotoxicological tests.

Control and Reference Sediments

All solid phase tests measure toxicity relative to a negative control or reference sediment (ASTM 2003). For bulk sediment tests a negative control will be a control sediment that is essentially free of contaminants. Such a negative control provides evidence of test organism health, which is in most acute tests defined as mortality lower than 10% and for chronic tests a survival of more than 80% is sufficient. Control sediments can be provided from field collected sediments or from artificial or formulated sediments. The physicochemical properties such as grain size, TOC and background levels of contaminants should be determined. However, these properties could be different to those of the area studied.

Although sediment test organisms should tolerate a wide range of physicochemical sediment conditions, the contribution of those confounding factors may be assessed using a reference sediment as parallel test only. Reference sediments are ideally collected from sites near the contaminated site, representing the same sediment conditions exclusive of contaminants. The impact of the reference sediment has to be determined, inclusive the site-specific variability. Whereas the control sediment provides a reference point for interpreting effects from the test, the reference sediment can help estimate the relative contributions of natural and anthropogenic stress.

Positive Controls

A positive control uses a reference toxicant that affects the test organism in a reproducible manner. Reference toxicants provide a general measure of the precision of a toxicity test method over time. Acceptability limits are in general a ±2standard deviation of the EC50. The criteria for determining the sensitivity of the test organisms is an important component of good quality assurance.

Reference toxicants recommended by Environment Canada (1995) are copper and fluoranthene. Both chemicals are reference toxicants used to be preferentially in spiked control sediments for chronic whole-sediment tests. If the aim of the positive control is to measure the sensitivity of the test organisms in acute tests water only exposures may be used.

Data Management Procedures

The project proponent is responsible for the quality assurance review of data generated in any sediment investigation. There are two levels of quality assurance review applicable for sediment data. On the first level a review of bioassay data covers field and reporting elements and evaluates the acceptability of test results for positive controls, negative controls, reference sediment, replicates, and experimental conditions (temperature, salinity, pH, dissolved oxygen). Detailed guidance on review procedures is available from Ecology (2003).

The second level represents a more vigorous level of quality assurance review, and is appropriate for sediment data that are to be used for the development of numerical chemical criteria or the derivation of effect classes (Ahlf and Heise 2005). Such a review is also recommended in cases where the data may be used in litigation. We expect a more complex environmental scene investigation in future due to the fact that point sources are less important in comparison to diffuse ones and all lines of evidence have to be used to characterize the environmental impact (Wenning et al. 2004).

Uncertainty of Laboratory Toxicity Tests

Uncertainties of laboratory toxicity tests have been regarded as to fall into two categories: *(1)* uncertainties related to the phase tested, and *(2)* uncertainties related to the selection of endpoints measured in toxicity tests (Ingersoll et al. 1997). While those parameters that need to be controlled and monitored before and during a test have been mentioned above, this subsection will address intrinsic properties of tests which can not easily be overcome but have to be known and evaluated in order to interpret the results well.

Test Immanent Uncertainties Related to the Tested Phase

Toxicity investigations can comprise (from Ingersoll et al. 1997):

a whole sediment tests that are carried out incubating test organisms in direct contact with sediment and which should reflect the effects of in-place pollutants

b pore water test systems, assuming that the concentration in the pore water is in equilibrium with the sediment and the main exposure pathway is through contact with the pore water
c toxicity tests with organic extracts, simulating a worst case scenario in which even strongly bound contaminants may become available, and
d tests with elutriate samples and/or suspended solids to simulate resuspension events

For evaluation of toxicity data, information on a number of aspects is needed, that are influenced by the matrix. Among these are precision of the test system, standardization, sensitivity, and interference of the sediment matrix.

Results from round robin tests and inspection of variability of positive controls over time indicated that laboratory precision (i.e. precision not related to sampling collection, handling and storage) seemed to be high for whole sediment tests and in the same range as elutriate, organic extracts and porewater tests. Intra- and interlaboratory variability were low (Mearns et al. 1986; Ahlf and Heise 2005; Burton et al. 1996). Only tests on suspended matter showed high uncertainty due to the low standardization of this testing method (ASTM 1995).

Sensitivity of biotests in terms of un/certainty considerations refers to the potential of a test system to indicate or to predict correctly the effects of contaminants in the sediment, therewith minimizing false positive (non-toxic sample incorrectly classified as toxic) and false-negative (toxic sample incorrectly classified as non-toxic) results. Ingersoll et al. pointed out, that whole sediment tests with benthic organisms and acute measurement endpoints showed a high degree of certainty in this respect (Ingersoll et al. 1997). This has also been shown by Rönnpagel et al. (1998) for a bacteria contact test which correlated well with the autochthonous microbial activity in experiments with spiked sediments.

Interference of sediment compounds with measurement endpoints has been described for organic extracts in the luminescence bacteria test (Greene et al. 1992) when extracted pigments interfere with the luminescence, for the Algae Growth Inhibition in cases of shading effects of elutriated compounds, and for bacterial contact tests, where sediment properties reduced the measurable endpoint due to adsorption of organisms (solid phase microtox – Ringwood et al. 1997; Benton et al. 1995) or of the indicator substance (bacterial contact assay (Heise and Ahlf 2005)).

Uncertainties Related to the Selection of Endpoints

Measurement endpoints in toxicity tests are for example survival, growth, behavior, development, reproduction, metabolic activity, or biomarkers. Precision, ecological relevance, and sensitivity towards contaminants are among those parameters that affect the degree of certainty for the different endpoints (for more details see Ingersoll et al. 1997). Precision of these endpoints can be evaluated based on the replicability of responses.

Ecological relevance may influence how much weight is given to certain responses in decision making. Those endpoints with a linkage to ecological resources are considered to be of higher relevance. A comparison of responses from a contact assay with nitrifying bacteria with autochthonous nitrifying activity e.g., showed a highly correlated response in experiments with spiked material (Frühling 2003).

Sensitivity of toxicity tests is not only species and contaminant-specific, it also varies depending on the measurement endpoint. Nevertheless assessing the sensitivity of a biotest is crucial for the evaluation of responses and their translation into management decisions. If a biotest is little sensitive, only the strongest effects would be determined. Use of a highly sensitive test system, on the other side, may be overprotective. Information on the responsiveness of a test system is hence of high importance for the interpretation of biotest responses and the degree of confidence that would be involved in resulting management decisions.

Reduction and Assessment of Uncertainty in the Interpretation of Ecotoxicological Data

One approach to increase certainty of data has been suggested e.g., by Suter (1983): The use of several lines of evidence in order to make a best-judgment weight of evidence decision. In this context, the application of different biotests with different endpoints, exposure routes, and sensitivities towards contaminants could be regarded as different lines of evidence. If such a biotest battery, however, comprises e.g., sediment contact tests as well as elutriate tests, additional information could be drawn from the results on whether potential risks of contaminants are sediment focused or could also effect organisms in the water column upon resuspension.

Among the possibilities to interpret results from a biotest battery, there is an evaluation according to the maximal value after simple addition of biotest data. (e.g., inhibition values), statistical evaluation e.g., by Hasse Diagram Technique (Hollert et al. 2002) or by expert judgment on e.g., the basis of pattern analysis and its translation into rule-based expert systems (Ahlf and Heise 2005).

Following the concept of traceability, however, the next lower level of uncertainty which needs to be considered is the precision and sensitivity of different endpoints. Figure 10.2 depicts the results of two endpoints – inhibition of growth and egg production – in the nematode sediment contact test with *Caenorhabditis elegans*. Data were gained from the toxicity testing of 114 sediment samples from different rivers (Rhine and Elbe, among others), measuring both endpoints for every sample. Even though exposed to the same samples, the distribution of responses for the two endpoints are

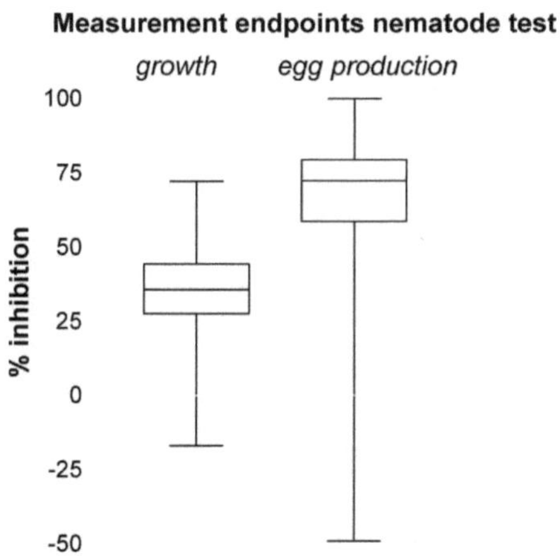

Fig. 10.2. Box-Whiskers plots of two measurement endpoints of the nematode test – inhibition of growth and egg production – determined for the same environmental sediment samples ($n = 114$)

shown to be very different, even though both are statistically closely correlated (Ahlf and Heise 2005). Considering these two endpoints, an inhibition in growth of e.g., 36% correlates well with an inhibition in egg production of about 70% and should therefore also lead to the same conclusion. If 36% inhibition (growth) were to be assessed as medium toxic, 70% (Egg production) should be as well because this endpoint seems to be more sensitive than growth inhibition.

In order to assess the extent of toxicity, which is indicated by a certain endpoint response, a characterization of the measurement endpoint in terms of sensitivity should be carried out.

Another parameter mentioned above, is the precision of biotest data depending on a variety of factors. These factors comprise the heterogeneity of the matrix, the preparation procedure, but also the number of organisms in the test system and the robustness of the bioassay. Precision can be estimated by reproducibility of e.g., positive controls or replicability of environmental samples.

Table 10.2 shows an attempt to estimate the test-immanent variability on the basis of standard deviation of positive controls, standard deviation of replicates, characteristics of the matrix and of the test organisms, and assessment of accuracy of 5 ecotoxicological assays of a biotest battery. The resulting degree of uncertainty needs to be taken into account any time, that biotest data are interpreted. If, for example, an inhibition of 20% is considered to mark the transition from "no effect" to "significant effect", a 15% inhibition can actually indicate already a significant response whereby 30% could also mean that no effect is apparent – as the uncertainty of data in this test is 15%.

A possibility to integrate this uncertainty into a decision framework is presented by fuzzy mathematics which defines overlapping areas between two sets in which more than one information can be valid (Babut et al. 2007, Heise et al. 2000; Hollert et al. 2002). Figure 10.3 demonstrates the interpretation of "no toxicity", "moderate toxicity" and "high toxicity" gained from 2 theoretical test systems *A* and *B*, whereby *A* has a low degree of uncertainty, the overlaps are small, and *B* has a high degree of uncertainty, with many inhibition values belonging to two result classses.

Table 10.2. Estimation of accuracy of test data for establishing the certainty toxicity, based on average standard deviation of replicates of negative controls (*column 1*) within a test and of positive control (*column 2*) between tests in % of inhibition value (Ahlf and Heise 2005)

Tests	SD_{av} of replicates	SD_{av} of positive control	Heterogenous matrix	Genetic variability	Resulting uncertainty
BCA	6[a]	12[e]	Yes	Low	±20%
AGI	4[b]	11–15[f]	No	Low	±15%
LBT (elutriate)	1[c]	10[g]	No	Low	±10%
LBT (extract)	1[c]	10[g]	No	Low	±10%
Nematode (eggs)	10–15[d]	n.d.	Yes	Low	±20%

The values were determined from different numbers of sediment samples ([a] *n* = 15, Hamburg Harbor; [b] *n* = 26, Hamburg Harbor, [c] *n* = 31, Hamburg Harbor and river Rhine) and comparing different test completions ([d] *n* = 23; [e] *n* = 10; Ahlf and Gratzer 1999).

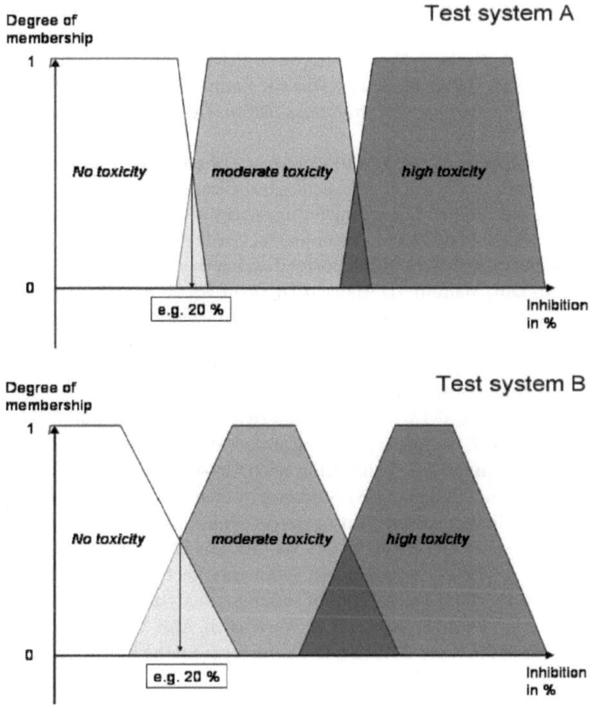

Fig. 10.3.
Demonstration of overlapping fuzzy sets, representing two theoretical test systems with **a** (*upper graph*) being a bioassay with a low degree of uncertainty and **b** (*lower graph*) showing a high overlap of toxicity classes due to high test-immanent variability

By using such mathematical models, the confidence in the interpretation of ecotoxicological data and finally the reliability of management decisions can be greatly enhanced. This may prove a valid possibility to include ecotoxicological data into environmental assessment – not despite their uncertainties but estimating and addressing them.

References

Ahlf W, Heise S (2005) Sediment toxicity assessment: Rationale for effect classes. JSS – J Soils and Sediments 5(1):16–20

Ahlf W, Gratzer H (1999) Erarbeitung von Kriterien zur Ableitung von Qualitätszielen für Sedimente und Schwebstoffe – Entwicklung methodischer Ansätze. UBA Texte 41/99:1–171

Anonymous (1987) GEMS/Water operational guide. World Health Organization (WHO). Geneva

Anonymous (2004) Evaluation of Current Gaps and Recommendations for further Actions in the Field of Environmental Analysis and Monitoring. METROPOLIS (Metrology in Support of EU Policies). Position Paper, March 2004, 8 p. Verneuil-en-Halatte/France

Apitz S, White S (2003) A conceptual framework for river-basin-scale sediment management. JSS – J Soils and Sediments 3(3):125–220

ASTM (1995) Standard Guide for Developing Conceptual Site Models for Contaminated Sites. E 1689-95

ASTM (2003) Standard guide for designing biological tests for sediments. E 1367-03. American Society for Testing and Materials. West Conshohocken, PA, USA, 2003

Babut M, Oen A, Hollert H, Apitz SE, Heise S, White S (2007) Prioritization at River Basin Scale, Risk Assessment at Local Scale: suggested approaches. In: Heise S (ed) Sediment Risk Management and Communication, Chapter 4. Elsevier, Amsterdam, pp 107–151

Benton MJ, Malott ML, Knight SS, Cooper CM, Benson WH (1995) Influence of sediment composition on apparent toxicity in a solid-phase test using bioluminescent bacteria. Environ Toxicol Chem 14: 411–414

Burton GA, Norberg-King TJ, Ingersoll CG, Ankley GT, Winger PV, Kubitz J, Lazorchak JM, Smith ME, Greer IE, Dwyer FJ, Call DJ, Day KE, Kennedy P, Stinson M (1996) Interlaboratory study of precision: *Hyalella azteca* and *Chironomus tentans* freshwater sediment toxicity assays. Environ Toxicol Chem 15:1335–1343

Carr RS, Nipper M (eds) (2001) Summary of a SETAC Technical Workshop "Porewater Toxicity Testing: Biological, Chemical, and Ecological Considerations with a Review of Methods and Applications, and Recommendations for Future Areas of Research, SETAC, Pensacola, FL

Crane M, Babut M (2006) Enironmental Quality Standards for Water Framework Directive Priority Substances: Challenges and Opportunities. International Environmental Assessment and Managment in press

DiToro DM, Mahony JD, Hansen DJ, Scott KJ, Hicks MB, Mayr SM, Redmond MS (1990) Toxicity of cadmium in sediments: the role of acid volatile sulfide. Environmental Toxicology and Chemistry 9:1487–1502

Dickson KL, Giesy JP, Parrish R, Wolfe L (1994) Summary and conclusions. In: Hamelink JL, Landrum PF, Bergman HL, Benson WH (eds) Bioavailability – Physical, Chemical and Biological Interactions. pp 221–230. SETAC Publication, Lewis Publ. Boca Raton

Ecology (2003) Sediment Sampling and Analysis Plan Appendix. Washington State Department of Ecology, Olympia, WA. Publication no. 03-09–043

Environment Canada (1995) Guidance document on measurement of toxicity test precision using control sediment spiked with a reference toxicant. Environment Canada Environmental Protection Series Report EPS 1/Rm/30, Ottawa, ON, Canada, 1995

Förstner U (1989) Contaminated Sediments, Lecture Notes in Earth Sciences. Springer-Verlag, Berlin

Förstner U (2004) Traceability of sediment analysis. Trends Anal Chem 23(3):217–236

Förstner U, Heise S, Schwartz R, Westrich B, Ahlf W (2004) Historical Contaminated Sediments and Soils at the River Basin Scale: Examples from the Elbe River Catchment Area. JSS – J Soils and Sediments 4(4):247–260

Frühling W(2003) Funktionale und strukturelle Untersuchungsparameter für mikrobielle Bodenbiozönosen und ihr ökotoxikologischer Nutzen. Dissertation at Umweltschutztechnik, Technische Universität Hamburg-Harburg Hamburg 167 pp

Greene MW, Bulich AA, Underwood SR (1992) Measurement of soil and sediment toxicity to bioluminescent bacteria when in direct contact for a fixed time period. Proceedings, 65th Annual Conference and Exposition of the Water Environment Federation, New Orleans, LA, USA, September 20–24, pp 53–63

Heise S, Ahlf W (2005) A new microbial contact assay for marine sediments. JSS – J Soils and Sediments 5(1):9–15

Heise S, Claus E, Heininger P, Krämer T, Krüger F, Schwartz R, Förstner U (2005) Studie zur Schadstoffbelastung der Sedimente im Elbeeinzugsgebiet. Commissioned by the Hamburg Port Authority, Hamburg, 181 pp

Heise S, Förstner U, Westrich B, Jancke T, Karnahl J, Salomons W (2004) Inventory of Historical Contaminated Sediment in Rhine Basin and its Tributaries. On behalf of the Port of Rotterdam Rep. Nr.: October 2004, 225 pp

Heise S, Maaß V, Gratzer H, Ahlf W (2000) Ecotoxicological Sediment Classification – Capabilities and Potentials – Presented for Elbe River Sediments. BfG- Mitteilungen Nr. 22 – Sediment Assessment in European River Basins: 96–104

Hollert H, Heise S, Pudenz S, Brüggemann R, Ahlf W, Braunbeck T (2002) Application of a Sediment Quality Triad and different statistical approaches (Hasse Diagrams and Fuzzy Logic) for the comparative evaluation of small streams. Ecotoxicology 11:311–321

Ingersoll CG, Ankley GT, Baudo R, Burton GA, Lick W, Luoma SN, MacDonald DD, Reynoldson TB, Solomon KR, Swartz RC, Warren-Hicks WJ (1997) Workgroup summary report on uncertainty evaluation of measurement endpoints used in sediment ecological risk assessment. In: Ingersoll CG, Dillon T, Biddinger GR (eds) 297. SETAC Pr.: Pensacola FL

Mearns AJ, Swartz RC, Cummins JM, Dinnel PA, Plesha P, Chapman PM (1986) Inter-laboratory comparison of a sediment toxicity test using the marine amphipod, *Rheposynius abronius.* Mar Environ Res 18:13–37

Meybeck M, Kimstach V, Helmer R (1992) Strategies for water quality assessment. In: Chapman D (ed) Water Quality Assessments. A Guide to the Use of Biota, Sediments and Water in Environmental Monitoring. Chapter 2, pp 19-50. Chapman & Hall, London
Mudroch A, Azcue JM (1995) Manual of Aquatic Sediment Sampling. Lewis Publ. Boca Raton
Quevauviller P (ed) (2002) Methodologies for Soil and Sediment Fractionation Studies. 180 p. The Royal Society of Chemistry, Cambridge UK
Quevauviller P (2004) Traceability of environmental chemical measurements. Trends Anal Chem 23(3):171–177
Ringwood AH, DeLorenzo ME, Ross PE, Holland AF (1997) Interpretation of Microtox® solid-phase toxicity tests: The effects of sediment composition. Environ Toxicol Chem 16:1135–1140
Rönnpagel K, Jansen E, Ahlf W (1998) Asking for the Indicator Function of Bioassays Evaluating Soil Contamination: Are Bioassay Results Reasonable Surrogates of Effects on Soil Microflora? Chemosphere 6:1291–1304
Simpson StL, Bateley GE, Chariton AA, Stauber JL, King CK, Chapman JC, Hyne RV, Gale SA, Roach AC, Maher WA (2005) Handbook for Sediment Quality Assessment, CSIRO, Bangar, NSW, pp 126
Suter GA (1993) Ecological Risk Assessment. Boca Raton Florida: Lewis
Tessier A, Campbell PGC (1987) Partitioning of trace metals in sediments: Relationship with bioavailability. In: Thomas R, Evans A, Hamilton A, Munawar M, Reynoldson T, Sadar H (eds) Ecological Effects of in situ Sediment Contaminants. Hydrobiologia 149:43–52
Wenning RJ, Batley GE, Ingersoll CG, Moore DW (2004) Use of Sediment Quality Guidelines and Related Tools for the Assessment of Contaminated Sediments. Society of Environmental Toxicology and Chemistry (SETAC). ISBN 1-880611-71-6 (2004)

Anett Matthäi · Pei-Chi Hsu · Wolfgang Ahlf

10.2 Evaluation of Sediment Toxicity in the Elbe River Basin

10.2.1 Introduction

Current sediment quality guidelines (SQGs) use chemical concentration criteria to evaluate sediment quality (Fig. 10.4). Different approaches for SQGs are linked to sediment toxicity by comparing concentrations in the environmental sample to effect levels that derive from tests with spiked water or sediments (Wenning and Ingersoll 2002; Giesy and Hoke 1990). In the field, however, mixtures of different contaminants cause effects that are difficult to estimate from the known toxicities of single substances and may deviate from ideal additive behavior, exhibiting synergistic or antagonistic effects, depending on their modes of action (Altenburger et al. 1996; Altenburger et al. 2004). Especially the large amount of organic compounds and their often low single compound concentration in natural sediments makes it difficult to relate toxicity to single substances. In addition biometabolization and biodegradation as well as abiotic degradation of organic compounds occur. Metabolites as well as their effects are mostly unknown. Potentials for chemical analyses are often limited and information on their effects insufficient.

Furthermore, bioavailability of substances, especially when adsorbed to sediment, is influenced by a variety of factors and difficult to conclude from chemical data. Characteristics of sediments and contaminants as well as the biodiversity in sediment all play a role in bioavailability (Ehlers and Luthy 2003) and can be altered with time. The time dependent increase of bonding strength of contaminants to particle surfaces, resulting in their decreased availability to organisms has been described as aging pro-

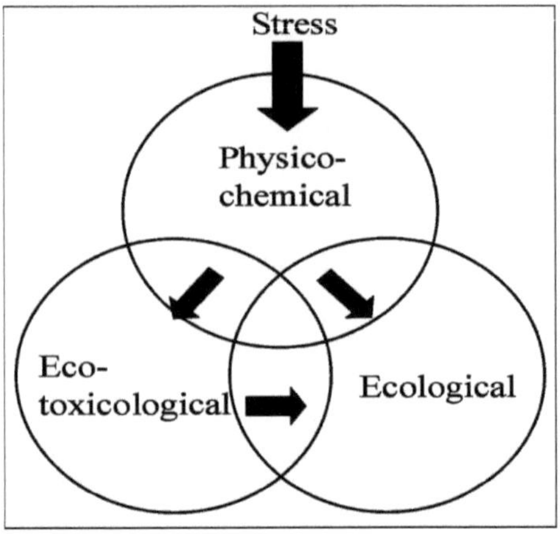

Fig. 10.4.
Conceptual model of the sediment quality triad based on Chapman et al. 1992

cess (Hatzinger and Alexander 1995; Reid et al. 2004). On the other side, it has been shown that chemically not extractable substances from sediments have been available to organisms, consequently causing effects (Scheifler et al. 2003).

In conclusion, the information of chemical SQGs on a potential impact on the protection aim "to prevent the impairments in living environment" has to be regarded with care, considering potential synergistic or antagonistic effects of chemicals, effectivity of their metabolites and bioavailability of compounds in different matrices (see also Heise and Ahlf 2002).

As the impact of mixture toxicity in sediments can only insufficiently be described by chemical analysis, to measure direct effects of environmental samples on organisms gives additional information. This ecotoxicological testing uses bioassays as tools to indicate bioavailability and the adverse effect of environmental samples by exposing them directly to test organisms. It interlinks between measured chemical concentration and ecological impairments observed, and points to a hazard which could be present in the ecosystem depending on exposure.

Biotesting, however, can only be done with a limited number of species, tested under controlled laboratory conditions, thus limiting their ecological relevance.

In an integrative approach to describe the sediment quality that has become known as the "sediment quality triad" (Chapman 1990), the limitations of the single methods are overcome by using chemical, ecotoxicological and in situ ecological data as different "lines of evidence", hence integrating both physical-chemical and biological processes. In such a "Weight of evidence"-concept, the more lines of evidence support the conclusion, the stronger the weight of evidence. It could mean a quantitative, semi quantitative, or qualitative estimate of the degree to which the evidence supports or undermines the conclusion (Burton et al. 2002).

This study presents ecotoxicological data as one line of evidence in order to indicate the ecological risk and transport of hazardous substances during/after flood events downstream. Recently it was shown that sediment dynamics may serve as additional line of evidence (Chapman and Hollert 2006). Sediment dynamics and re-mobilization of highly contaminated sediments influence both sediment toxicity of fresh water and marine systems (Förstner 2004; Leipe et al. 2005). Regarding to that it was of great

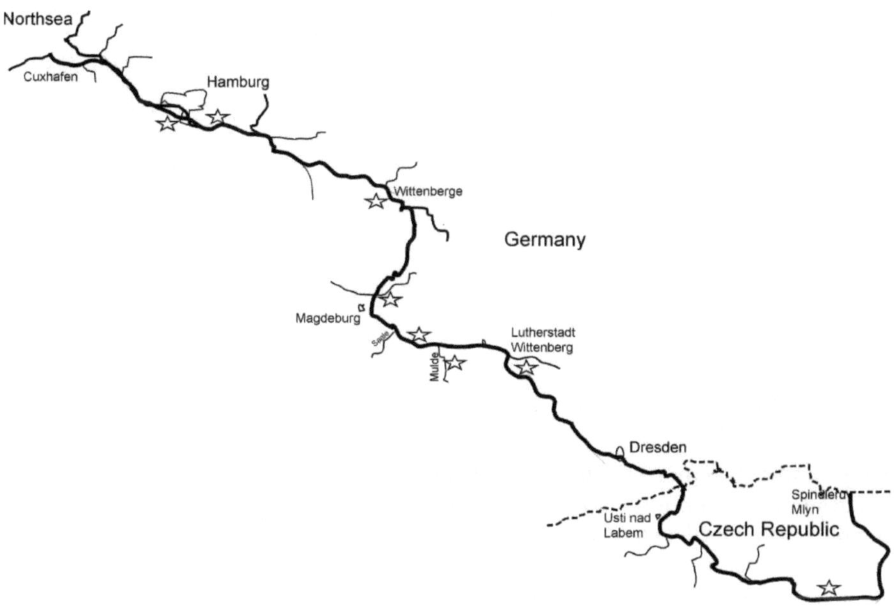

Fig. 10.5. Elbe River basin (sampling sites are marked with *stars*)

importance for this work to screen hotspots and monitor trends over time. Two main questions built the basis for this work. The first was to look for sources of highly contaminated material upstream in the Elbe River which might be transported downstream during flood events. For that reason sediment samples were taken along the Elbe River from spring to river mouth (Fig. 10.5).

The second was to find changes in toxicity without strong flood events to determine if contaminated material was transported downstream under normal conditions. For that concern sediment and water samples were taken monthly at a groyne field of Elbe River in Oortkaten near the city of Hamburg. Water samples were taken to answer the question how contaminants would be transported: dissolved or attached to suspended particles. All samples were analyzed with a set of four bioassays (Ahlf et al. 2002). It is hypothesized that is it not sufficient to look only on the transfer of contaminants to the water phase to determine the hazard potential. To have a closer look on the bioavailability of particle bound contaminants two bioassays testing elutriates of sediments and water phases were combined with two bioassays testing the sediment in direct contact to the organisms. Freshly deposited particles can have different effects compared to historically contaminated sediments due to aging effects even if they show similar contents of toxic compounds. Hence, it seemed important to differentiate between freshly deposited sediments and consolidated sediments. Samples from two depths were taken accordingly. The collected data will provide a spatial as well as a seasonal survey of toxicity and quality occurring in the Elbe River. The toxic impact on the test organisms should describe the different exposure conditions in the sediments and be considered as a characteristic of the sediment. Sediment as well as water phase samples will be rated in a refined system of effect classes where emphasis will be laid on the transport.

Table 10.3. The used bioassays for ecotoxicological classification

Test organism	Endpoint	Exposure route	Standard number
Algae: *P. subcapitata*	Growth (fluorescence)	Elutriate	DIN 38412-33
Bacteria: *V. fischeri*	Enzymatic activity (bioluminescence)	Elutriate and methanol extract	DIN EN ISO 11348-2
Bacteria: *A. globiformis*	Enzymatic activity (dehydrogenase activity)	Solid phase	DIN 38412-48
Nematode: *C. elegans*	Growth (NTG), fertility (NTF), reproduction (NTR)	Solid phase	In prep.

10.2.2 Sampling and Methods

Samples were taken in Prelouc (Czech Republic), Schöneberg, Barby, Saale, Rosslau, Muldestein, in Hamburg Harbor and Oortkaten (Germany) along the Elbe River. Samples from Oortkaten were taken from December 2004 to September 2005. Sediment cores (diameter 10 cm) from all sampling sites were spliced into layers and a upper layer sample was taken from a depth of 0–5 cm; a lower layer from 15 to 20 cm. Water phase samples were taken from the surface. To determine toxicity a test battery of four bioassays was used comprising two sediment contact assays and two elutriate assays (see Table 10.3). A dilution series of elutriates and water phase samples was performed for algal growth inhibition test (AGI) (DIN 38412-33) and luminescence bacteria assay (LBT) (DIN EN ISO 11348-2). LBT was performed with elutriates (LBT EL) and methanol extracts (LBT EX). Elutriates were prepared with sediment and VE water in a ratio of 1:4 shaken over head over night for 24 hours. For bacteria contact assay (BCA)(DIN 38412-48) and nematode test (NT) the samples were tested undiluted (Traunspurger et al. 1997). BCA was carried out in a modified way according to (Rönnpagel et al. 1998).

For the tests the endpoints as percentage of inhibition compared to a control that consists of medium and de-ionized water was determined, the latter replacing the elutriate component. In BCA quartz sand is used to replace the natural sediment sample and for the NT an artificial sediment works as control.

10.2.3 Results

Spatial Distribution of Sediment Toxicity in the Elbe River

The inhibition of algal growth was higher in elutriates of the upper layer than of the lower layer of the sediments among all investigated sites. The inhibition of bioluminescence was in most of the samples weak (~10%). Only samples from Muldestein had a higher toxic impact on the bioluminescence (40–50%). Both contact assays showed for most of the samples higher toxicity in the upper layer sediments than in the lower layer sediments. To give an example of the ecotoxicological profile of one sample, results from sampling in March in Oortkaten near Hamburg are shown (Fig. 10.6).

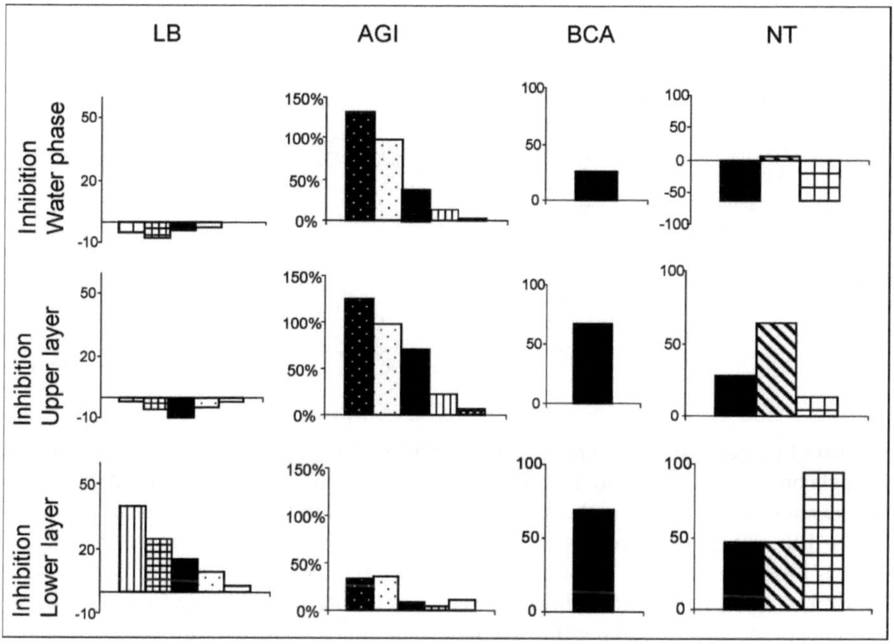

Fig. 10.6. Results of the bioassays for samples from Oortkaten (March 2005) (*columns from left to right:* Test with luminescence bacteria: G1, G2, G4, G8; algal growth inhibition test: G1, G2, G4, G8, G16; bacterial contact test: only G1; nematode test: growth, fertility, reproduction)

Temporal Changes of Sediment Toxicity at Sampling Site Oortkaten

In Oortkaten the water phases and elutriates of the upper layer inhibited the algal growth rate much stronger than the elutriates of the lower layer. In contrast the activity of the luminescent bacteria were strongly inhibited by lower layer elutriates but upper layer elutriates and water phases had low impact. The described pattern noticed in the two elutriate assays was observed for all samples during the sampling period. The similar toxicity in the water phase and in the elutriate prepared from the upper layer sediment indicated a high exchange between these two phases in opposite to a strong difference between the two sediment layers. Both sediment contact assays showed a common pattern; sediment toxicity is slightly increasing with the depth. For the nematode test, growth (first column), fertility (second column) and reproduction (third column) were determined. The water phase showed only minor toxic effect on the nematodes and the bacteria. Principle component analysis showed that for a defined area the quality can be well described conducting not all biotests (Ahlf et al. 2002).

From Toxicity Data to Classification of Effects

The classification used was based on (Ahlf and Heise 2005) where response patterns are used to define different classes. Interpretation and scaling of classes (e.g., class 1: low toxicity, class 3: high toxicity) can explain the level of toxicity as well as the effect

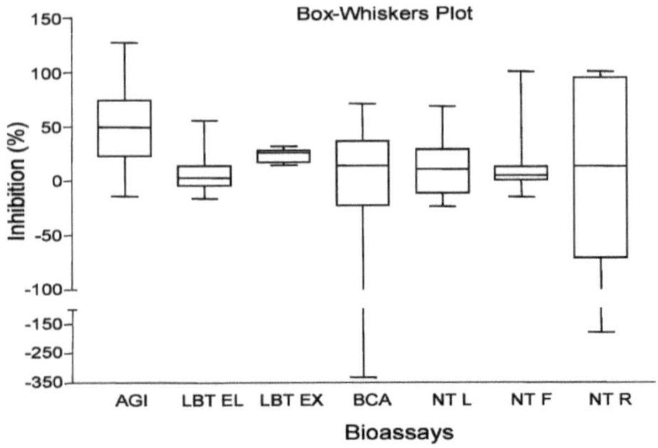

Fig. 10.7. Box-Wiskers plot of bioassays response distribution from spatial sampling in Elbe River: *x-axis* describes different bioassays (*AGI:* results from algal test; *LBT EL/EX:* luminescence bacteria; *BCA:* bacteria contact assay; *NT F/G/R:* nematode test). The inhibition in % is shown on the *y-axis*

pathways like being bound to particles or dissolved in the water phase. Interpretation of single biotest data took into account the following observations: test organisms have different sensitivities among the mixture of substances in the environmental samples. Via the different exposure routes; the test organisms show a different range of responses towards the same samples. Figure 10.7 shows the data distribution in Box-Wisker plot. The different response ranges show the necessity to interpret the data from the different biotests specifically; taking into account their characteristics (Ahlf 2005).

Due to the heterogeneity of the matrix (sediment) and the genetic variability of test organisms, the results of the bioassays also have different variabilities. To avoid the inflexibility of the rigid numerical separation of the toxicity categories, the variation and standard deviation of each bioassay were considered to create transient zones between the toxicity categories. Important is the gradient between the categories which is dependent on the deviation of the test results. Instead of belonging to one class or not, fuzzy logic describes the degree of membership of data to a class. The responses of the ecotoxicological bioassays are divided into three fuzzy sets indicating low toxicity (up to 25th percentile), moderate toxicity (25th to 75th percentile), and high toxicity (higher than 75th percentile). With cluster analysis and *K*-means analysis the grouping behavior of all samples' test responses were considered to determine the number of classes describing total toxicity. Three classes were selected to communicate the risk according to the characteristics and transport abilities of the pollutants in the sediments. The first class was described as having no negative effect on the test organisms. The second class indicated contaminations with mostly particle bound compounds and the third class contained samples contaminated with water soluble compounds. Contaminants in samples from the latter class will easily be transported with the water phase. Contaminants from samples of the second class will only be transported in the particle bound form when particles are eroded from the sediment and transported as suspended matter. From this categorization following three descriptive classes had been created:

- *class 1* is referring to no or low toxicity
- *class 2* is referring to particle bound toxicity, and
- *class 3* is referring to water soluble toxicity

Table 10.4. Results of the classification (spatial sampling)

	HH Harbor	Schöneberg	Barby	Saale	Rosslau	Muldestein	Prelouc
Upper layer	3	3	3	3	2	3	3
Lower layer	3	3	3	3	2	1	3

Table 10.5. Ecotoxicological classification for Elbe River and pH value of the water phase in 2004/2005

	Dec	Feb	Mar	Apr	May	Jul	Sept
pH value	7.68	7.78	6.88	9.44	8.16	8.55	8.73
Water phase class	3	3	3	3	1	3	3
Upper layer class	3	3	3	3	2	3	2
Lower layer class	3	2	2	2	2	3	2

Most sediment samples along the river were classified as class 3 indicating water soluble contaminants. Only samples from Rosslau fit in class 2 according to particle bound contaminations. The lower layer of the Muldestein sample taken in the Elbe tributary Mulde showed no toxicity (Table 10.4).

The classification approach was also applied to the samples taken during one year at Oortkaten. Table 10.5 shows the classes compared to the pH value of the water phase. In winter and beginning of spring the toxicity pattern remained stable. Most of the lower layer samples were classified in class two according to particle bound toxicity, whereas upper layer and water phase were grouped in class three which indicated more water soluble contaminants. From late spring on, changes in this pattern were observed. The toxicity in the water phase decreased in May (class 1) and now both sediment layers showed particle bound toxicity (class 2) which indicated an increased exchange/transport of material. This ecotoxicological pattern was compared to abiotic factors like discharge of the river, redox potential and pH value. A high discharge ($>2\,000\ m^3\ s^{-1}$) in the end of March/begin of April (Data Arge Elbe) which was accompanied by an increase of the pH-value in the water phase was observed. The increase of the pH-value indicated the first algae bloom in Elbe River.

10.2.4 Discussion

Sediment quality is strongly connected with sediment transport and sediment-water interaction. Water being the fastest transportation system in the river is able to carry soluble contaminants as well as particle bound contaminants attached to suspended particles. With higher discharges more suspended particles are transported in the Elbe River. In years of lower discharges 144 kt of suspended material is transported. During years of higher discharges this amount is almost doubled (Heise et al. 2005).

During average annual flood events both the quantity of river water and the quantity of suspended particles increase and contaminants are transported downstream to a larger extent. The connection between transport and toxicity in Elbe River was investigated for this work. A new classification system was created on the basis of the management tool for contaminated sediments and dredged material from Ahlf and Heise (2005). This approach could particularly differentiate between the transport of toxicity either via water phase or via suspended sediments. It was applied to classify the samples along Elbe River to find sources of different contaminated material. For Rosslau the toxicity of both sediment layers (class 2) were considered only to be transported via suspended material during higher discharges. Toxicity of all other samples fitting in class 3 would easily be transported even with the water phase during lower discharge.

The classification system was also applied to the samples taken in Oortkaten. Seasonal changes especially in spring/early summer could be detected with the approach mostly for the water phase and the upper layer sediment. One possible reason for this change may be the resuspension and transportation of sediments along the river caused by the high discharge (>2 000 $m^3 s^{-1}$) in April. Contaminated material may be transported further downstream, changing the classification of the sediment samples taken in May from class three to class two. This would also explain the low toxicity of the water sample taken in May. Before May similar toxicity pattern in water phase and upper layer sediments could be detected. It is hypothesized that with the high discharge water soluble contaminants were washed out of the sediment and were more diluted in the water phase. Table 10.6 shows the discharge of the Elbe River in 2005 compared to the classification. Although there was an increase in discharge already in end of February/begin of March, changes in toxicity pattern were not detected before May. It is possible that the first discharge peak in March transported only superficial sediment layers downstream. The second discharge peak in April eroded lower layer sediments with a larger amount of contaminants. This phenomena was already described for the Rhine by (Heise et al. 2004). It is also likely that the changes in toxicity after the second peak of discharge were due to changing input ratios of Elbe tributaries with different historical contaminations (Barborowski et al. 2006).

A third possible explanation for the changes in the toxicity classes may be the first algae bloom which was indicated by an increase of the pH value in the water phase from 6.8 in March to 9.4 in April (Table 10.5). The beginning of an algal bloom is dependent on water temperature and light regime. It is assumed that the algae were

Table 10.6. Discharge and ecotoxicological classification for Elbe River in 2005

	Jan	Feb	Mar	Apr	May	Jul	Sept	Discharge ($m^3 s^{-1}$)
								2 500
Water phase		3	3	3	1	3	3	
Upper layer		3	3	3	2	3	2	1 500
Lower layer		2	2	2	2	3	2	500

working as a filter for the water phase. Especially organic compounds would be accumulated by the algae and eventually accumulate on top of the sediment, when the algae material settles on top of the sediment at the end of the bloom.

10.2.5 Conclusion

The classification system proved to be valuable for monitoring local seasonal changes. It was possible to detect spatial distribution of different qualities along the Elbe River as well as temporal quality trends. The accumulation of particle bound toxicity between the upper and lower sediment layer could be demonstrated, but also high dynamic exchanges between the upper layer and the water phase. As depicted in Fig. 10.6, different results from elutriate assays and sediment contact tests are obvious. Data showed, that it is not enough to describe sediment quality with elutriate assays only; a test battery combining elutriate assay and contact assays is recommended. The bioavailability of toxic compounds is highly coupled with its chemical characteristics, and so is the transport. It is assumed that for water soluble contaminants, best described by elutriate assays, there is an equilibrium between water phase and the upper sediment layer. So the quality of the water is a factor of great importance for the transport of sediment toxicity during normal discharge. During higher discharge the depth of the mobilized sediment layers is important. To prove this hypothesis resuspension experiments should follow. Re-suspending both sediment layers with different sheer stresses over time and analyzing the samples with elutriate assays as well as sediment contact assays will give the information about the transport of sediment toxicity. It is expected that only weak changes in toxicity pattern between water phase and upper layer sediment will occur. By resuspending the lower layer sediment it is assumed to find stronger changes in sediment and water toxicity over time. The modified classification system worked well to detect changes in spatial and temporal sediment quality. It was suitable to distinguish differences in toxicity caused by the intense sedimentation resuspension cycles occurring in natural sediments of the Elbe River over time.

Acknowledgments

We thank G. Fengler (University of Greifswald) for successful sampling, R. Siepmann for experimental realization and M. Bergemann (ARGE Elbe) for providing technical data for Elbe River. Special thanks to S. Heise (BIS) for scientific discussion and support in classification. The study was supported by BMBF (FKZ 02WF0471).

References

Ahlf W, Braunbeck T, Heise S, Hollert H (2002) Sediment and soil quality criteria. In: Burden FR, McKelvie I, Förstner U, Günther H (eds), Environmental Monitoring Handbook.McGraw-Hill, New York 17.1-17.18

Ahlf W (2005) Trends in sediment research – dedicated to Prof. Dr. Ulrich Förstner on his 65th birthday. J Soils and Sediments 5:1

Ahlf W, Heise S (2005) Sediment toxicity assessment – rationale for effect classes. J Soils and Sediments 5:16–20

Altenburger R, Boedeker W, Faust M, Grimme LH (1996) Regulations for combined effects of pollutants: Consequences from risk assessment in aquatic toxicology. Food Chem Toxicol 34: 1155–1157

Altenburger R, Walter H, Grote M (2004) What contributes to the combined effect of a complex mixture? Environ Sci Technol 38:6353–6362

Baborowski M, Lobe I, Krüger F, v Tümpling W, Rupp H, Büttner O, Morgenstern P, Guhr H (2006) Transport and fate of dissolved and suspended matter in the Middle Elbe region during floods. SEDYMO Intern Symp, Hamburg, March 26–29, 2006. Abstract L 13, p 25, Hamburg

Burton GA, Chapman PM, Smith EP (2002) Weight of evidence approaches for assessing ecosystem impairment. Human and Ecological Risk Assessment 8:1657–1673

Chapman PM (1990) The Sediment Quality Triad approach to determining pollution induced degradation. Sci Tot Environ 97:815–825

Chapman PM, Hollert H (2006) Should the sediment quality triad become a tetrad, a pentad, or possibly even a hexad? J Soils and Sediments 6:4–8

Chapman PM, Power E, Burton G (1992) Interactive assessment in aquatic ecosystems. In: Burton GA (ed) Sediment toxicity assesseement. pp 313–340. Lewis Publ, Boca Raton, Fl

Ehlers LJ, Luthy RG (2003) Contaminant bioavailability in soil and sediment. Environ Sci Technol 37:295A-302A

Förstner U (2004) Sediment dynamics and pollutant mobility in rivers: An interdisciplinary approach. Lakes and Reservoirs: Research and Management 9:25–40

Giesy JP, Hoke RA (1990) Freshwater sediment quality criteria: Toxicity bioassessment. In: Baudo R, Giesy JP, Muntau H (eds) Sediments: chemistry and toxicity of in-place pollutants. CRC Press, BocaRaton, pp 265–348

Hatzinger PB, Alexander M (1995) Effect of aging of chemicals in soil on their biodegradability and extractability. Environ Sci Technol 29:537–545

Heise S, Ahlf W (2002) The Need for new concepts in risk management of sediments historical developments, future perspectives and new approaches. J Soils and Sediments 2:4–8

Heise S, Förstner U, Westrich B, Salomons W, Karnahl J, Jancke T, Schönberger H (2004) Inventory of historical contaminated sediments in Rhine Basin and its tributaries. On behalf of the Port of Rotterdam, October 2004, Hamburg, 225 p

Heise S, Claus E, Heininger P, Krämer T, Krüger F, Schwartz R, Förstner U (2005) Studie zur Schadstoffbelastung der Sedimente im Elbeinzugsgebiet: Ursachen und Trends. Hamburg Port Authority. December 2005, Hamburg, 169 p

Leipe T, Kersten M, Heise S, Pohl C, Witt G, Liehr G, Zettler M, Tauber F (2005) Ecotoxicity assessment of natural attenuation effects at a historical dumping site in the western Baltic Sea. Mari Poll Bull 50:446–459

Reid BJ, Stokes JD, Jones KC, Semple KT (2004) Influence of hydroxypropyl-beta-cyclodextrin on the extraction and biodegradation of phenanthrene in soil. Environ Toxicol Chem 23:550–556

Rönnpagel K, Janssen E, Ahlf W (1998) Asking for the indicator function of bioassays evaluating soil contamination: Are bioassay results reasonable surrogates of effects on soil microflora? Chemosphere 36:1291–1304

Scheifler R, Schwartz C, Echevarria G, De Vaufleury A, Badot PM, Morel JL (2003) "Nonavailable" soil cadmium is bioavailable to snails: Evidence from isotopic dilution experiments. Environmental Science and Technology 37:81–86

Traunspurger W, Haitzer M, Hoss S, Beier S, Ahlf W, Steinberg C (1997) Ecotoxicological assessment of aquatic sediments with Caenorhabditis elegans (nematoda) – A method for testing liquid medium and whole-sediment samples. Environ Toxicol Chem 16:245–250

Wenning RJ, Ingersoll CG (2002) Summary of the SETAC Pellston Workshop on Use of Sediment Quality Guidelines and Related Tools for the Assessment of Contaminated Sediments. Pensacola FL, USA 17–22 August 2002, 48 pp

*Henner Hollert · Matthias Dürr · Ingo Haag · Jan Wölz · Klara Hilscherova
Ludek Blaha · Sabine Ulrike Gerbersdorf*

10.3 Influence of Hydrodynamics on Sediment Ecotoxicity

10.3.1 Role of Sediments in Freshwater Quality

There is general agreement that sediment-bound substances are of major importance for the fate and effects of trace contaminants as well as water quality in aquatic systems. Sediments can act as sinks for various pollutants but could also become a contamination source under certain circumstances such as dredging or flood events (Ahlf et al. 2002a,b; Förstner and Müller 1974; Hollert et al. 2000a, 2003a). Contaminated sediments are known to cause various adverse effects on organisms even when contaminant levels in the overlying water are low (Chapman 1989). Thus, monitoring and assessment of sediment quality is of prime significance for national legislation in general and for the implementation of the European Water Framework Directive in particular (SedNet 2004).

Especially through the activities of SETAC North America (Wenning et al. 2005; Wenning and Ingersoll 2002) and the European SedNet network (Salomons and Brils 2004; SedNet 2004) sediment related issues were given increasing attention in both science and the public. While water quality has notably improved over the past three decades, the sediments in many European river basins still retain the toxic heritage from the past era of uncontrolled industrial production, and which will continue to influence the quality of waters significantly for many years to come (Salomons and Brils 2004; SedNet 2004).

Since the 1970s several chemical analytical studies revealed elevated concentrations of dominant environmental contaminants such as heavy metals and organic pollutants in marine and riverine sediments using chemical analyses (Foerstner et al. 2004; Förstner and Müller 1974; Giger et al. 1974; Haag et al. 2001; Stoffers et al. 1977). These hazardous contaminants are often accumulated in deeper layers covered by relatively unpolluted sediments, and thus are sequestered from the bioavailable oxic sediment surface zone (Haag et al. 2001; Ziegler 2002). However, these chemicals are mostly persistent in the natural environment, and can enter the oxic water column after an erosion events such as bioturbation (Chapman et al. 1992), flood events (Hollert et al. 2000a, 2003b) or dredging and relocation of sediments (Koethe 2003). Consequently, toxicants can become bioavailable (Calmano et al. 1993; Simpson et al. 1998; Ziegler 2002) and may result in detrimental effects on aquatic organisms at various trophic levels. Furthermore, downstream transport and deposition of contaminated particles in inundated areas may also result in negative effects on biota in these regions (Japenga and Salomons 1993).

With a delay of more than one decade to the first geochemical studies, the assessment of biological consequences of particle-bound pollutants has become a major topic in international water research (Burton 1991; Giesy and Hoke 1989; Power and Chapman 1992). To date, most studies focused on the development of suitable bioanalytical methods and the assessment of their potential to investigate sediment-bound contaminants. However, the role of sediment remobilization and possible ecotoxicological effects of contaminants bound to suspended material has been scarcely investigated.

10.3.2 Factors Affecting Mobilization of Sediments and (Bio)Availability of Contaminants

In many river systems, hazardous contaminants are predominantly transported in association with suspended particulate matter. The suspended particles and the sediment-bound pollutants accumulate in regions of low turbulence, such as groyne fields, harbours, and river reservoirs forming sites with high levels of contamination.

An important issue related to the role of sediments in water quality is their potential to be subject to remobilization, transport and redistribution during certain environmental events such as floods. Although these processes increase accessibility and bioavailability of contaminants, the conditions under which these processes occur, their amplitude and possible role in contaminant accessibility and effects are still poorly understood. The complexity of cohesive sediments, which are biologically active and chemically reactive, precludes the definition of a general analytical theory for their resuspension behavior. Moreover, the sediment properties of cohesive sediments vary on a number of spatial, temporal and vertical scales (Gerbersdorf et al. 2005, 2007) and empirically based field and laboratory experiments are needed to elucidate the mechanisms which govern the erosion resistance of cohesive sediments. As well, interdisciplinary studies are needed, to obtain better and realistic conceptual understanding of natural sediments and their inherent physical and biological complexity (Black et al. 2002). However, either physico-chemical or biological sediment properties have been in the focus of research on their impact on sediment stability, and only recently, the first comprehensive investigative approach to derive master-variables affecting sediment stability was published (Gerbersdorf et al. 2005, 2007).

Over the past decades numerous studies have been focused on primarily isolated aspects of sediment pollution issues. Recently the fate of particle-bound pollutants and hydrodynamic transport processes has been addressed increasingly in interdisciplinary joint projects. These studies documented that particle-bound priority pollutants (e.g., EPA-PAHs) are major contributors to both the overall contamination and transport of lipophilic pollutants in rivers. Work that significantly contributed to these findings were, among others, the DFG-Research Group 371 or the interdisciplinary BMBF-funded joint project, SEDYMO (Förstner et al. 2004; Förstner and Westrich 2005). However, the questions regarding physico-chemical surface properties of suspended particles, chemical mobilization and biological degradation of pollutants as well as regarding the related bioavailability of contaminants and their toxicity have not been satisfactorily addressed to date. Especially the important link between the erosion potential and hazard potential of sediments/distinctive sediment horizons originating from contaminated riverine sites, need to be addressed in future studies if a realistic risk assessment is to be derived.

The fate of the contaminants associated with sediments is strongly influenced by the amount and type of the sedimentary organic matter, which reflects the environmental evolution in the drainage area and fluviative or lake depositional systems (Martínek et al. 2006; Stout et al. 2002). The geochemical parameters of organic matter are controlled by the interplay of biomass productivity, weathering during transport, and microbial reworking during and shortly after deposition (Peters et al. 2005). Valuable monitoring data have been collected on the contaminants in sediments. However, only limited data exist on the associated organic matter and the role of different organic matrices for the fate of pollutants is insufficiently documented and not well understood (Stout et al. 2004). Fresh sedimentary particles behave in a different way when compared to the re-deposited older sediments, even if the content of pollutants is similar. It is, therefore, highly desirable to integrate the role of natural organic matter of different biological origin, mainly terrestrial plants, woody material, bacteria and algae (Gonzáles-Vila et al. 2003; Meyers 2003) into the ecotoxicological assessment of complex sedimentary systems of rivers and their relevance for potential contaminant bioavailability.

The extracellular polymeric substances (EPS) excreted by microorganisms such as microalgae or bacteria, can be a significant part of the total organic pool. These polymeric substances have received more and more attention over the last years due to their role in biostabilization of sediments (e.g., De Brouwer et al. 2000; Paterson et al. 2000). Only recently, the importance of EPS for the erosion resistance could be shown for several contaminated freshwater sites (Gerbersdorf et al. 2005, 2007). Concerning the fate of the contaminants, these polymeric substances may influence the nature of the eroded material, but this work is at an early stage (Perkins et al. 2004). The polymeric substances may alter the adsorption/degradation of contaminants to/within the eroded material, but may change as well their lateral transport through altering floc characteristics such as floc size and floc strength (Droppo 2004). Thus, the binding capacity of the polymeric substances, as well as their influence on the nature of the erodable material should be addressed in order to contribute to the questions on the bioavailability and re-deposition of contaminants.

10.3.3 Ecotoxicological Methods to Assess Sediment Contamination

As discussed in the previous paragraphs, decreased stability of cohesive sediments and their mobilization leads to increased bioavailability of hazardous contaminants. Sediment mobilization is affected by numerous physico-chemical, geochemical and biological parameters that are poorly understood and that have been scarcely investigated by complex interdisciplinary research projects. In spite of intensive research and development of numerous model testing systems, still is little known about possible ecotoxicological consequences of mobilized sediment contaminants. To evaluate adverse effects on ecosystems, neither biotests nor chemical-analytic techniques alone are sufficient. In contrast, a combination of biotests and chemical methods allows comprehensive insights into the hazard caused by sediment contamination.

To monitor the sediment quality, ecotoxicological bioassays are first applied to screen if contamination had significant effects on biological functions of the model organisms/systems. A broad spectrum of test batteries of standardized bioassays has been used to assess the possible hazardous effect of particulate matter and elutriate. The bioassays included *in vivo* tests at different levels of the aquatic food chain and *in vitro* tests. Various microbiological toxicity tests have been developed and validated for use in sediment risk assessment during the past 20 years (Ahlf et al. 1989; van Beelen 2003). It was shown that contamination correlates with the shift in microorganism communities towards toxicant-resistant species and that persistent toxic effects on the microflora caused for example by zinc, cadmium and copper often occur at concentrations lower than European Community limits (van Beelen 2003). Other assays for ecotoxicological studies include the algae growth inhibition assay, the bacterial bioluminescence bioassay, and the Daphnia assay (den Besten et al. 2003; Koethe 2003). Since fish are representing vertebrates, and can be linked via bioaccumulation to humans, large efforts have been undertaken to develop fish-based test systems for the assessment of sediment bound substances (Chen and White 2004; Davoren et al. 2005; Hilscherova et al. 2000; Hollert et al. 2000a, 2005; Kammann et al. 2005a; Kosmehl et al. 2004; USEPA 2002). In addition to *in vivo* sediment exposure tests with fish, a number of sub-organismal assays are in use such as cell-based *in vitro* systems (Davoren et al. 2005; Hollert et al. 2000a; Kosmehl et al. 2004; Segner 1998) and the fish egg assay with *Danio rerio* (Hallare et al. 2005; Hollert et al. 2003b).

While acute toxicity was of major concern in the last decades, recently for many river basins a change in focus to more subtle specific chronic non-lethal effects occurred (Brack et al. 2005a). While these effects are difficult to assess using *in vivo* tests, they can be relatively easily determined by *in vitro* techniques that allow to predict toxic potentials of complex environmental mixtures (Janošek et al. 2006). The *in vitro* bioassay approach serves as efficient, fast and cost effective screening for evaluation of the receptor-mediated activities of the complex mixtures (Hilscherova et al. 2002). We have successfully used this approach to prioritize contaminated sediment sites (Hilscherova et al. 2003; Hollert et al. 2002a) and to study novel endocrine disruptive effects observed in situ (Blaha et al. 2006). A further advantage of a bioassay approach is, that the combination of different bioanalytical methods allows to investigate multiple endpoints such as genotoxic or mutagenic (Chen and White 2004; Kosmehl et al. 2006), dioxin-like (Hilscherova et al. 2002; Hilscherova et al. 2001; Hilscherova et al. 2000), or various endocrine effects (Ankley et al. 1998; Sumpter and Johnson 2005) in parallel in the same sample.

10.3.4 Combined Approaches to Investigate the Influence of Hydrodynamics on Sediment Ecotoxicity

Recently, in several studies toxicity has been evaluated at various sediment depths (Burton Jr. et al. 2001; Hollert et al. 2003a; Kosmehl et al. 2004), showing for at least some of the locations a dramatically increase of chemical contamination and toxicity with the sediment depth. For several European river basis, including Neckar, Rhine and Elbe, highly contaminated old sediments can be described as "potential chemical time bombs" (Cappuyns et al. 2006; Japenga and Salomons 1993). An important process which may remobilize such sediments and which is still of increasing importance in relationship to the global climate change is more often occurrence of stronger floods in Europe as well as in other parts of the world. To understand and predict possible toxicological and ecotoxicological consequences of contaminants mobilized from sediments by flood events it is necessary to develop scientific approaches for the assessment of regularly flooded rivers. The combination of hydrodynamics and ecotoxicological investigations is devolving to an emerging field of research. Recently, it was shown that hydrodynamic aspects can be involved as additional line-of-evidence in Weight-of-evidence studies assessing the impact of sediments (Chapman and Hollert 2006). In the last five years several studies were published *(i)* addressing the ecotoxicological impact of flood events (Brack et al. 2002; Grote et al. 2005; Hollert et al. 2000a; Oetken et al. 2005; Matthaei et al. 2006, Sect. 10.2) or *(ii)* using combined approaches for evaluating flood events and the risk of erosion (Babut et al. 2006; Haag et al. 2001; Hollert et al. 2000b, 2003a). In this context, studies on the Elbe flood in 2002 indicated elevated effects in bioassays (Heise et al. in prep). Moreover, cellular changes could be found in livers from flounder (*Platichthys flesus* L.) and digestive glands of blue mussels (*Mytilus edulis*), 5 month after the flood disaster in the Elbe Estuary and the Wadden Sea (Einsporn et al. 2005). In comparison to earlier data from long-term studies at the same stations, a significant impairment in the function of cell organelles (lysosomes), involved in the detoxification and elimination of pollutants in fish liver, was found. In addition, in a long time study, EROD activity was measured in livers of dab (*Limanda limanda*) from the German Bight (North Sea) from 1995 to 2003 (Kammann et al. 2005b). In autumn 2002, significantly elevated EROD activities were detected, possibly related to effects of the river Elbe flood event.

These findings support the hypothesis that extreme flood events can affect not only freshwater ecosystems but also marine systems and have deleterious effects on animal health. Furthermore, flood events can influence floodplains and wetlands negatively (Schwartz et al. 2006; Ulrich et al. 2002). Consequently, the risk of extreme flood events for drinking water supply will be an emerging topic in the future (Maier et al. 2006).

In conclusion, research should consider the potential of sediments to serve as sources of contamination for the aquatic ecosystem, for drinking water supply but also for the floodplain soils and other flooded areas. In the following case studies, two examples for such integrated approaches addressing the risk of erosion are presented briefly.

Case Study River Neckar (Germany)

During the seventies, the *river Neckar in Southern Germany* ranged among the most strongly contaminated rivers in Germany with high loads of both organic pollutants and heavy metals (Förstner and Müller 1974). For instance, cadmium loads were increased by a factor of up to 300, when compared to pre-industrial clay stone sediments. As a consequence of sewage treatment, the quality of water and sediments improved significantly, and today the Neckar can be classified among Germany's moderately contaminated rivers, however, with heavily loaded old sediments at some sites (Hollert et al. 2000a). Hence, earlier studies within the Neckar catchment area or the river Neckar itself, revealed moderate to strong ecotoxicological effectiveness in several bioassays for mutagenic, genotoxic, endocrine, teratogenic, and dioxin-like responses as well as correlations between biological effects and concentrations of organic pollutants (Hollert et al. 2005, 2002a,b, 2003b).

The objective of the presented study was to develop a combined ecotoxicological and hydraulic approach by the cooperation between the Universities Stuttgart and Heidelberg to elucidate the ecotoxicological implications associated with the risk of erosion of contaminated sediments (Hollert et al. 2000b, 2003a). This integrated strategy was applied to the lock-regulated river Neckar in Southern Germany (Haag et al. 2002, 2001, Hollert et al. 2000b, 2003a; Knauert et al. 2004). For this purpose, sediment cores of the heavily contaminated Lauffen reservoir/river Neckar were investigated (A) as well as suspended particulate matter during a flood event in the river Neckar (B) in order to give the potential and effective pollution risk under different hydraulic scenarios (Fig. 10.8).

Methods

a Two undisturbed sediment cores (13.5 cm in diameter and 150 cm in length) were taken from one location in the backwater region of the Lauffen reservoir/river Neckar in south-west of Germany (in total 7 locations and 16 sediment cores). In both cores, vertical profiles of bulk densities were measured in 1 cm steps non-intrusively by using a γ-ray-densitometer. Thus, similar sediment layering within the parallel cores was ensured as well as subsequent sampling of the appropriate sediment layers (Haag et al. 2001). If, on the basis of the density profiles, parallel cores were considered to be similar, one of them served to experimentally determine the critical shear stress of mass erosion ($\tau_{c,e}$) as a function of sediment depth. Erosion experiments were carried out in a rectangular water flume, the so called SETEG-system (Kern et al. 1999). The second one of the parallel cores was sectioned into layers of almost uniformly texture, thus, the core was cut at depths of significant bulk density changes. From this material, concentrations of heavy metals and PCBs were identified by

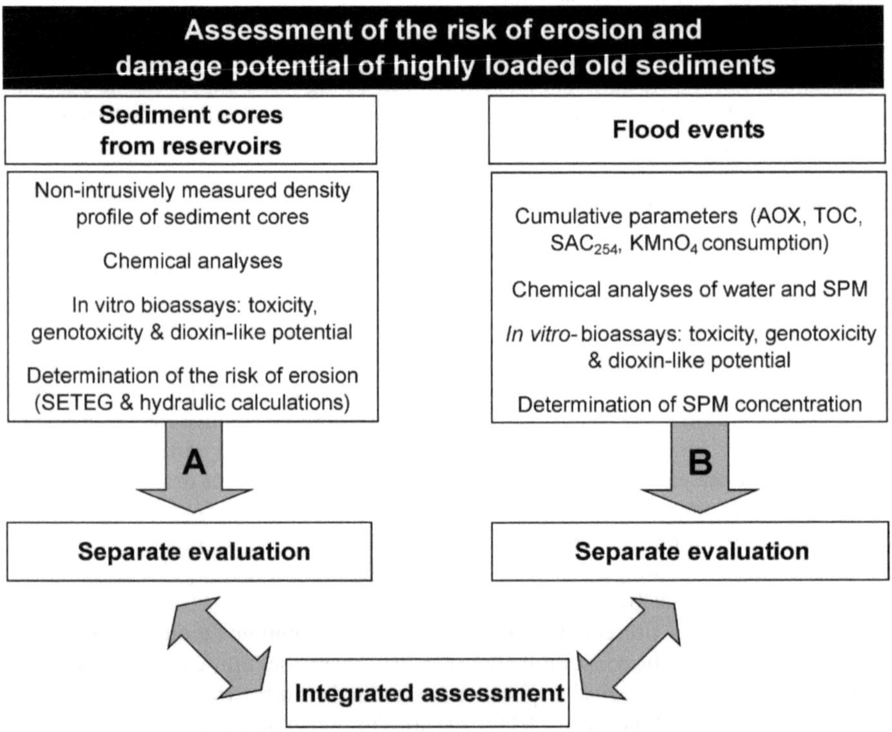

Fig. 10.8. Test strategies for examination and evaluation of the remobilization risk of old sediments in lock-regulated river systems (redrawn from Hollert et al. 2000)

chemical analyses while the cytotoxicity, dioxin-like activity and mutagenicity were investigated by bioanalytical methods (Kosmehl et al. 2004; Seiler et al. 2006). By comparison of the critical shear stress /sediment stability of the investigated sediment cores with the natural occurring bottom shear stresses, calculated by the 1-D flow and transport model COSMOS (Kern and Westrich 1997), the possible resuspension risk of contaminated sediment layers could be predicted.

b In order to gain insight into the ecological effects of a possible remobilization of heavily contaminated old sediments, suspended particulate matter (SPM) was collected in SPM traps from two sites of the lock-regulated section of the river Neckar: downstream the Lauffen reservoir with its high cadmium contaminations and downstream the less polluted Heidelberg reservoir (reference site). Parameters investigated are presented in Fig. 10.8.

Results and Discussion

The combined hydraulic and ecotoxicological approach revealed the high risk of erosion down to depth of 70 cm as well as an ecotoxicological hazard potential of the associated contaminants (Haag et al. 2002; Hollert et al. 2000, 2003).

Clear cut changes in bulk densities, the percentage of particles size $d < 20$ µm and ^{137}Cs content support the hypothesis of an erosional unconformity (Fig. 10.9).

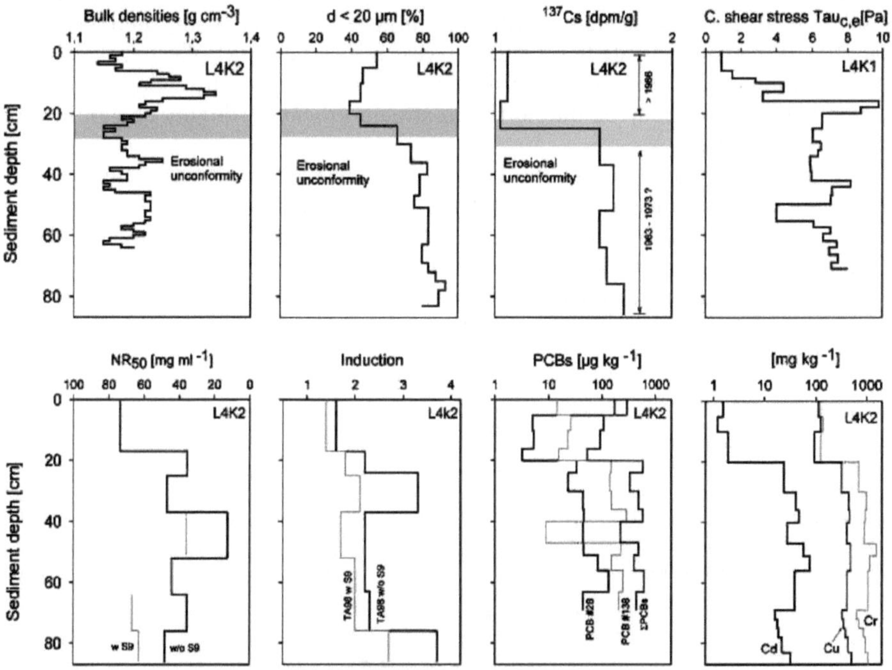

Fig. 10.9. Non-intrusively measured density profile, $d < 20$ µm, critical shear stress, cytotoxicity, mutagenicity, heavy metals and PCBs of core LN4K2 from the Lauffen Reservoir on the Neckar River depending on the depth (according to Hollert et al. 2003)

An erosional unconformity is the result of a flood event, where fine grained sediments are resuspended and non-cohesive particles are re-deposited (Haag et al. 2000). In the vertical sediment profiles, layers with coarse particles, low TOC and consequently increased bulk densities could be detected mostly below 25 cm depth. Often these layers were also characterized by sudden decreases of $\tau_{c,e}$ in the corresponding parallel core, indicating the predominance of non-cohesive particles (Fig. 10.9). Bioanalytical and chemical investigations (Fig. 10.9) were showing clear-cut changes of the ecotoxicological hazard potential below the depth of the erosional unconformity. The younger sediments within the top 25 cm depth revealed neither strong cytotoxicity nor mutagenicty. In contrast, for the older sediments below that zone, a strong cytotoxicity, dioxin-like and mutagenic potential could be determined. PCBs and anthropogenic influenced heavy metals such as Cd and Pb showed up to 100 times higher concentrations in the sediment layers below the erosional unconformity. Concentrations above 10.8 mg kg^{-1} of cadmium and 193 mg kg^{-1} copper, respectively, allowed the classification of these sediment layers to the older, highly contaminated sediments (HCS). In contrast, the upper layers represented low contaminated sediment layers (LCS, Haag et al. 2001; Hollert et al. 2003a). Since this unconformity happens in a transition zone between younger, less contaminated and older, heavily contaminated sediment layers, the last flood must have exposed not only deeper sediment layers but also their contamination load.

The suspended matter of the high discharge (return periods of 15 to 20 years (Hollert et al. 2003a)) exerted significantly higher cytotoxicity and mutagenicity (Fig. 10.10) than a moderate flood with a 1-year return period (Hollert et al. 2000a). These findings supported the conclusion that the observed ecotoxicological effects during major floods may be due to the in-stream erosion of highly contaminated bottom sediments.

Recently, SPM of a flood event at the Neckar in 2004 with a recurrence interval of five year was sampled using a sediment trap. Highest EROD activities of the extracts could be found for the peak of the flood, with a ten time higher Dioxin-like activity when compared to other SPM samples (details will be shown in an additional paper by Wölz et al. presented in the forthcoming Sedymo-special issue in Journal of Soils and Sediments, 2007). The two samples with the highest effects have been used for effect directed analyses.

Fig. 10.10.
Time-course of the mutagenicty during the flood event of Oct/ Nov 1998. Since for a moderate flood event (HQ_1) no mutagenicity could be found, several SPM extracts revealed genotoxic effects in the Ames test without S9 mix

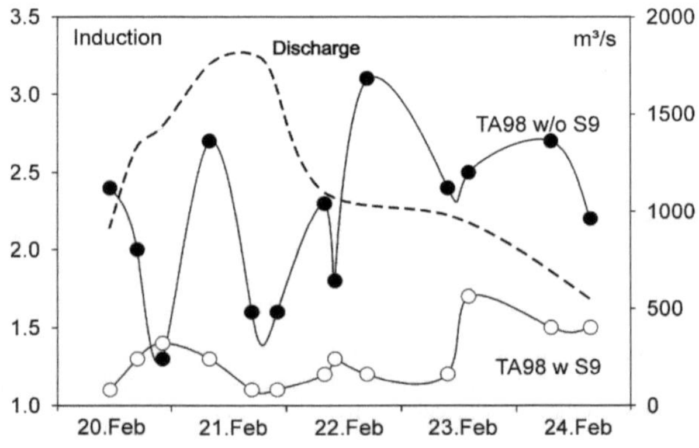

Using the shown strategy it is possible to investigate the risk of erosion. However, the identity of the pollutants causing effects in the bioassay is still unknown.

Effect directed analyses is a strategy to gain insight into the character of the noxious substances (Brack 2003; Brack et al. 2005b). Organic extracts from SPM sampled during the 2004 flood events was fractionated for polarity and aromaticity according to an previously developed methodology (Brack et al. 2005b). Only the fractions revealing high toxicity on bioassays are used for chemical analyses in order to identify the toxic substance class or substance. Using this approach, it was possible to elucidate PCBs and Dioxins/Furans to contribute only for less than 1% of the biologically derived EROD activities. The EROD activities of the fractions with PAHs explained the major part of the Dioxin-like potential of the crude extracts. However, the measured US EPA priority PAHs contributed less than 20% to the total EROD activities (Wölz et al. 2007).

Case Study Morava Catchment Area (Czech Republic)

Major flooding events also occur regularly in the *catchment area of the river Morava (Czech Republic)*. Water and sediment quality in this area has been impacted by historical industrial activities within the watershed. In July 1997 the region was affected by disastrous floods caused by two periods of exceptionally heavy rainfalls that resulted in great material and ecological damages. Extensive rainfall plagued the north part of the Morava River basin and the situation was even more complicated by the second flood wave within 10 days period. In historical context it was very rare event but due to human landscape interventions it is possible to expect similar events still more frequently. During the flooding period lasting for several days, older sediments were washed away and new silt materials were deposited up to several centimeters layer. Because of our earlier monitoring of this area, the situation brought unique opportunity to evaluate the changes in contaminant levels and the toxic effects in relation to flood events. Initial evaluations of the target contaminant profiles in sediment and water samples from several sites revealed that there was a gradient of concentrations along the Morava River from upstream to downstream, and suggested that the tributary of the little stream Drevnice serves as a source of pollution to the Morava River (Hilscherova et al. 2001; Holoubek et al. 1998). There are no limit values for sediment contaminants in the Czech Republic but the concentrations of polycyclic aromatic hydrocarbons (PAHs) as well as other organic compounds were above the maximal permissible limits that apply for instance in the Netherlands, as were the concentrations of Cd and Zn for soils. Most studies have been performed with the freshly contaminated top sediment layers, but still there is only little information on the deeper layers that might be mobilized during frequent floods.

Previous investigations have also shown the impact of floods on the periodically flooded soils with significantly elevated contaminated levels namely with persistent organic compounds (Hilscherova et al. 2001; Holoubek et al. 1998), and some heavy metals (recent unpublished data). The most obvious changes related to major floods in 1997 were observed for PAHs – the dominant contaminants in the area. The results clearly showed that in some regions there was significant decrease in PAHs concentrations in riverine sediments after the floods while the concentrations in the

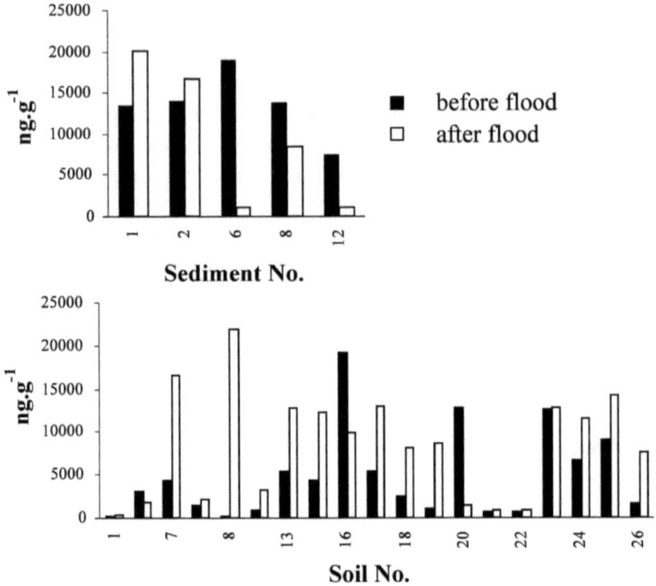

Fig. 10.11. Effect of floods in 1997 on concentrations of PAHs (sum of 16 US EPA PAHs) in sediments and floodplain soils of the river Morava Catchment area

surrounding soils at most sites within the flood affected area significantly increased (Fig. 10.11).

Application of *in vitro* biotests has shown significant toxic, genotoxic, dioxin-like, and estrogenic potentials in sediments collected from numerous sites (Hilscherova et al. 2002, 2001) and the bioassay results confirmed significant effects of floods. Both dioxin-like and estrogenic activities in sediments were generally either unaffected or significantly decreased after floods (Fig. 10.12, 10.13) showing removal of upper contaminated layers and their transport downstream by the flood water.

The greatest added value of *in vitro* assays is that they provide an integrative measure of the potential of the complex mixture of compounds within the sample that may cause a negative effect through the specific mechanism of action. They serve as rapid, sensitive and relatively simple screening systems evaluating the presence of chemicals and their mutual interactions with specific mode of action. Fractionation of extracts enables separation of compounds present in the complex mixture and allows determination of the most active classes of compounds. In the study in part of the Morava catchment area, the simple fractionation procedure also revealed the important role of mediate polar PAHs and pesticides for both the estrogenic and dioxin-like effects (Hilscherova et al. 2001, 2002) which was confirmed by the mass balance calculations (Hilscherova et al. 2002, 2001). Further, mechanism-specific bioassays were confirmed to be an effective tool in initial screening of river sediments compared to the more time- and cost-demanding instrumental analyses.

In vitro biotests have shown significant toxic, genotoxic, dioxin-like, and estrogenic potentials in sediments collected from numerous waters (Hilscherova et al. 2002, 2001). However, most studies have been performed on the freshly contaminated top sediment layers, and there are no information regarding the deeper layers of sediments that might be mobilized during frequent floods.

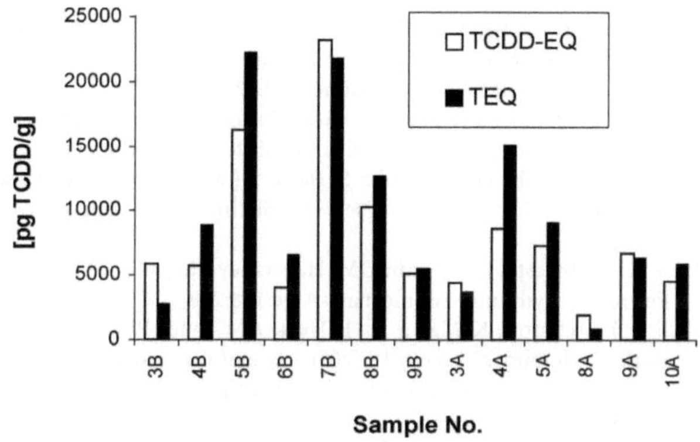

Fig. 10.12.
Dioxin-like equivalents determined in bioassay with H4IIE.luc cells (TCDD-EQ) and by chemical analysis (TEQ) of organic extracts from sediments sampled in Morava Catchment area before ("*B*" samples) and after ("*A*") the major floods in 1997 (Hilscherova et al. 2001)

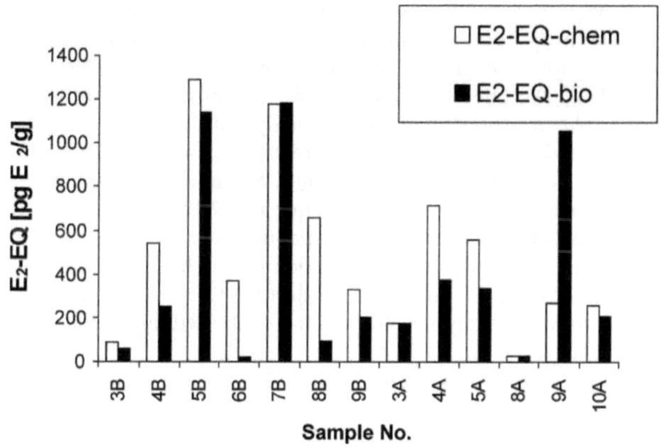

Fig. 10.13.
Estrogenic equivalents determined in bioassay with MVLN cells (E2-EQ-bio) and by chemical analysis (E2-EQ-chem) of organic extracts from sediments sampled in Morava Catchment area before (*B*) and after (*A*) the major floods in 1997 (according to Hilscherova et al. 2002)

Conclusion

The present article features the urgent need to cross disciplinary boundaries in order to derive a realistic assessment regarding the erosion risk of old deposited sediment layers as well as the bioavailablitiy and hazard potential of their associated contaminants at different aquatic sites. Especially the combination of hydrodynamic and ecotoxicological methods will give *(i)* comprehensive insights into the effects of flood events on biota and ecosystems and *(ii)* allow evaluation of sediment and thus water quality with regard to the global change and the expectations of more severe floods in the near future.

References

Ahlf W, Calmano W, Erhard J, Forstner U (1989) Comparison of 5 Bioassay Techniques for Assessing Sediment-Bound Contaminants. Hydrobiologia 188:285–289

Ahlf W, Braunbeck T, Heise S, Hollert H (2002a) Sediment and Soil Quality Criteria. In: Burton F, McKelvie I, Förstner U, Guenther A (eds) Environmental Monitoring Handbook. McGraw-Hill, New York, pp 17–18

Ahlf W, Hollert H, Neumann-Hensel H, Ricking M (2002b) A Guidance for the Assessment and Evaluation of Sediment Quality: A German Approach Based on Ecotoxicological and Chemical Measurements. Journal of Soils and Sediments 2:37–42

Ankley G, et al. (1998) Overview of a workshop on screening methods for detecting potential (anti-) estrogenic/androgenic chemicals in wildlife. Environmental Toxicology and Chemistry 17:68–87

Babut M, Oen A, Hollert H, Apitz S, Heise S (2007) From priority setting to risk ranking: suggested approaches, Chapter 8. In: Heise S (ed) Sediment – Risk management and communication, Elsevier, in press

Black KS, Tolhurst TJ, Paterson DM, Hagerthey SE (2002) Working with natural cohesive sediments, Journal of Hydraulic Engineering-Asce 128:2–8

Blaha L, Hilscherova K, Mazurova E, Hecker M, Jones PD, Newsted JL, Bradley PW, Gracia T, Duris Z, Horka I, Holoubek I, Giesy JP (2006) Alteration of steroidogenesis in H295R cells by organic sediment contaminants and relationships to other endocrine disrupting effects. Environment International 32:749–57

Brack W, Altenburger R, Dorusch F, Hubert A, Moder M, Morgenstern P, Moschutz S, Mothes S, Schirmer K, Wennrich R, Wenzel KD, Schuurmann G (2002) Hochwasser 2002: Chemische und toxische Belastung überschwemmter Gemeinden im Raum Bitterfeld Umweltwissenschaften and Schadstoffforschung – Z. Umweltchem. Ökotox 14:213–220

Brack W (2003) Effect-directed analysis: a promising tool for the identification of organic toxicants in complex mixtures? Analytical Bioanalytical Chemistry 377:397–407

Brack W, Bakker J, de Deckere E, Deerenberg C, van Gils J, Hein M, Jurajda P, Kooijman B, Lamoree M, Lek S, López de Alda MJ, Marcomini A, Munoz I, Rattei S, Segner H, Thomas K, von der Ohe PC, Westrich B, de Zwart D, Schmitt-Jansen M (2005) MODELKEY. Models for assessing and forecasting the impact of environmental key pollutants on freshwater and marine ecosystems and biodiversity Environmental Science and Pollution Research 12:252–256

Brack W, Schirmer K, Erdinger L, Hollert H (2005b) Effect-directed analysis of mutagens and ethoxyresorufin-O-deethylase inducers in aquatic sediments. Environmental Toxicology and Chemistry 24:2445–2458

Burton GA (1991) Assessing the toxicity of freshwater sediments. Environmental Toxicology and Chemistry 10:1585–1627

Burton Jr. GA, Baudo R, Beltrami M, Rowland C (2001) Assessing sediment contamination using six toxicity assays. Journal of Limnology 60:263–267

Calmano W, Hong J, Forstner U (1993) Binding and Mobilization of Heavy-Metals in Contaminated Sediments Affected by Ph and Redox Potential. Water Science and Technology 28:223–235

Cappuyns V, Swennen R, Devivier A (2006) Dredged river sediments: Potential chemical time bombs? A case study. Water, Air, and Soil Pollution 171:49–66

Chapman PM (1989) Current Approaches to Developing Sediment Quality Criteria. Environmental Toxicology and Chemistry 8:589–599

Chapman PM, Power EA, Burton Jr. GA (1992) Integrative Assessments in Aquatic Ecosystems. In: Burton GA (ed) Sediment Toxicity Assessment. Boca Raton, USA, pp 313–340

Chapman PM, Hollert H (2006) Should the sediment quality triad become a tetrad, a pentad, or possibly even a hexad? Journal of Soils and Sediments 6:4–8

Chen GS, White PA (2004) The mutagenic hazards of aquatic sediments: a review. Mutation Research-Reviews in Mutation Research 567:151–225

Davoren M, Ni-Shuilleabhain S, Hartl MGJ, Sheehan D, O'Brien NM, O'Halloran J, Van Pelt F, Mothersill C (2005) Assessing the potential of fish cell lines as tools for the cytotoxicity testing of estuarine sediment aqueous elutriates. Toxicology in Vitro 19:421–431

De Brouwer J, Bjelic S, de Deckere E, Stal L (2000) Interplay between biology and sedimentology in a mudflat (Biezelingse Ham, Westerschelde, The Netherlands). Continental Shelf Research 20:1159–1177

Den Besten PJ, de Deckere E, Babut MP, Power B, Angel DelValls T, Zago C, Oen AMP, Heise S (2003) Biological effects-based sediment quality in ecological risk assessment for European waters. Journal of Soils and Sediments 3:144–162

Droppo, I (2004) Structural controls on floc strength and transport. Canadian Journal of Civil Engineering 31:569–578

Einsporn S, Broeg K, Koehler A (2005) The Elbe flood 2002-toxic effects of transported contaminants in flatfish and mussels of the Wadden Sea. Marine Pollution Bulletin 50:423–429

Förstner U, Müller G (1974) Schwermetalle in Flüssen und Seen. Spinger-Verlag, Heidelberg, 225 pp

Förstner U, Westrich B (2005) BMBF coordinated research project SEDYMO (2002–2006): Sediment dynamics and pollutant mobility in river basins. Journal of Soils and Sediments 5:134–138

Förstner U, Heise S, Schwartz R, Westrich B, Ahlf W (2004) Historical Contaminated Sediments and Soils at the River Basin Scale. Journal of Soils and Sediments 4:247–260

Gerbersdorf SU, Jancke T, Westrich B (2005) Physico-chemical and biological sediment properties determining erosion resistance of contaminated riverine sediments – Temporal and vertical pattern at the Lauffen reservoir/river Neckar, Germany. Limnologica 35:132–144

Gerbersdorf SU, Jancke T, Westrich B (2006) Biostabilisation by polymeric substances in riverine sediments. Environmental Microbiology, under review

Gerbersdorf SU, Jancke T, Westrich B (2007) Sediment properties for assessing the erosion risk of contaminated riverine sites. Journal of Soils and Sediments, under review

Giesy JP, Hoke RA (1989) Fresh-Water Sediment Toxicity Bioassessment – Rationale for Species Selection and Test Design. Journal of Great Lakes Research 15:539–569

Giger W, Reinhard M, Schaffner C (1974) Petroleum-derived and indigenous hydrocarbons in recent sediments of Lake Zug. Environmental Science Technology 8:454–455

Gonzáles-Vila FJ, Polvillo O, Boski T, Moura D (2003) Biomarker patterns in a time-resolved Holocene/terminal Pleistocene sedimentary sequence from the Guadiana River estuarine area (SW Portugal/Spain border), Organic Geochemitry, 1601–1613

Grote M, Altenburger R, Brack W, Moschütz S, Mothes S, Michael C, Narten GB, Paschke A, Schirmer K, Walter H, Wennrich R, Wenzel KD, Schuurmann G (2005) Ecotoxicological profiling of transect river Elbe sediments. Acta Hydrochimica et Hydrobiologica 33:555–569

Haag I, Kern U, Westrich B (2001) Erosion investigation and sediment quality measurements for a comprehensive risk assessment of contaminated aquatic sediments. Science of the Total Environment 266:249–257

Haag I, Hollert H, Kern U, Braunbeck T, Westrich B (2002) Flood Event Sediment Budget for a Lock-Regulated River Reach and Toxicity of Suspended Particles. Proceedings 3rd International Conference on Water Resources and Environment Research (ICWRER), Dresden, 33–37

Hallare A, Kosmehl T, Schulze T, Hollert H, Koehler H-R, Triebskorn R (2005) Assessing The Severity of Sediment Contamination in Laguna Lake, Philippines Using A Sediment Contact Assay with Zebrafish (*Danio rerio*) Embryos. Science of the Total Environment 347:254–71

Hilscherova K, Machala M, Kannan K, Blankenship AL, Giesy JP (2000) Cell bioassays for detection of aryl hydrocarbon (AhR) and estrogen receptor (ER) mediated activity in environmental samples. Environmental Science and Pollution Research 7:159–171

Hilscherova K, Kannan K, Kang YS, Holoubek I, Machala M, Masunaga S, Nakanishi J, Giesy JP (2001) Characterization of dioxin-like activity of sediments from a Czech river basin. Environmental Toxicology and Chemistry 20:2768–2777

Hilscherova K, Kannan K, Holoubek I, Giesy JP (2002) Characterization of estrogenic activity of riverine sediments from the Czech Republic. Archives Environmental Contamamination Toxicology 43:175–85

Hilscherova K, Kannan K, Nakata H, Hanari N, Yamashita N, Bradley PW, McCabe JM, Taylor AB, Giesy JP (2003) Polychlorinated dibenzo-p-dioxin and dibenzofuran concentration profiles in sediments and flood-plain soils of the Tittabawassee River, Michigan. Environmental Science Technology 37:468–74

Hollert H, Dürr M, Erdinger L, Braunbeck T (2000a) Cytotoxicity of settling particulate matter (SPM) and sediments of the Neckar River (Germany) during a winter flood. Environmental Toxicology and Chemistry 19:528–534

Hollert H, Dürr M, Haag I, Winn N, Holtey-Weber R, Kern U, Färber H, Westrich B, Erdinger L, Braunbeck T (2000b) A combined hydraulic and in vitro bioassay approach to assess the risk of erosion and ecotoxicological implications of contaminated sediments in a lock-regulated river system. In: BfG (ed) Sediment assessement in European River Basins, Reihe: Mitteilungen der Bundesanstalt für Gewässerkunde, Koblenz, Berlin, pp 156–160

Hollert H, Dürr M, Olsman H, Halldin K, Bavel Bv, Brack W, Tysklind M, Engwall M, Braunbeck T (2002a) Biological and chemical determination of dioxin-like compounds in sediments by means of a sediment triad approach in the catchment area of the Neckar River. Ecotoxicology 11:323–336

Hollert H, Heise S, Pudenz S, Brüggemann R, Ahlf W, Braunbeck T (2002b) Application of a sediment quality triad and different statistical approaches (Hasse diagrams and fuzzy logic) for the comparative evaluation of small streams. Ecotoxicology 11:311–321

Hollert H, Haag I, Dürr M, Wetterauer B, Holtey-Weber R, Kern U, Westrich B, Färber H, Erdinger L, Braunbeck T (2003a) Untersuchungen zum ökotoxikologischen Schädigungspotenzial und Erosionsrisiko von kontaminierten Sedimenten in staugeregelten Flüssen. Umweltwissenschaften und Schadstoffforschung, Z Umweltchem Ökotox 15:5–12

Hollert H, Keiter S, König N, Rudolf M, Ulrich M, Braunbeck T (2003b) A New Sediment Contact Assay to Assess Particle-bound Pollutants Using Zebrafish (*Danio rerio*) Embryos. Journal of Soils and Sediments 3:197–207

Hollert H, Dürr M, Holtey-Weber R, Islinger M, Brack W, Färber H, Erdinger L, Braunbeck T (2005) Endocrine disruption of water and sediment extracts in a non-radioactive dot blot/RNAse protection-assay using isolated hepatocytes of rainbow trout – How explain deficiencies between bioanalytical effectiveness and chemically determined concentrations? Environmental Science and Pollution Research 12:347–360

Holoubek I, Machala M, Štaffová K, Helešic J, Ansorgová A, Schramm KW, Kettrup A, Giesy JP, Kannan K, Mitera J (1998) PCDD/Fs in sediments from Morava River catchment area. Organochlorine compounds 39:261–266

Janošek J, Hilscherová K, Bláha L, Holoubek I (2006) Environmental xenobiotics and nuclear receptors-Interactions, effects and in vitro assessment. Toxicology in Vitro 20:18–37

Japenga J, Salomons W (1993) Dyke-protected floodplains: a possible chemical time bomb? Land Degradation and Rehabilitation 4:373–380

Kammann U, Danischewski D, Vobach M, Biselli S, Theobald N, Reineke N, Hühnerfuss H, Wosniok W, Kinder A, Sierts-Herrmann A, Steinhart H, Vahl HH, Westendorf J (2005a) Bioassay-directed fractionation of organic extracts of marine surface sediments from the north and Baltic Sea part II: Results of the biotest battery and development of a biotest index. Journal of Soils and Sediments 5:225–232

Kammann U, Lang T, Vobach M, Wosniok W (2005b) Ethoxyresorufin-O-deethylase (EROD) activity in dab (*Limanda limanda*) as biomarker for marine monitoring. Environmental Science and Pollution Research 12:140–145

Kern U, Westrich B (1997) Sediment budget analysis for river reservoirs. Water, Air, and Soil Pollution 99:105–112

Knauert S, Dürr M, Haag I, Braunbeck T, Hollert H (2004) Dioxin-ähnliche Wirksamkeit in der permanenten Fischzelllinie RTL-W1-Tiefenprofile von Sedimentbohrkernen am Neckar. ALTEX 21:162

Koethe F (2003) Existing sediment management guidelines: An overview. What will happen with the sediment/dredged material? Journal of Soils and Sediments, pp 139–143

Kosmehl T, Krebs F, Manz W, Erdinger L, Braunbeck T, Hollert H (2004) Comparative genotoxicity testing of Rhine River sediment extracts using the permanent cell lines RTG-2 and RTL-W1 in the comet assay and Ames assay. Journal of Soils and Sediments 4:84–94

Kosmehl T, Hallare AV, Reifferscheid G, Manz W, Erdinger L, Braunbeck T, Hollert H (2006) Development of a new contact assay for testing whole sediment genotoxicity in zebra fish larvae. Environmental Toxicology and Chemistry 25:2097–2106

Maier M, Kühlers D, Brauch HJ, Fleig M, Maier D, Jirka GH, Mohrlok U, Bethge E, Bernhart HH, Lehmann B, Hillebrand G, Wölz J, Hollert H (2006) Flood retention and drinking water supply – Preventing conflicts of interest. Journal of Soils and Sediments 6:113–114

Martínek K, Blecha M, Daněk V, Francù J, Hladíková J (2006) Record of palaeoenvironmental changes in a Lower Permian organic-rich lacustrine succession: Integrated sedimentological and geochemical study of the Rudník member, Krkonoše Piedmont Basin, Czech Republic. Palaeogeography, Palaeoclimatology, Palaeoecology 230:85–128

Meyers PA (2003) Applications of organic geochemistry to paleolimnological reconstructions: a summary of examples from the Laurentian Great Lakes. Organic Geochemistry 34:261–289

Oetken M, Stachel B, Pfenninger M, Oehlmann J (2005) Impact of a flood disaster on sediment toxicity in a major river system – the Elbe flood 2002 as a case study. Environmental Pollution 134:87–95

Paterson D, Tolhurst T, Kelly J, Honeywill C, de Deckere E, Huet V, Shayler S, Black K, de Brouwer J, Davidson I (2000) Variations in sediment properties, Skeffling mudflat, Humber Estuary, UK. Continental Shelf Research 20:1373–1396

Perkins RG, Sun H, Watson J, Player MA, Gust G, Paterson DM (2004) In-line laser holography and video analysis of eroded floc from engineered and estuarine sediments. Environmental Science Technology 38:4640–4648

Peters KE, Walters CC, Moldowan JM (2005) The biomarker guide; I, Biomarkers and isotopes in the environment and human history. II, Biomarkers and isotopes in petroleum systems and Earth history. Cambridge University Press, Cambridge, 155 pp

Power EA, Chapman PM (1992) Assessing sediment quality. In: Burton GA (ed) Sediment toxicity assessment. Lewis-Publishers, Boca Raton, pp 1–18

Salomons W, Brils J (2004) Contaminated Sediments in European River Basins – European Sediment Research Network SedNet booklet. online: www.sednet.org

Schwartz R, Gerth J, Neumann-Hensel H, Bley S, Forstner U (2006) Assessment of highly polluted fluvisol in the Spittelwasser floodplain – Based on national guideline values and MNA-Criteria. Journal of Soils and Sediments 6:145–155

SedNet (2004) Sediment, a valuable resource that needs Europe's attention; SedNet recommendations for sediment research priorities related to the soil research clusters. http://www.sednet.org/materiale/Sediment_a_valuable_resource.pdf

Segner H (1998) Fish cell lines as a tool in aquatic toxicology. In: Braunbeck T, Hinton DE, Streit B (eds) Fish ecotoxicology – Experientia Supplement, vol. 86. Birkhäuser, Basel/Switzerland, pp 1–38

Seiler TB, Rastall AC, Leist E, Erdinger L, Braunbeck T, Hollert H (2006) Membrane dialysis extraction (MDE): A novel approach for extracting toxicologically relevant hydrophobic organic compounds from soils and sediments for assessment in biotests. Journal of Soils and Sediments 6:20–29

Simpson SL, Apte SC, Batley GE (1998) Effect of short term resuspension events on trace metal speciation in polluted anoxic sediments. Environmental Science and Technology 32:620–625

Stoffers P, Summerhayes C, Forstner U, Patchineelam S (1977) Copper and Other Heavy Metal Contamination in Sediments from New Bedford Harbor, Massachusetts: A Preliminary Note. Environmental Science Technology 11:819–821

Stout SA, Uhler AD, McCarthy KJ, Emsbo-Mattingly S (2002) Chemical fingerprinting of hydrocarbons. In: Murphy BL, Morrison D (eds) Introduction to environmental forensics. Academic Press, San Diego, pp 139–260

Stout SA, Uhler AD, Emsbo-Mattingly SD (2004) Comparative evaluation of background anthropogenic hydrocarbons in surficial sediments from nine urban waterways. Environmental Science Technology 38:2987–2994

Sumpter JP, Johnson AC (2005) Lessons from endocrine disruption and their application to other issues concerning trace organics in the aquatic environment. Environmental Science and Technology 39:4321–4332

Ulrich M, Schulze T, Leist E, Glaß B, Maier M, Maier D, Braunbeck T, Hollert H (2002) Ökotoxikologische Untersuchung von Sedimenten und Schwebstoffen: Abschätzung des Gefährdungspotenzials für Trinkwasser und Korrelation verschiedener Expositionspfade (acetonischer Extrakt, natives Sediment) im Bakterienkontakttest und Fischeitest Umweltwissenschaften und Schadstoffforschung. Z Umweltchem Ökotox 14:132–137

USEPA (2002) A Guidance Manual to Support the Assessment of Contaminated Sediments in Freshwater Ecosystems, Volume I – An Ecosystem-Based Framework for Assessing and Managing Contaminated Sediments. http://www.cerc.usgs.gov/pubs/sedtox/VolumeI.pdf, 149 p

Van Beelen P (2003) A review on the application of microbial toxicity tests for deriving sediment quality guidelines. Chemosphere 53:795–808

Wenning R, Ingersoll C (2002) Summary of the SETAC Pellston Workshop on Use of Sediment Quality Guidelines and Related Tools for the Assessment of Contaminated Sediments; 17–22 August 2002; Fairmont, Montana, USA. Society of Environmental Toxicology and Chemistry (SETAC). Pensacola FL, USA. online: http://www.setac.org/files/SQGSummary.pdf

Wenning R, Batley G, Ingersoll C, Moore D (2005) Use of Sediment Quality Guidelines and Related Tools for the Assessment of Contaminated Sediments. SETAC, Pensacola, FL

Wölz J, Olsman H, Hagberg J, Brack W, Möhlenkamp C, vanBavel B, Engwall M, Claus E, Manz W, Braunbeck T, Hollert H (2007) Effect-directed Analysis to identify AH-Receptor agonists in suspended particulate matter during flood events. to be submitted to Environmental Toxicology and Chemistry

Ziegler CK (2002) Evaluating sediment stability at sites with historic contamination. Environmental Management 29:409–427

Appendix

Table A.1. List of contributions of the "Sedymo 2006 Symposium, March 27–29, 2006 in Hamburg" which are not reproduced in this volume (addresses are associated to the "List of Authors" in the beginning)

Lecture	Contributor
Opening	
L1 Contaminated Sediment Transfers in River Basins	Kevin G. Taylor
Session II: Transport Modeling	
L6 Physics of Fine-Sediment Dynamics in Low-Energetic Open Water Systems	Johan C. Winterwerp
L10 Bed Form and Flow Resistance During Motion of Partly Cohesive Sediments	Robert Banasiak, Ronny Verhoeven
Session III: Catchment Modeling	
L14 Conceptual Framework and Desk-Based Study of Sediment Inputs to a Catchment for a Sediment Management Strategy: The Norfolk Broads	Sue White, Sabine E. Apitz, Andrea Kelly, Trudi Wakelin
Session V: Transport Indicators	
L23 Contribution of Combined Sewer Overflow Emissions to Pollutant Fluxes in the Urbanised Alzette-Catchment in Luxembourg	Tom Gallé, Ulli Leopold, Anna-Marie Solvi, Kai Klepiszewski, Andreas Kurtenbach, Joëlle Welfring, Pol Schosseler
Session VI: Fine Sediment Particles	
L29 Mobility and Retention of Natural Particles in Rivers as a Function of Flow Heterogeneity and Biofilm Growth Studied in Mesoscaled Flumes	Frank von der Kammer, Thilo Hofmann, Montserrat Roura, Tom J. Battin
Session VII: Sediment Toxicity	
L31 Sediment and Water Quality in a Mixed Land Use Scottish River Basin: Linking Chemical and Ecological Indicators	Marc Stutter, Simon Langan, Richard Cooper, Lynn Clark, David Lumsdon

Table A.1. (*Continued*)

Session VII: Sediment Toxicity (*continued*)		
L32	The SeKT Joint Research Project: Definition of Refefrence Conditions, Control Sediments and Toxicity Thresholds for Limnic Sediment Contact Tests	Ute Feiler, Wolfgang Ahlf, Christiane Fahnenstich, Daniel Gilberg, Monika Hammers-Wirtz, Sebastian Höss, Henner Hollert, Michael Meller, Kerstin Melbye, Helga Neumann-Hensel, Hans-Toni Ratte, Thomas-Benjamin Seiler, Jürgen Weber, Peter Heininger
L33	Miniaturized Test Kit with *A. globiformis* for the Risk Assessment of Soils and Sediments	Kerstin Melbye, Helga Neumann-Hensel
Session VIII: Sediments in European River Basins		
L36	Soil-Catchment-Coast Continuum	Wim Salomons
L37	Fluxes and Storage of Sediment in Rivers: Basin- and European-Scale Perspectives	Philip Owens
L38	Evaluation of Contaminated Sites Along the Elbe and Rhine River Basins	Susanne Heise, Evelyn Claus, Marc Eisma, Ulrich Förstner, Peter Heininger, Thomas Jancke, Joachim Karnahl, Thomas Krämer, Frank Krüger, Axel Netzband, Wim Salomons, René Schwartz, Tiedo Vellinga, Bernhard Westrich
Session IX: Sediment Risk Assessment		
L39	Sediment Risk Assessment: A Matter of Scale	Marc Babut, Sabine Apitz, Amy Oen, Pieter J. den Besten, Susanne Heise
L40	Sediment Dynamics and Risk Assessment	Sue White
Session X: Sediment Management		
L41	Sediment: The Natural Basis of the Water and Marine Framework Directive	Jos Brils, Marc Eisma, Axel Netzband, Pieter J. den Besten

Index

A

B

F

G

H

I

K

L

M

N

O

P

T

U

V

W

MIX
Papier aus verantwortungsvollen Quellen
Paper from responsible sources
FSC® C105338

If you have any concerns about our products,
you can contact us on
ProductSafety@springernature.com

In case Publisher is established outside the EU,
the EU authorized representative is:
Springer Nature Customer Service Center GmbH
Europaplatz 3, 69115 Heidelberg, Germany

Printed by Libri Plureos GmbH
in Hamburg, Germany